ANNUAL REVIEW OF FLUID MECHANICS

ANNUAL REVIEW OF FLUID MECHANICS

MILTON VAN DYKE, *Co-Editor*
Stanford University

J. V. WEHAUSEN, *Co-Editor*
University of California, Berkeley

JOHN L. LUMLEY, *Associate Editor*
Cornell University

VOLUME 12

1980

ANNUAL REVIEWS INC. 4139 EL CAMINO WAY PALO ALTO, CALIFORNIA 94306

ANNUAL REVIEWS INC.
Palo Alto, California, USA

REPRINTS The conspicuous number aligned in the margin with the title of each article in this volume is a key for use in ordering reprints. Available reprints are priced at the uniform rate of $1.00 each postpaid. The minimum acceptable reprint order is 5 reprints and/or $5.00 prepaid. A quantity discount is available.

International Standard Serial Number: 0066-4189
International Standard Book Number: 0-8243-0712-7
Library of Congress Catalog Card Number: 74-80866

Annual Reviews Inc. and the Editors of its publications assume no responsibility for the statements expressed by the contributors to this Review.

FILMSET BY TYPESETTING SERVICES LTD, GLASGOW, SCOTLAND
PRINTED AND BOUND IN THE UNITED STATES OF AMERICA

Annual Review of Fluid Mechanics
Volume 12, 1980

CONTENTS

INDEXES

ANNUAL REVIEWS INC. is a nonprofit corporation established to promote the advancement of the sciences. Beginning in 1932 with the *Annual Review of Biochemistry*, the Company has pursued as its principal function the publication of high quality, reasonably priced Annual Review volumes. The volumes are organized by Editors and Editorial Committees who invite qualified authors to contribute critical articles reviewing significant developments within each major discipline.

Annual Reviews are published in the following sciences: Anthropology, Astronomy and Astrophysics, Biochemistry, Biophysics and Bioengineering, Earth and Planetary Sciences, Ecology and Systematics, Energy, Entomology, Fluid Mechanics, Genetics, Materials Science, Medicine, Microbiology, Neuroscience, Nuclear and Particle Science, Pharmacology and Toxicology, Physical Chemistry, Physiology, Phytopathology, Plant Physiology, Psychology, Public Health, and Sociology. In addition, four special volumes have been published by Annual Reviews Inc.: *History of Entomology* (1973), *The Excitement and Fascination of Science* (1965), *The Excitement and Fascination of Science, Volume Two* (1978), and *Annual Reviews Reprints: Cell Membranes, 1975–1977* (published 1978). For the convenience of readers, a detachable order form/envelope is bound into the back of this volume.

The Rockefeller University, 1961

George E. Uhlenbeck

Ann. Rev. Fluid Mech. 1980. 12: 1–9

SOME NOTES ON THE RELATION BETWEEN FLUID MECHANICS AND STATISTICAL PHYSICS

G. E. Uhlenbeck
The Rockefeller University, New York, New York 10021

When your Editor asked me to write a prefatory chapter on my contributions to fluid mechanics, I thought he had made a mistake! As I told him, I have never written a paper on bona fide fluid mechanics, and although I am quite interested in the field I am certainly not an expert. In fact as a student of Ehrenfest at the University of Leiden and later as a teacher at the University of Michigan (from 1927 till 1935) I had never taken or given a course in fluid mechanics. In checking over my old notes, I noticed that Ehrenfest never even mentioned the hydrodynamical equations. The reason, of course, was that at that time the so-called "frontier of physics" was the development of the quantum theory and the problem of the structure of the atomic nuclei. All other topics were "desperation physics" as Pauli called them.

However, since your Editor assured me that a description of my work in statistical physics in so far as it touches some of the basic problems of fluid mechanics might be of interest to the readers of the Annual Review, I will try to do just that.

Since my student days, I have been deeply interested in the kinetic theory of matter, and, as I have often said, I am proud to be, through Ehrenfest, a "grandchild pupil" of Boltzmann. My dissertation of 1927 was on the statistical methods in the quantum theory and at Ann Arbor I always gave the course in statistical mechanics. However, during all this time the main emphasis was on the explanation (or better, interpretation) of the laws of thermodynamics and of the equilibrium properties of material systems. I knew, of course, that the founders of the kinetic theory of gases (Clausius, Maxwell, and Boltzmann) had explained the transport properties of gases and that this was in fact the main success of the theory.

0066-4189/80/0115-0001$01.00

I was, however, not very well acquainted with the more general and exact work of Sidney Chapman and David Enskog and as a result the connection with fluid mechanics was rather neglected in my lectures. This changed in the early thirties when I started to work with E. A. Uehling.

1 *The Uehling-Uhlenbeck Equation*

The problem, as Uehling and I saw it, was to generalize the Boltzmann equation and the Chapman-Enskog theory to the Bose-Einstein and Fermi-Dirac statistics.[1] The first part was easy. It only required the combination of the familiar streaming terms with the modification of the so-called Stosszahl-Ansatz required by the change in statistics, which was discussed in the literature.[2] The resulting so-called U-U equation has all the required conservation laws. It leads to an H-theorem, so that also all the thermodynamic consequences of the Boltzmann equation remain valid. Of course, the equilibrium velocity distribution now becomes either the Bose or the Fermi distribution.

The second part of the problem was more complicated because we had to understand in detail the Enskog successive-approximation method for solving the Boltzmann equation close to equilibrium and for finding the hydrodynamical equations in successive orders. I will spare the reader all details. Let me only say that we succeeded in showing that the U-U equation also leads to the Euler equations in zeroth order, the Navier-Stokes equations in first order, and so on. At least at this stage, one may conclude that the quantum theory leads to the same macroscopic description of fluid motion as the classical theory. Of course, the quantum theory affects the equation of state and the transport coefficients of the gas, and we had hoped that this would lead to observable effects, especially to a density dependence of the transport coefficients. But this was an illusion. We had not clearly realized that both the Boltzmann and the U-U equation take into account only *binary collisions*. In this sense the Boltzmann equation consists of the first two terms of the expansion in powers of nd^3, where n is the particle density and d the range of the intermolecular force. However, in the U-U equation the Bose or Fermi statistics produce a quasi triple and quadruple interaction, which has nothing to do with the intermolecular force. In reality they are at best a correction to the actual triple and higher-order collisions, which would extend the Boltzmann equation with terms of higher order in nd^3, which at that time were quite unknown and which even now are still in dispute. As a result,

[1] E. A. Uehling and G. E. Uhlenbeck, *Phys. Rev.* 43:552–61 (1933). See also E. A. Uehling, *Phys. Rev.* 46:917–29 (1934).

[2] See L. S. Ornstein and H. A. Kramers, *Zs f. Phys.* 42:481–86 (1927) and especially the article by W. Pauli in the Sommerfeld Festschrift (*Collected Papers*, Vol. I, p. 549).

there is still no theory for the transport coefficients of a Bose gas like He_4 at very low temperature.

In 1935 I went back to Holland where I succeeded Kramers as professor in theoretical physics *and* mechanics at the University of Utrecht. There, already in the first semester, a student came to see me to ask for an oral examination in fluid mechanics! I heard that Kramers in his three-year cycle of six courses always devoted at least one semester to bona fide classical hydrodynamics, and each student for his doctoral degree could choose four topics for the final examination. So all I could do was to postpone the examination of the student for a couple of weeks, and to study *very* hard the notes of Kramers' lectures and some books (especially Prandtl and some of Lamb) in order to know at least a bit more than the student. Naturally, the boy passed with quite a good grade. From this experience I concluded that I had better follow Kramers' example, and as a result in 1936 I gave for the first time a regular course in fluid mechanics. It made a great impression on me, and I think that every physicist should have some familiarity with the field. It will do even the high energy physicists some good. Personally, I had the satisfaction that I finally could explain to my young nephew Harm Buning (who was at the time a teenage airplane buff) why an airplane does not fall down and what the motion of the air is that keeps it up. Buning is now professor of aeronautical engineering at the University of Michigan and became, of course, a real expert.

My work in Utrecht, and later during the war in Ann Arbor and in the Radiation Laboratory at Massachusetts Institute of Technology, was not related in any way to fluid mechanics. This changed when after the war I started to work with Mrs. C. S. Wang Chang on various problems in the kinetic theory of gases.[3]

2 *The Problem of the Higher-Order Hydrodynamical Equations*

Why do the Navier-Stokes equations describe fluid motion so satisfactorily? I have already indicated that in the Chapman-Enskog theory the hydrodynamical equations are derived from the Boltzmann equation by a successive-approximation method. The expansion parameter is $\lambda\nabla$, where λ is the mean free path and the gradient operator ∇ acts on the five

[3] Mrs. C. S. Wang Chang worked with me as a post doctoral fellow until she returned in 1956 with her husband and child to China where she joined the Institute of Physics in Peking. Our work was supported by a grant from the Office of Naval Research and was published in a series of reports, of which the first five were later reprinted in *Studies in Statistical Mechanics*, Vol. V, Part A, pp. 1–100 (North-Holland Publishing Company). The first paper, especially, gives a survey of the problems which were going to occupy us.

macroscopic variables (density, ρ, velocity $\mathbf{u}$, and temperature T). This makes a plausible[4] explanation of why, for liquids where λ is of the same order as the diameter d of the molecules, the Navier-Stokes equations, which are of the *first* order in $\lambda\nabla$, are almost exact. But for very dilute gases higher approximations should play a role.[5] This was the main problem Mrs. Chang and I tried to attack.

The second-order hydrodynamical equations were already derived by J. Burnett[6] but the consequences were not discussed. Mrs. Chang and I worked on a number of special problems always hoping to find observable deviations from the predictions of the usual Navier-Stokes equations. Our main success came from the application of the theory to the dispersion and absorption of sound, for which Mrs. Chang developed the third and even in part the fourth-order hydrodynamical equations. The deviations from the classical theory were appreciable, but the comparison with experiment took quite a while. This was done in 1967 by J. Foch in his dissertation at the Rockefeller University.[7] The agreement with the Burnett and higher-order equations was quite remarkable.

Mrs. Chang and I also tried to discuss for flow problems the transition from the normal density or Clausius-gas regime to the very dilute or Knudsen-gas regime. It turned out that for such problems the higher-order hydrodynamical equations were useless because of the *basic* difficulty that there was *no* successive-approximation method to find the boundary conditions for the equations. They are still unknown and this limits severely the use of the higher-order equations. At present it seems that the only way to discuss the transition from the Clausius- to the Knudsen-gas regimes for flow problems with solid boundaries is to go back to the Boltzmann equation. Mrs. Chang and I tried to do this, for instance, for Couette flow but we had only limited success. For more recent work on these problems see especially the book of Cercignani.[8] Some progress has been made, but I believe that one is still far from a complete theory that can be checked experimentally.

Finally, one may ask whether we understand now why the Navier-Stokes equations are so satisfactory. Quantitatively, some of the predic-

[4] The explanation is only "plausible," because for a liquid the Boltzmann equation is of course not valid since the higher-order collisions must then be taken into account.

[5] Note that λ is inversely proportional to the density ρ. The Chapman-Enskog development is therefore an expansion in *inverse powers of* ρ.

[6] See S. Chapman and T. G. Cowling, *The Mathematical Theory of Non-Uniform Gases*, 2nd edition, p. 162 (Cambridge University Press, Cambridge, 1959).

[7] The dissertation was not published but the results can be found in the article by G. W. Ford and J. D. Foch (*Studies in Statistical Mechanics*, Vol. V, Part B, pp.101–231).

[8] Carlo Cercignani, *Mathematical Methods in Kinetic Theory* (Plenum Press, New York, 1969).

tions from the equations surely deviate from experiment, but the *very* remarkable fact remains that *qualitatively* the Navier-Stokes equations *always* describe the physical phenomena sensibly. This is *not* true for the high-order equations. For instance, the Burnett equations will lead for very-high-frequency sound to a negative absorption coefficient. The mathematical reason for this virtue of the Navier-Stokes equations is completely mysterious to me. It is, I think, a deep question.

The surge of interest in the kinetic theory of dense gases which is still going on today began in the late forties. The problem was how to generalize the Boltzmann equation to a higher density of the gas by taking into account triple and higher-order collisions of the molecules. To do this in a systematic fashion, it seemed that one had to go back to the basic Liouville equation of statistical mechanics and the first question was then whether one could develop a successive-approximation procedure in powers of the density (or better nd^3), which in first approximation would give the Boltzmann equation.

At that time I was so absorbed by the recent advances in quantum electrodynamics that I had not followed in detail the attempts of M. Born and H. S. Green and of J. G. Kirkwood and his co-workers to develop a consistent theory of dense gases.[9] This changed in the early fifties when I got hold of the treatise of N. N. Bogolubov entitled "Dynamical Theory in Statistical Mechanics" .(Moscow, 1946). Unfortunately, it was in Russian, but luckily Mark Kac was in Ann Arbor at the time and made a rough translation of the book.[10] It made a great impression on me. I have on various occasions[11] tried to explain the ideas of Bogolubov, so let me only say that his basic view, that the approach to equilibrium of a dynamical system may go in successive stages depending on the magnitudes of the basic relaxation times, seemed to me to unify the whole theory of nonequilibrium phenomena. It lead to my work with S. T. Choh.

3 *The Choh-Uhlenbeck Equation*

For a gas of moderate density Bogolubov distinguishes three widely separated ranges of relaxation times clustered around $\tau = d/v$ (= time

[9] The papers by M. Born and H. S. Green were collected in a book called *A General Kinetic Theory of Fluids* (Cambridge University Press, Cambridge, 1949). The work of J. G. Kirkwood and co-workers appeared in a series of papers in *J. Chem. Phys.*, 1946–1953.

[10] A careful translation was made later by E. K. Gora and appeared in *Studies in Statistical Mechanics*, Vol. 1 (1962), pp. 1–118.

[11] The first time was in the Higgins Lectures, which I gave in Princeton in 1954, and which were not published. The last time was during a summer symposium at the University of Colorado in Boulder in 1960, which led to the book (with G. W. Ford as co-author) entitled *Lectures in Statistical Mechanics* (American Mathematical Society, Providence, Rhode Island, 1963).

of a collision), around $t_0 = \lambda/v$ (= time between collisions), and around $\theta = L/v$ (= macroscopic relaxation time), where v is some average velocity. Since $\tau \ll t_0 \ll \theta$, Bogolubov argues that for a time $t > \tau$ the temporal development of the gas should be determined by the one-particle distribution $f_1(r,p,t)$, since this is the only distribution function which varies slowly over a time of order t_0. One has reached the so-called *kinetic stage*. Bogolubov showed from the Liouville equation that in the first order of the small parameter τ/t_0, f_1 fulfills the Boltzmann equation and he suggested that in higher orders one would get the corrections due to the triple and higher-order collisions. Note that $\tau/t_0 \sim nd^3$ (since $\lambda \sim 1/nd^2$), which is the parameter we expected. Then, for time $t > t_0$, Bogolubov expects that the temporal development of the gas can be described by the five macroscopic variables ρ, $\mathbf{u}$, and T since these are the only variables that vary slowly over a time of order θ. One has reached the so-called *hydrodynamical stage*, and in the first order of the small parameter t_0/θ one should get the Navier-Stokes equations. Since $t_0/\theta = \lambda/L$, this is of the order of the relative change of the macroscopic variables over a mean free path, which is just the Chapman-Enskog parameter.

These ideas of Bogolubov clearly contain a whole program. Choh and I began to work on it by first deriving the triple-collision correction to the Boltzmann equation, which leads to the so-called Choh-Uhlenbeck equation,[12] and then by applying the Chapman-Enskog method to this equation we again obtained the Navier-Stokes equations but now with transport coefficients that were density dependent. *Formally*, our results seemed reasonable, but unfortunately the equations were so complicated that, even for simple molecular models, we were unable to get numerical results. There was also a bothersome question. In the kinetic stage the successive-approximation procedure, although formally quite similar to the Enskog method in the hydrodynamical stage, started from the Liouville equation which is *reversible* in time, while in the Enskog method one starts from the *irreversible* Boltzmann equation. One *must* therefore in the kinetic stage postulate an initial condition, for which Bogolubov chose the absence of all molecular correlations at $t = 0$. Choh and I followed his example but we then ran into the difficulty that we could not prove the analogue of the H-theorem for our equation. The initial condition is *not* explained! Is it sufficient to ensure ergodic phase mixing in the sense of Boltzmann and Gibbs? I think it is still an open question.

More serious difficulties were discovered in the sixties by J. R. Dorfman and E. G. D. Cohen. They showed that at some stage of the development

[12] See the dissertation of S. T. Choh (University of Michigan, 1958), which was not published but is readily available. I gave a short summary in Chapter VII of the book with G. W. Ford (see footnote 11).

one always runs into divergent integrals, so that a power-series expansion in nd^3 is not possible. This work is still in flux.[13] It raises a similar question asked earlier for thc Navier-Stokes equations. Why is the Boltzmann equation so satisfactory in describing at least qualitatively the physical phenomena?

4 *The Hilbert Paradox and the Fluctuation Problem*

In the famous book of Hilbert on the theory of integral equations[14] there is a chapter devoted to the foundations of the kinetic theory of gases, in which Hilbert claims to have *proved* from the Boltzmann equation that the state of a gas is completely determined if one knows initially the space dependence of the five macroscopic variables ρ, $\mathbf{u}$, and T. I havc always been intrigued and puzzled by this so-called causality theorem. On the one hand it couldn't be true, because the initial-value problem for the Boltzmann equation (which supposedly gives a better description of the state of the gas) requires the knowledge of the initial value of the distribution function $f_1(r,\mathbf{v},t)$ of which ρ, $\mathbf{u}$, and T are only the first five moments in $\mathbf{v}$. But on the other hand the hydrodynamical equations surely give a *causal* description of the motion of a fluid. Otherwise how could fluid mechanics be used? Clearly, one has here to do not with a theorem but with a paradox, which I propose to call the *Hilbert paradox.*

A similar paradoxical situation occurs in the kinetic stage, where one describes the state of the gas by the *causal* Boltzmann equation for $f_1(r,\mathbf{v},t)$, which is a function of six variables (besides t) and which is supposed to be derivable from the *really causal* Liouville equation, which describes the state of the gas by a function of $6N$ variables if N is the number of molecules. How is this possible?

I believe that the ideas of Bogolubov give at least a hint of how these paradoxes could be resolved. Consider an ensemble of conservative mechanical systems of many degrees of freedom, like a dense gas. The temporal development is then governed by the Liouville equation. Let us now make the following assumptions: (*a*) the system fulfills all the requirements (metrical intransitivity, etc.) that ensure that in the coarse-grained Boltzmann-Gibbs sense *any* initial state will relax to an equilibrium state; (*b*) in the spectrum of relaxation times one relaxation time t_0 is very much longer than all the others, which are, say, of order τ. It is then clear that for *any* initial state of the system after a short time of order τ, the whole temporal development will be determined by the

[13] For a review see J. R. Dorfman and H. van Beyeren, *The Kinetic Theory of Gases*, Part B of *Statistical Mechanics* edited by Bruce J. Berne (Plenum Press, 1977).

[14] D. Hilbert, *Grundzüge einer allgemeinen Theorie der linearen Integralgleichungen* (Teubner Verlag, Leipzig, 1912); see Chapter 22.

temporal development of the few variables that correspond to the relaxation time t_0. I have expressed this by saying that after the time τ one can *contract* the description of the temporal evolution of the system.[15] However, this contracted description is *not causal* but *stochastic*. Only if one averages over the initial ensemble does one get a causal description, but each member of the ensemble will show fluctuations around the average behavior, which are due to all the fast varying variables that have been contracted away.

I hope that this view clarifies the ideas of Bogolubov. For a gas of moderate density with the three relaxation times, $\tau \ll t_0 \ll \theta$, we shall now say that after a time τ a contraction of the description is possible in which the only variable is the first distribution function $f_1(r,\mathbf{v},t)$. This is the kinetic stage, which after a time of order t_0 can be contracted further to the hydrodynamical stage with fewer variables, namely ρ, $\mathbf{u}$, and T. But *both* the kinetic equation for $f_1(r,\mathbf{v},t)$ and the hydrodynamical equations for ρ, $\mathbf{u}$, and T are not causal but stochastic. One *must* take into account the fluctuations.

Clearly, this generalization of the ideas of Bogolubov avoids the Hilbert paradox, but it makes the general program much more complicated. At any stage of contraction the fluctuations may change character, and it is also far from clear how the successive-approximation methods can be extended to include the fluctuations.

In one case significant progress has been made by R. F. Fox.[16] If the system is close to equilibrium so that the relaxation equations in the various stages can be taken to be *linear*, then one may expect that the fluctuations can be described by a Gaussian Markov process and, by appealing to the general fluctuation-dissipation theorem, one can complete the relaxation equations by adding known purely random forces. For the hydrodynamical equations this had been done already by Landau and Lifshitz.[17] Fox extended this to the linearized Boltzmann equation, and he also showed that at this level, at least, the contraction procedure also connects properly the fluctuating terms. In fact from the fluctuating hydrodynamical equations Fox derived the Langevin equation for the Brownian motion of an immersed particle using, so to speak, a further contraction.

What is still lacking is the derivation of the fluctuating terms in the

[15] I used the notion of a contracted description for the first time in my Boulder lectures of 1957. See Appendix I in the book by Mark Kac (*Probability and Related Topics in Physical Sciences*, Interscience, 1959).

[16] R. F. Fox, Dissertation (Rockefeller University, 1969). In part published in the papers by R. F. Fox and G. E. Uhlenbeck, *Phys. Fluids* 13:1893–1902, 2881–90 (1970).

[17] L. D. Landau and E. M. Lifshitz, *Fluid Mechanics* (Pergamon Press, 1959), Chapter 17.

Boltzmann equation from the appropriately coarse-grained Liouville equation. This would be an extension of the ergodic theorems and may therefore be a real challenge.

I am also disappointed that so far the fluctuating hydrodynamical equations have not lead to new and interesting phenomena. I have advocated the study of light scattering in a moving fluid just before it has reached an instability point, since one may expect that it will be similar to the scattering near a critical point.

Finally, I should say that if the fluid is *not* close to equilibrium, the theory becomes part of the problem of how to describe *nonlinear* stochastic processes, which is still a wide-open field.

I hope that these notes give an impression of the relation of my work in the kinetic theory of gases to some of the problems in fluid mechanics. Clearly, they show that I am *not* a fluid mechanician. I still remember that once after a talk I had given about these questions, an aeronautical engineer insisted that in fluid mechanics it was *not* necessary to know that the air or the water consisted of molecules. He may be right, but the behavior of fluids is also a part of physics, and the molecular structure is no longer just an hypothesis.

Finally, I want to make a few remarks about turbulence. In my lectures on fluid mechanics, which I gave regularly in Ann Arbor, "das grosse Problem der Turbulenz" was for me always the high point. Because the description of turbulent flow clearly must be stochastic, I had hopes that it could be looked upon as a further contraction of the hydrodynamical description. But this is no doubt an illusion, because the fluctuations are *not* due to a simplification of the description, but in the present view (due to Landau) they are due to successive instabilities. I should therefore also warn the reader not to confuse the hydrodynamical fluctuations discussed in Section 4 with turbulence. The former are due to the temperature motion and are more or less understood, while the latter are due to macroscopic instabilities in the fluid flow and their description is in my opinion still a deep riddle.

Ann. Rev. Fluid Mech. 1980. 12: 11–43

SOLITARY WAVES ×8151

John W. Miles
Institute of Geophysics and Planetary Physics, University of California, San Diego, La Jolla, California 92093

> I believe I shall best introduce this phaenomenon by describing the circumstances of my own first acquaintance with it. I was observing the motion of a boat which was rapidly drawn along a narrow channel by a pair of horses, when the boat suddenly stopped—not so the mass of water in the channel which it had put in motion; it accumulated round the prow of the vessel in a state of violent agitation, then suddenly leaving it behind, rolled forward with great velocity, assuming the form of a large solitary elevation, a rounded, smooth and well-defined heap of water, which continued its course along the channel apparently without change of form or diminution of speed. I followed it on horseback, and overtook it still rolling on at a rate of some eight or nine miles an hour, preserving its original figure some thirty feet long and a foot to a foot and a half in height. Its height gradually diminished, and after a chase of one or two miles I lost it in the windings of the channel. Such, in the month of August 1834, was my first chance interview with that singular and beautiful phaenomenon. . . .
>
> John Scott Russell (1845)

1 INTRODUCTION

John Scott Russell discovered and named the solitary wave in 1834, carried out extensive experiments in 1834 and 1835, and reported his investigations at the British Association meetings of 1837 and 1844 (Russell 1838, 1845).[1] His description of the wave as a solitary elevation of finite amplitude and permanent form, which appeared to contradict Airy's (shallow-water-theory) prediction that a wave of finite amplitude cannot propagate without change of form, stimulated considerable scientific interest at the time and later in the nineteenth century. Interest waned after the resolution of the Airy paradox by Boussinesq and Rayleigh (see below) and was sporadic prior to Zabusky & Kruskal's (1965) discovery that solitary waves typically dominate the asymptotic solution of the Korteweg-deVries equation (see Section 3). Current interest stems from that discovery and is intense.

[1] Lamb (1932, Section 252) and many other writers identify Russell as "Scott Russell," but the correct surname is simply Russell (*Encyc. Brit.*, 11th Ed.).

0066-4189/80/0115-0011$01.00

The present review deals primarily with solitary waves on the surface of a homogeneous, nonrotating liquid of finite depth (see Keulegan & Patterson 1940, Munk 1949, and Wiegel 1964a for earlier reviews of solitary surface waves). Internal solitary waves, which may be associated with any or all of stratification, shear, compressibility, and rotation are considered in Section 9. Solitary waves in nonfluid media are not discussed explicitly, nor are envelope solitons (Cohen, Watson & West 1976, Lake, Yuen & Ferguson 1978, Yuen & Lake 1975), which are relevant for both deep-water and atmospheric waves and are sufficiently important to merit a separate review.

Russell's Experiments

Russell, following the encounter described in the above quotation, produced solitary waves in a laboratory channel by either releasing an impounded elevation of water or dropping a weight at one end of the channel (Figure 1). He concluded from his experiments that: the volume of the wave is equal to that of the initial displacement (although the wave does not comprise the same particles of water); a wave of amplitude a in water of depth d advances with the speed

$$c = [g(d+a)]^{\frac{1}{2}}; \tag{1.1}$$

breaking occurs for $a \doteqdot d$. Bazin (1865) repeated Russell's experiments, confirmed (1.1), and concluded that breaking occurs for a somewhat less than d.

Russell did not determine an explicit relation between the height and characteristic length of a solitary wave [see (1.2) and (1.4) below], but he did observe that an initial elevation might, depending on the relation between its height and length, evolve into any of a pure solitary wave, a single solitary wave plus a residual wave train, or two or more solitary waves with or without a residual wave train. He also attempted to produce negative solitary waves but found that an initial depression is transformed into an oscillatory wave train of gradually increasing length and decreasing

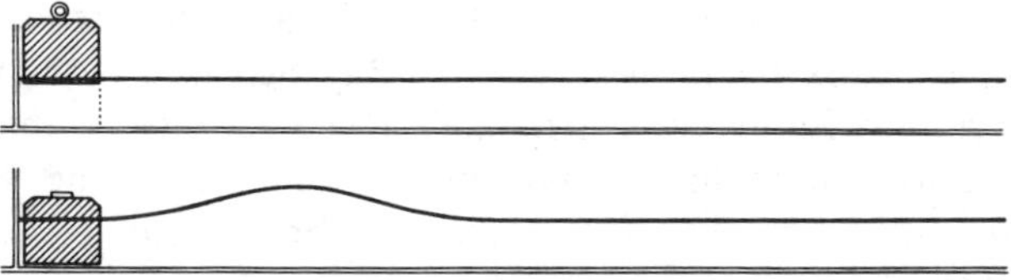

Figure 1 Russell's (1845) observation of the evolution of a solitary wave in a wave tank (reproduced from the original drawing).

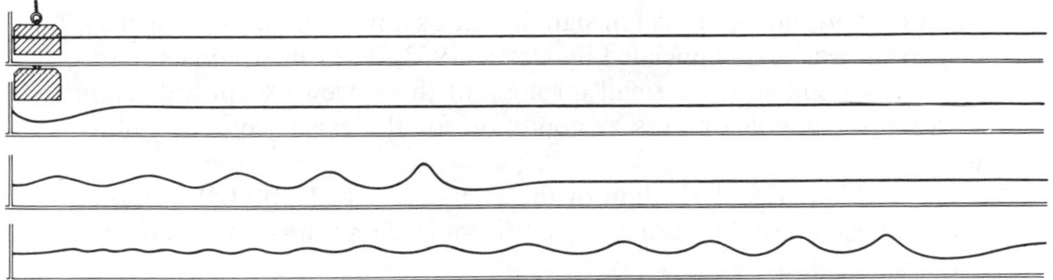

Figure 2 Russell's (1845) observations of the evolution from an initial depression in a wave tank (reproduced from the original drawing).

amplitude (Figure 2). [A solitary wave of depression is possible if the surface tension T is such that $d < (3T/\rho g)^{\frac{1}{2}} \doteqdot 0.5$ cm for water; see Section 3.]

Basic Theoretical Results

The conflict between Russell's observations and Airy's shallow-water theory (see Airy 1845, Stokes 1891, Rayleigh 1876, Lamb 1932, Section 252, Ursell 1953, Stoker 1957, Section 10.9) was resolved independently by Boussinesq (1871a,b, 1872, 1877) and Rayleigh (1876), who showed that appropriate allowance for vertical acceleration (which is ultimately responsible for dispersion but is neglected in the shallow-water theory), as well as for finite amplitude, leads to the solution

$$y = a \operatorname{sech}^2 [(x-ct)/l], \quad \alpha \equiv a/d \ll 1, \quad \beta \equiv (d/l)^2 = O(\alpha) \tag{1.2}$$

(y is the free-surface displacement, x the horizontal coordinate, and t the time). The wave speed is given by (1.1) or, equivalently,

$$\mathscr{F}^2 \equiv c^2/gd \doteqdot 1+\alpha, \tag{1.3}$$

where $\mathscr{F}$ is the Froude number (solitary waves are *supercritical*); the characteristic length l is determined by

$$\mathscr{U} \equiv 3al^2/4d^3 = 3\alpha/4\beta = 1 \tag{1.4}$$

[Rayleigh obtained $l^2 = 4d^2(d+a)/(3a)$, which reduces to (1.4) for $\alpha \ll 1$]. The essential quality of the solitary wave, then, is the balance between nonlinearity, which tends to steepen the wave front in consequence of the increase of wave speed with amplitude and is measured by α, and dispersion, which tends to spread the wave front (the wave speed of any spectral component decreases with increasing wave number) and is measured by β. The dimensionless parameter $\mathscr{U}$ appears in the work of

Stokes (1849); however, its full significance as a measure of nonlinearity/dispersion was first enunciated by Ursell (1953). The dimensionless parameter α is assumed to be small throughout this review except in Section 4, but $\alpha \ll 1$ is not a necessary condition for the existence of a solitary wave.

Rayleigh's (1876) derivation of the equivalents of (1.2)–(1.4) is reproduced by Lamb (1932, Section 252). It is more direct but less penetrating than that of Boussinesq.

2 BOUSSINESQ THEORY

Boussinesq's first (1871a) paper on the solitary wave only sketches the derivation of (1.2), and it is necessary to look to his 1871b supplement and to either his 1872 paper or his 1877 essay to obtain a fuller appreciation of his contributions [Lamb (1932) refers only to the 1871a paper and, at least in retrospect, appears to have underestimated the significance of Boussinesq's work]. The following summary is from the 1872 paper.

Boussinesq begins by reducing the equations governing two-dimensional, irrotational motion of an inviscid liquid in a uniform rectangular channel to

$$y_{tt} = c_0^2[y_{xx} + \tfrac{3}{2}(y^2/d)_{xx} + \tfrac{1}{3}d^2 y_{xxxx}] \qquad (c_0^2 \equiv gd), \tag{2.1}$$

which is known as Boussinesq's equation (*singular*).[2] It rests on the assumptions that: (*a*) $d/l \ll 1$, where l is a characteristic length, such that $y_x = O(a/l)$ if $y = O(a)$; (*b*) $a/d = O(d^2/l^2)$; (*c*) the disturbance changes only slowly in a reference frame moving to the right with the speed c_0, such that $y_t \doteq -c_0 y_x$ in the derivation of the second and third terms on the right hand side of (2.1), which represent nonlinearity and dispersion, respectively. It should be emphasized that (2.1) admits both right- and left-running solutions, but only the former are consistent with (*c*); this restriction has occasionally been overlooked (see remarks by Lin & Clark 1959, Long 1964a, and Byatt-Smith 1971). The Korteweg-deVries equation, (3.1) below, incorporates (*c*) more completely and admits only right-moving solutions.

Boussinesq next shows that any local section of a unidirectional solution of (2.1) moves with the approximate speed

[2] The Boussinesq equations, which, in their conventional form, are evolution equations for the free-surface displacement and the mean horizontal velocity and are not restricted to unidirectional propagation, do not appear explicitly in the 1871 and 1872 papers. However, Boussinesq's (19) in the 1872 paper or (280) in the 1877 essay are, after dropping several higher-order terms, equivalent equations for the free-surface displacement and the horizontal velocity at the bottom of the channel (cf Whitham 1974, Section 13.11).

$$c = c_0[1+\tfrac{3}{4}(y/d)+\tfrac{1}{6}(d^2 y_{xx}/y)], \tag{2.2}$$

wherein c_0 is the speed of infinitesimal long waves and the second and third terms within the brackets represent nonlinearity and dispersion, respectively. He infers from (2.2) that an initial elevation of water for which al^2 is significantly in excess of the value determined by (1.4) would tend to disintegrate into two or more solitary waves (plus, in most cases, a residual wave train) and that an initial depression would tend to decay into an oscillatory wave train, all in conformity with Russell's observations.

He then goes on to show that the solutions of (2.1) are characterized by the invariants

$$Q = \int_{-\infty}^{\infty} y\,dx, \quad E = \int_{-\infty}^{\infty} y^2 dx, \quad M = \int_{-\infty}^{\infty} [y_x^2 - 3(y/d)^3]\,dx \tag{2.3a,b,c}$$

provided that y vanishes with sufficient rapidity as $x \to \pm\infty$. Q and E evidently represent the volume and energy of the wave; Boussinesq designates M as the *moment of instability* and demonstrates that the solitary wave is the unique solution of the conditional variational problem $\delta M = 0$ with E fixed (he omits the implicit constraint that Q be fixed, but this has no effect on the end result, for which the corresponding Lagrange multiplier vanishes).

He also shows that the amplitude and volume of a solitary wave of prescribed energy are given by

$$a = \tfrac{3}{4}E^{\frac{2}{3}}d^{-1}, \quad Q = 2E^{\frac{1}{3}}d \tag{2.4a,b}$$

and remarks that

> When a wave propagates along the canal of which the depth $[d]$ is slowly decreasing from one point to the next . . . the bottom of the canal must continuously reflect a small part of the movement, in a manner such that the volume and energy of the [wave] must divide between the direct wave and this reflected wave. The latter being of an increasing length and of a height which is evidently very small, its volume will become finite without its energy, which is at once proportional both to this volume and to its height, ceasing to remain very small. The direct wave thus will conserve, approximately, all the energy of the [wave], and, as it retains effectively the form of a solitary wave, . . . its height $[a]$ and . . . its volume Q . . . will be obtained at any particular instant by means of [(2.4)], where E will be invariable: one sees that the wave will become higher, shorter, and consequently less stable, until finally it lacks a base and breaks. The opposite would occur if the depth increased.

These predictions appear to have been overlooked in most of the current literature (although not by Keulegan & Patterson 1940), and the result $a \propto 1/d$ has been rediscovered by various writers through much more indirect analyses, as well as through Boussinesq's energy-conservation argument (see Section 6).

3 KORTEWEG-DEVRIES THEORY

Korteweg-deVries Equation

The Boussinesq equation (2.1) is reduced to the Korteweg-deVries (KdV) equation[3]

$$y_t + c_0[y_x + \tfrac{3}{2}(y/d)y_x + \tfrac{1}{6}d^2 y_{xxx}] = 0 \tag{3.1}$$

by factoring the operator $c_0^2\partial_x^2 - \partial_t^2$, invoking the prior assumption of unilateral propagation, and integrating with respect to x. Korteweg & deVries (1895) derive a somewhat more general equivalent of (3.1), in which they allow for any uniform translation of the reference frame and incorporate a surface tension T, such that the coefficient d^2 in the coefficient of y_{xxx} is replaced by $d^2 - 3(T/\rho g)$. They then obtain a family of periodic solutions of the form $y = f(x - ct)$, which they call *cnoidal waves*; see W13.12[4] for details. [Boussinesq (1877) also discussed periodic solutions of (2.1) but did not obtain explicit integrals.] This family comprises the Boussinesq solitary wave (1.2) in the limit of infinite period, although allowance for surface tension implies that l in (1.2) must be replaced by

$$l_* = l[1 - (d_*/d)^2]^{\frac{1}{2}}, \quad d_* = (3T/\rho g)^{\frac{1}{2}}. \tag{3.2a,b}$$

If $d < d_*$ (0.5 cm for water), l must be replaced by $|l_*|$ and the sign of η must be changed, so that the solitary wave becomes one of depression rather than elevation. Dispersion vanishes in the present approximation if $d = d_*$.

Current interest in the KdV equation stems in general from the corresponding interest in nonlinear wave theory and in particular from the discoveries that (3.1) or its equivalent arises in a wide variety of physical contexts, that its solutions may exhibit Fermi-Pasta-Ulam (1955) recurrence (Zabusky & Kruskal 1965), and that it can be solved exactly for appropriately restricted initial data (Gardner et al 1967, 1974). There have been several recent reviews (Jeffrey & Kakutani 1972, Scott et al 1973, Miura 1974, 1976, Kruskal 1974, 1975, Makhankov 1978) with varying emphases, and it is dealt with extensively by Whitham (1974) and Karpman (1975); the present review is written within the context of

[3] Curiously, none of Lamb (1932), Stoker (1957), or Wehausen & Laitone (1960) mention the KdV equation explicitly, although each of them cites the original paper in connection with cnoidal waves.

[4] The prefix W, followed by the appropriate section number, is used throughout this section to refer to Whitham's (1974) treatise.

fluid mechanics (although it is worth emphasizing that the neglect of dissipation may be rather more realistic in other contexts).

It is convenient, before proceeding further, to nondimensionalize (3.1) in a reference frame moving with the basic wave speed c_0 by introducing

$$\xi = (x - c_0 t)/l, \quad \tau = \tfrac{1}{6}(d/l)^2 (c_0 t/l), \quad \eta = y/a, \tag{3.3a,b,c}$$

where l and a now are appropriate length scales. The end result is

$$\eta_\tau + 12\mathscr{U}\eta\eta_\xi + \eta_{\xi\xi\xi} = 0, \tag{3.4}$$

where $\mathscr{U} \equiv 3al^2/4d^3$, as in (1.4). This is Whitham's (W17.1) normal form of the KdV equation, although he uses $\sigma \equiv 12\mathscr{U}$. Equations in which the nonlinear term $\eta\eta_\xi$ is replaced by $\eta^p\eta_\xi$ ($p = 2, 3, \ldots$) are known as *modified* KdV equations (although that description is sometimes reserved for $p = 2$). The solitary-wave solution (1.2)–(1.4) reduces to

$$\eta = \operatorname{sech}^2(\xi - 4\tau), \quad \mathscr{U} = 1. \tag{3.5a,b}$$

Stability

The stability of (3.5) as a solution of (3.4) has been established by Benjamin (1972); see also Bona (1975). Jeffrey & Kakutani (1970) had earlier demonstrated stability with respect to small perturbations; see also Berryman (1976). All of these references deal with one-dimensional perturbations; Kadomtsev & Petviashvili (1970) and Oikawa et al (1974) have shown that the one-dimensional solitary wave (3.5) is neutrally stable with respect to small, transverse perturbations. Observation strongly supports the conclusion that a solitary wave in a uniform channel is stable with respect to any moderate disturbance.

Interacting Solitons

A crucial discovery in the development of KdV theory was that unidirectional, Boussinesq solitary waves of different amplitudes, and hence of different speeds, pass through one another without any permanent loss of identity and suffer only phase shifts, even though nonlinear distortion is quite significant during the interaction. This phenomenon was discovered from numerical solutions of the KdV equation by Zabusky & Kruskal (1965), who coined the term *soliton* for any solitary wave having this quality. [Higher-order solitary waves, such as those discussed in Section 4 below, may not be solitons in this sense. Solitary waves in two dimensions, such as those discussed in Section 5, may or may not be solitons.] Subsequently, Gardner et al (1974), Hirota (1971, 1976), and Whitham (W17.2) constructed analytical solutions of (3.1) that describe the interaction among N solitons for any N; see also Zabusky (1967), Lax (1968), Wadati & Toda (1972), Kruskal (1974), and the film made by Zabusky et al (1965). Experimental confirmation has been provided by Zabusky & Galvin (1971), Hammack & Segur (1974), and Weidman & Maxworthy (1978).

Integral Invariants

Closely associated with the existence of these N-soliton solutions is the existence of an infinite number of integral invariants of the KdV equation. The first three of these are equivalent to those constructed by Boussinesq, although his discovery of (2.3c) and his associated variational principle for the solitary wave have been widely overlooked. Zabusky & Kruskal (1965) discovered two more, and Miura et al (1968) demonstrated the existence, and gave an algorithm for the construction, of an infinite sequence.

Kruskal (1974) and Zabusky (1967) subsequently rediscovered Boussinesq's variational principle and then went on to show that the conditional variational problem $\delta I_{N+2} = 0$ with $I_1, \ldots, I_{N+1}$ (I_n is the nth invariant) fixed would yield an N-soliton solution of (3.4). See also Lax (1975, 1976).

Inverse Scattering Theory

Perhaps the most remarkable property of the KdV equation is the existence of an algorithm for the reduction to a linear integral equation of the corresponding initial-value problem for initial data that vanish at $x = \pm\infty$. The details are given in the original papers (Gardner et al 1967, 1974) and in W17.3. Let

$$\eta(\xi,0) = \eta_0(\xi), \quad \langle \eta_0 \rangle \equiv \int_{-\infty}^{\infty} \eta_0(\xi)\, d\xi, \tag{3.6a,b}$$

and N be the number of solitons that evolve from η_0; then $N \geqq 1$ if $\langle \eta_0 \rangle > 0$, $N = 1$ if $\langle \eta_0 \rangle = 0$ and $\mathscr{U} \ll 1$, and $N = 0$ if $\eta_0 \leqq 0$ for all ξ. An upper bound to N is given by Segur (1973).

The solitons are determined by solving the eigenvalue problem (a subproblem of the complete inverse scattering algorithm)

$$\{(d/d\xi)^2 - \kappa^2 + 2\mathscr{U}\eta_0(\xi)\}\psi(\xi) = 0 \quad (-\infty < \xi < \infty), \tag{3.7}$$

$$\psi \sim e^{-\kappa\xi} \quad (\xi \uparrow \infty), \tag{3.8}$$

to obtain the discrete eigenvalues $\kappa_1 > \kappa_2 > \ldots > \kappa_N > 0$ ($N \equiv 0$ if the discrete spectrum is empty) and the corresponding normalizing parameters $\gamma_n = 1/\langle\psi_n^2\rangle$. The asymptotic ($\tau \uparrow \infty$) solution for $N \geqq 1$ then is given by (W17.4, 17.5)

$$\eta \sim \mathscr{U}^{-1} \sum_{n=1}^{N} \kappa_n^2 \operatorname{sech}^2 (\kappa_n \xi - 4\kappa_n^3 \tau + \delta_n), \quad \delta_n = \tfrac{1}{2} \ln (2\kappa_n/\gamma_n). \tag{3.9a,b}$$

It follows from (3.7) and (3.9) that the maximum amplitude of a solitary wave cannot exceed twice the amplitude of the initial profile from which it evolves (Peregrine 1972).

Perhaps the simplest example is described by (Landau & Lifshitz 1958, p. 69, Problem 4)

$$\eta_0 = \operatorname{sech}^2 \xi : \kappa_n = \tfrac{1}{2}\{(1+8\mathscr{U})^{\frac{1}{2}} - (2n-1)\}, \quad N = [\tfrac{1}{2}(1+8\mathscr{U})^{\frac{1}{2}} + \tfrac{1}{2}], \tag{3.10}$$

where [ ] signifies *the integral part of*. There is one soliton if $0 < \mathscr{U} < 1$, and an additional soliton appears as each of the critical values $\mathscr{U} = 1, 3, 6, 10, \ldots$ is exceeded; $\kappa_N = 0$ at a critical value of $\mathscr{U}$ and therefore is trivial in (3.9).

The eigenvalue problem for $\mathscr{U} \ll 1$ may be solved by perturbation theory for an arbitrarily prescribed initial displacement and yields the first approximations (Landau & Lifshitz 1958, p. 155, Problem 1; Miles 1978b)

$$\kappa_1 = \mathscr{U}\langle\eta_0\rangle + O(\mathscr{U}^2) \qquad (\langle\eta_0\rangle > 0) \tag{3.11a}$$

or

$$\kappa_1 = 2\mathscr{U}^2 \left\langle \left(\int_{\xi}^{\infty} \eta_0 d\xi \right)^2 \right\rangle + O(\mathscr{U}^3) \quad (\langle\eta_0\rangle = 0). \tag{3.11b}$$

Ursell Paradox

The limit $\mathscr{U} \downarrow 0$ presumably leads to linear theory; however, as both Russell's observations and (3.11a) imply, an initial mound of water in a wave tank, no matter how small, ultimately produces a solitary wave ($N = 1$), whereas linear theory yields no such wave (Ursell 1953). In fact, (see Hammack & Segur 1978, Miles 1978b) linear theory is valid only for $\mathscr{U}\tau^{\frac{1}{3}} \ll 1$, whereas the soliton is fully evolved only for $\mathscr{U}\tau^{\frac{1}{3}} \gg 1$ if $\langle\eta_0\rangle > 0$ or $\mathscr{U}^2\tau^{\frac{1}{3}} \gg 1$ if $\langle\eta_0\rangle = 0$ (zero initial volume).

Perturbed KdV Equation

The inverse-scattering theory may be perturbed to obtain solutions of perturbations of (3.1) or (3.4), such as (6.2) below; see Karpman & Maslov (1977) and Kaup & Newell (1978).

4 HIGHER-ORDER APPROXIMATIONS

McCowan (1891, 1894) attempted to improve Boussinesq's approximation (1.2) to the solitary wave, but his solution does not satisfy the requirement of constant pressure along the free surface. Korteweg & deVries (1895) developed a series expansion in $\alpha = a/d$ and obtained the second approximation. Subsequent expansions in α were developed by Gwyther (1900a,b), Levi-Cività (1907), Weinstein (1926) [Weinstein's paper contains a significant numerical error (see Long 1956)], Keller (1948), Long (1956), Laitone (1960), Grimshaw (1971), Fenton (1972), Epstein (1974), and Longuet-Higgins & Fenton (1974); of these, only the last is useful for $0.9\alpha_{max} < \alpha < \alpha_{max} = 0.83$ (see below), although Gwyther (1900a) obtained the somewhat fortuitous estimate $\alpha_{max} = 5/6 = 0.833$. An integral-equation formulation was initiated by Nekrasov (1921) and developed by Levi-Cività (1925), Packham (1952), Hunt (1955), Yamada (1957, 1958), Milne-Thomson (1968, Section 14.75), Lenau (1966), Yamada et al (1968), Strelkoff (1971), and Witting (1975); generically related is the integral-equation formulation of Byatt-Smith (1970, 1971) and Byatt-Smith & Longuet-Higgins (1976). It should perhaps be emphasized that the principal interest of all of this work is theoretical and that capillarity and viscosity must be expected to have significant effects as $\alpha \uparrow \alpha_{max}$; indeed, laboratory measurements (Daily & Stephan 1953a,b) suggest that it may be difficult to exceed $\alpha = 0.6$–0.7 for a steady solitary wave. Moreover, Boussinesq's profile fits both observation and the more accurate theoretical calculations rather well for $\alpha \leqq 0.5$ and appears to be superior to the approximations of Rayleigh (1876), McCowan (1894), and Packham (1952); see Yamada (1958).

It suffices here to consider further only the work of Longuet-Higgins & Fenton (1974), who also summarize much of the earlier work, and of Yamada (1957) and Witting (1975).

The failure of the earlier series-expansion formulations in the neighborhood of $\alpha = \alpha_{max}$ stems essentially from the fact that the mass, momentum, energy, and speed of a solitary wave all exhibit maxima at values of $\alpha < \alpha_{max}$, in consequence of which convergence is irregular in some finite neighborhood of α_{max}. [Some improvement may be obtained by expanding in powers of $\mathscr{F}^2 - 1$ or some equivalent thereof, as in the work of Friedrichs & Hyers (1954), but there remains the difficulty that two different solitary waves exist for a finite interval of $\mathscr{F}$ bounded by the waves of maximum speed, $\mathscr{F} = \mathscr{F}_{max}$, and maximum amplitude, $\alpha = \alpha_{max}$.] Moreover, convergence presumably is adversely affected by the singularity associated with the sharp (120°) crest of the wave of maximum height. (The term *convergence* is used loosely here, and it appears likely that at least some of the series in question are only asymptotic for any finite value of α.) Longuet-Higgins & Fenton (1974) circumvent these difficulties by introducing the expansion parameter $\omega = 1-(U/c_0)^2$, where U is the velocity at the wave crest in a reference frame moving with the wave speed; $U = c_0$ and $\omega = 0$ for $\alpha = 0$, $U = 0$, and $\omega = 1$ for $\alpha = \alpha_{max}$, and ω varies monotonically from 0 to 1 as α varies from 0 to α_{max}. They also improve convergence by introducing Padé approximants for the series and invoke various identities (Longuet-Higgins 1974) to check the numerical significance of their results. The details of their formulation are too complicated to recapitulate here, where it must suffice to state the results $\alpha_{max} = 0.827$ ($\mathscr{F} = 1.286$) and $\mathscr{F}_{max} = 1.294$ at $\alpha = 0.790$.

The prediction $\alpha_{max} = 0.827$ is confirmed by the earlier work of Yamada (1957), who obtained 0.827 ± 0.008 [subsequently modified to 0.8262 by Yamada et al (1968)] through an integral-equation formulation, and Lenau (1966), who obtained 0.828 using essentially the same procedure. Witting (together with J. Bergin) also has repeated Yamada's (1957) calculation and reports $\alpha_{max} = 0.8332$. The discrepancy between this value and Yamada, Kimura & Okabe's value of 0.8262, which presumably was obtained by the same method, remains unexplained. Moreover, Witting states that his solution (and, by implication, that of Longuet-Higgins & Fenton) is incomplete and that "the functions needed to complete the solution may shift singularities" such that the uncertainty in α_{max} could be of the order of 0.005; whatever the reason, his conclusion that "the exact (to 3 decimal places) amplitude of the highest wave cannot be considered known" appears to remain true in 1979. [It should perhaps be emphasized that Longuet-Higgins and his colleagues have published rather full details of their work; Witting & Bergin, on the other hand, have not published any details of their calculation, nor has Witting published any details on the possible incompleteness of any of the solutions in question.]

Longuet-Higgins & Fenton's (1974) theoretical prediction of $\mathscr{F}$ vs α falls slightly below that of Boussinesq and slightly above the measured values of Daily & Stephan (1953a,b) in the range $0 < \alpha < 0.62$ (the upper limit of the measurements). The difference between the two theoretical predictions is less than 1% for $\alpha < 0.5$ and exceeds the experimental scatter only for $\alpha \gtrsim 0.4$ (although the Longuet-Higgins & Fenton approximation is consistently closer to observation than is that of Boussinesq). This, together with the aforementioned comparison of Yamada (1958), suggests that higher-order approximations offer little or no practical advantage over Boussinesq's approximation for $0 < \alpha < 0.5$.

5 INTERACTING SOLITARY WAVES

The simplest theoretical interaction problem appears to be the collision of two solitary waves moving in opposite directions. The interaction then is of relatively brief duration, and the waves emerge unchanged within the KdV approximation (Gwyther 1900c); a second-order calculation yields a relative phase shift that is $O(\alpha)$ as $\alpha \to 0$. If the waves are moving in the same direction, the interaction is of relatively long duration, and the waves emerge unchanged in profile but with a relative phase shift that is $O(1)$ (Zabusky & Kruskal 1965).

The first type of interaction may be generalized to include oblique interactions between any two solitary waves (*solitons* in the present context) for which $\kappa \equiv \sin^2 \frac{1}{2}\psi_2 - \psi_1) \gg 3\alpha$, where $\psi_2 - \psi_1$ is the angle between the wave normals; it may be characterized as *weak* in the sense that the first approximation to the solution is obtained by superimposing the solutions of the KdV equation for the individual solitons. The second-order interaction term has been obtained by Byatt-Smith (1971) for normal reflection and by Miles (1977b) for oblique reflection, wherein the interaction is between the incident wave and its image. The predicted maximum amplitude for normal reflection has been experimentally confirmed by Maxworthy (1976), but the predicted phase shift disagrees with his measurements.

The second type of interaction may be generalized to include oblique interactions for which $\kappa = O(\alpha)$ and may be analytically characterized as essentially nonlinear or *strong* (the adjective *strong* is used here in the sense of scattering theory; the actual nonlinearity is assumed to remain weak). Only strong interactions are considered further in this section; however, the results for $\kappa = O(\alpha)$ do predict a phase shift that is $O(\alpha)$ for $\kappa \gg \alpha$ and, to that extent, are consistent with the corresponding results for weak interactions.

One-Dimensional Interactions

It now is convenient to adopt the description

$$\alpha\eta_i = \alpha_i \operatorname{sech}^2 \theta_i, \quad \theta_i = (3a_i/4d^3)^{\frac{1}{2}}(\mathbf{n}_i \cdot \mathbf{x} - c_i t) + \theta_{i0} \tag{5.1a,b}$$

for a soliton of strength $\alpha_i \equiv a_i/d$ ($\eta_i \equiv y_i/a$, where a is the reference amplitude) propagating in the direction $\mathbf{n}_i$ in the $\mathbf{x}$ plane with the wave speed $c_i = c_0(1 + \frac{1}{2}\alpha_i)$; θ_{i0} is a phase constant, which may be chosen arbitrarily for one of a set of, but not for the remaining, solitons [(5.1) reduces to (3.5) for $\alpha_i = \alpha$, $\mathbf{n}_i = \{1,0\}$, and $\theta_{i0} = 0$]. It also is convenient to introduce

$$\alpha\eta_{i\phi} \equiv \alpha_i \operatorname{sech}^2 (\theta_i - \phi) \tag{5.2}$$

for a soliton that differs from η_i only by the phase shift ϕ.

The solution (W17.2) for the interaction between two solitons of strengths α_1 and α_2, $\alpha_2 > \alpha_1$, may be asymptotically described by

$$\{\eta_1, \eta_{2\phi}\} \to \{\eta_{1\phi}, \eta_2\}, \tag{5.3}$$

where

$$\phi = \ln [(\alpha_2^{\frac{1}{2}} - \alpha_1^{\frac{1}{2}})/(\alpha_2^{\frac{1}{2}} + \alpha_1^{\frac{1}{2}})] \qquad (\alpha_2 > \alpha_1) \tag{5.4}$$

and the left- and right-hand sides of (5.3) refer to the solitons in the limits $t \downarrow -\infty$ and $t \uparrow +\infty$, respectively. The only remnant of the interaction then is a forward phase shift of the stronger soliton and a corresponding backward phase shift of the weaker soliton; moreover, the interaction is *phase-conserving* in that the sum of the phases of the incoming waves is equal to the sum of the phases of the outgoing waves. Experimental confirmation of both the phase shift and the predicted description of the collision has been reported by Weidman & Maxworthy (1978). Detailed, numerical-graphical descriptions of the interaction of KdV and modified-KdV solitons have been given by Fornberg & Whitham (1978).

It is worth noting that one-dimensional interactions between solitary waves appear to be relevant to the evolution and structure of undular bores; see Keulegan & Patterson (1940), Peregrine (1966), and Witting (1975).

Regular Oblique Interactions

Strong, oblique interactions, for which $\mathbf{x} = \{x, z\}$,

$$\mathbf{n}_i = \{\cos \psi_i, \sin \psi_i\} \quad (i = 1,2), \qquad \kappa = \sin^2 \tfrac{1}{2}(\psi_2 - \psi_1) = O(\alpha), \tag{5.5a,b}$$

are governed by the two-dimensional KdV equation (Kadomtsev & Petviashvili 1970, Oikawa et al 1974)

$$(\eta_\tau + 12\mathscr{U}\eta\eta_\xi + \eta_{\xi\xi\xi})_\xi + 12\mathscr{U}\eta_{\zeta\zeta} = 0, \tag{5.6}$$

where $\zeta = (3\alpha)^{\frac{1}{2}}(z/l)$ is a dimensionless transverse coordinate and ξ, η, and $\mathscr{U}$ are defined as in Section 3 [note that the characteristic lengths in the x and z directions are $O(d/\alpha^{\frac{1}{2}})$ and $O(d/\alpha)$, respectively]. The solution of (5.6) for interacting solitons has been obtained by Chen (1975), Satsuma (1976), Hirota (1976), and Miles (1977b) through a generalization of the corresponding solution of the KdV equation (3.4), and the subsequent results in this section follow from that generalization.

If the soliton $\eta_1 = \operatorname{sech}^2 \theta_1$ [$\alpha_1 = \alpha$ in (5.1)] is incident on the rigid plane $z = 0$, the reflected wave is given by $\eta_{2\phi} = \operatorname{sech}^2 (\theta_2 - \phi)$, where

$\alpha_2 = \alpha$, $\psi_2 = -\psi_1$ (the angle of reflection equals the angle of incidence), $\theta_{20} = \theta_{10} (= 0$ without loss of generality), and

$$\phi = \tfrac{1}{2} \ln [\kappa/(\kappa - 3\alpha)]. \tag{5.7}$$

It follows that $\phi = O(\alpha)$ if $\kappa = O(1)$ or $\phi = O(1)$ if $\kappa = O(\alpha)$, as anticipated above. Moreover, the solution is singular, and regular reflection is impossible, if $\kappa < 3\alpha$; see below.

Regular reflection may be regarded as a symmetrical interaction between the incident wave η_1 and its image η_2 that yields the outgoing waves $\eta_{1\phi}$ and $\eta_{2\phi}$; in brief

$$\{\eta_1, \eta_2\} \to \{\eta_{1\phi}, \eta_{2\phi}\}. \tag{5.8}$$

The interaction, in contrast to that described by (5.3), is not phase-conserving.

Oblique interactions between solitary waves of unequal strength are asymmetric and require two parameters, measuring relative obliquity and relative strength, for their description; these may be conveniently chosen as $\psi = \frac{1}{2}(\psi_2 - \psi_1)$, $\varepsilon = \frac{1}{4}(\alpha_2 - \alpha_1)$. The resolution among incoming and outgoing waves is less direct than for the reflection problem in consequence of a spatial nonuniformity of the asymptotic limits $t \to \pm\infty$ with θ_n fixed; however, this difficulty is avoided by adopting a reference frame in which the interaction is stationary. The asymptotic description then is of the form (5.3), and therefore phase-conserving, if $\varepsilon > \tan^2 \psi$ and similarly, with $1 \leftrightarrow 2$, if $\varepsilon < -\tan^2 \psi$; it is of the form (5.8), and therefore not phase-conserving, if $|\varepsilon| < \tan^2 \psi$. Satsuma (1976) overlooked the spatial nonuniformity of the asymptotic limits $t \to \pm\infty$ with θ_n fixed and concluded that all interactions are phase-conserving, like those of (5.3) but unlike those of (5.8).

The phase shift for $\alpha_2 \neq \alpha_1$ is given by [cf (5.7)]

$$\phi = \tfrac{1}{2} \ln \{(\kappa - \kappa_-)/(\kappa - \kappa_+)\}, \quad \kappa_\mp = \tfrac{3}{4}(\alpha_2^{\frac{1}{2}} \mp \alpha_1^{\frac{1}{2}})^2. \tag{5.9a,b}$$

The interaction is regular if either $\kappa < \kappa_-$ or $\kappa > \kappa_+$, but it is singular if $\kappa_- < \kappa < \kappa_+$. Both Chen (1975) and Satsuma (1976) appear to have overlooked the existence of the singular regime $\kappa_- < \kappa < \kappa_+$, although it is implicit in Satsuma's solution, as also are other such regimes for interactions among more than two solitons.

Resonant Interactions

The condition $\kappa = \kappa_\pm$ is precisely that necessary for a resonant (or phase-locked) interaction among three solitary waves with

$$\mathbf{k}_3 = \mathbf{k}_2 \pm \mathbf{k}_1, \quad k_3 c_3 = k_2 c_2 \pm k_1 c_1, \tag{5.10a,b$_\pm$}$$

where

$$\mathbf{k}_i = k_i \mathbf{n}_i, \quad k_i = (3a_i/4d^3)^{\frac{1}{2}} l = (\alpha_i/\alpha)^{\frac{1}{2}}, \tag{5.11a,b}$$

and, here and subsequently, the alternative signs are vertically ordered and refer to alternatives. The phases θ_1 and θ_2, and hence the solitons η_1 and η_2, are stationary in a reference frame R_* moving with a velocity $\mathbf{c}_* \equiv c_* \{\cos \psi_*, \sin \psi_*\}$, where

$$c_* = c_0[1 + \tfrac{1}{3}(\alpha_1 + \alpha_2 + \alpha_3)], \quad \psi_* = \tfrac{1}{3}(\psi_1 + \psi_2 + \psi_3). \tag{5.12a,b}$$

The asymptotic description of the resonant interaction $(5.10)_-$ is given by (Miles 1977c)

$$\{\eta_1, \eta_2\} \rightarrow \eta_3 \quad (k_2 < \tfrac{1}{2}k_1), \tag{5.13a}$$

$$\eta_2 \rightarrow \{\eta_1, \eta_3\} \quad (\tfrac{1}{2}k_1 < k_2 < k_1), \tag{5.13b}$$

$$\eta_1 \rightarrow \{\eta_2, \eta_3\} \quad (k_1 < k_2 < 2k_1), \tag{5.13c}$$

and

$$\{\eta_1, \eta_2\} \rightarrow \eta_3 \quad (k_2 > 2k_1), \tag{5.13d}$$

where the left/right-hand sides correspond to the incoming/outgoing waves at large distances from the interaction zone and the x axis is now aligned with $\mathbf{c}_*$.

The asymptotic description of the resonant interaction $(5.10)_+$ is given by

$$\eta_1 \rightarrow \{\eta_2, \eta_3\} \quad (k_2 < k_1), \quad \{\eta_1, \eta_3\} \rightarrow \eta_2 \quad (k_2 > k_1). \tag{5.14a,b}$$

Higher-order resonant interactions have been discussed by Anker & Freeman (1978). Newell & Redekopp (1977) have remarked that resonant interactions also are possible for all those multi-dimensional systems to which Zakharov-Shabat theory is applicable and imply a breakdown of that theory for such systems.

Mach Reflection

Observation and experiment (Perroud 1957, Chen 1961, Wiegel 1964a,b) reveal that regular reflection of a solitary wave at a rigid wall (for which $k_2 = k_1, \psi_2 = -\psi_1 \equiv \psi$, and $\psi_* = 0$ in the present notation) is impossible for sufficiently small angles of incidence and is replaced by "Mach reflection" (geometrically similar to the corresponding shock-wave reflection). The apex of the incident and reflected waves then moves away from the wall at a constant angle, say ψ_*, and is joined to the wall by a third solitary wave (the "Mach stem"), as shown in Figure 3. Moreover, the strength of the reflected wave decreases to zero with the angle of incidence. There is some question as to the stability of the resulting waves, and the observed profiles may depart significantly from that of (1.2), but the available data are not definitive. It seems, nevertheless, that there exists a parametric regime in which the Mach-reflection pattern is realized and that the pattern is asymptotically stationary (but non-stationary effects may be significant near the leading edge at which the reflection is initiated).

The asymptotic configuration may be modelled by the resonant interaction described above with η_1 as the incident wave, η_2 as the reflected wave, and η_3 as the stem wave (Miles 1977c). It can be shown that both mass and energy are conserved if the reflected wave PT (see Figure 3) is terminated by the ray OT that extends from the leading edge O and is perpendicular to PT. The available observations (Perroud 1957, Melville, unpublished) yield Mach-stem amplitudes that are smaller than these predicted by this model, but the reasons for this discrepancy are not clear at this time.

This same resonant interaction also provides an asymptotic solution for the diffraction of a solitary wave at a concave corner of internal angle $\pi-\mu(3\alpha)^{\frac{1}{2}}$, $0<\mu<1$, or, equivalently, by a wedge of angle $2\mu(3\alpha)^{\frac{1}{2}}$. The corresponding result for $-1<\mu<0$ provides a solution for diffraction at a convex corner of internal angle $\pi-\mu(3\alpha)^{\frac{1}{2}}$ and suggests that a solitary wave cannot turn through an angle greater than $(3\alpha)^{\frac{1}{2}}$ at a convex corner without separating or otherwise losing its identity (see also last paragraph in Section 7).

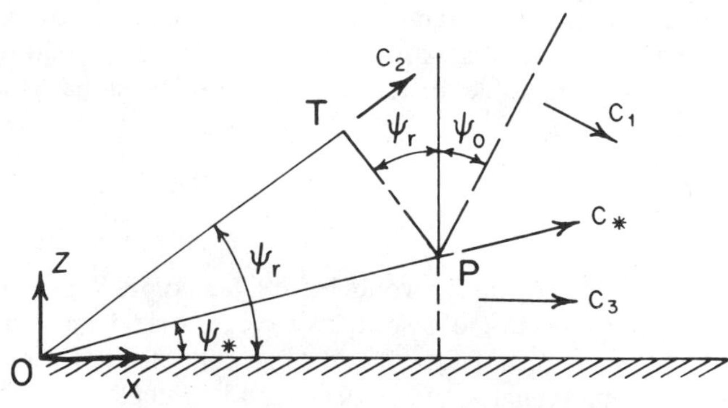

Figure 3 Mach-reflection pattern (the angular scale is exaggerated), showing the incident (— — —), reflected (— - —), and stem (- - -) waves. The entire pattern is expanding uniformly from the leading edge O, such that the apex P is moving with speed c_*, and the reflected wave is terminated by the perpendicular OT.

6 QUASI–ONE-DIMENSIONAL SOLITARY WAVES

The one-dimensional solitary wave (1.2) may be regarded as propagating in a uniform rectangular channel. It therefore is natural to study the effects of changing the breadth, depth, or cross-sectional shape of the channel. Solitary waves in channels of nonrectangular cross section have been considered by Peters (1966), Peregrine (1968, 1969, 1972), Fenton (1973), and Grimshaw (1978); they are of considerable engineering interest and pose some still unresolved problems but will not be considered further here.

Gradually Varying Channel

Boussinesq, as noticed in the last paragraph of Section 2, inferred from conservation of energy that the amplitude of a solitary wave in a channel of gradually varying depth d would vary inversely as d; see also Ostrovsky & Pelinovsky (1970), Grimshaw (1970, 1971), Kakutani (1971), Johnson (1972, 1973a,b), Shuto (1973), and Reutov (1976). Saeki et al (1971) appear to have been the first to consider a channel of gradually varying breadth $b(x)$, for which they predicted that $a \propto b^{-\frac{2}{3}}d^{-1}$; see also Shuto (1974), Ostrovsky & Pelinovsky (1975), and Miles (1977a). The corresponding generalization of (1.2) is

$$y = a \operatorname{sech}^2 \left[(3ga)^{\frac{1}{2}}(2d)^{-1}\left(\int c^{-1}dx - t\right)\right], \quad a = a_0(b/b_0)^{-\frac{2}{3}}(d/d_0)^{-1}, \tag{6.1a,b}$$

where $c(x)$ is given by (1.1) with $a = a(x)$ and $d = d(x)$ therein, and the subscript 0 implies $x = x_0$, at which $b \equiv b_0$ and $d \equiv d_0$. It is convenient, albeit not necessary, to assume that the channel is uniform in $x < x_0$.

Shuto (1973, 1974) compares the prediction $a \propto d^{-1}$ for a channel of constant breadth and constant bottom slope with the experimental results of Ippen & Kulin (1954), Street & Camfield (1967), Camfield & Street (1969), and Saeki et al (1971). It appears that the approximation is valid for sufficiently small values of bottom slope, but that Green's law, $a \propto d^{-\frac{1}{4}}$, provides a better approximation for sufficiently large bottom slope or sufficiently small amplitude (cf Peregrine 1967 and Svendsen 1976). Recent experiments in a channel of constant depth and linearly varying breadth (Chang et al 1979) confirm the prediction $a \propto b^{-\frac{2}{3}}$ in a channel with an expansion angle of 0.02 rad for $a/d = 0.05 - 0.4$, but corresponding experiments in a contracting channel imply an exponent of approximately -0.4. The latter discrepancy appears to be a consequence of nonlinear distortion of the characteristics (which is much stronger for the diverging channel).

Generalized KdV Equation

The joint assumptions that $a/d \ll 1$, $d^2/l^2 \ll 1$, and $l/L \ll 1$, where L is the scale of the channel variation, lead to the generalized KdV equation (Shuto 1974, Ostrovsky & Pelinovsky 1975)

$$y_x + \tfrac{3}{2}g^{-\frac{1}{2}}d^{-\frac{3}{2}}yy_s + \tfrac{1}{6}g^{-\frac{3}{2}}d^{\frac{1}{2}}y_{sss} + [\ln (b^{\frac{1}{2}}d^{\frac{1}{4}})]_x y = 0, \tag{6.2}$$

where

$$s = \int_{x_0}^{x} (gd)^{-\frac{1}{2}}dx - t \tag{6.3}$$

is a characteristic variable. The transformation [cf (3.3) with $c_0 t$ replaced by x in (3.3b)]

$$\xi = (gd_0)^{\frac{1}{2}}s/l, \quad \tau = \tfrac{1}{6}(d_0^2/l^3)\int_{x_0}^{x} (d/d_0)^{\frac{1}{2}}dx, \quad \eta = (b/b_0)^{\frac{1}{2}}(d/d_0)^{\frac{1}{4}}(y/a_0) \tag{6.4a,b,c}$$

reduces (6.2) to the KdV form (3.4) with

$$\mathscr{U} = \mathscr{U}(\tau) = \mathscr{U}_0(b/b_0)^{-\frac{1}{2}}(d/d_0)^{-\frac{9}{4}} \tag{6.5}$$

therein, where $\mathscr{U}_0 = 3a_0l^2/4d_0^3$. The following results may be rendered more compact if transformed according to (6.4), but their physical interpretation is more direct in the original variables. It is noteworthy that $\mathscr{U} \equiv \mathscr{U}_0$ if b^2d^9 is constant, in which special case (6.1) is an exact similarity solution of (6.2).

In contrast to the KdV equation (3.1), (6.2) appears to admit only the two integral invariants (Miles 1979a)

$$I = b^{\frac{1}{2}}d^{\frac{1}{4}}\int_{-\infty}^{\infty} y\,ds, \quad E = b(gd)^{\frac{1}{2}}\int_{-\infty}^{\infty} y^2ds \tag{6.6a,b}$$

unless b^2d^9 is constant. The integral E is a measure of energy, which is conserved, but I is not a measure of mass unless b^2d is constant. The volumetric (or mass) integral [cf (2.4a)]

$$Q = b(gd)^{\frac{1}{2}}\int_{-\infty}^{\infty} y\,ds = g^{\frac{1}{2}}b^{\frac{1}{2}}d^{\frac{1}{4}}I \tag{6.7}$$

is not invariant unless b^2d is constant.

If $l/L \ll a/d \equiv \alpha$ and $d^2/l^2 = O(\alpha)$, the coefficients in (6.2) are slowly varying functions of x, and (6.1) is an adiabatic approximation that conserves E, but neither I (unless b^2d^9 is constant) nor Q (unless b^2d is constant). An approximation that conserves both I and Q may be developed in the form (Miles 1979a)

$$y = y_1 + y_+ + y_-, \quad y_\pm = b^{-\frac{1}{2}} d^{-\frac{1}{4}} f_\pm \left[\mp t + \int_{x_0}^{x} (gd)^{-\frac{1}{2}} dx \right] \tag{6.8a,b}$$

(subscripts are vertically ordered), where y_1 is the adiabatic approximation (6.1) and $y_\pm$ is a right/left moving secondary wave of length scale L. Conservation of I and Q determine f_+ and f_-, respectively.

An alternative method of improving the adiabatic approximation is to regard (6.1) as the first term in an inner expansion. It is relatively straightforward to construct higher-order terms in this inner expansion, but the results are ambiguous in the absence of a matched outer expansion, the construction of which poses significant difficulties (Johnson 1973b).

Fission of Solitary Wave

The restriction $l/L \ll \alpha \ll 1$ is severe. If $l/L = O(\alpha)$ the solution of (6.2) appears to be available only through numerical integration. If $\alpha \ll l/L \ll 1$ the second (nonlinear) and third (dispersion) terms in (6.2) are dominated by the fourth term, and Green's law, $a \propto b^{-\frac{1}{2}} d^{-\frac{1}{4}}$, holds over distances of $O(L)$; however, the approximation is not uniformly valid (unless $b^2 d^9$ is constant) and must be expected to evolve further. In particular, as Madsen & Mei (1969) discovered through numerical integration and confirmed through experiment, a solitary wave may split into two or more solitary waves (plus, in general, a residual wave train) on advancing into shallower water.

Suppose that the breadth and depth of a channel change continuously from b_0 and d_0 at $x = x_0$ to b and d at $x = x_0 + L$ and then remain uniform in $x > x_0 + L$, that the solitary wave (1.2) is incident from $x < x_0$, and that $\alpha \ll l/L \ll 1$ with $l^2 = 4d_0^3/3a_0$ ($\mathscr{U}_0 = 1$). Green's law then may be invoked to determine the wave at $x = x_0 + L$ and the subsequent evolution determined through inverse scattering theory. This procedure, which was suggested independently by Johnson (1972, 1973a), Ono (1972), Peregrine (1972), and Tappert & Zabusky (1971), is tantamount to the transformation (6.4) and (6.5) followed by the solution of (3.4) with the initial condition $\eta_0 = \operatorname{sech}^2 \xi$. The asymptotic solution in $x - (x_0 + L) \gg l$ then is determined by (3.9) and (3.10) and comprises N solitons of amplitudes

$$a_n = \tfrac{1}{4} a_0 (d/d_0)^2 \left\{ (1 + 8\mathscr{U})^{\frac{1}{2}} - (2n-1) \right\}^2 \quad (n = 1, 2, \ldots, N), \tag{6.9}$$

where $\mathscr{U} = (b/b_0)^{-\frac{1}{2}} (d/d_0)^{-\frac{9}{4}}$ and N is the largest integer for which the bracketed quantity in (6.9) is positive. For example, if $b = b_0$ and $d = \frac{1}{2} d_0$, as in Madsen & Mei's (1969) experiments, $N = 3$, and the predicted relative amplitudes are 1.72, 0.66, and 0.10, which compare favorably with the observed amplitudes of 1.67, 0.75, and 0.17. Note that $N = 1$ if $b^2d^9 > b_0^2d_0^9$ and that fissioning is much more sensitive to a decrease in depth than to a decrease in breadth; e.g. the transition depth for $N = 3$ is $0.614d_0$ if $b = b_0$, whereas the corresponding transition breadth is $b_0/9$ if $d = d_0$.

The effects of changes of depth, including one or more discontinuities, also have been considered by Pelinovsky (1974) on the basis of either conservation laws or (6.2). He allows for a reflected wave at a discontinuity, presumably by analogy with the long-wave approximation of linear theory (Lamb 1932, Section 176); however, his hypotheses are not explicitly stated.

The phenomenon of fission, even where the change of depth takes place over a very gentle slope (as small as 0.05 in Madsen & Mei's experiments), provides further evidence that the adiabatic approximation (6.1) may not be uniformly valid and that the model of a slowly varying solitary wave may be inadequate for determining the effects of topography if the cumulative change is sufficiently large. Kaup & Newell (1978) have suggested that the secondary wave y_+ (their "shelf") in (6.8) may be regarded as an initial displacement for the solution of the KdV equation and hence, if positive, ultimately should be resolved into one or more solitons, in qualitative agreement with the preceding results.

Cylindrical Solitary Wave

The preceding results apply to axisymmetric wave propagation if x is taken to be the radial coordinate and $b \equiv x$. The asymptotic expansion, for which (6.1) is the first term in the inner approximation, then may be developed in some detail if d is constant (Miles 1978a), but nonconservation of mass obscures the physical significance of the results. Nevertheless, numerical integrations of both the Boussinesq equations (Chwang & Wu 1976) and the KdV equation (Cumberbatch 1978, Ko & Kuehl 1979) suggest that (6.1) does provide a rather good description of a converging, axisymmetric solitary wave, at least over limited distances. [The numerical solution of the counterpart of (6.2) for spherically symmetric ion-acoustic waves (Maxon & Viecelli 1974) provides similar, albeit indirect, support for (6.1).] On the other hand, Chwang & Wu's numerical solution implies that the reflected wave is a dispersive wave train rather than a solitary wave.

7 RAY THEORY

Analytical solutions of the two-dimensional KdV equation (5.6), other than those for the soliton interactions described in Section 5 and the two-dimensional, "lump" soliton recently discovered by Ablowitz & Satsuma (1978) and Zakharov (unpublished lecture referenced by Ablowitz & Satsuma), appear to have been obtained only through nonlinear, geometrical-optics approximations of the type developed originally by Whitham (1974, Sections 8.3–7) for shock waves and applied subsequently by Ostrovsky & Shrira (1976) and Miles (1977d) to solitary waves; see also Gorschkov et al (1974) and Ostrovsky (1976).

Let the successive positions of a solitary wave (more precisely, the wave peak) and the corresponding orthogonal rays be described by the families of curves $\mu =$ constant and $v =$ constant, respectively, such that $\mathscr{F} d\mu$ is the incremental advance of a wave element in the temporal increment $d\mu/(gd)^{\frac{1}{2}}$ and $\mathscr{A} dv$ is the width of a hypothetical channel between the rays v and $v+dv$ ($\mathscr{A}$ is dimensionless, v has the dimensions of length, and $d\mu$ and dv are differentials); then θ, the local inclination between a ray and a fixed direction, and $\mathscr{F} \equiv c/(gd)^{\frac{1}{2}}$, the dimensionless local wave speed, satisfy

$$\mathscr{A}\theta_{\mu} = -\mathscr{F}_{v}, \quad \mathscr{F}\theta_{v} = \mathscr{A}_{\mu}, \tag{7.1a,b}$$

where subscripts imply partial differentiation. This system of first-order, nonlinear, partial differential equations is completed by a relation of the form $\mathscr{A} = \mathscr{A}(\mathscr{F})$, which describes the evolution of the wave in a channel of gradually varying breadth; the system is hyperbolic if, as in the present instance, $d\mathscr{A}/d\mathscr{F} < 0$.

The required result for $\mathscr{A}(\mathscr{F})$ may be expressed in the parametric form

$$\mathscr{A} = (\alpha_0/3)^{\frac{1}{2}}(a/a_0)^{-\frac{3}{2}}, \quad \mathscr{F} = 1+\tfrac{1}{2}\alpha_0(a/a_0), \tag{7.2a,b}$$

by invoking (1.3) and (6.1b) and normalizing $\mathscr{A}$ such that the distance between the rays v and $v+dv$ is $(\alpha_0/3)^{\frac{1}{2}}dv$ for $a = a_0$ (this scaling is convenient for most applications); a_0 is a reference value of a, $\alpha_0 = a_0/d$, and d is assumed to be constant. The formulation could be generalized to allow for variable depth, but then the relation between $\mathscr{A}$ and $\mathscr{F}$ would

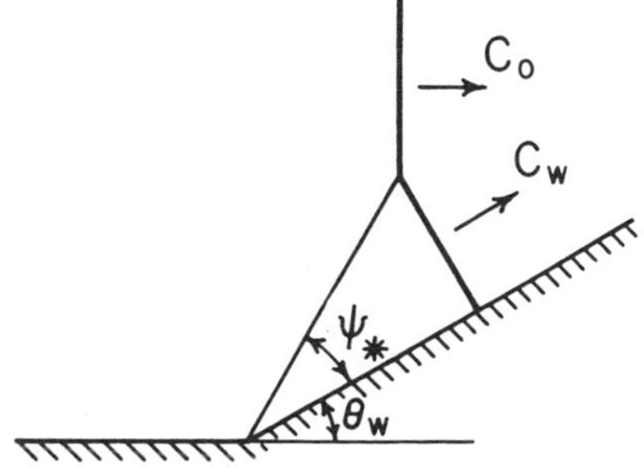

Figure 4 Diffraction at a concave corner of turning angle θ_w.

depend on local position, which would render the resulting partial differential equations analytically intractable.

The hyperbolic system (7.1) and (7.2) may be treated by characteristics theory. Envelopes of the characteristics may form and give rise to discontinuities in the transverse slope of the solitary wave; these discontinuities are naturally described as *shocks*.

A particularly simple example is diffraction of a solitary wave at a corner (Miles 1977d). The results for a concave corner (Figure 4) of turning angle θ_w are compared with those provided by the Mach-reflection solution of Section 5 in Figures 5a,b. The ray approximation to the amplitude at the wall is surprisingly good for $\theta_w < (3\alpha_0)^{\frac{1}{2}}$, but that for the Mach-stem angle ψ_* deteriorates as $\theta_w \uparrow (3\alpha_0)^{\frac{1}{2}}$ [ψ_* should vanish for $\theta_w = (3\alpha_0)^{\frac{1}{2}}$]; the approximation fails for $\theta_w > (3\alpha_0)^{\frac{1}{2}}$, in which regime Mach reflection should give way to regular reflection. The corresponding

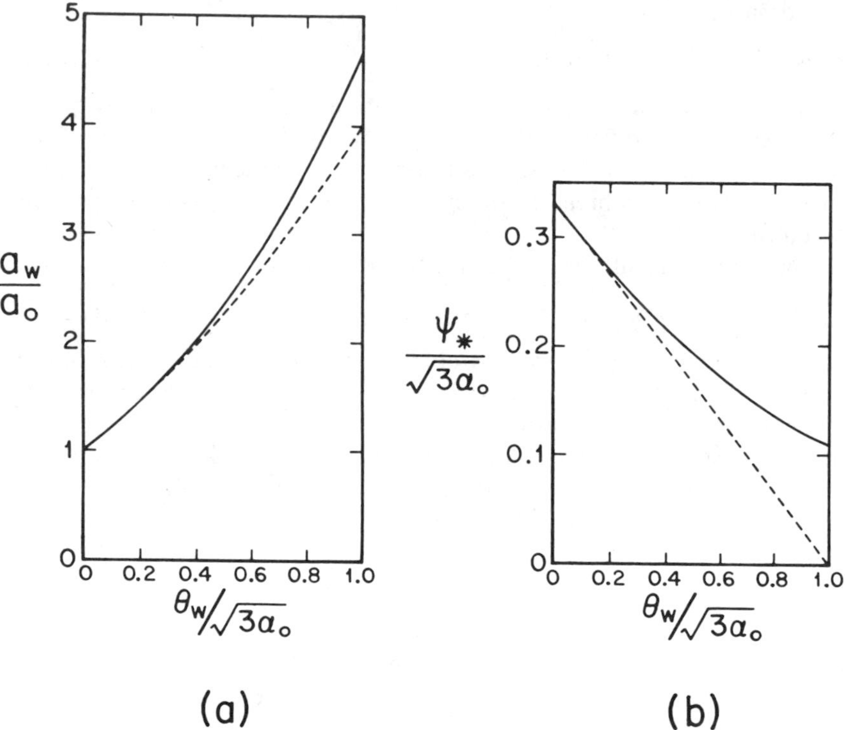

Figure 5a The amplitude of the diffracted wave at a concave corner, as predicted by ray theory (———) and the Mach-reflection solution of Section 5 (- - -).

Figure 5b The angle ψ_* (see Figure 4), as predicted by ray theory (———) and the Mach-reflection solution of Section 5 (- - -).

results for a convex corner imply separation of the diffracted wave for $\theta_w < -(3\alpha_0)^{\frac{1}{2}}$ (see last paragraph in Section 5). It should perhaps be emphasized that these predictions have not been tested experimentally (see penultimate paragraph in Section 5).

8 DISSIPATION

Dissipation of a solitary wave may occur through any of boundary-layer viscosity, surface contamination, capillary hysteresis, parasitic capillary waves, and scattering by bottom roughness. All but the last may be significant for laboratory measurements, whereas only the first and last are likely to be significant in the field. Moreover, boundary-layer damping is likely to be laminar in the laboratory but turbulent in the field.

The predicted decay of the amplitude of a solitary wave in a rectangular channel with laminar boundary layers is given by (Keulegan 1948)

$$a(t) = a_0\{1+0.237(a_0/d)^{\frac{1}{4}}(\nu c_0/8d^3)^{\frac{1}{2}}[1+(2d/b)]t\}^{-4}, \tag{8.1}$$

where ν is the kinematic viscosity and $a_0 \equiv a(0)$. The factor $1+2(d/b)$ may be augmented by an additional term $\mathscr{C}$ to allow for surface contamination; the data required to evaluate $\mathscr{C}$ are seldom available, at least in most laboratory experiments, but a rough estimate is $\mathscr{C} = 1$ (Miles 1967).

Ostrovsky (1976) and Shuto (1976) invoke the empirical form $C_*\rho u|u|$ for the wall stress under a turbulent boundary layer to obtain

$$a(t) = a_0[1+(8C_*/15)(a_0/d^2)c_0t]^{-1}. \tag{8.2}$$

Shuto gives estimates of C_* based on wave-tank measurements.

Damping due to bottom scattering from random depth variations of mean-square amplitude $d^2\langle\delta^2\rangle$ and correlation length $L \ll (d^3/a)^{\frac{1}{2}}$ implies

$$a(t) = a_0[1+0.396\langle\delta^2\rangle(L/d)^2(a_0/d)^{\frac{3}{4}}(c_0t/d)]^{-\frac{4}{3}}. \tag{8.3}$$

[Ostrovsky (1976) gives a different result, but it is valid, after inserting a missing factor of L in the coefficient of $\langle\zeta^2\rangle$, only for *one-dimensional* depth variation (Ostrovsky, private communication).]

It is noteworthy that the asymptotic form of $a(t)$ implied by each of (8.1)–(8.3) as $t\uparrow\infty$ is independent of the initial value a_0.

Weidman & Maxworthy (1978) discuss various measurements designed to confirm (8.1) and conclude that it typically (although not always) underestimates, but nevertheless comprehends the dominant contributions to, the observed damping in laboratory wave tanks and that capillary hysteresis (Miles 1967) provides the most plausible explanation of the observed discrepancies. It seems reasonably certain, on the other hand, that (8.1) is not valid for solitary waves in natural channels or open bodies of water, for which either turbulent damping or bottom scattering might be expected to dominate and for which the decay might be expected to be at least qualitatively described by (8.2); however, definitive measurements do not appear to be available at this time.

9 INTERNAL SOLITARY WAVES

The preceding sections deal exclusively with solitary waves on the free surface of a homogeneous liquid; however, the hallmark of the solitary wave is the balance between nonlinearity and dispersion, and that balance may be possible in the absence of a free surface by virtue of any or all of stratification, shear, compressibility, or rotation.

Two-Layer Model

The simplest model that supports an internal solitary wave is a two-layer liquid with a small discontinuity in density. It provides at least a qualitative description of the dominant internal mode in a typical thermocline but suppresses the higher modes, which depend on the detailed structure of the thermocline. (It also comprises a surface wave if the upper boundary is free.)

Let $\mathscr{d}d$ and $(1-\mathscr{d})d$ be the depths of the lower and upper layers and ρ_0 and $\rho_0(1-\sigma)$, $0<\sigma\ll 1$, the corresponding densities. The resulting internal solitary wave then is approximately independent of whether the upper boundary is rigid or free and is a surface of elevation/depression for $\mathscr{d}\lesssim\frac{1}{2}$. The interfacial displacement is described by (1.2), with (1.3) and (1.4) replaced by (Keulegan 1953)

$$\mathscr{F}^2=\sigma\{\mathscr{d}(1-\mathscr{d})+\alpha(1-2\mathscr{d})+O(\alpha^2)\}+O(\sigma^2) \tag{9.1}$$

and

$$\mathscr{d}^2(1-\mathscr{d})^2\beta=\tfrac{3}{2}(\tfrac{1}{2}-\mathscr{d})\alpha+O(\alpha^2,\alpha\sigma),\quad \operatorname{sgn}\alpha=\operatorname{sgn}(\tfrac{1}{2}-\mathscr{d}). \tag{9.2a,b}$$

Keulegan (1952) has reported experimental confirmation of the wave of elevation for $\mathscr{d}<\frac{1}{2}$ (he presumably did no experiments for $\mathscr{d}>\frac{1}{2}$).[5]

If the upper surface is free, the streamline displacement is reversed in a

[5] Keulegan does not appear to have published the details of his experimental work. Some experimental results for a two-fluid medium have been published by Walker (1973), who observed wave speeds inferior to, and profiles narrower than, those predicted by theory; however, his measurements are far from definitive.

thin layer of depth $\mathscr{F}^2 d$ just below the free surface, and the ratio of the free-surface displacement to the interfacial displacement is $-\sigma\mathscr{d}$. This reversal constitutes the principal effect of the upper boundary on the internal-wave solution vis-à-vis that for a rigid boundary.

The appropriate measure of (quadratic) nonlinearity in Keulegan's model is $(\frac{1}{2}-\mathscr{d})\alpha$, rather than simply α, and vanishes for $\mathscr{d}=\frac{1}{2}$. Cubic nonlinearity is comparable with quadratic nonlinearity if $\mathscr{d}-\frac{1}{2}=O(\alpha)$ and implies that $|\alpha|<|\alpha_1|$, where (Long 1956)

$$\alpha_1 = \mathscr{d}(1-\mathscr{d})\left[\mathscr{d}^3+(1-\mathscr{d})^3\right]^{-1}(\tfrac{1}{2}-\mathscr{d}). \tag{9.3}$$

[Long assumes a rigid upper boundary and does not invoke the approximation $\sigma \ll 1$; however, his results are equivalent to those cited above if $\sigma \ll 1$. See also Kakutani & Yamasaki (1978).]

The two-layer model may be generalized to a channel of arbitrary and gradually varying cross section, in which form it provides a model for internal surges in lakes (Grimshaw 1978).

Kaup & Newell (1978) have suggested that an interfacial solitary wave in water of variable depth could reverse polarity on passing through a transition region in which $\mathscr{d}-\frac{1}{2}$ changes sign, but the impossibility of a solitary wave for $\mathscr{d}=\frac{1}{2}$ casts doubt on this suggestion. If $l/L \ll \alpha$, where L is the scale of the depth variation, an adiabatic approximation with appropriate corrections can be developed as in Section 6; however, $a \to 0$ and $l \to \infty$ as $\mathscr{d} \to \frac{1}{2}$ for a solitary wave of prescribed mass, and the initial assumption breaks down. If, on the other hand, $\alpha \ll l/L \ll 1$ Green's law may be invoked to determine the change in amplitude through the transition, after which the wave at the end of the transition may be taken as the initial condition for the appropriate KdV equation (cf the corresponding surface-wave problem in Section 6.) Suppose, for example, that a solitary wave of amplitude a_0 moves from water of depth d_0 and depth ratio $\mathscr{d}_0 > \frac{1}{2}$ across a transition layer of length L into water of depth d_1 and depth ratio $\mathscr{d}_1 < \frac{1}{2}$. The incident wave then will be one of depression, and Green's law implies that the wave at the end of the transition will be of similar (Boussinesq) form with amplitude $a_0(c_0/c_1)^{\frac{1}{2}}$, where c_0 and c_1 are determined by (9.1). Thus, the initial displacement for the KdV equation will be negative in a domain in which solitons are intrinsically positive, and the wave will be transformed into an oscillatory wave train of gradually increasing length and decreasing amplitude (cf Figure 2). A similar argument holds if $\mathscr{d}_0 < \frac{1}{2}$ and $\mathscr{d}_1 > \frac{1}{2}$. It therefore appears (although the preceding arguments do not prove) that a reversal of polarity in consequence of a gradual change in depth is impossible for an interfacial solitary wave. Djordjevic & Redekopp (1978) agree.

Stratified Shear Flows

Continuously stratified flow, which supports an infinite, discrete set of internal solitary waves, has been considered by Peters & Stoker (1960), Long (1965), Benjamin (1966), Benney (1966), Djordjevic & Redekopp (1978), Weidman (1978), and Miles (1979b). All but Miles include only quadratic nonlinearity, although the formulations of Benney and Long are readily extended to include the effects of cubic nonlinearity, which are comparable with those of quadratic nonlinearity if and only if $\beta = O(\alpha^2)$ (α and β are amplitude and dispersion parameters, as in Section 1). Benjamin, Benney, and Miles also allow for shear, although not for critical layers (at which $U = c$); see also Long (1956), Benjamin (1962), Ter-Krikorov (1962, 1963), Freeman & Johnson (1970), and Leonov & Miropol'skiy (1975). Long (1965) and Benjamin (1966) remark that the conventional Boussinesq approximation, in which the inertial effects of stratification are neglected, is untenable in this context and, if invoked, suppresses internal solitary waves.

The form of the internal solitary waves is given by (Miles 1979b)

$$y = a\,\{\cosh^2 \mathscr{x} - [\alpha/(2\alpha_n - \alpha)]\,\sinh^2 \mathscr{x}\}^{-1}\phi_n(\mathscr{y}) \quad (|\alpha| < |\alpha_n| \ll 1) \tag{9.4a}$$

or

$$y = \tfrac{1}{2}a(1+\tanh \mathscr{x})\phi_n(\mathscr{y}) \quad (|\alpha| = |\alpha_n| \ll 1), \tag{9.4b}$$

where: $\mathscr{x} = (x - c_n t)/l$ is a dimensionless coordinate in a reference frame moving with the wave, y is the vertical displacement of the streamline that originates at the dimensionless (with d as the unit of depth) elevation $\mathscr{y}$ in the primary flow; ϕ_n is a solution of a linear eigenvalue problem for prescribed density $\rho(\mathscr{y})$ and horizontal velocity $U(\mathscr{y})$; the parameters α_n and $\beta = (d/l)^2$ and the wave speed c_n are (for given α) functionals of ρ, U, and ϕ_n; $c_0 > c_1 > c_2 \ldots$, and only the free-surface ($n = 0$) and dominant internal ($n = 1$) modes are likely to be important. Cubic nonlinearity is negligible if $|\alpha_n| \gg |\alpha|$, and the individual streamlines then have the Boussinesq profile (1.2). The parameter α_n, which reduces to (9.3) for Keulegan's model, limits the amplitude if $|\alpha_n| \ll 1$; it also is a measure of the volume Q associated with the wave. The limit $|\alpha_n| \downarrow |\alpha|$ implies $Q \to \infty$, whereas the limit $|\alpha_n| \to 0$ with Q fixed implies $|a| \to 0$ and $l \to \infty$. The solution (9.4b) implies $Q = \infty$, and its physical significance (perhaps as an approximation to a bore) has not yet been explored.

Djordjevic & Redekopp (1978) consider internal waves in water of variable depth with special reference to fissioning (cf Section 6).

Embedded Layer

The preceding results are for waves that are long compared with the depth of the fluid and exclude the important case of an inhomogeneous layer of depth d embedded in a fluid of great depth D, such that $d \ll l \ll D$. Solitary waves in such a layer were originally studied both analytically and experimentally by Benjamin (1967) and Davis & Acrivos (1967) and have since been observed in the atmosphere (Christie et al 1978).

The assumption $D \gg 1$ does not preclude boundaries (see Benjamin 1967), but the simplest, and perhaps the most interesting, case is that of a layer embedded in an unbounded fluid ($D = \infty$). A simple model is described by

$$\rho = \rho_0(1-\sigma \tanh \mathscr{y}) \qquad (\sigma \ll 1), \quad U = 0, \tag{9.5}$$

for which (Benjamin 1967)

$$y = a(\mathscr{x}^2+1)^{-1} \tanh \mathscr{y}, \quad \mathscr{F}^2 = \tfrac{1}{2}\sigma(1+\tfrac{3}{5}\alpha), \quad al/d^2 = 5/2, \tag{9.6a,b,c}$$

all within $1+O(\sigma)$, for the dominant mode. Davis & Acrivos (1967) and Hurdis & Pao (1975) have reported experimental confirmation of both the profile and the wave speed.

Joseph (1977) has generalized Benjamin's results to arbitrary D/l. The resulting profile is proportional to $\operatorname{sech}^2 \mathscr{x}$ for $D/l \downarrow 0$ and to $1/(\mathscr{x}^2+1)$ for $D/l \uparrow \infty$.

Compressibility

The incorporation of compressibility in a stratified flow leads to an additional family of internal solitary waves. These waves have speeds that are of the same order as those for the waves that are (essentially) independent of compressibility, but their lengths tend to infinity as the compressibility tends to zero (Long & Morton 1966).

Rotation

Rotation has significant or dominant effects on atmospheric and oceanic waves of planetary scale. Solitary Rossby waves in a zonal flow (westerly current) appear to have been discovered (analytically) by Long (1964b) and have been studied subsequently by Larsen (1965), Benney (1966), Clarke (1971), and Redekopp (1977) [see also Maxworthy & Redekopp (1976), Maxworthy et al (1978), and Redekopp & Weidman (1978)]. All invoke Rossby's β-plane model, in which the northerly gradient of the vertical component of the Earth's rotation is constant, and all but Redekopp assume that the wave speed relative to the local flow never vanishes (there are no critical layers).

Long, Larsen, and Benney neglect vertical variation of the flow (the "non-divergent" approximation) and assume that the shear, which is essential for the existence of solitary Rossby waves, is weak (the zonal flow is almost uniform) but not so weak as to preclude a balance between dispersion and quadratic nonlinearity. The streamline displacement is northerly/southerly, and the wave represents a pressure ridge/trough, if the westerly current increases/decreases with latitude. The details of the analysis are rather similar to those for a stratified shear flow.

Clarke (1971) and Redekopp (1977) have extended Long's model to allow for vertical variation of the flow in a model with constant Väisälä frequency. Clarke also considers the effects of topographic variation transverse to the wave. Redekopp (1977) invokes matched asymptotic expansions to allow for critical layers; the details are complicated.

Pritchard (1970) has produced solitary waves in swirling flows in a cylindrical container with a free surface (so that the basic flow is a Rankine vortex) and in cylindrical tubes. These waves are counterparts of those discussed by Benjamin (1967) and Davis & Acrivos (1967), and Pritchard's observed wave forms are close to those calculated from an appropriate modification of Benjamin's theory.

Leibovich & Randall (1972) have carried out numerical solutions of a nonlinear integro-differential equation governing finite-amplitude wave propagation on concentrated vortices and obtained solitary waves that are qualitatively similar to that of Boussinesq and are related to those observed by Pritchard (1970).

ACKNOWLEDGMENTS

This work was partly supported by the Physical Oceanography Division, National Science Foundation, NSF Grant OCE77-24005, and by a contract with the Office of Naval Research. I am indebted to Michael Longuet-Higgins, W. K. Melville, Robert Miura, Alan Newell, Howell Peregrine, Alwyn Scott, Harvey Segur, and Ib Svendsen for various references and suggestions.

Literature Cited

Ablowitz, M. J., Satsuma, J. 1978. Solitons and rational solutions of nonlinear evolution equations. *J. Math. Phys.* 19:2180–86

Airy, G. B. 1845. Tides and waves, Section 392. *Encyc. Metropolitana*, Vol. 5, pp. 241–396

Anker, D., Freeman, N. C. 1978. Interpretation of three-soliton interactions in terms of resonant triads. *J. Fluid Mech.* 87:17–31

Bazin, H. 1865. Recherches expérimentales sur la propagation des ondes. *Mém. présentés par divers Savants à L'Acad. Sci. Inst. France* 19:495–644

Benjamin, T. B. 1962. The solitary wave on a stream with an arbitrary distribution of vorticity. *J. Fluid Mech.* 12:97–116

Benjamin, T. B. 1966. Internal waves of finite amplitude and permanent form. *J. Fluid Mech.* 25:241–70

Benjamin, T. B. 1967. Internal waves of permanent form in fluids of great depth. *J. Fluid Mech.* 29:559–92

Benjamin, T. B. 1972. The stability of solitary waves. *Proc. R. Soc. London Ser. A* 328: 153–83

Benney, D. J. 1966. Long non-linear waves in fluid flows. *J. Math. & Phys.* 45:52–63

Berryman, J. G. 1976. Stability of solitary waves in shallow water. *Phys. Fluids* 19: 771–77

Bona, J. 1975. On the stability theory of solitary waves. *Proc. R. Soc. London Ser. A* 344:363–74

Boussinesq, M. J. 1871a. Théorie de l'intumescence liquide, appelée onde solitaire ou de translation, se propageant dans un canal rectangulaire. *Acad. Sci. Paris, Comptes Rendus* 72:755–59

Boussinesq, M. J. 1871b. Théorie générale des mouvements qui sont propagés dans un canal rectangulaire horizontal. *Acad. Sci. Paris, Comptes Rendus* 73:256–60

Boussinesq, M. J. 1872. Théorie des ondes et des remous qui se propagent le long d'un canal rectangulaire horizontal, en communiquant au liquide contenu dans ce canal des vitesses sensiblement pareilles de la surface au fond. *J. Math. Pures Appl.* (2) 17:55–108; transl. A. C. J. Vastano, J. C. H. Mungall, Texas A&M Univ., Ref. 76-2-T (March, 1976)

Boussinesq, M. J. 1877. Essai sur la théorie des eaux courantes. *Mém. présentés par divers Savants à L'Acad. Sci. Inst. France* (séries 2) 23:1–680 (see also 24:1–64)

Byatt-Smith, J. G. B. 1970. An exact integral equation for steady surface waves. *Proc. R. Soc. London Ser. A* 315:405–18

Byatt-Smith, J. G. B. 1971. An integral equation for unsteady surface waves and a comment on the Boussinesq equation. *J. Fluid Mech.* 49:625–33

Byatt-Smith, J. G. B., Longuet-Higgins, M. S. 1976. On the speed and profile of steep solitary waves. *Proc. R. Soc. London Ser. A* 350:175–89

Camfield, F. E., Street, R. L. 1969. Shoaling of solitary waves on small slopes. *J. Waterway Harb. Div., Proc. ASCE* 95:1–22

Chang, P., Melville, W. K., Miles, J. W. 1979. On the evolution of a solitary wave in a gradually varying channel. *J. Fluid Mech.* In press

Chen, H.-H. 1975. A Bäcklund transformation in two dimensions. *J. Math. Phys.* 16:2382–84

Chen, T. C. 1961. Experimental study on the solitary wave reflection along a straight sloped wall at oblique angle of incidence. *Tech. Mem. No. 124*. Dep. Army, Corps of Engineers, Beach Erosion Board, Office of the Chief of Engineers

Christie, D. R., Muirhead, K. J., Hales, A. L. 1978. On solitary waves in the atmosphere. *J. Atmos. Sci.* 35:805–25

Chwang, A. T., Wu, T. Y. 1976. Cylindrical solitary waves. *Proc. IUTAM Symposium on water waves in water of varying depth, Canberra, Australia, 1976*, pp. 80–90

Clarke, R. A. 1971. Solitary and cnoidal planetary waves. *Geophys. Fluid Dyn.* 2:343–54

Cohen, B. I., Watson, K. M., West, B. J. 1976. Some properties of deep water solitons. *Phys. Fluids* 19:345–54

Cumberbatch, E. 1978. Spike solution for radially symmetric solitary waves. *Phys. Fluids* 21:374–76

Daily, J. W., Stephans, S. C. Jr. 1953a. The solitary wave. *Proc. Conf. Coastal Eng., 3rd, October 1952*, pp. 13–30

Daily, J. W., Stephans, S. C. Jr. 1953b. Characteristics of the solitary wave. *Trans. ASCE* 118:575–87

Davis, R. E., Acrivos, A. 1967. Solitary internal waves in deep water. *J. Fluid Mech.* 29:593–607

Djordjevic, V. D., Redekopp, L. G. 1978. The fission and disintegration of internal solitary waves moving over two-dimensional topography. *J. Phys. Oceanogr.* 8:1016–24

Epstein, B. 1974. A formal expansion procedure for the solitary wave problem. *Quart. Appl. Math.* 32:89–95

Fenton, J. D. 1972. A ninth-order solution for the solitary wave. *J. Fluid Mech.* 53:257–71

Fenton, J. D. 1973. Cnoidal waves and bores in uniform channels of arbitrary cross-section. *J. Fluid Mech.* 58:417–34

Fermi, E., Pasta, J., Ulam, S. 1955. Studies of nonlinear problems. I. *Los Alamos Rep. LA 1940* (1955). Reprinted in Newell (1974), pp. 143–56

Fornberg, B., Whitham, G. B. 1978. A numerical and theoretical study of certain nonlinear wave phenomena. *Phil. Trans. R. Soc. London Ser. A* 289:373–404

Freeman, N. C., Johnson, R. S. 1970. Shallow water waves on shear flows. *J. Fluid Mech.* 42:401–9

Friedrichs, K. O., Hyers, D. H. 1954. The existence of solitary waves. *Comm. Pure Appl. Math.* 7:517–50

Gardner, C. S., Greene, J. M., Kruskal, M. D., Miura, R. M. 1967. Method for solving the Korteweg-deVries equation. *Phys. Rev. Lett.* 19:1095–97

Gardner, C. S., Greene, J. M., Kruskal, M. D., Miura, R. M. 1974. Korteweg-deVries equation and generalizations. VI. Methods for exact solution. *Comm. Pure Appl. Math.* 27:97–133

Gorschkov, K. A., Ostrovsky, L. A.,

Pelinovsky, E. N. 1974. Some problems of asymptotic theory of nonlinear waves. *Proc. IEEE* 62:1511–17

Grimshaw, R. 1970. The solitary wave in water of variable depth. *J. Fluid Mech.* 42:639–56

Grimshaw, R. 1971. The solitary wave in water of variable depth. Part 2. *J. Fluid Mech.* 46:611–22

Grimshaw, R. 1978. Long nonlinear internal waves in channels of arbitrary cross-section. *J. Fluid Mech.* 86:415–31

Gwyther, R. F. 1900a. The classes of progressive long waves. *Phil. Mag.* 50:213–16

Gwyther, R. F. 1900b. An appendix to the paper on the classes of progressive long waves. *Phil. Mag.* 50:308–12

Gwyther, R. F. 1900c. The general motion of long waves, with an examination of the direct reflexion of the solitary wave. *Phil. Mag.* 50:349–52

Hammack, J. L., Segur, H. 1974. The Korteweg-deVries equation and water waves. Part 2. Comparison with experiments. *J. Fluid Mech.* 65:289–314

Hammack, J. L., Segur, H. 1978. Modelling criteria for long water waves. *J. Fluid Mech.* 84:359–73

Hirota, R. 1971. Exact solution of the Korteweg-deVries equation for multiple collisions of solitons. *Phys. Rev. Lett.* 27:1192–94

Hirota, R. 1976. Direct method of finding exact solutions of nonlinear evolution equations. *Lecture Notes in Mathematics* 515, ed. A. Dold, B. Eckmann (*Bäcklund Transformations, the Inverse Scattering Method, Solitons, and Their Applications*, ed. R. M. Miura), pp. 40–68. Berlin: Springer

Hunt, J. N. 1955. On the solitary wave of finite amplitude. *La Houille Blanche* 10: 197–203

Hurdis, D. A., Pao, H.-P. 1975. Experimental observation of internal solitary waves in a stratified fluid. *Phys. Fluids* 18:385–86

Ippen, A. T., Kulin, G. 1954. The shoaling and breaking of the solitary wave. *Proc. Conf. Coastal Eng., 5th*, pp. 27–49

Jeffrey, A., Kakutani, T. 1970. Stability of the Burgers shock wave and the Korteweg-deVries soliton. *Indiana Univ. Math. J.* 20:463–68

Jeffrey, A., Kakutani, T. 1972. Weak nonlinear dispersive waves: a discussion centered around the Korteweg-deVries equation. *SIAM Rev.* 14:582–643

Johnson, R. S. 1972. Some numerical solutions of a variable-coefficient Korteweg-deVries equation (with applications to solitary wave development on a shelf). *J. Fluid Mech.* 54:81–91

Johnson, R. S. 1973a. On the development of a solitary wave moving over an uneven bottom. *Proc. Camb. Phil. Soc.* 73:183–203

Johnson, R. S. 1973b. On an asymptotic solution of the Korteweg-deVries equation with slowly varying coefficients. *J. Fluid Mech.* 60:813–24

Joseph, R. J. 1977. Solitary waves in a finite depth fluid. *J. Phys. A: Math. Gen.* 10:L225–27

Kadomtsev, B. B., Petviashvili, V. I. 1970. On the stability of solitary waves in weakly dispersing media. *Sov. Phys.-Dokl.* 15:539–41

Kakutani, T. 1971. Effect of an uneven bottom on gravity waves. *J. Phys. Soc. Jpn.* 30:272–76, 593

Kakutani, T., Yamasaki, N. 1978. Solitary waves on a two-layer fluid. *J. Phys. Soc. Jpn.* 45:674–79

Karpman, V. I. 1975. *Non-linear Waves in Dispersive Media.* Oxford: Pergamon. 186 pp.

Karpman, V. I., Maslov, E. M. 1977. A perturbation theory for the Korteweg-deVries equation. *Phys. Lett.* 60A:307–8

Kaup, D. J., Newell, A. C. 1978. Solitons as particles, oscillators, and in slowly changing media: a singular perturbation theory. *Proc. R. Soc. London Ser. A* 361:413–46

Keller, J. B. 1948. The solitary wave and periodic waves in shallow water. *Comm. Pure Appl. Math.* 1:323–39. [cf 1948, same title, *Ann. NY Acad. Sci.* 51:345–50]

Keulegan, G. H. 1948. Gradual damping of solitary waves. *J. Res. Natl. Bur. Stand.* 40:487–98

Keulegan, G. H. 1952. The characteristics of internal solitary waves. *Gravity Waves, Natl. Bur. Stand. Circ. 521*, p. 279. See also Keulegan 1953

Keulegan, G. H. 1953. Characteristics of internal solitary waves. *J. Res. Natl. Bur. Stand.* 51:133–40

Keulegan, G. H., Patterson, G. W. 1940. Mathematical theory of irrotational translation waves. *J. Res. Natl. Bur. Stand.* 24:47–101

Ko, K., Kuehl, H. H. 1979. Cylindrical and spherical Korteweg-deVries solitary waves. *Phys. Fluids.* 22:1343–48

Korteweg, D. J., deVries, G. 1895. On the change of form of long waves advancing in a rectangular canal and on a new type of long stationary waves. *Phil. Mag.* 39:422–43

Kruskal, M. D. 1974. The Korteweg-deVries equation and related evolution equations. In Newell (1974), pp. 61–84

Kruskal, M. D. 1975. Nonlinear wave equations. *Dynamical Systems Theory and*

Applications, ed. J. Moser, pp. 310–54. New York: Springer. 624 pp.

Laitone, E. V. 1960. The second approximation to cnoidal and solitary waves. *J. Fluid Mech.* 9:430–44

Lake, B. M., Yuen, H. C., Ferguson, W. E. 1978. Envelope solitons and recurrence in nonlinear deep water waves: theory and experiment. *Rocky Mountain Math. J.* 8:105–16

Lamb, H. 1932. *Hydrodynamics*. Cambridge: Univ. Press. 738 pp. 6th ed.

Landau, L. D., Lifshitz, E. M. 1958. *Quantum Mechanics, Non-relativistic Theory*. London: Pergamon 506 pp.

Larsen, L. H. 1965. Comments on (Long 1964b). *J. Atmos. Sci.* 22:222–24

Lax, P. D. 1968. Integrals of nonlinear equations of evolution and solitary waves. *Comm. Pure Appl. Math.* 21:467–90

Lax, P. D. 1975. Periodic solutions of the KdV equation. *Comm. Pure Appl. Math.* 28:141–88

Lax, P. D. 1976. Almost periodic solutions of the KdV equation. *SIAM Rev.* 18:351–75

Leibovich, S., Randall, J. D. 1972. Solitary waves in concentrated vortices. *J. Fluid Mech.* 51:625–35

Leibovich, S., Seebass, A. R., eds. 1974. *Nonlinear Waves*. Ithaca: Cornell Univ. Press. 331 pp.

Lenau, C. W. 1966. The solitary wave of maximum amplitude. *J. Fluid Mech.* 26: 309–20

Leonov, A. I., Miropol'skiy, Y. Z. 1975. Toward a theory of stationary nonlinear internal gravity waves. *Atmospheric and Oceanic Physics* 11:298–304

Levi-Cività, T. 1907. Sulle onde progressive di tipo permanente. *Atti Accad. Lincei, Rend. Cl. Sci. Fis. Mat. Nat.* (5) 16^{II}: 770–90

Levi-Cività, T. 1925. Détermination rigoureuse des ondes permanentes d'ampleur finie. *Math. Ann.* 93:264–314

Lin, C. C., Clark, A. 1959. On the theory of shallow water waves. *Tsing Hua J. Chinese Studies* 1:54–62

Long, R. R. 1956. Solitary waves in one- and two-fluid systems. *Tellus* 8:460–71

Long, R. R. 1964a. The initial-value problem for long waves of finite amplitude. *J. Fluid Mech.* 20:161–70

Long, R. R. 1964b. Solitary waves in the Westerlies. *J. Atmos. Sci.* 21:197–200

Long, R. R. 1965. On the Boussinesq approximation and its role in the theory of internal waves. *Tellus* 17:46–52

Long, R. R., Morton, J. B. 1966. Solitary waves in compressible, stratified fluids. *Tellus* 18:79–85

Longuet-Higgins, M. S. 1974. On the mass, momentum, energy and circulation of a solitary wave. *Proc. R. Soc. London Ser. A* 337:1–13

Longuet-Higgins, M. S., Fenton, J. D. 1974. On the mass, momentum, energy and circulation of a solitary wave. II. *Proc. R. Soc. London Ser. A* 340:471–93

Madsen, O. S., Mei, C. C. 1969. The transformation of a solitary wave over an uneven bottom. *J. Fluid Mech.* 39:781–91

Makhankov, V. G. 1978. Dynamics of classical solitons (in non-integrable systems). *Phys. Rep.* 35:1–128

Maxon, S., Viecelli, J. 1974. Spherical solitons. *Phys. Rev. Lett.* 32:4–6

Maxworthy, T. 1976. Experiments on collisions between solitary waves. *J. Fluid Mech.* 76:177–85

Maxworthy, T., Redekopp, L. G. 1976. A solitary wave theory of the Great Red Spot and other observed features in the Jovian atmosphere. *Icarus* 29:261–71

Maxworthy, T., Redekopp, L. G., Weidman, P. D. 1978. On the production and interaction of planetary solitary waves: applications to the Jovian atmosphere. *Icarus* 33:388–409

McCowan, J. 1891. On the solitary wave. *Phil. Mag.* (5) 32:45–58

McCowan, J. 1894. On the highest wave of permanent type. *Phil. Mag.* (5) 38:351–58

Miles, J. W. 1967. Surface-wave damping in closed basins. *Proc. R. Soc. London Ser. A* 297:459–75

Miles, J. W. 1977a. Note on a solitary wave in a slowly varying channel. *J. Fluid Mech.* 80:149–52

Miles, J. W. 1977b. Obliquely interacting solitary waves. *J. Fluid Mech.* 79:157–69

Miles, J. W. 1977c. Resonantly interacting solitary waves. *J. Fluid Mech.* 79:171–79

Miles, J. W. 1977d. Diffraction of solitary waves. *ZAMP* 28:889–902

Miles, J. W. 1978a. An axisymmetric Boussinesq wave. *J. Fluid Mech.* 84:181–91

Miles, J. W. 1978b. On the evolution of a solitary wave for very weak nonlinearity. *J. Fluid Mech.* 87:773–83

Miles, J. W. 1979a. On the Korteweg-deVries equation for a gradually varying channel. *J. Fluid Mech.* 91:181–90

Miles, J. W. 1979b. On internal solitary waves. *Tellus*. In press

Milne-Thomson, L. M. 1968. *Theoretical Hydrodynamics*. London: Macmillan. 743 pp. 5th ed.

Miura, R. M. 1974. The Korteweg-deVries equation: a model equation for nonlinear dispersive waves. In Leibovich & Seebass (1974), pp. 212–34

Miura, R. M. 1976. The Korteweg-deVries equation: a survey of results. *SIAM Rev.* 18:412–59

Miura, R. M., Gardner, C. S., Kruskal,

M. D. 1968. Korteweg-deVries equation and generalizations. II. Existence of conservation laws and constants of motion. *J. Math. Phys.* 9:1204–9

Munk, W. H. 1949. The solitary wave theory and its application to surf problems. *Ann. NY Acad. Sci.* 51:376–424

Nekrasov, A. I. 1921. On waves of permanent type. *Izv. Ivan. Voznesensk. politekh. Inst.* 3:52–65 (in Russian)

Newell, A. C., ed. 1974. *Nonlinear Wave Motion. Lectures in Applied Mathematics*, Vol. 15. Providence, R.I.: Am. Math. Soc. 229 pp.

Newell, A. C., Redekopp, L. G. 1977. Breakdown of Zakharov-Shabat theory and soliton creation. *Phys. Rev. Lett.* 38:377–80

Oikawa, M., Satsuma, J., Yajima, N. 1974. Shallow water waves propagating along undulation of bottom surface. *J. Phys. Soc. Jpn.* 37:511–17

Ono, H. 1972. Wave propagation in an inhomogeneous anharmonic lattice. *J. Phys. Soc. Jpn.* 32:332–36

Ostrovsky, L. A. 1976. Short-wave asymptotics for weak shock waves and solitons in mechanics. *Int. J. Non-Linear Mech.* 11:401–16

Ostrovsky, L. A., Pelinovsky, E. N. 1970. Wave transformation on the surface of a fluid of variable depth. *Atmospheric and Oceanic Physics* 6:552–55

Ostrovsky, L. A., Pelinovsky, E. N. 1975. Refraction of nonlinear ocean waves in a beach zone. *Atmospheric and Oceanic Physics* 11:37–41

Ostrovsky, L. A., Shrira, V. I. 1976. Instability and self-refraction of solitons. *Sov. Phys.-JETP* 44:738–43

Packham, B. A. 1952. The theory of symmetrical gravity waves of finite amplitude. II. The solitary wave. *Proc. R. Soc. London Ser. A* 213:238–49

Pelinovsky, E. N. 1974. The evolution of a solitary wave in a nonhomogeneous medium. *J. Appl. Mech. Tech. Phys.* 12:853–58 (transl. from Russian)

Peregrine, D. H. 1966. Calculations of the development of an undular bore. *J. Fluid Mech.* 25:321–30

Peregrine, D. H. 1967. Long waves on a beach. *J. Fluid Mech.* 27:815–27

Peregrine, D. H. 1968. Long waves in a uniform channel of arbitrary cross-section. *J. Fluid Mech.* 32:353–65

Peregrine, D. H. 1969. Solitary waves in trapezoidal channels. *J. Fluid Mech.* 35:1–6

Peregrine, D. H. 1972. Long waves in two and three dimensions. *Proc. Symp. on Long Waves*, Univ. Delaware, Sept. 10–11, 1970, pp. 63–90

Perroud, P. H. 1957. *The solitary wave reflection along a straight vertical wall at oblique incidence.* PhD thesis. Univ. Calif., Berkeley. 93 pp.

Peters, A. S. 1966. Rotational and irrotational solitary waves in a channel with arbitrary cross section. *Comm. Pure Appl. Math.* 19:445–71

Peters, A. S., Stoker, J. J. 1960. Solitary waves in liquids having nonconstant density. *Comm. Pure Appl. Math.* 13:115–64

Pritchard, W. G. 1970. Solitary waves in rotating fluids. *J. Fluid Mech.* 42:61–83

Rayleigh, Lord. 1876. On waves. *Phil. Mag.* 1:257–79. *Sci. Pap.* 1:251–71

Redekopp, L. G. 1977. On the theory of solitary Rossby waves. *J. Fluid Mech.* 82:725–45

Redekopp, L. G., Weidman, P. D. 1978. Solitary Rossby waves in zonal shear flows and their interactions. *J. Atmos. Sci.* 35:790–804

Reutov, V. A. 1976. Motion of a solitary wave over a submerged ridge. *Fluid Dyn. (USSR)* 10:604–10; *Izv. Akad. Nauk SSSR, Mekh. Zhidk. Gaza* 1975, no. 4:79–85

Russell, J. S. 1838. Report of the Committee on Waves. *Rep. Meet. Brit. Assoc. Adv. Sci., 7th*, Liverpool, 1837, pp. 417–496. London: John Murray

Russell, J. S. 1845. Report on waves. *Rep. Meet. Brit. Assoc. Adv. Sci., 14th*, York, 1844, pp. 311–390. London: John Murray

Saeki, H., Takagi, K., Ozaki, A. 1971. Study on the transformation of the solitary wave (2). *Proc. Conf. Coastal Eng. in Japan, 18th*, pp. 49–53 (in Japanese; cited by Shuto 1974)

Satsuma, J. 1976. N-soliton solution of the two-dimensional Korteweg-deVries equation. *J. Phys. Soc. Jpn.* 40:286–90

Scott, A. C., Chu, F. Y. F., McLaughlin, D. W. 1973. The soliton: a new concept in applied science. *Proc. IEEE* 61:1443–83

Segur, H. 1973. The Korteweg-deVries equation and water waves. Solutions of the equation. Part I. *J. Fluid Mech.* 59:721–36

Shuto, N. 1973. Shoaling and deformation of non-linear long waves. *Coastal Eng. Jpn.* 16:1–12

Shuto, N. 1974. Non-linear long waves in a channel of variable section. *Coastal Eng. Jpn.* 17:1–12

Shuto, N. 1976. Transformation of nonlinear long waves. *Proc. Conf. Coastal Eng., 15th*, pp. 1–18

Stoker, J. J. 1957. *Water Waves.* New York: Interscience. 567 pp.

Stokes, G. G. 1849. On the theory of oscillatory waves. *Trans. Camb. Phil. Soc.* 8:441–55. *Math. Phys. Pap.* 1:197–229

Stokes, G. G. 1891. Note on the theory of

the solitary wave. *Phil. Mag.* 31:314–16. *Math. Phys. Pap.* 5:160–62

Street, R. L., Camfield, F. E. 1967. Observations and experiments on solitary wave deformation. *Proc. Conf. Coastal Eng., 10th*, pp. 284–301

Strelkoff, T. 1971. An exact numerical solution of the solitary wave. *Proc. Int. Conf. Num. Meth. Fluid Dyn., 2nd*, Springer, pp. 441–46 (cited by Longuet-Higgins & Fenton 1974)

Svendsen, I. A. 1976. A direct derivation of the KdV-equation for waves on a beach, and discussion of its implications. *Prog. Rep. Inst. Hydrodyn. Hydraul. Eng., Tech. Univ. Denmark* 39:9–16

Tappert, F. D., Zabusky, N. J. 1971. Gradient-induced fission of solitons. *Phys. Rev. Lett.* 27:1774–76

Ter-Krikorov, A. M. 1962. The solitary wave on the surface of a turbulent liquid [*sic*; "turbulent liquid" presumably should be *liquid with vorticity*]. *USSR Comp. Math. Math. Phys.* 1:1253–64

Ter-Krikorov, A. M. 1963. Théorie exacte des ondes longues stationnaires dans un liquide hétérogène. *J. Méc.* 2:351–76

Ursell, F. 1953. The long-wave paradox in the theory of gravity waves. *Proc. Camb. Phil. Soc.* 49:685–94

Wadati, M., Toda, M. 1972. The exact N-soliton solution of the Korteweg-deVries equation. *J. Phys. Soc. Jpn.* 32:1403–11

Walker, L. R. 1973. Interfacial solitary waves in a two-fluid medium. *Phys. Fluids* 16:1796–1804

Wehausen, J. V., Laitone, E. V. 1960. Surface waves. *Encyc. Phys.* 9:446–778

Weidman, P. D. 1978. Internal waves in a linearly stratified fluid. *Tellus* 30:177–84; and Corrigendum. In press

Weidman, P. D., Maxworthy, T. 1978. Experiments on strong interactions between solitary waves. *J. Fluid Mech.* 85:417–31

Weinstein, A. 1926. Sur la vitesse de propagation de l'onde solitaire. *Accad. Naz. Lincei, Cl. Sci. Fis., Mat. Nat. Rendiconti* (6) 3:463–68

Whitham, G. B. 1974. *Linear and Nonlinear Waves*. New York: Wiley. 636 pp.

Wiegel, R. L. 1964a. *Oceanographical Engineering*. Englewood Cliffs, N.J.: Prentice-Hall. 551 pp.

Wiegel, R. L. 1964b. Water wave equivalent of Mach-reflection. *Proc. Conf. Coastal Eng., 9th, Lisbon, Portugal, June 1964*, pp. 82–102

Witting, J. 1975. On the highest and other solitary waves. *SIAM J. Appl. Math.* 28:700–19

Yamada, H. 1957. On the highest solitary wave. *Rep. Res. Inst. Appl. Mech. Kyushu Univ.* 5:53–67

Yamada, H. 1958. On approximate expressions of solitary wave. *Rep. Res. Inst. Appl. Mech. Kyushu Univ.* 6:35–47

Yamada, H., Kimura, G., Okabe, J.-I. 1968. Precise determination of the solitary wave of extreme height on water of a uniform depth. *Rep. Res. Inst. Appl. Mech. Kyushu Univ.* 16:15–32

Yuen, H. C., Lake, B. M. 1975. Nonlinear deep water waves: theory and experiment. *Phys. Fluids* 18:956–60

Zabusky, N. J. 1967. A synergetic approach to problems of nonlinear dispersive wave propagation and interaction. *Nonlinear Partial Differential Equations*, pp. 223–58. New York: Academic

Zabusky, N. J., Galvin, C. J. 1971. Shallow-water waves, the Korteweg-deVries equation and solitons. *J. Fluid Mech.* 47:811–24

Zabusky, N. J., Kruskal, M. D. 1965. Interaction of "solitons" in a collisionless plasma and the recurrence of initial states. *Phys. Rev. Lett.* 15:240–43

Zabusky, N. J., Kruskal, M. D., Deem, G. S. 1965. *Formation, Propagation and Interaction of Solitons* (*Numerical Solutions of Differential Equations Describing Wave Motion in Nonlinear Dispersive Media*). A 16 mm, 35 min, silent, black-and-white film available on loan from the Bell Telephone Laboratories, Inc., Film Library, Murray Hill, N.J. 07971

Ann. Rev. Fluid Mech. 1980. 12:45–76

TOPOGRAPHICALLY TRAPPED WAVES

×8152

Lawrence A. Mysak

Departments of Mathematics and Oceanography, University of British Columbia, Vancouver, British Columbia V6T 1W5, Canada

1 INTRODUCTION

At the beach we have all enjoyed watching the incessant approach of surface waves that have traveled vast distances across the ocean. Because this is such a familiar sight, it perhaps comes as a surprise to discover that along this same beach considerable amounts of wave energy can also travel long distances parallel to the coast. Such "edge" waves have amplitudes that decrease away from the coast and have periods and wavelengths that are generally far larger than those associated with surface waves. The first mathematical solution of the appropriate equations of motion for such topographically trapped waves was given over a century ago by Stokes (1846). For decades this solution was regarded as a hydrodynamical curiosity; indeed, Horace Lamb (1945, p. 447) once wrote: ". . . it does not appear that the motion here referred to is very important." However, recently it has been established that edge waves are an essential ingredient of our understanding of such phenomena as rip currents (Figure 1), beach cusps, crescentic bars, and nearshore sediment transport (see LeBlond & Mysak 1977, 1978).

The main purpose of this article is to review the basic dynamics of edge waves, and also of two other types of coastal trapped waves that have even larger periods and wavelengths: Kelvin waves and shelf waves. Unlike edge waves, Kelvin and shelf waves depend crucially upon the earth's rotation. Further, in a given hemisphere they can travel in only one direction. In the northern (southern) hemisphere, their phase propagates with the coast to the right (left). Shortly after their discovery near the end of the last century, it was recognized that Kelvin waves could serve as propagators of the tides along the coast (see Lamb 1945, p. 319). Shelf waves, on the other hand, are, relatively, newcomers to the field,

0066-4189/80/0115-0045$01.00

Figure 1 Strong rip currents (pointed narrow grey strips) periodically spaced along Hell's Mouth Bay, North Wales. Photo courtesy of A. J. Bowen.

having been discovered about twenty years ago. Yet since that time it has been found that shelf waves play an important role in such phenomena as Gulf Stream meanders, coastal upwelling (for example, see J. S. Allen's article in this volume), shelf circulation, and maritime climate.

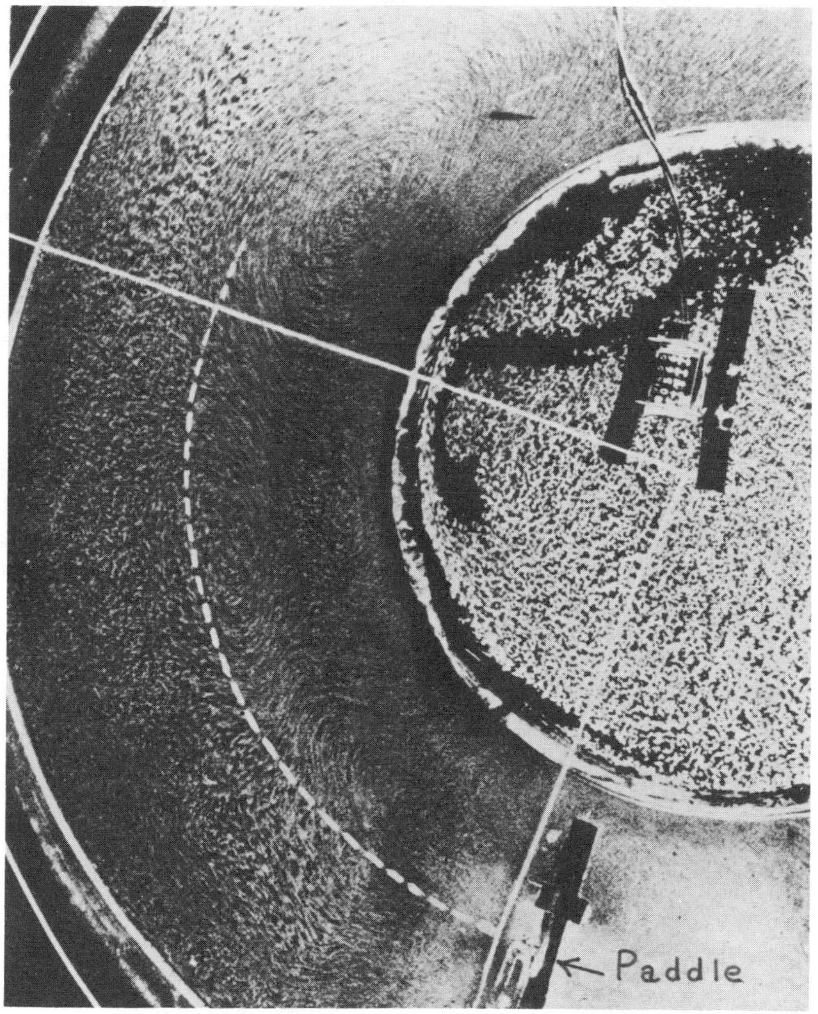

Figure 2 A streak photograph showing a laboratory model of long shelf waves propagating along an exponential shelf profile (see Figure 3) toward the top of the picture (Caldwell, Cutchin & Longuet-Higgins 1972). The waves are generated by the paddle (at bottom) which moves back and forth across the depth contours. The solid lines are basin radii and the broken line marks the foot of the continental slope region. Three large eddies centered near the broken lines are clearly visible.

Edge waves can be thought of as long gravity waves over a sloping beach that are trapped near the coast by refraction. A straightforward understanding of this trapping mechanism can be obtained from the methods of ray theory (see, for example, LeBlond & Mysak 1977). Further, because these waves have relatively high frequencies (periods ranging from hours to minutes) and consequently are only slightly modified by the earth's rotation, they are often referred to as "first class" oscillations. Longuet-Higgins (1972) gave a lucid account of the trapping mechanism for shelf waves, which is based on the fact that these waves conserve potential vorticity, defined as $(\zeta+f)/h$ where ζ is the vertical component of the relative vorticity, f is the Coriolis parameter, and h is the total fluid depth. Thus shelf waves depend both on depth variations and the earth's rotation and are highly rotational. Their associated particle motions consist of large vortices (eddies) of alternating spin that drift along the shelf-slope region (Figure 2). Also, having much larger scales than edge waves, the amplitude of a shelf wave extends well into the deep-sea region (Figure 3). Further, their frequencies are always subinertial ($\omega < f$), and as $f \to 0$ the waves reduce to steady currents. For these reasons they are sometimes called "second class" oscillations. Finally, we remark that a Kelvin wave, in its idealized form, is a type of hybrid wave, depending both on the earth's rotation and on gravity. Kelvin waves are exponentially trapped against a vertical wall that acts as a side boundary to an ocean of constant depth. They are maintained by an exact balance between the Coriolis force and a pressure gradient normal to the wall. Yet the speed of their phase propagation is simply given by

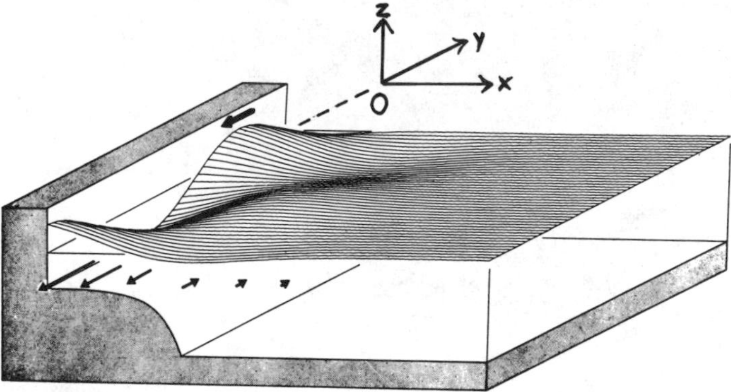

Figure 3 A computer-assisted artist's conception of a gravest-mode shelf wave (Cutchin & Smith 1973). The heavy arrow along the coast indicates direction of phase propagation in the northern hemisphere and the lighter arrows over the shelf-slope region indicate the water velocity under the crest. The vertical displacement of the sea surface is greatly exaggerated—in the ocean, shelf-wave amplitudes are only a few centimeters.

$(gH)^{\frac{1}{2}}$ (H = equilibrium depth), which is the speed of a long gravity wave in a nonrotating system! In the presence of realistic coastal topography, a Kelvin wave does get slightly modified, in a manner that will be discussed in the next section.

Ocean waves can also be trapped around islands and seamounts, and the classification of these trapped waves into first class oscillations, second class oscillations, or Kelvin waves generally carries over to these nonrectilinear topographies (see LeBlond & Mysak 1978, Mysak 1980). However, there is one new feature of these waves: because the waves are circularly traveling, the longshore wave number is now quantized, there being an integral number of wavelengths around the island or seamount. As a consequence there is a discrete spectrum for both the frequency and wave number of island or seamount trapped waves. In principle, this should make it easy to identify such waves from observed current or sea level spectra. However, in practice there has been limited success in this direction. On the other hand, there have been many observations of trapped waves of all types along rectilinear coasts. For this reason, we shall focus most of our attention on the dynamics of waves trapped along extensive straight coastlines whose mean properties (e.g. topography, stratification) do not vary in the longshore direction. Further, the reader will probably notice a slight bias in the paper toward low-frequency trapped waves. This is because, over the last fifteen years or so, much of the author's research has centered on various aspects of shelf waves and, to a lesser extent, Kelvin waves.

2 COASTAL TRAPPED WAVES IN A HOMOGENEOUS OCEAN

We now give a more quantitative discussion of the different types of barotropic trapped waves that can propagate along a straight continental shelf-slope region with a monotonic depth profile (Figure 4). The analysis will be based on the unforced linearized shallow water equations for a homogeneous, uniformly rotating fluid. The effects of stratification, mean flows, longshore variations, nonuniform rotation, and nonlinear interactions will be discussed in the subsequent sections (Sections 3–7).

Governing Equations

The nonlinear unforced shallow-water equations for a depth profile $H(x)$ are given by

$$\left.\begin{aligned} u_t - fv + g\eta_x &= -uu_x - vu_y \\ v_t + fu + g\eta_y &= -uv_x - vv_y \end{aligned}\right\} \tag{2.1}$$

$$(Hu)_x + Hv_y + \eta_t = -(\eta u)_x - (\eta v)_y. \tag{2.2}$$

The notation is standard (see Figure 4), with $f = 2\Omega \sin \theta$ (Ω = angular velocity of the earth and θ = latitude) denoting the constant Coriolis parameter. At middle and higher latitudes, the effects of a variable f are negligible for typical magnitudes of the bottom slope (dH/dx) characterizing the shelf-slope region. However, near the equator, where f is rapidly changing, it is important that both depth and f variations be taken into account (for example, see Mysak 1978a,b, Geisler & Mysak 1978).

For long-wave motions characterized by a frequency ω, wave number k, and particle speed U, the nonlinear inertial terms on the right side of (2.1) are small compared with the local acceleration terms, provided

$$U \ll c, \tag{2.3}$$

where $c = \omega/k$ is the phase speed. For the types of waves to be discussed below, the restriction (2.3) generally holds. The nonlinear terms on the right side of (2.2) can be neglected, provided

$$\eta \ll H \tag{2.4}$$

everywhere. Clearly, the criterion (2.4) breaks down in the neighborhood of the mean coastline ($x = 0$) if the mean depth tends to zero (as in Figure 4). For such a depth profile the linearized form of (2.2) is strictly valid only for x in the range of $0 < x_0 \leqq x < \infty$. In practice this has motivated the use of shelf profiles with a small vertical wall at the coast. It turns out that the use of the linearized form of (2.2) together with a depth

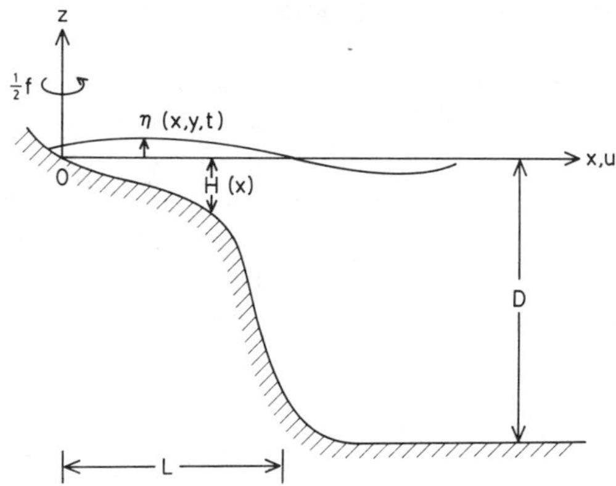

Figure 4 The coordinate system. The y axis and y velocity component v are directed into the plane at 0. L is the characteristic shelf-slope width and D is the deep-sea depth far from the coast.

profile with $H(0) = 0$ imposes important restrictions on the nature of the dispersion curves, $\omega = \omega(k)$, for shelf waves.

From the linearized form of (2.1) and (2.2) we readily obtain

$$H\Delta\eta_t + H_x\eta_{xt} + fH_x\eta_y - (M/g)\eta_t = 0, \tag{2.5}$$

where $\Delta = \partial_{xx} + \partial_{yy}$ and $M = \partial_{tt} + f^2$. For waves traveling parallel to the coast we let

$$\eta = F(x)\exp(iky + i\omega t), \tag{2.6}$$

where we suppose that $k > 0$ and ω can have either sign. Thus for $\omega > 0$ Equation (2.6) implies that the wave phase travels in the negative y direction, with the coast on the right (as in Figure 3). We shall call such waves right-bounded. Similarly, for $\omega < 0$ the waves travel with the coast on the left and shall be called left-bounded. Substitution of (2.6) into (2.5) yields

$$(HF')' + [(\omega^2 - f^2)/g - k^2H + fkH'/\omega]\,F = 0, \quad 0 < x < \infty \tag{2.7}$$

where a prime indicates d/dx. The appropriate boundary conditions are $Hu = 0$ at $x = 0$ and $\eta \to 0$ as $x \to \infty$ (for trapped waves):

$$H(fkF + \omega F') = 0 \quad \text{at} \quad x = 0, \tag{2.8}$$

$$F \to 0 \quad \text{as} \quad x \to \infty. \tag{2.9}$$

If $H(0) \neq 0$, (2.8) can be written as

$$fkF(0) + \omega F'(0) = 0, \tag{2.10a}$$

whereas if $H(0) = 0$, (2.8) will hold provided F is well-behaved (differentiable) at $x = 0$:

$$|F'(0)| < C_0 \text{ (a constant).} \tag{2.10b}$$

Qualitative Results

The qualitative theory of the system (2.7), (2.9), (2.10a) has been studied in some detail by Huthnance (1975). He showed that for any monotonic depth profile that tends to a constant as $x \to \infty$, there are three types of trapped waves:

1. An infinite discrete set of high-frequency edge waves that can travel in both directions along the shelf;
2. An infinite discrete set of low-frequency ($\omega < f$) continental shelf waves that are right-bounded in the northern hemisphere;
3. A single low-frequency, low–wave-number "Kelvin wave" that is effectively trapped against the continental slope. Further, it is also right-bounded in the northern hemisphere. At high wave numbers

(and high frequencies) it becomes a right-bounded, fundamental-mode edge wave.

The unidirectional propagation property of shelf waves and the Kelvin wave when $\omega < f$ follows immediately from (2.7) upon multiplying by F, integrating over all x, and invoking the boundary conditions.

At sufficiently low wave numbers the edge waves cannot be trapped, owing to the breakdown of the refraction mechanism; at these long length scales only the continuum of "leaky" Poincaré waves is possible (see Figure 5 below). Shelf waves, however, are trapped at all wave numbers.

Finally, we note that Huthnance (1975) also established the following important results regarding the group velocity ($c_g = \partial\omega/\partial k$) of each shelf wave mode:

1. $c_g \to c = \omega/k$ as $k \to 0$ (long shelf waves are nondispersive); (2.11)
2. If H'/H is bounded for all x, then $c_g < 0$ for some range of $k > 0$. (2.12)

Since the dispersion curve $\omega(k)$ of each mode is continuous, (2.11) and (2.12) imply that at some intermediate wave number, the shelf wave has a zero group velocity. This is to say, the phase and energy (which propagates with the group velocity) of shelf waves propagate in the same direction at low wave numbers and in opposite directions at (generally) high wave numbers *provided* H'/H is always bounded.

It is interesting to note here that similar qualitative results for long waves in a variable-depth channel were obtained at about the same time by Odulo (1975). In place of the trapping condition (2.9), he imposed another condition of the form (2.10a) at a vertical wall at $x = x_0$. Because of the presence of two walls, Odulo showed that two Kelvin waves are possible. Finally, we mention that the decomposition of the energy density and fluxes into the different wave types has been discussed by Huthnance (1975) and by Odulo (1977).

Explicit Results

As an example we now give the solution for trapped waves over a sloping beach of finite width that drops off to a flat deep-sea region. For this topography, $H(x)$ has the form

$$\begin{aligned} H(x) &= dx/L, \qquad 0 < x < L \\ &= D, \qquad L < x < \infty. \end{aligned} \tag{2.13}$$

For the profile (2.13), the solution of (2.7), (2.9), and (2.10b) takes the form (Mysak 1968a)

$$F(x) = Ae^{-kx}L_\nu(2kx), \qquad 0 < x < L$$
$$= Be^{-Kx}, \qquad L < x < \infty \tag{2.14}$$

where $L_\nu(z)$ is the Laguerre function, $\nu = (-1+\mu)/2$ and

$$\mu = f/\omega + (\omega^2 - f^2)/\alpha g k \qquad (\alpha = d/L) \tag{2.15}$$

$$K = [k^2 + (f^2 - \omega^2)/gD]^{\frac{1}{2}}. \tag{2.16}$$

K must be real and positive for trapped waves. Examination of (2.14) and (2.16) reveals that high-frequency edge waves ($\omega^2 \gg f^2$) cannot be trapped when k is very small, i.e. there exists a long-wave cutoff for trapped edge waves defined by the hyperbolas (obtained by setting $K^2 = 0$)

$$\sigma = \omega/f = \pm[1 + (gD/f^2L^2)\kappa^2]^{\frac{1}{2}}, \tag{2.17}$$

where $\kappa = kL$. When $K^2 < 0$, the spectrum is continuous and consists of topographically modified "leaky" Poincaré waves. For shelf waves, which have $\omega^2 < f^2$, $K > 0$ for all $k > 0$ and trapping occurs at all wavelengths.

By making η and Hu continuous at $x = L$, we obtain two homogeneous equations for the unknowns A and B, which in turn yield the implicit dispersion relation

$$L_\nu(2\kappa)\{[1 + \delta\Delta(1-\sigma^2)\kappa^{-2}]^{\frac{1}{2}} - \sigma^{-1} - \Delta(1-\sigma^{-1})\} + 2\Delta L'_\nu(2\kappa) = 0, \tag{2.18}$$

where here $\delta = f^2L^2/gd$, $\Delta = d/D$, and ν is related to σ, κ, and δ by

$$\sigma^3 - [1 + (2\nu+1)\delta^{-1}\kappa]\sigma + \delta^{-1}\kappa = 0. \tag{2.19}$$

For small Δ, (2.18) can be approximated by

$$L_\nu(2\kappa) = 0. \tag{2.20}$$

Physically, this means that at the edge of the shelf ($x = L$), the waves have a node. Since typically $\Delta = 4 \times 10^{-2}$ ($d = 200$ m, $D = 5000$ m) for the ocean, this approximation introduces an error of only a few percent. However, to determine the Kelvin-wave dispersion curve at low frequencies as well as the details of the edge-wave dispersion curves near the long-wave cutoff, the full relation (2.18) together with (2.19) must be solved. Csanady (1976) showed that for large lakes such as Lake Ontario, $\Delta = 1$ applies and (2.18) can be approximated by

$$L'_\nu(2\kappa) = 0 \tag{2.21}$$

provided also $\delta \ll 1$.

For a given κ, (2.20) is satisfied for a countably infinite number of discrete

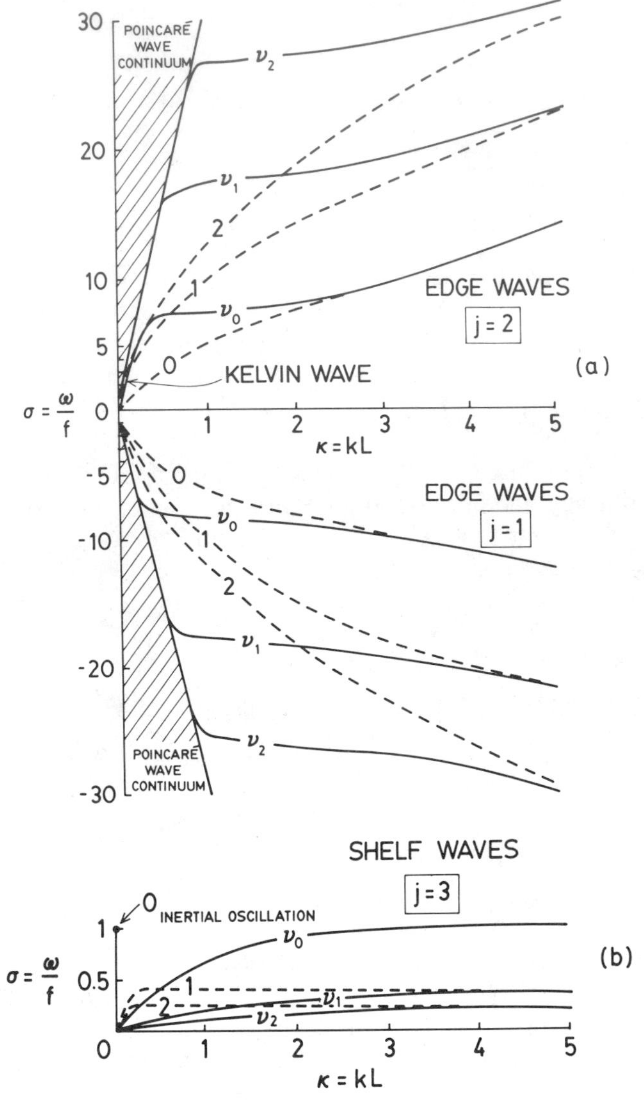

Figure 5 (*a*) Edge wave ($j = 1$ and 2), and (*b*) shelf wave ($j = 3$) dispersion curves for a sloping shelf of finite width (solid lines) and semi-infinite width (dashed lines) with $\delta = f^2L^2/gd = 0.027$ [$f = 0.73 \times 10^{-4}$ rad s^{-1}, $d = 200$ m, $L = 10^5$ m (or $\alpha = 2 \times 10^{-3}$)] (adapted from Mysak 1968a). The different modes v_n or n are indicated on the curves. The shaded region corresponds to the continuous spectrum of topographically modified Poincaré waves and is bounded by the hyperbolas $\omega/f = \pm(1 + gDk^2/f^2)^{\frac{1}{2}}$ where $D = 5 \times 10^3$ m (deep-sea depth).

values of $\nu = \nu_0, \nu_1, \nu_2, \ldots$, the first three of which have been tabulated as a function of κ by Mysak (1968a). Then, corresponding to each value of ν_n, the cubic (2.19) can be solved for the frequency functions (dispersion relations) $\sigma = \sigma[j, \nu_n(\kappa), \kappa]$, $j = 1, 2, 3$. For each value of ν_n the roots $j = 1$ and 2 correspond to the nth-mode edge-wave dispersion relations (see Figure 5*a*) whereas the root $j = 3$ corresponds to the nth-mode shelf-wave dispersion relation (Figure 5*b*). In the limit of a wide shelf, or equivalently, as $\kappa \to \infty$, $\nu_n \to n = 0, 1, 2, \ldots$, which are the values obtained by Reid (1958) for the semi-infinite sloping beach, with depth profile

$$H(x) = \alpha x. \tag{2.22}$$

The corresponding dispersion curves are shown as dashed lines in Figure 5. For the sloping beach (2.22), Reid (1958) showed that in the presence of rotation, there are two types of trapped long waves: high-frequency edge waves that are slightly modified by f, and low-frequency quasi-geostrophic waves that are highly rotational. Further, the gravest-mode quasi-geostrophic wave is simply an inertial oscillation ($\sigma = 1$ at $\kappa = 0$).

In computing the values of ν_n from (2.20), κ must always be chosen so that $K > 0$ (trapped wave condition). Thus the edge-wave dispersion curves end abruptly on the hyperbolas $\sigma = \pm[1 + (gD/f^2L^2)\kappa^2]^{\frac{1}{2}}$, inside of which only the leaky Poincaré-wave continuum exists. However, the one exception is the curve for the gravest-mode edge wave for $j = 2$, which at small κ turns into the dispersion relation for a nondispersive Kelvin wave trapped against the vertical wall at $x = L$: $\omega = (gD)^{\frac{1}{2}}k \operatorname{sgn} f$.

From Figure 5*b* we note that the phase and group velocities of each shelf-wave mode have the same sign for all κ. This does not contradict Huthnance's result (2.12) since for the profile (2.13), H'/H is unbounded at both $x = 0$ and $x = L$. Thus the hypothesis of his theorem does not hold. However, we note that as $\kappa \to 0$, the shelf waves are nondispersive: $\sigma \propto \kappa$ as $\kappa \to 0$ and therefore $c_g \to c$ as $\kappa \to 0$, in agreement with (2.11). Finally we note that as $\kappa \to \infty$, $\sigma \to 1/(2n+1)$, so that the group velocity tends to zero at high wave numbers. That is, short shelf waves over a sloping beach are rather inert (cannot propagate energy).

For other shelf-slope topographies used in the literature, the dispersion curves of the different types of trapped waves are generally similar in character to those shown in Figure 5. In particular, for the Ball (1967) profile

$$H(x) = H_0(1 - e^{-ax}), \qquad a > 0, \qquad 0 < x < \infty \tag{2.23}$$

and the flat-shelf model

$$H(x) = d, \qquad 0 < x < L$$
$$\quad = D, \qquad L < x < \infty \tag{2.24}$$

the dispersion curves have been computed for a variety of parameter values by Munk, Snodgrass & Wimbush (1970). However, we note that because of the singular nature of $H'(x)$ for the flat-shelf model, there is only one (the gravest) shelf-wave mode, the higher modes having been filtered out.

An important topographic model, which has been widely used in the shelf-wave literature, is the exponential depth profile first used by Buchwald & Adams (1968),

$$H(x) = de^{2bx}, \qquad 0 < x < L$$
$$\quad = D, \qquad L < x < \infty \tag{2.25}$$

where $D = d \exp(2bL)$, so that the depth is continuous at $x = L$. Since it follows from (2.25) that H'/H is bounded for all x, it is anticipated on the basis of (2.12) that shelf waves over this profile have a negative group velocity at high wave numbers. This was indeed found from the numerical computations of Buchwald & Adams (1968) who also invoked the rigid-lid approximation. In using the latter approximation the linearized form of (2.2) is simply

$$(Hu)_x + Hv_y = 0, \tag{2.26}$$

which allows the use of a mass-transport stream function ψ:

$$Hv = \psi_x, \qquad Hu = -\psi_y. \tag{2.27}$$

Then, in place of the amplitude Equation (2.7), one obtains

$$(H^{-1}\phi')' - [fk\omega^{-1}(H^{-1})' + k^2H^{-1}]\phi = 0, \qquad 0 < x < \infty, \tag{2.28}$$

where $\phi(x)$ is defined by $\psi = \phi(x) \exp(iky + i\omega t)$. From (2.28) it is clear that there is only one class of waves possible, namely the shelf waves. The Kelvin and edge waves are filtered out of the problem by the rigid-lid approximation. The dispersion curves for the shelf-wave modes that result from solving (2.28) with H given by (2.25) are shown in Figure 7*a*.

3 COASTAL TRAPPED WAVES IN A STRATIFIED OCEAN

In a stratified fluid of variable depth, it is not generally possible to separate the dependent variables describing the wave motion into vertical and horizontal parts, corresponding to the normal modes and horizontally propagating waves, respectively. However, because of the importance of stratification in the slope and deep-sea regions, many theoretical papers have attempted to assess quantitatively the effects of stratification in coastal trapped waves, especially those of subinertial frequency.

The problem of edge waves in a stratified fluid has only been briefly examined in the literature. This is because edge waves are generally confined close to the coast where the water is fairly homogeneous. Indeed, it was shown by Mysak (1968b) and Iida (1974) that in two-layer models of the deep-sea stratification (see Figure 6*a*) edge waves are only weakly coupled to long internal waves in the deep ocean. For specialized edge-wave solutions over a sloping beach [see (2.22)] supporting a continuously stratified fluid, we refer the reader to Greenspan (1970) and Odulo (1974).

Low-Frequency Trapped Waves in Two-Layer Models

Most of the earlier studies dealing with low-frequency coastal trapped waves in the presence of stratification involved the use of two-layer models. Further, two types of models have been considered, according

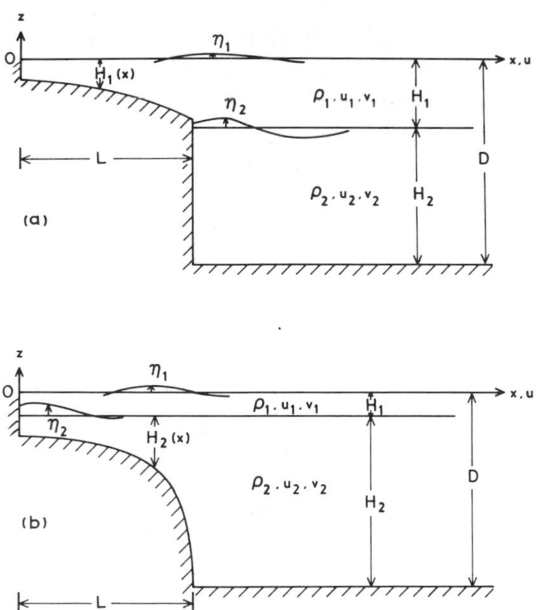

Figure 6 Two-layer models used in the theory of coastal trapped waves: (*a*) deep-sea stratification model, (*b*) on-shelf stratification model.

to whether the stratification is confined to the deep-sea region (Figure 6*a*) or extends onto the shelf region (Figure 6*b*).

In addition, the rigid-lid approximation has usually been invoked, which filters out the edge waves and the external Kelvin wave from the problem. Generally speaking, the governing two-layer long-wave equations for these models can be combined into a pair of coupled equations for the mass-transport stream function ψ [see (2.27)] and the quantity $h = \eta_2 - \eta_1$ (see Figure 6). Symbolically, these equations take the form (see Allen 1975 for details)

$$L_1\psi = L_2 h \tag{3.1}$$

$$L_3 h = L_4\psi, \tag{3.2}$$

where L_2 and L_4 are each proportional to (d/dx) (total depth) and L_1 and L_3 are third-order linear differential operators. Because of the form of L_2 and L_4, over a region of variable depth, the barotropic and baroclinic parts of the motion, ψ and h respectively, are coupled. If there is no stratification ($\rho_1 = \rho_2$) it follows that $h \equiv 0$ and (3.1) simply reduces to the vorticity equation for barotropic shelf waves [effectively (2.28)]. If, however, stratification is present but the bottom is flat, (3.1) and (3.2) become uncoupled and represent the so-called normal mode equations for a two-layer fluid on an f-plane. In this case, $L_1\psi = 0$ reduces to Laplace's equation, implying that only steady barotropic currents exist, which is a direct consequence of the rigid-lid approximation. The equation $L_3 h = 0$ is the governing equation for a coastally trapped internal Kelvin wave that is nondispersive. Further, its current structure is described by $u_1 \equiv 0 \equiv u_2$, $v_1/v_2 = -H_2/H_1$ (the upper and lower layer longshore velocities are 180° out of phase and have no net mass transport). Recently it has been shown (McCreary 1976) that internal Kelvin waves appear to be closely connected with El Niño, a well-known oceanographic phenomenon characterized by the appearance of abnormally warm water off the coast of Peru.

From the above discussion it might now be anticipated that in the combined case of stratification and topography, which includes a vertical wall, two types of low-frequency coastal trapped waves will exist in a two-layer system: 1. shelf waves modified by the presence of stratification, and 2. internal Kelvin waves modified by the presence of topography. We shall next discuss the properties of these waves as found by a number of investigators.

Mysak (1967a) determined the phase speed of long (nondispersive) shelf waves in a deep-sea stratification model (Figure 6*a*) in which $H_1(x) = dx/L$ and $H_2 \gg H_1$. He found that for typical deep-sea parameter values, the stratification increases the phase speed of the gravest-mode shelf wave approximately twofold. The significant increase in the shelf-wave phase speed due to deep-sea stratification was confirmed later by Kajiura (1974) using a flat shelf model. However, when the stratification lies over the shelf (Figure 6*b*), Wang (1975) and Wright & Mysak (1977) found that the (long) shelf-wave phase speeds increased by only a few percent. These results explain why stratification is important for shelf waves in regions like the east Australian coast (where the deep-sea stratification model is most relevant) but not so important in coastal upwelling regions like Oregon (for example, see Mysak 1967a, Wright & Mysak 1977, respectively).

The first thorough studies of the coupled Equations (3.1) and (3.2) were carried out independently by Allen (1975) and Wang (1975) for the case of an exponential shelf profile (as illustrated in Figure 6*b*). Allen (1975) showed that the strength of the coupling between shelf waves and the internal Kelvin wave trapped against the coast is measured by the parameter $\lambda = r_{i2}/L_B$, where $L_B = H/(dH/dx)$ (here $H = H_1 + H_2$) is the characteristic length scale of the bottom topography and r_{i2} is the internal Rossby radius for a two-layer fluid. For smoothly varying profiles such as the exponential shape (2.25), λ is usually small, i.e. $\lambda = O(10^{-1})$, in which case a regular perturbation expansion in λ can be used to analyze the coupled motions. For long wavelengths, Allen (1975) showed that an $O(1)$ shelf wave is coupled to a rather weak $O(\lambda^2)$ baroclinic motion, whereas an $O(1)$ internal Kelvin wave is coupled to an $O(\lambda)$ barotropic motion. For intermediate wavelengths, on the other hand, Allen (1975) showed that because of their comparable phase speeds the shelf waves are coupled at lowest order to the internal Kelvin wave. At each point where the dispersion curves appear to cross in the (ω,k) plane, there is in fact a change of modal structure of the waves (Allen 1975, Wang 1975). Figure 7*a* shows the dispersion curves for the first four shelf-wave modes and the internal Kelvin wave superposed, whereas Figure 7*b* shows how the curves switch families at the intersections of these curves. For example, imagine, in Figure 7*b*, following along the internal Kelvin-wave dispersion curve with κ increasing toward κ_0. Further, suppose that this wave has an amplitude of $O(1)$. When κ is very near κ_0, both an internal Kelvin wave and a shelf wave can now exist and oscillate with the same frequency. As κ increases further past κ_0, the former Kelvin-wave solution becomes a pure $O(1)$ shelf wave. Wang (1975) called this phenomenon a "resonant coupling" between the two types of wave modes; however, such a terminology is somewhat misleading since no unlimited amplification of the motion occurs. Also, when continuous stratification is used, this phenomenon apparently does not occur (see below). Finally, Allen

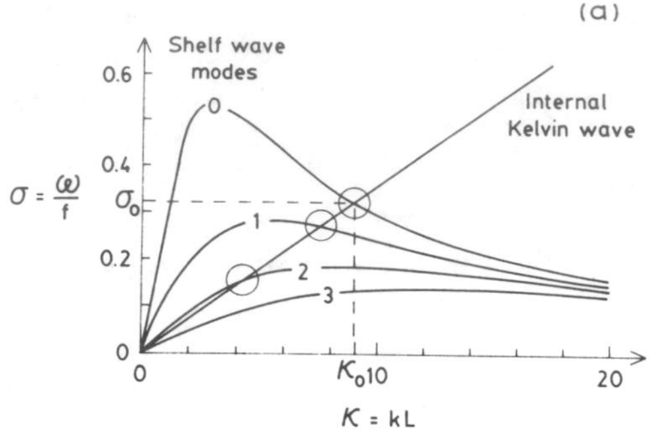

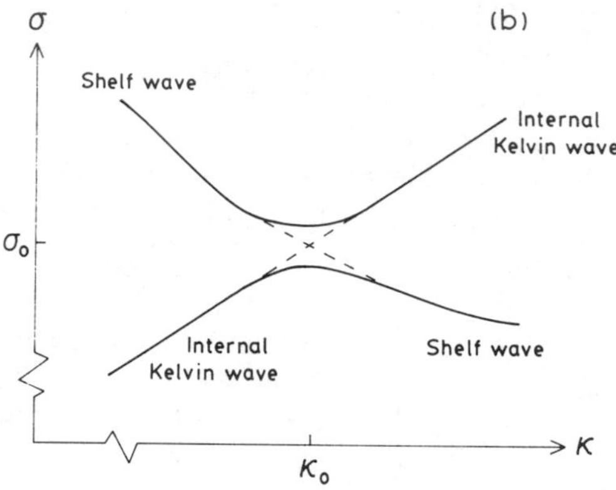

Figure 7 (*a*) Superposition of typical dispersion curves for shelf waves (first four modes) and the internal Kelvin wave for the stratification and exponential depth profile shown in Figure 6*b* (Allen 1975). When coupling is taken into account, the curves do not actually intersect at the circled points but change families, as shown in (*b*).

(1975) showed that for very short wavelengths, the shelf-wave motion is primarily confined to the bottom layer and the waves are bottom-trapped. These very short waves are analogous to Rhines' (1970) bottom-intensified topographic Rossby waves in a continuously stratified ocean without lateral boundaries.

Low-Frequency Trapped Waves in Continuously Stratified Models

The obvious limiting feature of two-layer models is that they lack sufficient vertical resolution to determine accurately the vertical structure of coastal trapped waves. Also, the discretization of the mean density field into two layers results in having to solve a coupled system of equations. In a continuously stratified model, a single governing equation results, albeit a partial differential equation. On the basis of a qualitative analysis of this equation together with appropriate boundary conditions, Huthnance (1978) has shown that over any monotonic depth profile $H(x)$ there is only one infinite sequence of trapped subinertial modes with frequencies decreasing to zero as mode number increases. As in the case of shelf waves and Kelvin waves, they are right-bounded in the northern hemisphere. In the special case of a vertical wall at the coast, Clarke (1977a) showed that at long wavelengths, these waves are nondispersive and represent a hybrid between shelf waves and internal Kelvin waves.

The first study of subinertial coastal trapped waves in a continuously stratified ocean was carried out by Wang & Mooers (1976). Under the rigid-lid approximation, they determined numerically the vertical and offshore structure of the velocity field over a variety of depth profiles. In particular, the phenomena of both coastal trapping and bottom trapping were studied in some detail. Clarke (1977a) focused his attention on the long-wave limit of these equations, in which case each wave mode satisfies a first-order wave equation. The most complete analysis of these equations, however, has been given by Huthnance (1978) who did not a priori make the rigid-lid approximation. However, he showed that divergence effects are generally very small. By invoking the rigid-lid approximation, the eigenfrequencies of the subinertial waves are increased by only a few percent.

For a discussion of the solutions of the governing equations for the continuously stratified problem, we refer the reader to Huthnance (1978) or Mysak (1980). For intermediate values of the stratification and wavelengths, the results are fairly complex. However, in certain limiting cases, the solutions simplify as follows. For weak stratification, the motions become depth-independent and the waves reduce to barotropic shelf waves, of the kind discussed in Section 2. For strong stratification, on the other hand, the current structure and wave characteristics are akin to those of internal Kelvin waves. Finally, for very short wavelengths, the waves become bottom-trapped topographic Rossby waves. In closing this section we note that over a wide range of parameters, Huthnance (1978) never found any evidence of the mode-coupling phenomenon (Figure 7) that was obtained in the two-layer models of Allen (1975) and Wang (1975). Thus it appears that this phenomenon may be essentially due to the two-layer approximation to the basic continuous density profile usually observed in the shelf-slope regions.

4 EFFECTS OF MEAN FLOWS

Generally speaking, all the studies of coastal trapped waves in the presence of longshore mean flows can be divided into two categories, according to whether the flows are barotropic or baroclinic. Thus our discussion in this section will also be divided in this manner.

Barotropic Flows

Because edge waves and barotropic Kelvin waves travel with relatively large phase speeds, they are generally unaffected by barotropic longshore mean flows. However, Kenyon (1972) showed that if such a flow possesses a strong lateral shear, then low-mode, low-frequency edge waves are strongly affected by the current. In particular, no trapped-wave solutions are possible in the frequency band $f^2 - V'^2 \leqq \omega^2 < f^2$, where $V' = dV/dx$ is the (constant) current shear. Also, Bowen and his collaborators (see Bowen 1977 for references) have proposed that the interaction between two edge-wave modes and a longshore mean flow could play an important role in a number of spatially periodic coastal processes, such as rip currents (Figure 1) and the formation of crescentic bars.

For nondivergent barotropic shelf waves, on the other hand, mean flows appear to be rather important in many coastal regions. The appropriate wave equation for shelf waves (under the rigid-lid approximation) can be derived by linearizing the following equation expressing conservation of potential vorticity:

$$\frac{D}{Dt}\left(\frac{\zeta+f}{H}\right)=0, \tag{4.1}$$

where $D/Dt = \partial_t + u\partial_x + v\partial_y$ and $\zeta = v_x - u_y$ is the relative vorticity of the fluid. Equation (4.1) shows that changes in ζ are compensated by changes in the background potential vorticity f/H. Suppose now that in addition to the shelf-wave currents u, v, there is also present a laterally sheared, alongshore mean flow $V(x)$. In this case ζ takes the form

$$\zeta = v_x - u_y + V'(x) \tag{4.2}$$

and thus the background potential vorticity is now given by

$$P(x) = [f + V'(x)]/H. \tag{4.3}$$

In regions where intense western-boundary currents occur (e.g. the east coasts of the United States, Australia, and Japan), the mean shear V' can be comparable in magnitude to f. If this is the case, three new effects may arise: 1. shelf waves can be significantly advected by the current,

2. shelf waves can become amplified, extracting kinetic energy from the mean flow through the process of barotropic instability, 3. a new class of shear waves can exist.

The relevant amplitude equation for trapped waves in the presence of $P(x)$ as given by (4.3) was first derived by Niiler & Mysak (1971); a slightly more general form, which allows for free surface divergence, was obtained by Tareyev (1971). In both cases, however, the authors showed that if unstable waves are present, then $P'(x) = 0$ (P has an extremum) at least once for $0 < x < \infty$. Equivalently, if $P'(x) \neq 0$ for $0 < x < \infty$, then only neutral waves exist. These statements are the counterpart of Rayleigh's stability criterion for shear flows in a nonrotating homogeneous fluid of constant depth. Many other qualitative results concerning the stability of the mean flow in the nondivergent case were obtained by Grimshaw (1976).

From the shelf-wave amplitude equation it is found that critical layers for neutral shelf waves can also exist at those points x_c where $c + V(x_c) = 0$. Thus at or near critical layers, shelf waves become left-bounded waves in the nothern hemisphere if $V(x_c) > 0$. The topic of critical layers was first explored by McKee (1977) who showed, using the flat-shelf profile (2.24) and a mean flow $V = V_0 x/L$ for which $P' \neq 0$, that in addition to the discrete spectrum of shelf-wave modes there is also a continuous spectrum of critical-layer solutions whose phase speeds lie in the range of the current speed. Further, these solutions have discontinuous derivatives at the critical layers. A detailed discussion of shelf-wave critical layers for any monotonic depth profile has also been given by Grimshaw (1976).

Niiler & Mysak (1971) proposed that Gulf Stream meanders may be due to barotropically unstable shelf waves over the Blake Plateau. In the Florida Strait, however, it appears that long-period shelf waves are stable (Brooks & Mooers 1977) whereas short-period shelf waves are unstable (Helbig 1980) in the presence of the Florida current. For a review of these works, we refer the reader to Mysak (1980).

Baroclinic Flows

All the studies of coastal trapped waves in the presence of baroclinic mean flows have been based on various two-layer models (see Figure 8). The choice of model in any particular study has usually been motivated by a desire to understand such phenomena as time-dependent coastal upwelling (e.g. off Oregon) or the temporal and spatial variability (i.e. meandering) of intense boundary currents (e.g. the Gulf Stream).

The first study of shelf and edge waves in the presence of a baroclinic flow appears to have been made by Iida (1970). For the basic state he

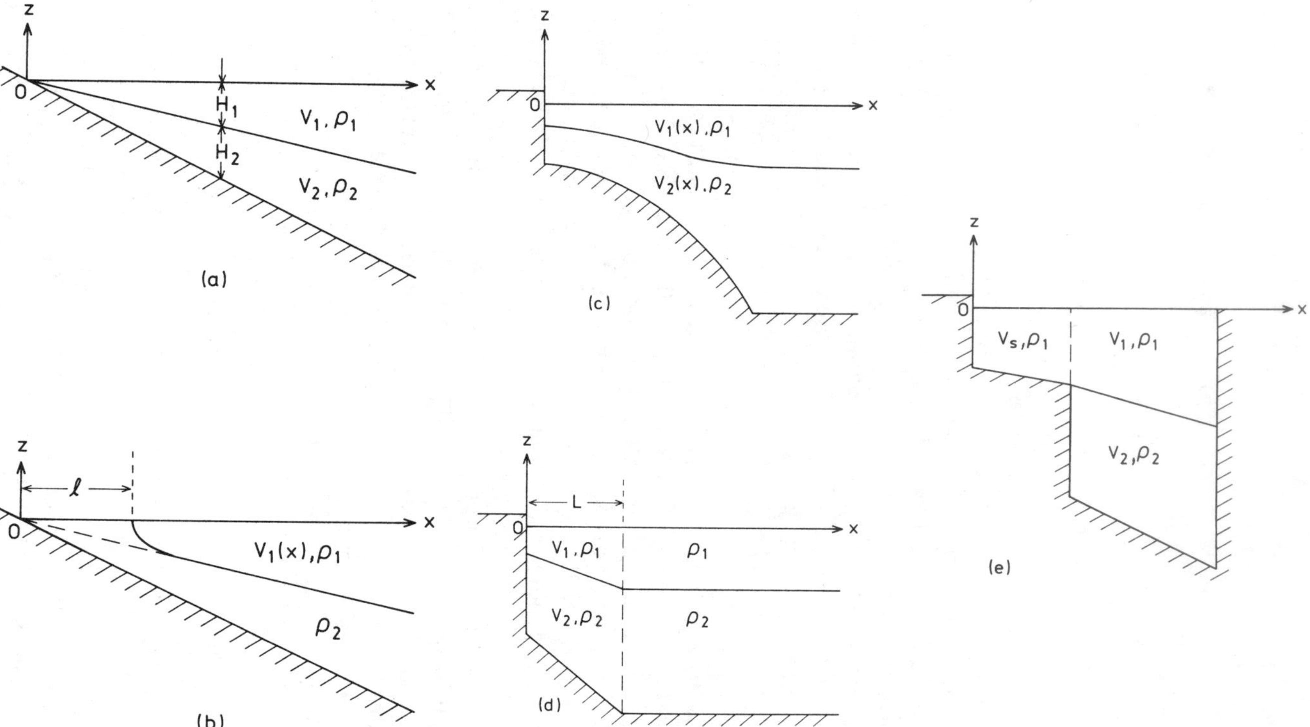

Figure 8 Two-layer models of baroclinic mean flows used in the coastal-trapped-wave literature.

used a two-layer geostrophic flow over a sloping beach, as depicted in Figure 8*a*. Because the interface has a constant slope and emanates from the coastline, exact solutions for stable trapped waves are possible. He showed that in the long-wave limit ($k \to 0$), all the trapped waves can be classified as being either barotropic or baroclinic in nature; further, each wave type is advected along with the velocity $(H_1 V_1 + H_2 V_2)/(H_1 + H_2)$. For any finite but small k, this classification is only approximate, and when k is large compared with $g'\alpha/f$ [$g' = g(1-\rho_1/\rho_2)$, $\alpha =$ bottom slope], Iida (1970) showed that the two types of motions are strongly coupled. Bane & Hseuh (1980) studied low-frequency trapped waves in the "after upwelling" model shown in Figure 8*b*. In this model the interface has risen up to the surface due to upwelling from below; thus a front now exists at $x = l$. Off the coast of Oregon, typically $l = 0(10$ km). In this "after upwelling" configuration, the low-frequency barotropic waves are very similar to those in the "before upwelling" configuration. However, now there can also exist a set of baroclinic waves trapped offshore of the front at $x = l$ which co-oscillate with a set of barotropic waves that occur in the homogeneous band of water in $0 < x < l$. When taken together, the motion of these waves for a given offshore mode resembles a narrow near-shore oscillatory barotropic jet that is coherent with vertical pycnocline motions further offshore.

An important numerical study concerning the stability of baroclinic coastal jets was carried out a decade ago by Orlanski (1969). He used the model in Figure 8*c* with $V_2 = 0$ to study the baroclinic instability with respect to quasi-geostrophic waves of a surface flow that modelled the Gulf Stream south of Cape Hatteras. Cutchin & Rao (1976) carried out a recent numerical study, which follows on from Orlanski. They used the model in Figure 8*c* with free-surface divergence as well to determine the dispersion relations of six wave types: external (barotropic) and internal (baroclinic) edge waves, external and internal Kelvin waves, and external and internal shelf waves. For bottom and mean current profiles along the Depoe Bay line off Oregon, they generally found only stable wave solutions; thus the main effect of the current is to advect the waves along. This result is not unexpected for the quasi-geostrophic shelf waves since the gradients of the mean potential vorticities for the upper and lower layers in their model are negative. On the other hand, Cutchin & Rao (1976) found that there were a few isolated points in the (ω,k) plane where the dispersion curves for different wave types intersected (for example, see their Figure 4). At such points there exists the possibility of hybrid-type instabilities, which were first discussed by Rao & Simons (1970) for internal and rotational waves in a two-layer channel of constant depth.

Finally, it is of interest to mention here the recent studies of Kubota (1977) and Johnson & Mysak (1980) who independently studied low-frequency wave propagation superimposed on the flows shown in Figure 8*d*. Kubota allowed for free-surface divergence and identified six types of subinertial trapped waves in his numerical solutions: external and internal Kelvin waves, a shear wave (associated with the discontinuity in V_i at $x = L$), a topographic Rossby wave, a surface-trapped wave, and a bottom-trapped wave. In this paper only stable wave solutions were obtained, although Kubota (1977) did offer the speculation that baroclinic instability takes place when the surface- and bottom-trapped waves are "resonantly coupled," in the manner shown in Figure 7*b* for shelf waves and internal Kelvin waves. In Johnson & Mysak (1980) the rigid-lid approximation was used, permitting the authors to obtain the dispersion relation analytically. They found that there are two shear modes, one of which is unstable at all wavelengths due to barotropic instability; the other is stable at very long wavelengths but unstable at shorter wavelengths. Also, they showed that the surface-trapped waves of different modes coalesce for a small range of wave numbers and became unstable due to baroclinic instability. Johnson & Mysak (1980) also investigated the stability of the flow shown in Figure 8*e*, again under the rigid-lid approximation and under the assumptions of quasi-geostrophy. Similar stability results were obtained, with one unstable shear mode now arising because of the discontinuity in the upper-layer velocity at $x = L$. Because Johnson & Mysak assumed that the shelf was "deep" in their analysis, no true shelf waves arose in their solution. To obtain such waves in a model like Figure 8*e*, ageostrophic effects would have to be taken into account.

From the above discussion it is evident that in the presence of two-layer baroclinic coastal flows, there is a plethora of subinertial trapped waves. In some models, one of these waves can be identified as a shelf wave. Is the picture so complex because of the two-layer approximation for the stratification? Clearly, a next useful step in the study of shelf-wave dynamics would be an investigation of low-frequency coastal trapped waves in a continuously stratified ocean with a mean flow containing vertical and lateral shears.

5 GENERATION OF COASTAL TRAPPED WAVES

Kelvin Waves

In coastal regions Kelvin waves make up a significant part of the barotropic tidal motions which are directly forced by the tidal potential (cf LeBlond & Mysak 1978, Section 52). Also, Kelvin waves can be generated by storm surges diffracted by extensive isolated land masses

(Crease 1956) and by scattering from irregular coastlines (Mysak & Howe 1978). Finally, meteorological forces can excite barotropic Kelvin waves. In particular, Kajiura (1962) and Thomson (1970) showed that Kelvin waves are efficiently generated by the longshore component of the wind.

Gill & Clarke (1974) considered the wind generation of internal Kelvin waves in a continuously stratified ocean. The purpose of their study was to show how wind-induced upwelling at the coast is modulated by the propagation of low-frequency internal Kelvin waves. They used the familiar normal-mode technique (see LeBlond & Mysak 1978, Section 10) to find the response as a sum of vertical modes; the amplitude of each mode in the low-frequency limit ($\omega^2 \ll f^2$) satisfies a first-order wave equation. Such equations can be easily integrated by the method of characteristics and the results, they suggest, may be of use in forecasting coastal upwelling.

Edge Waves

In contrast to Kelvin waves, atmospheric pressure disturbances generally appear to be more important than the wind stress in the generation of edge waves on a sloping beach. Detailed scaling arguments to show why this is the case are given in LeBlond & Mysak (1978, Section 55). Further, observations of hurricane-generated edge waves along the eastern US coast confirm this generation mechanism (for example, see Munk, Snodgrass & Carrier 1956).

The generation of edge waves by nonmeteorological forces has been receiving an increasing amount of attention in the literature. For example, Kajiura (1972) and Morris (1976) have discussed the generation of edge waves by transient and sinusoidal sources located near the coast. Kajiura (1972) presented detailed calculations for the response of a semi-infinite ocean bounded on one side by a flat shelf due to a source that is located either on or just off the shelf. Such a forcing mechanism is characteristic of broad crustal deformations due to earthquakes near the coast. He found that the proportion of energy trapped on the shelf as edge waves relative to that radiated into the deep water increases with the longshore dimension of the source. Also, edge waves are generated more efficiently when the source is close to the shoreline.

Other recent edge-wave generation studies are based on the concept of resonant wave-wave interactions. For example, Guza & Davis (1974) showed that a standing gravity wave normally incident on a sloping beach can transfer energy to edge waves through a weak resonant interaction resulting from the instability of the incident wave with respect to perturbation by edge waves. Their results suggest that subharmonic, lowest-mode edge waves are preferentially excited. Guza & Bowen (1975) extended this theory to the case of arbitrary incidence. Fuller & Mysak (1977), on the other hand, showed that when a long gravity wave is incident upon a flat shelf whose coastline has an irregular shape, one or more edge waves can be generated on the shelf through a resonant interaction of the incident wave with the wave number spectrum of the coastal irregularities.

Shelf Waves

At midlatitudes, shelf waves are usually generated by the surface wind stress associated with large-scale weather systems that pass over the shelf region. Such systems have periods of about a week, wavelengths of about 10^3 km and sea-surface pressure and wind-stress fluctuations of order 5 mbar and 1 dyne/cm^2, respectively. Consequently, shelf waves in midlatitude oceans typically have long wavelengths ($\lambda \gg L$, where $L =$ shelf width), low frequencies ($\omega \ll f$), and small amplitudes (a few centimeters).

It was originally proposed (Robinson 1964, Mysak 1967a,b) that shelf waves are generated when the sea surface near the coast responds resonantly to the atmospheric pressure fluctuations. However, these pressure fluctuations also tend to be highly correlated with changes in the wind stress (Hamon 1966). Further, the wind stress can effectively add vorticity to the ocean on the scale needed to drive the highly rotational shelf waves. Thus it was suggested by Adams & Buchwald (1969) that the wind stress is mainly responsible for the generation of shelf waves. A more extensive account of this generation mechanism was later given by Gill & Schumann (1974). In particular, they clearly showed that the vortex stretching over topography due to an oscillating wind produces large alternating vortices characteristic of shelf waves (see Figure 2). Their formulation of the generation problem was then used by Hamon (1976) to show that along the east Australian coast, shelf waves are most likely generated by the longshore component of the wind stress. For a recent account of this generation theory see Mysak (1980).

Shelf waves can also be generated by processes not related to meteorological forcing. For example, they can be excited by scattering from coastal or bottom topographic irregularities, by nonlinear interactions, and by slow modulations in the shelf-slope wave guide. These possibilities will be explored in Sections 6 and 7.

6 EFFECTS OF LONGSHORE VARIATIONS

So far, our treatment of coastal trapped waves has been based on the fundamental assumption of spatial homogeneity in the direction of wave propagation. In practice, however, there are usually longshore variations present in the shelf-slope wave guide. These may be due to 1. coastline curvature or coastal irregularities, 2. offshore topographic features (e.g. sea mounts and canyons), 3. the variation of the Coriolis parameter with latitude, and 4. downstream spatial modulation of the mean flow. In the coastal-trapped-wave literature, only the first three types of longshore variations have been examined.

Edge Waves and Kelvin Waves

It appears that the study of the effects of longshore variations on edge-wave propagation has hardly been touched upon in the literature. In a recent paper Fuller & Mysak (1977) showed that a coherent edge-wave mode traveling along a flat shelf with small coastal irregularities slowly loses energy to other edge-wave modes and also radiates energy out to the deep-sea region in the form of long gravity waves. The analysis in this paper is based on the theory of wave propagation in extensive random media (Mysak 1978c).

The problem of Kelvin waves in the presence of longshore variations has been much more extensively studied. These studies generally fall into one of three categories, according to whether the variations consist of 1. an isolated bend in the coastline (e.g. Packham & Williams 1968), 2. extensive, small-scale irregularities (Mysak & Tang 1974), or 3. slow variations in the coast (Clarke 1977b, Killworth 1978). The latter two studies by Clarke (1977b) and Killworth (1978) have been particularly useful for helping to understand the nature of coastal upwelling in the presence of capes and bays.

Shelf Waves

In transient coastal upwelling events, shelf waves are generated when the flow interacts with isolated coastal or bathymetric features (see Kishi & Suginohara 1975, Peffley & O'Brien 1976, Mysak 1980). Also, shelf waves of various scales can be generated when a single shelf-wave mode is incident upon such an isolated feature. In particular, if the depth profile allows for shelf waves with a negative group velocity at high wave numbers, such a scattering process will produce backscattered wave energy (Buchwald 1977) which will travel much slower than the incident wave energy. Thus in addition to producing a cascade toward small-scale motions, the presence of topographic irregularities leads to the possibility of significant nonlinear interactions between different shelf-wave modes since these shorter waves have phase speeds that are comparable with the current speed.

When the variations in the coastline, bottom topography, and Coriolis parameter occur on a scale much greater than the shelf width L, which is assumed to be an appropriate scale for the shelf waves, WKB or multiple-scale techniques can be used to describe the slow evolution of shelf waves in such a wave guide. Grimshaw (1977a) provided a unified treatment of all three types of variations, whereas Smith (1975) and Allen (1976) looked only at the effects of slow variations in the topography and coastline. Allen & Romea (1979), on the other hand, looked at the poleward propagation of shelf waves at low latitudes in which the slow variations in f are taken into account. In this paper they show how an internal Kelvin wave emanating from the equator is gradually transformed into a poleward-propagating barotropic shelf wave whose frequency is the same as that of the Kelvin wave. Their theory thus shows how coastal-trapped baroclinic disturbances may propagate efficiently from the equatorial region to midlatitudes where they take the form of barotropic shelf waves.

7 NONLINEAR EFFECTS

Barotropic Kelvin waves that propagate along a steeply sloping oceanic boundary are weakly dispersive, the correction to the "vertical-wall" phase speed $c = (gD)^{\frac{1}{2}}$ being of order $\delta = f^2L^2/gD$, the divergence parameter (Smith 1972). When this phase dispersion is balanced against amplitude dispersion, a modified Korteweg-de Vries (KdV) equation for nonlinear barotropic Kelvin waves is obtained (Smith 1972). The influence of nonlinearity is, however, detectable only over extremely long distances. Studies of nonlinear internal Kelvin waves have been presented by Bennett (1973) and Clarke (1977b).

Nonlinear corrections to the linear edge-wave solutions over a sloping beach have recently been presented by Guza & Bowen (1976) and independently by Whitham (1976), who did not use the shallow-water (long-wave) approximation. In particular, he showed that the nonlinear shallow-water solution is not uniformly valid as $x \to \infty$ in that the third-order correction does not decay far from the coast. However, if the depth tends to a constant as $x \to \infty$, this secularity is eliminated (Minzoni 1976).

The study of the nonlinear interactions between edge waves and surface waves incident upon the beach has received considerable attention during the last decade. It was first proposed by Gallagher (1971) that these interactions could be responsible for the generation of surf beat. Since then, many refinements have been made to the theory, which has also been confirmed by experiment (for example, see Guza & Davis 1974, Minzoni & Whitham 1977, Bowen & Guza 1978, Rockliff 1978).

The effects on long shelf waves of small nonlinearities and weak phase dispersion due to topography have been examined in two complementary papers by Smith (1972) and Grimshaw (1977b). When these two effects are exactly in balance, the slow spatial and temporal evolution of the amplitude of a single-shelf-wave mode is governed by a form of the KdV equation. It is interesting to note that although the approaches taken by Smith and Grimshaw are different, both yield the same amplitude equation in the common parameter regime. Smith (1972) carried out an expansion in the small-divergence parameter δ, whereas Grimshaw (1977b) held this parameter fixed and expanded in the small parameter $\varepsilon_0 = L^2/\lambda^2$ (where L = shelf width and λ = wavelength). Grimshaw then examined the limit $\delta \to 0$ a posteriori and showed that both expansions gave the same amplitude equation when δ and ε_0 are both small but $\varepsilon_0 \ll \delta$.

Although of considerable theoretical interest, the above nonlinear theory appears to have little practical application because the nonlinearities are so weak for long shelf waves. Off the east Australian coast, a gravest-mode shelf wave with a period of five days has a wavelength (in the nondispersive regime) of 1200 km and thus a phase speed of 240 km day. However, according to Grimshaw (1977b) the time scale for the development of nonlinear effects in this region is 73 days, which is considerably longer than the time for the wave to travel the length of the coast (about 10 days).

Finally, we mention here three shelf-wave problems involving nonlinear resonant interactions that have recently been examined in the literature: side-band instability and long-wave resonance (Grimshaw 1977c) and wind-shelf wave nonlinear interactions (Barton 1977, Barton & Buchwald 1977). For a brief review of these papers, see Mysak (1980).

8 OBSERVATIONS

Edge waves and shelf waves have been observed along many continental margins around the world. Thus it is not possible to give a detailed review of all these observations here. For a discussion of some of the edge-wave observations, we refer the reader to LeBlond & Mysak (1977, 1978). Below, we shall make a few comments about shelf-wave observations.

Shelf waves, interestingly enough, were first detected along the east Australian coast by Hamon (1962, 1963). Subsequently they have been observed off the west Australian coast, the Oregon coast, the Carolina coast, the east Japanese coast, the west coast of Scotland, the north Mediterranean coast, Lake Ontario, the Florida Strait, the west United States–British Columbia coast, and the Peruvian coast. Equatorial shelf waves have been observed along the northern coast of the Guinea Gulf. Finally, island shelf waves have been reported off Bermuda. For the references discussing these observations, see Mysak (1980).

In most studies of shelf-wave observations, lengthy time series of sea-level records along the coast have been analyzed statistically for phase lags that are consistent with shelf-wave propagation. Further, these lags have been analyzed in conjunction with atmospheric forcing. In a number of more recent investigations direct current measurements have also been made which were then analyzed for longshore phase propagation and the velocity-field structure in the vertical and offshore directions. Recent novel observational approaches used to detect shelf waves involve wave-model fits based on current auto- and cross-spectra and satellite imagery.

As mentioned earlier, shelf waves appear to play an important role in coastal upwelling and Gulf Stream meanders. Off the coast of South Carolina they may also trigger off lee-wave disturbances, which could be responsible for the persistent offshore deflection of the Gulf Stream in this region (Brooks & Bane 1978). On the other hand, off the east coast of Japan, variations in the Kuroshio (probably meteorologically induced) cause anomalously high sea-level disturbances at the coast, which appear to propagate as shelf waves (Isozaki 1972).

9 CONCLUDING REMARKS

In this paper we have reviewed the basic dynamics of the different types of long waves that can be trapped over the continental shelf-slope region. We mentioned in Section 1 that similar types of waves can also be trapped around islands and seamounts. Here we note that long waves of either the first or second class can also be trapped over isolated rectilinear topographic features such as oceanic ridges (Buchwald 1968), escarpments (Rhines 1969), and trenches (Mysak, LeBlond & Emery 1979). However, except for the case of trench waves (Mysak, LeBlond & Emery 1979), there has been no observational evidence of these trapped waves in the open ocean.

During the next few years there will likely be further theoretical and observational studies of the interactions between edge waves and mean currents or other types of waves. So far most of these studies have been based on the theory of weak wave-wave interactions, whereas in practice these interactions are often strong in the sense that significant energy exchanges between modes take place over one wave period. Clearly, new theoretical work is required to handle such strong interactions.

The effects of dissipation and mixing on edge waves and shelf waves have been tacitly left out of the discussion in this paper. To date little is known about these effects, especially for the case of shelf waves. Recently it has been shown by Brink & Allen (1978) that bottom Ekman friction appears to cause a small longshore phase lag between shelf-wave current fluctuations off the Oregon coast. In other situations, however, it appears that bottom stresses are of sufficient strength to eliminate shelf-wave propagation altogether (Lee & Brooks 1979).

The study of shelf waves and their relation to and interaction with oceanic motions beyond the shelf will likely engage many theoreticians during the next decade. The propagation of shelf waves in the presence of baroclinic currents over the slope region needs further elucidation, especially for the case of continuous stratification. Also, the propagation of "leaky" shelf waves should be examined. How much shelf-wave energy leaks away from the coast in the form of deep-sea Rossby waves? Further, can a significant portion of shelf-wave energy be transferred through nonlinear interactions to other low-frequency modes trapped farther offshore? Finally, how do eddies that impinge upon the shelf (for example, see Lee & Brooks 1979) affect shelf-wave propagation and generation?

From answers to the above questions we may then be in a position to determine whether the "inverse" problem for shelf waves is soluble. This is to say, what can we learn about the ocean climate beyond the shelf on time scales of days and longer from observing the properties of shelf and related coastal trapped waves near the coast? For example, can we detect accurately seasonal variations in the transport of the Gulf Stream (Mysak 1967a, Mysak & Hamon 1969) or the Kuroshio (Isozaki 1972) from shelf-wave measurements? Can the theory of Gill & Clarke (1974) be used to measure variations in upwelling at the shelf break? Also, is it possible to estimate the eddy density over the slope and beyond from shelf-wave measurements?

In conclusion, it is clear that although we now have a basic understanding of edge waves, Kelvin waves, and shelf waves, there remain a number of interesting and worthwhile problems to examine in the future. Also, it is becoming increasingly evident that the successful attacks on these problems will involve the theoreticians and experimentalists working together.

Acknowledgments

The author is indebted to the Canadian National Scientific and Engineering Research Council which has supported his work on topographically trapped waves over the last decade. The comments of Professor Paul H. LeBlond on a first draft of this paper are also gratefully acknowledged.

Literature Cited

Adams, J. K., Buchwald, V. T. 1969. The generation of continental shelf waves. *J. Fluid Mech.* 35:815–26

Allen, J. S. 1975. Coastal trapped waves in a stratified ocean. *J. Phys. Oceanogr.* 5: 300–25

Allen, J. S. 1976. Continental shelf waves and alongshore variations in bottom topography and coastline. *J. Phys. Oceanogr.* 6:864–78

Allen, J. S., Romea, R. D. 1979. On coastal trapped waves at low latitudes in a stratified ocean. *J. Fluid Mech.* Submitted for publication

Ball, F. K. 1967. Edge waves in an ocean of finite depth. *Deep-Sea Res.* 14:79–88

Bane, J. M., Hseuh, Y. 1980. On the theory of coastal trapped waves in an upwelling frontal zone. *J. Phys. Oceanogr.* 10: In press

Barton, N. G. 1977. Resonant interactions of shelf waves with wind-generated effects. *Geophys. Astrophys. Fluid Dyn.* 9:101–14

Barton, N. G., Buchwald, V. T. 1977. The nonlinear generation of shelf waves. In *Lecture Notes in Physics*, ed. D. G. Provis, R. Radok, 64:194–201. Canberra: Austral. Acad. Sci.; Berlin: Springer. 231 pp.

Bennett, J. R. 1973. A theory of large-amplitude Kelvin waves. *J. Phys. Oceanogr.* 3:57–60

Bowen, A. J. 1977. Wave-wave interactions near the shore. See Barton & Buchwald 1977, pp. 102–13

Bowen, A. J., Guza, R. T. 1978. Edge waves and surf beat. *J. Geophys. Res.* 83:1913–20

Brink, K. H., Allen, J. S. 1978. On the effect of bottom friction on barotropic motion over the continental shelf. *J. Phys. Oceanogr.* 8:919–22

Brooks, D. A., Bane, J. M. Jr. 1978. Gulf Stream deflection by a bottom feature off Charleston, South Carolina. *Science* 201:1225–26

Brooks, D. A., Mooers, C. N. K. 1977. Free, stable continental shelf waves in a sheared, barotropic boundary current. *J. Phys. Oceanogr.* 7:380–88

Buchwald, V. T. 1968. Long waves on oceanic ridges. *Proc. R. Soc. London Ser. A* 308: 343–54

Buchwald, V. T. 1977. Diffraction of shelf waves by an irregular coastline. See Barton & Buchwald 1977, pp. 188–93

Buchwald, V. T., Adams, J. K. 1968. The propagation of continental shelf waves. *Proc. R. Soc. London Ser. A* 305:235–50

Caldwell, D. R., Cutchin, D. L., Longuet-Higgins, M. S. 1972. Some model experiments on continental shelf waves. *J. Mar. Res.* 30:39–55

Clarke, A. J. 1977a. Observational and numerical evidence for wind-forced coastal trapped long waves. *J. Phys. Oceanogr.* 7:231–47

Clarke, A. J. 1977b. Wind-forced linear and nonlinear Kelvin waves along an irregular coastline. *J. Fluid Mech.* 83:337–48

Crease, J. 1956. Long waves on a rotating earth in the presence of a semi-infinite barrier. *J. Fluid Mech.* 1:86–96

Csanady, G. T. 1976. Topographic waves in Lake Ontatio. *J. Phys. Oceanogr.* 6:93–103

Cutchin, D. L., Rao, D. B. 1976. Baroclinic and barotropic edge waves on a continental shelf. *Spec. Rep. No. 30.* Ctr. Great Lakes Studies, Univ. Wis. Milwaukee. 53 pp.

Cutchin, D. L., Smith, R. L. 1973. Continental shelf waves: low frequency variations in sea level and currents over the Oregon continental shelf. *J. Phys. Oceanogr.* 3:73–82

Fuller, J. D., Mysak, L. A. 1977. Edge waves in the presence of an irregular coastline. *J. Phys. Oceanogr.* 7:846–55

Gallagher, B. 1971. Generation of surf beat by nonlinear wave interactions. *J. Fluid Mech.* 49:1–20

Geisler, J. E., Mysak, L. A. 1978. Trapped coastal waves on an equatorial beta plane. *J. Phys. Oceanogr.* 8:665–75

Gill, A. E., Clarke, A. J. 1974. Wind-induced upwelling, coastal currents and sea-level changes. *Deep-Sea Res.* 21:325–45

Gill, A. E., Schumann, E. H. 1974. The generation of long shelf waves by the wind. *J. Phys. Oceanogr.* 4:83–90

Greenspan, H. P. 1970. A note on edge waves in a stratified fluid. *Stud. Appl. Math.* 49:381–88

Grimshaw, R. 1976. The stability of continental shelf waves in the presence of a boundary current shear. *Res. Rep. No. 43.* Sch. Math. Sci., Univ. Melbourne. 19 pp.

Grimshaw, R. 1977a. The effects of a variable Coriolis parameter, coastline curvature and variable bottom topography on continental shelf waves. *J. Phys. Oceanogr.* 7:547–54

Grimshaw, R. 1977b. Nonlinear aspects of long shelf waves. *Geophys. Astrophys. Fluid Dyn.* 8:3–16

Grimshaw, R. 1977c. The stability of continental shelf waves, I. Side band instability and long-wave resonance. *J. Austral. Math. Soc. Ser. B* 20:13–30

Guza, R. T., Bowen, A. J. 1975. The resonant instabilities of long waves obliquely incident on a beach. *J. Geophys. Res.* 80:4529–34

Guza, R. T., Bowen, A. J. 1976. Finite amplitude edge waves. *J. Mar. Res.* 34:268–93

Guza, R. T., Davis, R. E. 1974. Excitation of edge waves by waves incident on a beach. *J. Geophys. Res.* 79:1285–91

Hamon, B. V. 1962. The spectrums of mean sea level at Sydney, Coff's Harbour, and Lord Howe Island. *J. Geophys. Res.* 67:5147–55

Hamon, B. V. 1963. Correction to "The spectrums of mean sea level at Sydney, Coff's Harbour, and Lord Howe Island". *J. Geophys. Res.* 68:4635

Hamon, B. V. 1966. Continental shelf waves and the effects of atmospheric pressure and wind stress on sea level. *J. Geophys. Res.* 71:2883–93

Hamon, B. V. 1976. Generation of shelf waves on the east Australian coast by wind stress. *Mém. Soc. R. Sci. Liège. 6e sér.* 10:359–67

Helbig, J. A. 1980. On the propagation of shelf waves in the presence of randomly perturbed mean flows, with application to the Florida Current. *J. Phys. Oceanogr.* 10: In press

Huthnance, J. M. 1975. On trapped waves over a continental shelf. *J. Fluid Mech.* 69:689–704

Huthnance, J. M. 1978. On coastal trapped waves: analysis and numerical calculation by inverse iteration. *J. Phys. Oceanogr.* 8:74–92

Iida, H. 1970. Edge waves on the linearly sloping coast, I. Free waves. *Oceanogr. Mag.* 22:37–62

Iida, H. 1974. Remarks on the classification of long waves propagating in the marginal sea in a uniformly rotating two-layer ocean. *Oceanogr. Mag.* 25:89–100

Isozaki, I. 1972. Unusually high mean sea level in September 1971 along the south coast of Japan, I. Some aspects of high sea level with time scale more than one week. *Pap. Meterol. Geophys.* 23:243–57

Johnson, E. R., Mysak, L. A. 1980. Baroclinic and barotropic instabilities of coastal currents. *J. Phys. Oceanogr.* Submitted for publication

Kajiura, K. 1962. A note on the generation of boundary waves of the Kelvin type. *J. Oceanogr. Soc. Jpn.* 18:51–8

Kajiura, K. 1972. The directivity of energy radiation of the tsunami generated in the vicinity of a continental shelf. *J. Oceanogr. Soc. Jpn.* 28:260–77

Kajiura, K. 1974. Effect of stratification on long-period trapped waves on the shelf. *J. Oceanogr. Soc. Jpn.* 30:271–81

Kenyon, K. E. 1972. Edge waves with current shear. *J. Geophys. Res.* 77:6599–603

Killworth, P. D. 1978. Coastal upwelling and Kelvin waves with small longshore topography. *J. Phys. Oceanogr.* 8:188–205

Kishi, M. J., Suginohara, N. 1975. Effects of longshore variation of coastline geometry and bottom topography on coastal upwelling in a two-layer model. *J. Oceanogr. Soc. Jpn.* 31:48–50

Kubota, M. 1977. Long-period topographic trapped waves in a two-layer ocean with basic flow. *J. Oceanogr. Soc. Jpn.* 33:199–206

Lamb, H. 1945. *Hydrodynamics.* Cambridge: Univ. Press. 738 pp. 6th ed.

LeBlond, P. H., Mysak, L. A. 1977. Trapped coastal waves and their role in shelf dynamics. In *The Sea*, ed. E. D. Goldberg, I. N. Cave, J. J. O'Brien, J. H. Steele, 6:459–95. New York: Wiley

LeBlond, P. H., Mysak, L. A. 1978. *Waves in the Ocean.* Amsterdam: Elsevier. 602 pp.

Lee, T. N., Brooks, D. A. 1979. Initial observations of current, temperature and coastal sea level response to atmospheric and Gulf Stream forcing on the Georgia shelf. *Geophys. Res. Lett.* 6:321–24

Longuet-Higgins, M. S. 1972. Topographic Rossby Waves. *Mém. Soc. R. Sci. Liège. 6e Sér.* 2:11–16

McCreary, J. 1976. Eastern tropical response to changing winds: with application to El Nino. *J. Phys. Oceanogr.* 6:632–45

McKee, W. D. 1977. Continental shelf waves in the presence of a sheared geostrophic current. See Barton & Buchwald 1977, pp. 212–19

Minzoni, A. A. 1976. Nonlinear edge waves and shallow-water theory. *J. Fluid Mech.* 74:369–74

Minzoni, A. A., Whitham, G. B. 1977. On the excitation of edge waves on beaches. *J. Fluid Mech.* 79:273–87

Morris, C. A. N. 1976. The generation of surface waves over a sloping beach by an oscillating line source, III. The three-dimensional problem and the generation of edge waves. *Math. Proc. Cambridge Philos. Soc.* 79:573–85

Munk, W. H., Snodgrass, F. E., Carrier, G. F. 1956. Edge waves on the continental shelf. *Science* 123:127–32

Munk, W. H., Snodgrass, F. E., Wimbush, M. 1970. Tides offshore: Transition from California coastal to deep-sea waters. *Geophys. Fluid Dyn.* 1:161–235

Mysak, L. A. 1967a. On the theory of continental shelf waves. *J. Mar. Res.* 25:205–27

Mysak, L. A. 1967b. On the very low frequency spectrum of the sea level on a continental shelf. *J. Geophys. Res.* 72:3043–47

Mysak, L. A. 1968a. Edge waves on a gently sloping continental shelf of finite width. *J. Mar. Res.* 36:24–33

Mysak, L. A. 1968b. Effects of deep-sea stratification and current on edge waves. *J. Mar. Res.* 26:34–43

Mysak, L. A. 1978a. Long-period equatorial topographic waves. *J. Phys. Oceanogr.* 8:302–14

Mysak, L. A. 1978b. Equatorial shelf waves on an exponential shelf profile. *J. Phys. Oceanogr.* 8:458–67

Mysak, L. A. 1978c. Wave propagation in random media, with oceanic applications. *Rev. Geophys. Space Phys.* 16:233–61

Mysak, L. A. 1980. Recent advances in shelf wave dynamics. *Rev. Geophys. Space Phys.* 18: In press

Mysak, L. A., Hamon, B. V. 1969. Low-frequency sea level behavior and continental shelf waves off North Carolina. *J. Geophys. Res.* 74:1397–1405

Mysak, L. A., Howe, M. S. 1978. Scattering of Poincaré waves by an irregular coastline, 2. Multiple scattering. *J. Fluid Mech.* 86:337–63

Mysak, L. A., LeBlond, P. H., Emery, W. J. 1979. Trench waves. *J. Phys. Oceanogr.* 9: In press

Mysak, L. A., Tang, C. L. 1974. Kelvin wave propagation along an irregular coastline. *J. Fluid Mech.* 64:241–61

Niiler, P. P., Mysak, L. A. 1971. Barotropic waves along an eastern continental shelf. *Geophys. Fluid Dyn.* 2:273–88

Odulo, A. B. 1974. Edge waves in a rotating stratified fluid at an inclined shore. *Atmos. Oceanic Phys.* 10:188–89

Odulo, A. B. 1975. Propagation of long waves in a rotating basin of variable depth. *Oceanology* 15:11–14

Odulo, A. B. 1977. Energetics of the long waves in a rotating basin of varying depth. *Okeanologiya* 17:565–69 (In Russian)

Orlanski, I. 1969. The influence of bottom topography on the stability of jets in a baroclinic fluid. *J. Atmos. Sci.* 26:1216–32

Packham, B. A., Williams, W. E. 1968. Diffraction of Kelvin waves at a sharp bend. *J. Fluid Mech.* 34:517–29

Peffley, M. B., O'Brien, J. J. 1976. A three-dimensional simulation of coastal upwelling off Oregon. *J. Phys. Oceanogr.* 6:164–80

Rao, D. B., Simons, T. J. 1970. Stability of a sloping interface in a rotating two-fluid system. *Tellus* 22:493–503

Reid, R. O. 1958. Effect of Coriolis force on edge waves, I. Investigation of the normal modes. *J. Mar. Res.* 16:104–44

Rhines, P. B. 1969. Slow oscillations in an ocean of varying depth, I. Abrupt topography. *J. Fluid Mech.* 37:161–89

Rhines, P. B. 1970. Edge-, bottom-, and Rossby waves in a rotating stratified fluid. *Geophys. Fluid Dyn.* 1:273–302

Robinson, A. R. 1964. Continental shelf waves and the response of the sea level to weather systems. *J. Geophys. Res.* 69:367–68

Rockliff, N. 1978. Finite amplitude effects in free and forced edge waves. *Math. Proc. Cambridge Philos. Soc.* 83:463–79

Smith, R. 1972. Nonlinear Kelvin and continental-shelf waves. *J. Fluid Mech.* 52:379–91

Smith, R. 1975. Second-order turning point problems in oceanography. *Deep-Sea Res.* 22:837–52

Stokes, G. G. 1846. Report on recent researches in hydrodynamics. *Rep. 16th Meet. Brit. Assoc. Adv. Sci. Southampton, 1846*, pp. 1–20. London: John Murray

Tareyev, B. A. 1971. Gradient-vorticity waves on the continental shelf. *Atmos. Oceanic Phys.* 7:283–85

Thomson, R. E. 1970. On the generation of Kelvin-type waves by atmospheric disturbances. *J. Fluid Mech.* 42:657–70

Wang, D.-P. 1975. Coastal trapped waves in a baroclinic ocean. *J. Phys. Oceanogr.* 5:326–33

Wang, D.-P., Mooers, C. N. K. 1976. Coastal trapped waves in a continuously stratified ocean *J. Phys. Oceanogr.* 6:853–63

Whitham, G. B. 1976. Nonlinear effects in edge waves. *J. Fluid Mech.* 74:353–68

Wright, D. G., Mysak, L. A. 1977. Coastal trapped waves, with application to the Northeast Pacific Ocean. *Atmosphere* 15:141–50

Ann. Rev. Fluid Mech. 1980. 12:77–102

WATER TRANSPORT IN SOILS

✖8153

J.-Y. Parlange
School of Australian Environmental Studies, Griffith University, Brisbane, Queensland, 4111 Australia

INTRODUCTION

Many branches of science and technology use and study porous media. Hence the physical and chemical processes discussed within the context of soil physics are not limited to that particular branch of science. It is, rather, the applications treated here that are more specifically important for soil physics. Some branches of engineering, e.g. hydrology, soil mechanics, agricultural and petroleum engineering, use soil physics as one of their foundations. Although less recognized in many cases, there are also strong connections between soil physics and chemical engineering (e.g. fluidized beds, chemical reactors), chromatography, and the drying of fibres, grains, bricks, ceramics, wood, and paper. Readers familiar with any of the above fields will recognize both the concepts and the problems discussed here, with some differences in jargon and emphasis.

We call a "soil" a collection of grains forming the soil matrix, with interconnected pore spaces forming the channels available for water and solute transport (Emerson et al 1978). The dynamic interactions between water movement and soil matrix, which can lead to fluidized beds (quick sands) and even soil erosion, will not be considered. Only the milder static processes of swelling and shrinking for a soil with high clay content will be described.

In general the length scale of interest in soil physics problems is much larger than the grain or pore size. The importance of microscopic studies, i.e. on the grain scale, then, is primarily to gain a better understanding of physical and chemical processes, e.g. hysteresis, and also to justify, via statistics, the macroscopic postulates that are presently used in soil physics (e.g. Neuman 1977). It would seem, however, that

0066-4189/80/0115-0077$01.00

attempts to describe water movement in soils on the grain scale have had a conceptual rather than practical interest, so far.

All physical parameters used here are averages over a control volume including many grains. Hence the thermodynamic state of the soil is described by a few state variables and its static properties by a few equations of state. If the state variables are not uniform in space, mass and energy transport may take place. The constitutive relations between fluxes and the appropriate "potential" gradients are essentially empirical.

The descriptive physical and chemical background material, which can be found in standard textbooks, e.g. Baver et al (1972) and Bear (1972), will be kept to a minimum. The phenomena associated with water transport in soils will be illustrated by a series of concrete problems important for soil-water management.

Several reviews on the topic of water movement in soils have preceded this one, e.g. Philip (1970) and Wooding & Morel-Seytoux (1976) in this series, and Parlange (1974). Obviously the present review concentrates on newer approaches and more recent developments.

STATIC CONSIDERATIONS

Thermodynamic Potentials

In any thermodynamic description of a system, state variables must be defined. These include, here, the mass of each constituent, e.g. water, air, soil, solutes, temperature, and the volume of the sample. As long as adsorption-desorption processes of solutes on the soil grains do not affect the volume occupied by the grains significantly, extensive thermodynamic variables can be defined per unit volume occupied by the grains. For instance the moisture ratio, υ, is the corresponding volume occupied by water and the void ratio, e, is the volume not occupied by the grains, i.e. $e+1$ is the volume of the sample. For a non-rigid matrix the void ratio is a variable and the study of its changes is a thermodynamic problem as long as they are not affected by water movement. The thermodynamic state of the water in the soil is dependent upon temperature and solute concentrations on the one hand and moisture and void ratios on the other. Only these last two are soil specific; the others will not be kept explicitly as variables. This is not to say, of course, that processes in soils are isothermal or that the solutes are in contact with reservoirs at constant chemical potentials. Rather, if temperature and solute concentration changes affect the state of water significantly, then standard results on the thermodynamics of solutions have to be taken into account also (the approach of Sposito 1973, 1975 is followed here).

Call g the Gibbs potential per unit mass of soil for isothermal processes with solutes at a constant chemical potential as well as air (assumed to remain in contact with the atmosphere). Then the thermodynamic relation

$$dg = \rho_w \mu_w d\upsilon + (e+1)\, dP \tag{1}$$

holds, where ρ_w and μ_w are the density and chemical potential of the liquid and P is the load pressure acting on the sample. The two equations of state defining the system are $\rho_w \mu_w$ and e which must be given as function of υ, P, and the other independent variables like temperature.

The void ratio, e, is routinely measured as a function of v and P. Figure 1 gives a sketch of $e\ (\upsilon, P)$ for a loam (Groenevelt & Parlange 1974). Note that $e = \upsilon$, i.e. the soil is saturated for large values of e, and $e > \upsilon$, i.e. air is present for lower values of e. At the air entry point, A on the figure, a new phase appears and results in a discontinuity of $(\partial e/\partial \upsilon)$. At a first-order phase transition the volume is discontinuous. Here the new phase being infinitely small in extent at the air entry point, the discontinuity is only in the derivative of e. For this reason the phase transition is a second-order phase transition.

Knowing $e(\upsilon, P)$, the second equation of state, $\rho_w \mu_w$, satisfies

$$\partial^2 g/\partial \upsilon \partial P = \partial(\rho_w \mu_w)/\partial P = \partial e/\partial \upsilon \tag{2}$$

or

$$\rho_w \mu_w(\upsilon, P) = \rho_w \mu_w(\upsilon, 0) + \int_0^P (\partial e/\partial \upsilon)\, dP'. \tag{3}$$

Hence $\rho_w \mu_w$ is known from the measurement of $e(\upsilon, P)$ and $\rho_w \mu_w\ (\upsilon, P = 0)$. Note that for a nonswelling soil $\rho_w \mu_w\ (\upsilon, 0)$ defines the thermodynamic properties since e is constant, and the value of P is irrelevant (the load pressure is entirely taken by the grains assumed incompressible). In any case the measurement of $\rho_w \mu_w\ (\upsilon, 0)$ is carried out with a tensiometer. The liquid is brought into contact with a liquid reservoir through a permeable membrane keeping the soil out but not the solutes (to avoid osmotic effects). Once equilibrium is reached, water on both sides of the membrane has the same chemical potential. Hence the water level in

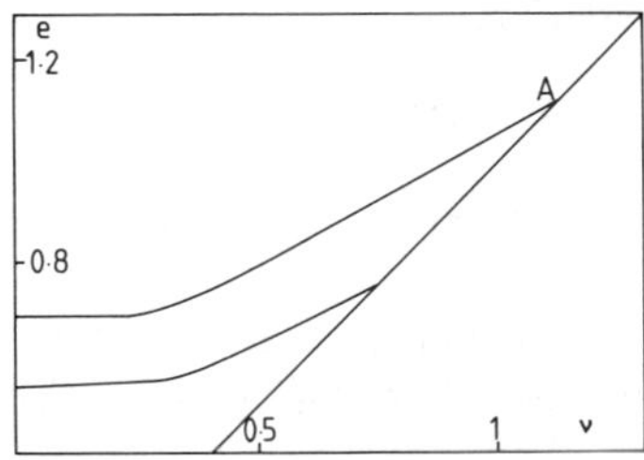

Figure 1 Sketch of void ratio (e) as a function of the moisture ratio (υ) for Lemnos loam. The top curve was measured under zero load pressure. The bottom curve corresponds to $P > 0$. At the air entry points (A) a second-order phase transition takes place.

the reservoir is a true measure of the chemical potential. By definition a saturated soil is said to be at the zero level. Other levels are measured in cm of water. To empty pores with a characteristic dimension, r, requires a suction of order σ/r, σ being the surface tension. Saturation corresponds to emptying the largest pores, i.e. $\sigma/r \to 0$ in the limit. Since suction is a negative pressure, the level of water in the reservoir is negative for an unsaturated soil. This level is usually denoted by ψ, and is often called matric or soil-water potential, $-\psi$ is also called the suction.

For a nonswelling soil, ψ only has to be measured, usually as a function of the water content, θ, instead of the moisture ratio, υ, with

$$\theta = \upsilon/(1+e). \tag{4}$$

Figure 2 shows typical $\psi(\theta)$ curves for a sandy loam and a coarse sand taken from two soil layers of the Connecticut Valley (Starr et al 1978). Note that for a suction of 40 cm the sand is almost dry since few pores are so small that they are not emptied for that value of the suction. However for $-\psi \simeq 40$ cm the sandy loam is still close to saturation since few pores are so large that they are emptied for that value of the suction.

Hysteresis

The $\psi(\theta)$ curves in Figure 2 are drawn as a unique curve. However, the relationship is often multivalued as a result of the "inkbottle" effect: A

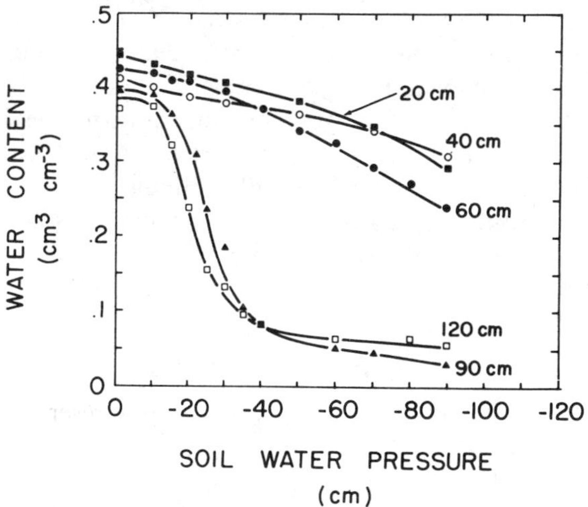

Figure 2 Soil-water suction or pressure ($-\psi$) as a function of the water content (θ) for five soil depths. The top curves (20, 40, and 60 cm) correspond to a sandy loam and the bottom ones (90 and 120 cm) to a coarse sand.

pore is wetted at a lower suction, $-\psi_w$, than the value, $-\psi_d$, necessary to dry it, since filling is determined by the radius of the pore which is larger than the radius of the neck through which it empties. The clearest conceptual model of soil hysteresis is the one developed by Philip (1964). A similar approach is followed here with some further simplifications (Parlange 1976).

The fundamental assumption is of the existence of a function $f(\psi_w, \psi_d)$ such that $f d\psi_w d\psi_d$ represents the volume of the pores that fill between ψ_w and $\psi_w + d\psi_w$, and empty between ψ_d and $\psi_d + d\psi_d$. Consider then a soil being wetted along a curve $\theta_w(\psi)$. At some potential, ψ, the soil is then dried along a curve $\theta_d(\psi, \psi_1)$, where ψ_1 indicates the potential at which drying first occurs. Then, by definition of f,

$$\theta_d(\psi, \psi_1) = \theta_w(\psi) + \int_{\psi_1}^{\psi} d\psi_w \int_{\psi}^{\psi_M} f(\psi_d, \psi_w)\, d\psi_d, \tag{5}$$

where $-\psi_M$ is the suction at which all pores are empty (in principle ψ_M can be infinite). In principle the determination of $f(\psi_d, \psi_w)$ would result from a microscopic analysis of the soil structure. Such a study does not exist at present. Instead the simplest assumption that still results in hysteresis will be made. It is assumed then that f is independent of ψ_w, i.e. we assume that necks are distributed independently of the pore size, and f is now written as $f(\psi_d)$. Such a conceptual soil may exist, in practice, only as an approximation to a real soil. The main question is whether the approximation is useful for the study of soil hysteresis.

By definition of f, it is clear that

$$d\theta_w/d\psi = -\int_{\psi}^{\psi_M} f(\psi_d) d\psi_d \tag{6}$$

and Equation (5) reduces to

$$\theta_d(\psi, \psi_1) = \theta_w(\psi) - (\psi - \psi_1) d\theta_w/d\psi, \tag{7}$$

which gives all possible drying curves issued from the wetting curve $\theta_w(\psi)$ at any point $\psi = \psi_1$. In turn, integration of Equation (7) yields $\theta_w(\psi, \psi_1, \psi_2)$ representing the wetting curves issued from $\theta_d(\psi, \psi_1)$ at $\psi = \psi_2$, where $\theta_w = \theta_d = \theta_2$, or

$$\theta_w(\psi, \psi_1, \psi_2) - \theta_2 = (\psi - \psi_1) \int_{\psi}^{\psi_2} [\theta_d(\psi^1, \psi_1) - \theta_2]/(\psi_1 - \psi^1)^2\, d\psi^1. \tag{8}$$

Figure 3 illustrates the application of Equations (7) and (8) for Caribou silt loam (Topp 1971). In the figure the solid lines represent averages of observed drying curve $\psi_1 = 0$ (upper curve) and wetting curve for

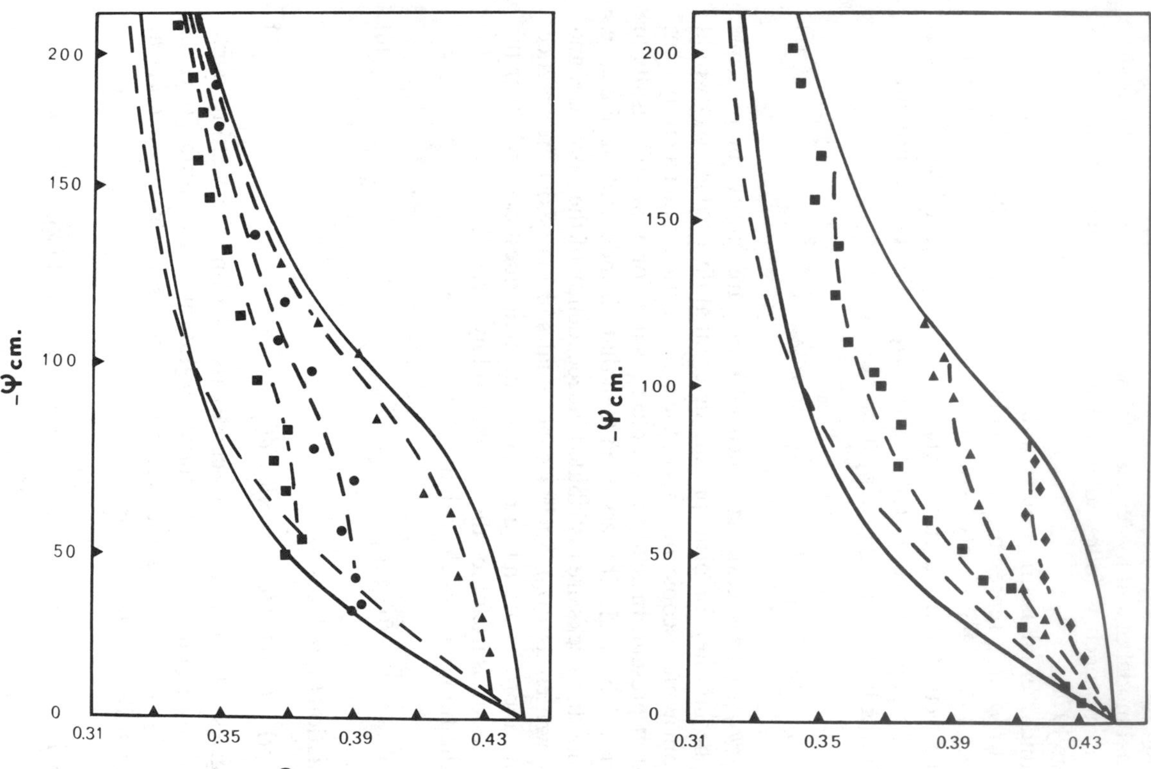

Figure 3 Wetting and drying curves for Caribou silt loam. Solid lines are measured wetting and drying boundaries. Dashed lines in left figure are wetting curves predicted by Equation (8), where θ_d is the measured drying boundary, and dots give the measured wetting curves. Dashed lines in right figure are drying curves predicted by Equation (7), where θ_w is the predicted wetting boundary given by Equation (8), and dots give the measured drying curves.

$\psi_1 = 0$, $\psi_2 \to \infty$ (lower curve). All wetting and drying curves being contained between those two, they are called wetting and drying boundaries. Together they also form the largest hysteresis loop. The left figure gives three wetting curves issued from the drying boundary; the agreement between experiments (dots) and the predictions of Equation (8) (dashes) is excellent. Note that the prediction is based on the knowledge of θ_d and the starting point of the wetting curve. As $\psi_2 \to \infty$, Equation (8) predicts a theoretical wetting boundary which differs somewhat from the observed average. However, the difference is within experimental scatter and in addition the wetting boundary is notoriously difficult to measure, whereas the drying boundary presents fewer problems. The right figure compares predicted and measured drying curves from Equation (7). However the *theoretical* wetting boundary was used for that purpose, because even small errors in θ_w are amplified by Equation (7), since the derivative of θ_w must be estimated. It is somewhat surprising that such a simple analysis of hysteresis in soils should be as accurate. The reason is undoubtedly that f enters the definition of θ_w and θ_d by its integral only. Hence even large errors in f will lead to smaller errors in θ_w and θ_d. Finally it must be pointed out that hysteresis is rarely considered under field situation, due to the difficulties in measurement. Hence even a crude theory is valuable to give an estimate of actual hysteresis to be expected when no measurements are available. This completes the static description of a soil and a discussion of the two most important problems in the thermodynamics of soils, swelling and hysteresis.

DYNAMIC CONSIDERATIONS

In this section no further explicit reference will be given to the static problems of swelling and hysteresis. This is not to say that they should be ignored in practice. Rather, if hysteresis occurs, it can usually be taken into account by taking the appropriate wetting or drying curve as $\psi(\theta)$ (Philip 1970). Similarly the dynamics of water in a swelling soil presents mathematical problems identical to those of a rigid soil as long as the flow is described relative to grain position (Philip 1970) and if grain orientation is a unique function of the void ratio (Smiles 1978).

Water movement in soils under a potential gradient is now considered.

Horizontal Water Movement

Horizontal is used here to mean that the effect of gravity is unimportant. The flux of water, J_w, is a function of the potential gradient, $\nabla\psi$, and assuming a linear relationship, Darcy's law is obtained, or

$$J_w = -K\nabla\psi \tag{9}$$

where K cm/sec is the soil-water conductivity. In the following discussion the dependence of K and ψ on the moisture content and its effect on the flow is considered.

Under somewhat exceptional circumstances, temperature, solute contents, etc., not only affect the value of $\nabla\psi$, but, in addition, terms proportional to the gradients of temperature, solute content, etc., must be included in Equation (9). For instance, even if ψ is uniform, water can be drawn to the surface if solid fertilizer is spread on it (Scotter 1974) or can be removed by evaporation if the soil surface is heated by the sun (Rose 1976). Those two effects are important in arid regions. Similarly, under freezing conditions temperature gradients control ice lens formation and permafrost movement. In that case, however, complications arise due to soil displacement by the ice and to the existence of a new phase (ice). At present two complementary approaches have provided some insight into that difficult problem, either within the framework of irreversible thermodynamics (Groenevelt & Kay 1977) or with the help of more mechanistic models (Miller et al 1975).

Equation (9) also implies that air movement which must accompany water movement does not affect it significantly. Indeed, in most cases the presence of air is relatively unimportant as will be shown later.

Conservation of mass requires that

$$\partial\theta/\partial t + \nabla \cdot J_w = 0 \tag{10}$$

where J_w obeys Equation (9), for instance, and t is the time. In the present case, when K and θ are functions of θ only, it is convenient to introduce the soil-water diffusivity D cm^2/sec, with

$$D = K\, d\psi/d\theta. \tag{11}$$

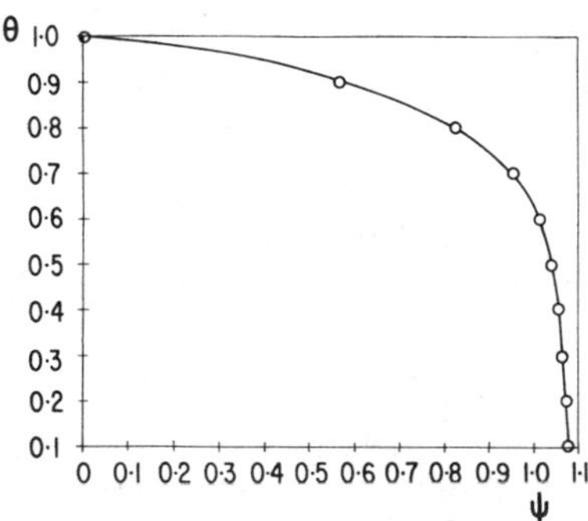

Figure 4 Water profile (ϕ) as a function of θ, yielding $D = 9 \times 10^{-4} \exp 8.36\theta$ from Equation (12).

Since $\psi(\theta)$ is known from tensiometer measurements, D (or K) must be measured. The standard method consists in taking a soil column of uniform moisture content, e.g. dry, and imposing a constant potential e.g. zero, at one end of the column. Dimensional analysis shows that a similarity variable, $\phi = xt^{-1/2}$, must exist since no length scale enters the problem (x cm is the distance along the column). Equation (10) then reduces to

$$D(\theta) = -\tfrac{1}{2}\int_0^\theta \phi(a)\, da\, d\phi/d\theta, \tag{12}$$

first introduced by Bruce & Klute (1956) in soil physics. This fundamental equation contains in compact form many of the mathematical problems encountered in soil physics, and for this reason will be discussed in some details.

Equation (12) indicates that $D(\theta)$ is known if $\phi(\theta)$ is measured. Figure 4 shows such a profile, $\phi(\theta)$, yielding $D = 9 \times 10^{-4} \exp 8.36\theta$ (Hayhoe 1978), where θ is here normalized to the pore volume, i.e. $\theta = 1$ at saturation. This typical example illustrates the basic difficulty of water movement in soil, i.e. D increases very rapidly with θ, corresponding to a profile where most of the change in θ occurs in a narrow zone. Ahead of that narrow zone or wetting front the soil is dry, and is nearly saturated behind it. This is totally different from a diffusion process with a near constant diffusivity; the curvatures of the profile in particular are of opposite signs. Hence standard linearization techniques are inapplicable to calculate the profile, $\phi(\theta)$, for a given $D(\theta)$.

Modern analytical methods, on the contrary, exploit directly the rapid variation of D with θ (Brutsaert 1974, Parlange 1975a,b, Babu 1976, Liu 1976). Although those methods are superficially different, they all yield solutions that are identical when considered as expansions in orders of a small parameter $\varepsilon = \int_0^1 (1-\theta) D\, d\theta / \int_0^1 D\, d\theta$ (Parlange & Babu 1978). Even the early study by Green & Ampt (1911), based on the existence of an abrupt wetting front, is of the same nature, as is the technique developed by Passioura (1976).

The easiest way to generate an approximate solution, $\phi(\theta)$, to Equation (12) is to observe that $d\phi/d\theta$ varies rapidly, i.e. like $D(\theta)$, while $\int_0^\theta \phi\, da$ is hardly varying in comparison. For instance, for θ small, $\int_0^\theta \phi\, da$ varies like θ, and approaches a constant, the sorptivity, $S = \int_0^1 \phi\, da$, as $\theta \to 1$. In general then, $\int_0^\theta \phi\, da$ will vary somewhat less rapidly than θ, which is indeed very slow compared to the exponential behavior of the diffusivity.

In general, if $\int_0^\theta \phi\, da$ is replaced by $\lambda^{-1} f(\theta)$, where λ is an unknown constant and $f(\theta)$ is an explicit function varying somewhat between 1 and θ, e.g. $f = \theta^{1/2}$ (Brutsaert 1976), then Equation (12) yields a first approximation

$$\phi_1 = \lambda \int_\theta^1 D/f\,da. \tag{13}$$

Whatever the choice of f is, the constant λ must also be given. For instance (for example, see Brutsaert 1976), one can impose that ϕ_1 satisfies Equation (12) at $\theta = 1$. Another possible condition (for example, see Babu 1976) is obtained by integrating Equation (12) between 0 and 1, or

$$\int_0^1 \phi^2\,d\theta = 2\int_0^1 D\,d\theta. \tag{14}$$

This last condition is normally preferable as it ensures that the resulting approximation, ϕ_1, yields an optimal estimate of the sorptivity and hence of the water intake. That is, the Euler equation associated with $I = \int_0^1 d\theta\,[4D \ln|d\phi/d\theta| + \phi^2]$ is Equation (12).

Approximating ϕ by ϕ_1, the condition $dI/d\lambda = 0$ then yields Equation (14). It can be shown that the resulting approximation for the sorptivity $\int_0^1 \phi_1\,d\theta$, is also stationary for small changes in f (Parlange 1975b) yielding the excellent estimate (Brutsaert 1976),

$$S^2 \simeq 2\int_0^1 \theta^{1/2} D\,d\theta. \tag{15}$$

For a given f, λ can, in principle, be calculated either from Equation (14) or from Equation (15). However, Equation (14) has the advantage of being exact while Equation (15) is only an approximation.

For the example considered by Hayhoe (1978), the exact value of the sorptivity is 0.9292. Taking $f = \theta^{1/2}$ as suggested by Brutsaert (1976) and Equation (14) to calculate λ yields $S_1 = \int_0^1 \phi_1\,d\theta = 0.9301$. By comparison Equation (15) yields an approximate value for S equal to 0.9277, which is almost as accurate, as expected.

To improve the accuracy of the solution it is possible to calculate a second approximate solution to Equation (12), ϕ_2, where $\int_0^\theta \phi\,da$ is calculated replacing ϕ by ϕ_1. In general two successive approximations ϕ_{n-1} and ϕ_n are related by

$$D = -\tfrac{1}{2}\int_0^\theta \phi_{n-1}\,da\,d\phi_n/d\theta, \tag{16}$$

which allows the calculation of ϕ_n when ϕ_{n-1} is known. In the present case two iterations are sufficient to reproduce Hayhoe's exact results.

It was first pointed out by Cisler (1974), see also Parlange (1975a),

Wooding & Morel-Seytoux (1976), and Hayhoe (1978), that the iterative method might not be a converging one. The resulting oscillation of two successive approximations can in principle be removed by various schemes, e.g. using $(\phi_n + \phi_{n-1})/2$ or $(\phi_n \phi_{n-1})^{1/2}$ as new approximations. In practice, however, the oscillation is irrelevant if f is chosen judiciously, e.g. $f = \theta^{1/2}$, and if λ is calculated with an optimal condition, e.g. Equation (14). Indeed, in the present example the oscillation affects the 5th decimal place only. In addition, in most cases, ϕ_1 is precise enough and no iteration is even necessary. Only when less appropriate choices for f are made, e.g. $f = 1$ or θ, and when nonoptimal conditions are used to calculate λ, does the lack of convergence necessitate some additional scheme (Hayhoe 1978).

The emphasis so far has been to calculate ϕ by analytical means. Soil physicists having to solve Equation (12) usually have relied on numerical solutions. The difficulty is clearly that the two boundary conditions, $\phi(\theta = 1) = 0$ and $\phi(\theta = 0) \to \infty$, do not apply at the same point. Forward numerical integration starting at $\phi = 0$ requires the knowledge of $(d\phi/d\theta)_{\theta=1}$ or of $\int_0^1 \phi \, d\theta$, as shown by Equation (12). It is then necessary to have a numerical iterative scheme as well. For instance, Equation (15) can be used to estimate the sorptivity and hence $(d\phi/d\theta)_{\theta=1}$. Because the estimate given by Equation (15), S_1, although accurate is not exact, numerical integration of Equation (12) will yield a value θ_∞ of θ as $\phi \to \infty$ different from zero. That is, S_1 would be the exact sorptivity if the initial water content was θ_∞ rather than zero. Replacing θ by $\theta - \theta_\infty$ in the analysis of Equation (15) shows that $S_1^2/(1-\theta_\infty)$ gives a new and better estimate of the sorptivity (Parlange 1975b, Smith & Parlange 1978). Taking this new estimate of the sorptivity allows the integration of Equation (12) numerically once more and the repeat of the procedure. Indeed two iterations are again sufficient to reproduce Hayhoe's solution. Hence it is elementary, either analytically or numerically, to generate solutions of Equation (12) when D increases rapidly with θ. This equation is the simplest one that can describe water movement in soils. However, the method of solution, which is based on the existence of a wetting front, can be extended easily to more complex systems, as is shown in the next two sections.

Horizontal Air and Water Movement

The previous section neglected air movement altogether; it is reintroduced and its effects discussed here. Some discussion of this problem can be found in Wooding & Morel-Seytoux (1976) based on the Buckley-Leverett theory, which has been very successful in modelling the displacement of oil and water. There are, however, fundamental differences between the two problems. With oil and water, surface tension, and hence diffusion effects, are negligible. The interpenetration of oil and water results from flux differences between the two fluids, under high pressure gradients. On the other hand with air and water, pressures caused by surface tension are not negligible and the front is more diffuse, and the Buckley-Leverett theory is less appropriate. The present analysis follows instead a fundamental theory developed by McWhorter (1971, 1975).

Applying Darcy's law to both fluids gives

$$J_{\mathrm{a}} = -K_{\mathrm{a}}\nabla p_{\mathrm{a}}; \quad J_{\mathrm{w}} = -K_{\mathrm{w}}\nabla p_{\mathrm{w}}, \quad \psi = p_{\mathrm{w}} - p_{\mathrm{a}}, \tag{17}$$

where fluxes result from gradients of the pressure in each phase (subscripts "a" and "w" refer to air and water respectively). The results of the previous section apply when it is legitimate to replace p_{a} by 1 atm. For simplicity, and following McWhorter (1971, 1975), air is taken as incompressible (it can be shown that air compressibility introduces much smaller corrections, Parlange & Hill 1979). Introducing a total flux V and an f-function by

$$V = J_{\mathrm{a}} + J_{\mathrm{w}}; \quad f = 1/(1 + K_{\mathrm{a}}/K_{\mathrm{w}}), \tag{18}$$

the condition of air and water compressibility gives $\nabla \cdot V = 0$. Mass conservation for water, Equation (10), becomes

$$\partial\theta/\partial t + f' V \cdot \nabla\theta = \nabla \cdot [D_{\mathrm{a}} \nabla\theta], \tag{19}$$

where $f' = df/d\theta$ and the diffusivity, D_{a}, is defined as $f K_{\mathrm{a}} d\psi/d\theta$. Air effect is expressed by the f' term. Hence, to estimate the influence of air the f-function must be measured. From its definition in Equation (18), f varies from zero for a dry soil to one for a fully saturated soil. It is clear that air movement is important when K_{a} is small and f is near one. Attempts to measure K_{a} directly are too imprecise to deduce f near one with any accuracy. The alternative is to devise two different experiments; in the first, air has great difficulties in moving but not in the second. Differences between those two experiments will then give a measure of f.

In the first experiment, for instance, water moves into a column sealed at the bottom and air has to escape through the surface. Such conditions can occur in the field when water ponds a flat area with a shallow water table. Bubbling of air can be observed and air movement greatly reduces water intake, e.g. by a factor two (Dixon & Linden 1972). In the second case the column is unsealed at the bottom and air can escape easily ahead of the wetting front without appreciable buildup of air pressure.

In both cases, V is independent of position, hence it must be zero in the first case, the value at the bottom of the column and equal to

$1/2St^{-1/2}$ in the second case, the value at the surface. It is paradoxical that Equation (19) has the form postulated in the previous section (when air was neglected altogether) in the first case (when air movement is in fact important). In the second case the equation has a different form since $V \neq 0$, even though air should not be as important in that case. The explanation of this paradox is that the diffusivities D and D_a are not the same.

As in the previous section, and for the same reason, Equation (19) has a similarity solution in terms of the variable, $\phi = xt^{-1/2}$. Writing $\phi = \phi_c$ for the closed column and $\phi = \phi_o$ for the open one, then ϕ_c obeys Equation (12) (replacing D by D_a) and ϕ_o obeys, from Equation (19),

$$D_a = 1/2(d\phi_o/d\theta)\left(\int_0^\theta \phi_o da - S_0 f\right). \tag{20}$$

Eliminating D_a between those two equations yields f, or

$$S_0 f = \int_0^\theta \phi_o da - (d\phi_c/d\phi_o)\int_0^\theta \phi_c da. \tag{21}$$

This equation shows that f can be measured from measurements of ϕ_o and ϕ_c. As an illustration, Figure 5 gives profiles ϕ_o and ϕ_c (Parlange & Hill 1979). Note that the primary qualitative difference between the two is that ϕ_c has more of a gradient in θ near the surface, i.e. there is more

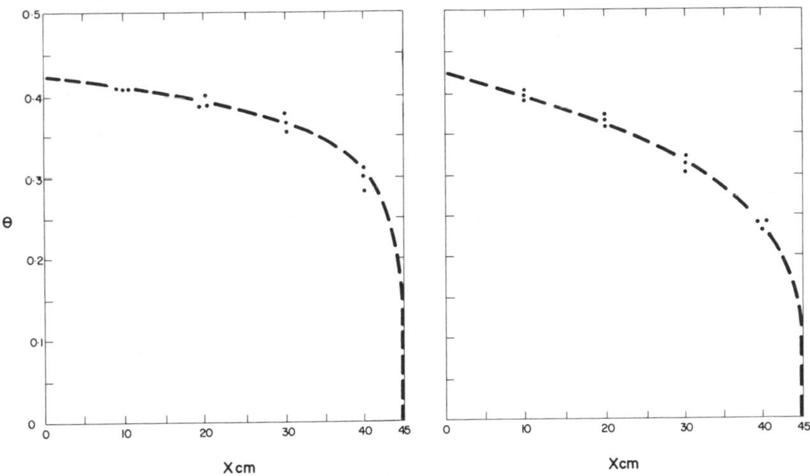

Figure 5 Observed water profiles giving (θ) as a function of distance x from the source for a fine sand. The left figure corresponds to the open column at $t = 0.87 \times 10^4$ sec., and $\phi_0 = xt^{-1/2}$. The right figure corresponds to the closed column at $t = 2 \times 10^4$ sec and $\phi_c = xt^{-1/2}$.

air present to allow for air movement. From those two profiles, f is calculated as indicated in Figure 6. As expected $f \to 0$ rapidly for a dry soil, and f increases toward one as θ increases. Note that $f < 1$ however, since at $f = 1$ air could not move to the soil surface.

From the estimate of f the following question can now be answered: In the usual case of air moving ahead of the wetting front, without appreciable pressure buildup, can air be ignored altogether (as was done in the previous section)?

In principle air would have no effect if it were possible to have each pore in direct contact with the atmosphere. Then water would diffuse into the soil with a diffusivity $D_w = K_w d\psi/d\theta$ or $D_w = D_a/(1-f)$ and, following Equation (14), $\int_0^1 \phi_o^2 \, d\theta$ would be equal to $2\int_0^1 D_w \, d\theta$. However, for the present example, they differ by 4%, i.e. air movement ahead of the wetting front affects water intake by about 2%. This correction is small, as expected, and certainly negligible under field conditions. Only careful laboratory experiments will show the effect. Usually, then Equation (12) is appropriate and its use is justified. However, because of the inherent error of Equation (12) in describing air-water movement, solving that equation with a precision much greater than 2% is of more mathematical than physical interest. It must also be remembered that the 2% error obtained here is only an order of magnitude. If the soil surface had been drier the error would have been less and more if the soil surface had been wetter.

Gravity Effects

The water intake by a soil is increased by the presence of gravity. In this section this increase is analyzed when the flow is one dimensional

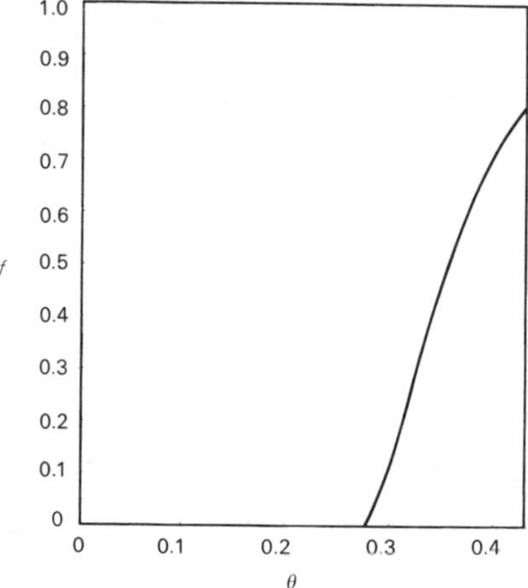

Figure 6 Measured f-function obtained from the profiles ϕ_o and ϕ_c given in Figure 5 and Equation (21).

in the z-direction, measured positively downwards. As we shall see later, in the section on solute movement, a second important effect of gravity is that under certain conditions the one dimensional flow becomes unstable.

With the effect of gravity added, Equation (9) is valid if the soil-water potential is replaced by the *total* potential $(\psi - z)$, and Equation (10) becomes, in one-dimension,

$$\partial\theta/\partial t = \partial[D\,\partial\theta/\partial z]/\partial z - (dK/d\theta)(\partial\theta/\partial z). \tag{22}$$

If the soil surface is ponded, the early stages of infiltration are hardly affected by gravity. As time progresses, however, gravity becomes more important (Philip 1970). Normally, under variable rainfall, ponding occurs only at a time $t > t_p$. The accurate determination of t_p is important to estimate the amount of runoff (which occurs for $t > t_p$).

Infiltration with gravity under a ponded surface was originally discussed by Green & Ampt (1911). His result can be reproduced by considering an exact solution of Equation (22): $z = g(t)\,dK/d\theta$ where $gK(\theta = 1)$ is the quantity of water in the soil. The solution is obviously exact if $D = -(S^2/2K_1)Kd^2K/d\theta^2$, and

$$t = g - (S^2/2K_1^2)\ln(1 + 2gK_1^2/S^2). \tag{23}$$

Equation (23) constitutes the Green & Ampt result. It is surprisingly accurate (Onstad et al 1973) even if D and K do not obey the relation that makes it exact. Equation (23) can be considered as an interpolation between the two general limits: $g = t$ as $t \to \infty$ and $K_1 g = S\sqrt{t}$ as $t \to 0$. To gain more understanding of Equation (23), integrate Equation (22) twice to obtain

$$\partial\left(\int_0^1 z^2\,d\theta\right)\Big/\partial t = 2\int_0^1 D\,d\theta - 2\int_0^1 K(\partial z/\partial\theta)\,d\theta, \tag{24}$$

which is the obvious extension of Equation (14) when gravity is taken into account. Now let D increase very rapidly with θ so that the profile has an almost abrupt front. This is the normal behavior for D. However, it appears that K may have two different limiting behaviors (Smith 1972, Smith & Parlange 1977, 1978). In one $dK/d\theta$ does not increase rapidly with θ; in the other, D and $dK/d\theta$ have the same behavior. In the first limiting case, of an abrupt front, with $dK/d\theta$ well-behaved, Equation (24) yields at once

$$g\,dg/dt = S^2/2K_1^2 - g, \tag{25}$$

where $S^2/2$ in Equation (25) stands for $\int_0^1 D\,d\theta$ in Equation (24), so that $gK_1 = S\sqrt{t}$ as $t \to 0$. Integration of Equation (25) yields Equation (23), which is not surprising since Equation (23) is exact when $D \sim d^2K/d\theta^2$, i.e. when $dK/d\theta$ grows less rapidly than D. The other limiting case of D and $dK/d\theta$ increasing similarly with θ gives, from Equation (24) (Smith & Parlange 1977, 1978),

$$g = (S^2/2K_1)\ln[1/(1-dt/dg)], \tag{26}$$

and by integration,

$$t = g+(S^2/2K_1)[\exp-(2gK_1/S^2)-1]. \tag{27}$$

The two limiting Equations (23) and (27) apply when the soil surface is ponded, i.e. for $t > t_p$. Figure 7 illustrates their validity for a particular soil (Smith & Parlange 1978). The surface flux, $K_1\, dg/dt$, is plotted as a function of time. The exact flux was obtained by direct numerical integration of Equation (22). The initial time in Equations (23) and (27) was adjusted so that the results are exact at $t = t_p$. As expected the exact result lies between the limiting approximations. Note also that the two approximations do not differ greatly from each other. Those two properties make the approximations useful in practice.

Equations (23) and (27) apply for $t > t_p$. For $t < t_p$ Equations (25) and (26) can still be used, with $g(t)$ now known from measurements of rainfall intensity. The equations then yield the ponding time, when $S^2/2K_1$ becomes equal to its value at saturation (Smith & Parlange 1977; in that paper, $S^2/2K_1$ is slightly modified to match the exact result as $t \to 0$, for a variable water content at the surface, rather than a constant one).

We have concentrated on the effect of gravity on the water intake. There is no difficulty in solving Equation (22) for the moisture profile, either by numerical or analytical iterative procedures. The method is essentially the same as in the horizontal case (for example, see Parlange 1975a) and will not be repeated here.

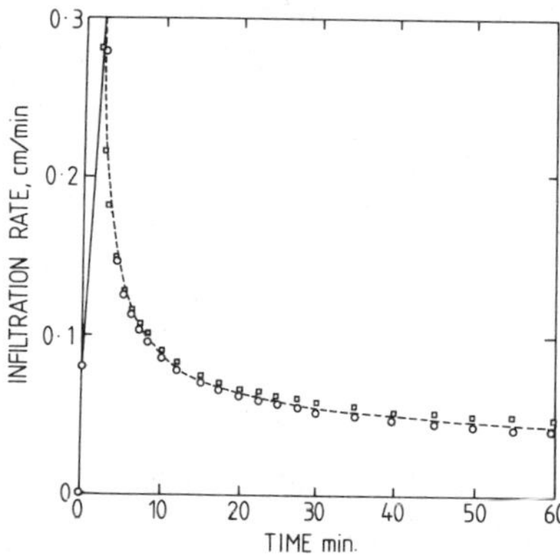

Figure 7 Infiltration rate for Nickel soil, as a function of time according to Equation (23), square points, and Equation (27), circular points, for $t > t_p$. The dashed line is obtained by direct integration of Equation (22). For $t < t_p$, the infiltration rate is equal to the rainfall rate equal to $(0.08+0.01016t)\,\text{cm min}^{-1}$, represented by the solid line, with $t_p = 1.54$ min.

SOLUTE TRANSPORT

In this last section the transport and dispersion of solutes in a soil is analyzed. Again, many of the questions associated with this problem occur in other sciences, e.g. chromatography. For this reason we shall restrict the study to recent advances, primarily related to soils.

Natural soils show a great variability in dynamic properties. Nielsen and co-workers (for examples, see Biggar & Nielsen 1976) have pioneered the statistical studies of soil properties. They have shown in particular that the soil-water conductivity is log normally distributed. Under those conditions large fluctuations occur in soil-water transport, and even the meaning of a proper averaging procedure is not quite clear at present. This problem will not be considered any further and all physical quantities in the following are considered as adequate averages.

Let us consider, then, the one-dimensional, e.g. downward, transport of a solute. The transport equation for the solute based on mass conservation is (for example, see Jury et al 1976)

$$\partial c/\partial t + V\partial c/\partial z = D\partial^2 c/\partial z^2 - \partial s/\partial t \qquad (28)$$

where c is the solute concentration and V the water velocity. Since this section is primarily concerned with solute transport, rather than with the details of the water profile described in the previous section, the water content is taken as constant. This simplifies the analysis without sacrificing the physics of solute transport [a good example of the simultaneous treatment of solute transport and detailed water movement can be found in Reeves et al (1977)]. D in Equation (28) is the dispersion coefficient of the solute (not to be confused with the soil-water diffusivity). Normally D depends on the magnitude of V, i.e. there is some hydrodynamic dispersion in addition to molecular diffusion [see Bear's (1972) excellent discussion for more details on the dispersion coefficient].

Finally, $\partial s/\partial t$ is the sink term which may represent adsorption, ion uptake by roots, biochemical reactions, diffusion into aggregates, etc. If s can be written as a function of c, the sink term represents a reversible reaction and the relation $s(c)$ describes the adsorption isotherm. For the linear case in particular, $s = Rc$, the only effect of the sink term is to reduce V and D by $(1+R)$. If $\partial s/\partial t$ can be written as a function of c, the reaction is irreversible. In that case, if $\partial s/\partial t$ is proportional to c^n, n is the order of the reaction. Under field conditions, so many factors control chemical and biochemical reactions that observations can often be fitted by reactions of an arbitrary order. This is particularly true when microbial populations have to be taken into account (for example, see Ardakani et al 1975, McLaren 1975, Starr & Parlange 1975, 1976a).

Three problems connected with Equation (28) are discussed in the following sections. The first investigates the influence of the terms on the right-hand side of the equation. The second looks at the boundary conditions associated with the equation. Finally, the third discusses conditions that make Equation (28) invalid because of flow instability.

Stagnant Phase Effect

It is the rule, rather than the exception, that solutes move rapidly, e.g. by convection, in part of the pore space (the mobile phase) and slowly, e.g. by molecular diffusion, in the other (the stagnant phase). This may occur because of the soil structure, e.g. due to the presence of cracks, aggregates, dead end pores, or particle size segregation (see, for instance, Skopp & Warrick 1974, Gaudet et al 1977, Starr & Parlange 1977, Ripple et al 1974). The division of the pore space in two phases may also be caused by density differences which enhance convection in large pores (for example, see Krupp & Elrick 1969, Biggar & Nielsen 1964, Starr & Parlange 1976b). The stagnant phase does not even have to be a physical part of the pore space. For instance a slow, i.e. nonequilibrium, reaction or adsorption on the grains creates an effect similar to the slow molecular diffusion into an aggregate (Murali, unpublished data).

In all those cases the stagnant phase effect is represented by $\partial s/\partial t$ in Equation (28). It is always characterized by two properties: first by its extent, i.e. the amount of solute in the stagnant phase, whenever equilibrium is reached; second, by the relaxation time, ζ, which characterizes the time necessary to reach equilibrium.

Consider a soil column of length L and change the concentration in the influent solution from zero to one, with the water velocity V remaining constant. If $\partial s/\partial t = 0$ in Equation (28) the breakthrough curve is the standard, $c = 1/2 \text{ erfc } (L - Vt)/\sqrt{4Dt}$, with the concentration rising rapidly when the pore volume $tV/L \simeq 1$, the dispersion around that time being a function of the magnitude of D. If, however, the stagnant fluid occupies a fraction b of the total pore volume and if $\zeta V/L \ll 1$, i.e. if the diffusion time is slow compared to the convection time, then the first rise in concentration occurs around the pore volume $(1-b)$. The value of b can be known a priori; for example, if it corresponds to the volume of aggregates, it can also depend on the experimental conditions. This is the case illustrated in Figure 8, where convection in the large pores is enhanced in proportion to the increasing density of the incoming solution.

Following the rapid rise when the pore volume is $(1-b)$, there is a much slower change during the time ζ. To conserve mass, when the slow rise occurs c must be less than one by an order $bL/\zeta V$, since bL/V is

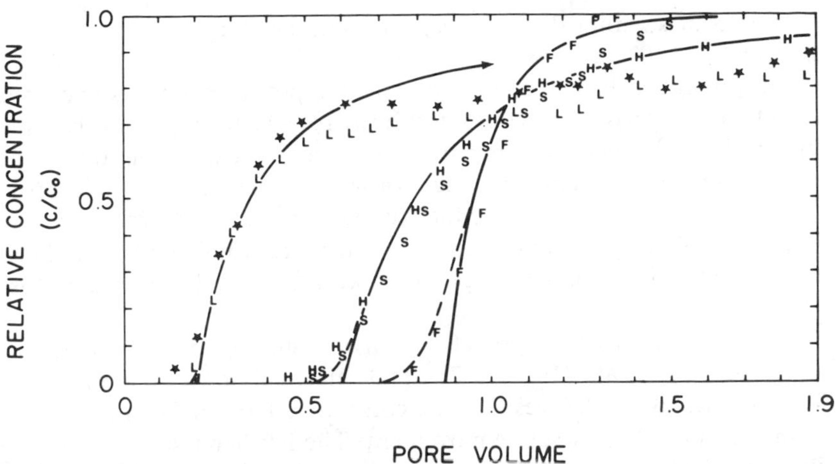

Figure 8 Breakthrough curves showing the earlier rise in concentration as the extent of the stagnant phase increases. Solid lines represent the theoretical result, without dispersion, the dashed lines show the correction due to the dispersion term in Equation (28).

the order of the mass of solute flowing out of the column ahead of one pore volume.

Although specific experiments may be analyzed more fully (see, for example, Gaudet et al 1977), b and ζ, or some equivalent parameters, govern the shape of the breakthrough curve. Indeed, it is striking that all stagnant phase experiments show very similar breakthrough curves even if the physical processes are quite different (for example, see Gaudet et al 1977, Skopp & Warrick 1974, Ripple et al 1974, Krupp & Elrick 1969, Biggar & Nielsen 1964, Starr & Parlange 1976b, 1977).

The primary discussion so far has been concerned with the influence of the $\partial s/\partial t$ term in Equation (28). Figure 8 also shows the effect of the dispersion term as the difference between the dotted and solid lines. This effect is negligible compared to the stagnant phase influence. This is often the case, particularly under field conditions, since the Péclet number, VL/D, which measures the relative importance of convection and diffusion, is normally much greater than one at any reasonable distance from the soil surface (usually $V \sim 1\,\mathrm{cm/hr}$; $D \sim 0.05\,\mathrm{cm^2/hr}$).

Variable Infiltration Rates

Neglecting the dispersion term altogether in Equation (28), we are now interested in the influence of the convective term, $V\partial c/\partial z$. Under field conditions, the infiltration rate $I(t)$, e.g. due to rainfall or irrigation, varies rapidly with time. This in turn produces rapid changes in the solute profiles within the soil (see, for example, Jury et al 1976, 1978; Saffigna et al 1977).

The present problem is of great practical importance. For instance, one might wish to predict the fate of fertilizer, i.e. which proportion is taken up by the plant and how much is eventually causing pollution. One might also wish to adjust the irrigation rate so that salinity build up in the soil does not make it unproductive. In both cases, the problems have implications for agricultural practices in large areas and affect the policies of national and international agencies, like FAO (see, for example, Rose 1979).

The most promising approach to solving such global problems was developed by Raats (1975, 1978) and his method is followed here. Basically Raats emphasizes that the convective term in Equation (28) is mainly responsible for solute movement. The left-hand side of Figure 9 illustrates the validity of that point. There, the rate of progress of the pulse is equal to the constant water velocity, on the *average*. That is, there is so much variability in soil properties that the scatter of the

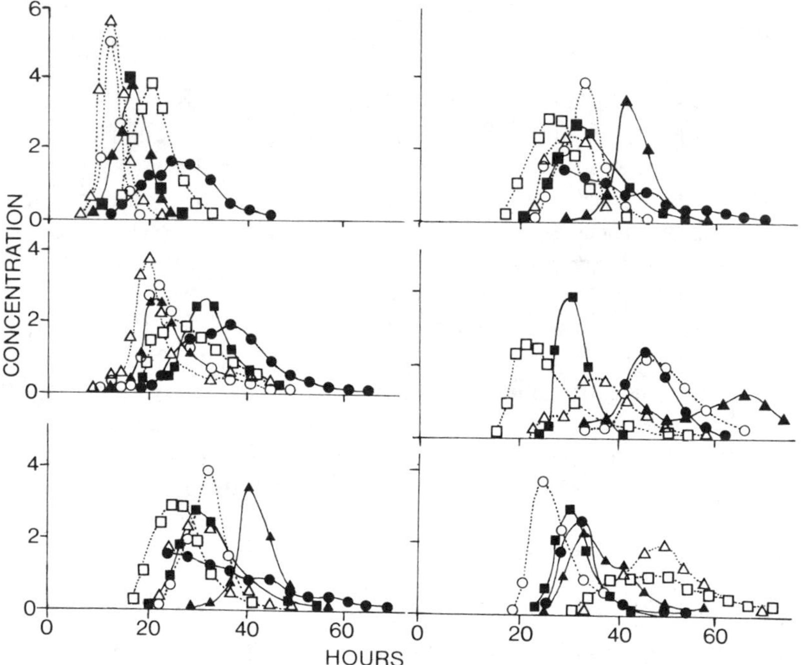

Figure 9 Salt concentration versus time, measured at five depths with six probes. The three depths on the left are at 20, 40, and 60 cm respectively and 60, 120, and 180 cm on the right.

data is typically large, as discussed earlier. Under those conditions it would not be meaningful to analyze the shape of the pulse in any great detail and consideration of convection as the primary factor for transport is quite justified.

Both the infiltration rate, I, and the evaporation rate, E, are known as a function of time, e.g. from micrometeorological observations. The evaporation sink has to be distributed in the root zone. Raats (1975) assumes the root zone to be exponentially decreasing with depth. However, it appears that water uptake by roots is primarily a function of water distribution (Herkelrath et al 1977). Hence it is more consistent to assume a uniform distribution of water uptake down to the root depth, L (the results are hardly changed, only slightly simpler; in general there is no difficulty in using Raats' method for any root distribution).

The convective velocity would be I/θ, in the absence of evaporation (here θ is the uniform moisture content, e.g. the holding capacity of the soil). Under condition of constant evaporation rate per unit depth, $E/L\theta$, the velocity is reduced at a depth z by $Ez/L\theta$. If the solute is taken by roots at the same rate as water then $\partial s/\partial t = 0$. On the other hand if the roots exclude the solute entirely then the concentration increases at a rate, $-\partial s/\partial t = cE/L\theta$. In general we shall take $-\partial s/\partial t = \alpha cE/L\theta$, with $0 < \alpha < 1$, depending on the amount of exclusion and also on the presence of other possible first-order reactions, e.g. microbial denitrification. Hence, Equation (28) becomes

$$\theta\partial c/\partial t + (I - zE/L)\partial c/\partial z = \alpha cE/L. \tag{29}$$

Following Raats (1975, 1978) the equation is solved by the method of characteristics, and yields at once

$$c \exp - \alpha \int_0^t E d\zeta/L\theta = F\left[\left(\exp - \int_0^t E d\zeta\right)\Big/(z_{\mathrm{f}} - z)L\theta\right] \tag{30}$$

where $t = 0$ is some arbitrary initial time and z_{f} is the solute front position, with

$$z_{\mathrm{f}}/L = \exp - \int_0^t E d\zeta/L\theta \left[\int_0^t d\zeta (I/L\theta) \exp\left(\int_0^\zeta E d\zeta^*/L\theta\right)\right]. \tag{31}$$

F is an unknown function determined for $z < z_{\mathrm{f}}$ by the boundary condition

$$c(z = 0, t) \exp - \alpha \int_0^t E d\zeta/L\theta = F\left[\left(\exp - \int_0^t E d\zeta\right)\Big/ L\theta z_{\mathrm{f}}\right]. \tag{32}$$

The most important conclusion to be drawn from the formulation is that Equation (30) predicts an exponential growth of c, proportional to

$\exp \int_0^t E\,d\zeta/L\theta$ at $z = z_f$. Hence the condition for continued agricultural use of the land is that at some time, T, when $c\,(z_f, T)$ has reached some critical value for crop survival, then we must have $z_f > L$, i.e. the critical level of salt must be below the root zone. Then, Equation (31) yields the condition,

$$\int_0^T dt\,(I/L\theta) \exp \int_0^t (E/L\theta)\,d\zeta > \exp \int_0^T (E/L\theta)\,dt. \qquad (33)$$

Obviously, Equation (33) is satisfied if I is large, but the interesting result, from a water management point of view, is that I is multiplied by $\exp \int_0^t (E/L\theta)\,d\zeta$. Hence, to conserve water, without letting salinity build up, enough water should be provided for plant needs, i.e. $I \simeq E$. For flushing purposes, however, irrigation water is more efficiently used as $\int_0^t (E/L\theta)\,d\zeta$ increases.

Dynamic Instability and Fingering

Equation (28) assumes that the flow of water and solute is one-dimensional at least in a reasonable average sense. However, under certain conditions the water flow may be unstable, and the water may move only along preferred paths or fingers. How quickly and how much of the solutes leave the root zone and reach the water table is then determined by the structure of the fingers. In the past, erratic observations of solute movement may well have been due to the presence of fingers, although explicit observations or discussions of fingering in soils is fairly recent (Wooding & Morel-Seytoux 1976, Quisenberry & Phillips 1976, Jury et al 1976, Starr et al 1978).

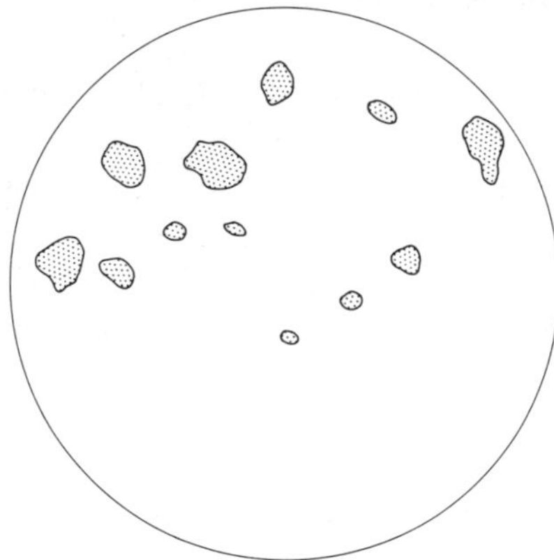

Figure 10 Cross sections of wetted areas (fingers) shown by shaded areas. Diameter of the column was 1.8 m.

Comparison of the left- and right-hand sides of Figure 9 shows a similar scatter of the data. However, the average progress of the pulse of solute is clearly controlled by the flux of water in the first 60 cm of soil but not below, even though the *average* water flux is constant. Observations showed (Starr et al 1978) that below 60 cm there was a coarse sand and a finer soil above. Also, removal of soil layers showed that water moved throughout the whole soil in the top 60 cm but only in a few zones, i.e. fingers, below (Figure 10). Clearly, the total cross section of the fingers being small, water moved quickly through the fingers, as shown in Figure 9. The instability, responsible for the formation of fingers, originates at the interface of the fine and coarse layers.

As in the case of water droplets hanging below some surface, the instability is gravity-driven and surface tension (here the soil-water diffusivity) is stabilizing. Some recent theoretical discussion of fingering in soils can be found in Hill & Parlange (1972), Raats (1973), Philip (1975), Wooding & Morel-Seytoux (1976), and Parlange & Hill (1976). The fundamental problem is the determination of the fingers cross section resulting from the balance between gravity and diffusivity.

Because of diffusivity effects, the velocity of water, U, is less than the average water velocity, V, if the water front moving downwards is convex. To determine the magnitude of the diffusive effect, water movement around a cylindrical source can be considered since the front is then curved (see, for example, Sawhney & Parlange 1976). Conservation of mass, Equation (10), written in cylindrical coordinates replaces Equation (12) by

$$D = -\tfrac{1}{2}\int_0^\theta \phi^2(a)\, da\, d\ln\phi/d\theta \tag{34}$$

where ϕ is the radial distance, r, from the source divided by $t^{1/2}$. If r_0 and r_1 are the positions of the wetting front ($\theta = 0$) and saturated front ($\theta = 1$), then dr_1/dt is the velocity of the saturated front and dr_0/dt would be the velocity of the front if diffusivity was negligible. The difference between the two velocities then represents the diffusive effect, i.e. $V-U$. Integrating Equation (34) shows that $\int_0^1 D\, d\theta$ is essentially proportional to $\ln(r_0/r_1)$ as long as D increases rapidly with θ. Differentiation with time then gives (Parlange & Hill 1976)

$$V-U \simeq \int_0^1 D\, d\theta/r_1 \tag{35}$$

which, as expected, shows that the reduction in velocity is greater as diffusivity increases and the front's curvature is smaller. Behind the saturated front, $\theta = 1$, hence $\partial\theta/\partial t$ in Equation (10) is zero, and the conductivity has its saturated value, K_1. Hence, from Equation (10), the Laplacian of the total potential is zero. Solving that equation behind the front, for a disturbance of wave length λ and frequency ω together with the condition of zero potential at the front and Equation (35), yields the dispersion relation $\omega(\lambda)$. The wave length of the fastest growing disturbance can be associated with the diameter d of a finger, or (Parlange & Hill 1976)

$$d \simeq 2\pi \int_0^1 D\,d\theta/K_1. \tag{36}$$

As expected, d is greater when the stabilizing influence of the diffusivity increases and is smaller when K_1, the value of the flux under gravity alone, is larger. The coarse sand, in the example reported in Figure 10, has properties corresponding to $d \simeq 18$ cm according to Equation (36) (Starr et al 1978), which is in rough agreement with the size of the wet zones as shown in Figure 10.

There are many regions where water moving through layered field soils can become unstable at textural interfaces. To predict and measure solute movement under those conditions becomes much more difficult because of fingering. It is perhaps a symptom of the difficulties of field measurements that recognition of this phenomenon is so recent in soil physics.

In conclusion, this review has attempted to present, through highly selected topics, some of the fundamental problems of water transport in soils. Concepts have been stressed, and hence only simple examples, which do not necessitate long mathematical developments, were considered, with many simplifying assumptions as well. It must be stressed, however, that the processes described here, in three separate and somewhat independent sections, must often be considered simultaneously, in practice. The resulting theoretical and numerical difficulties are such that in agricultural engineering, for instance, there is still a large component of empiricism.

Acknowledgment

This paper was prepared with the support of University Research Grant No. 35-0600-151.

Literature Cited

Ardakani, M. S., Volz, M. G., McLaren, A. D. 1975. Consecutive steady state reactions of urea, ammonium and nitrite nitrogen in soil. *Can. J. Soil Sci.* 55: 83–91

Babu, D. K. 1976. Infiltration analysis and perturbation methods. 2. Horizontal absorption. *Water Resour. Res.* 12:1013–18

Baver, L. D., Gardner, W. H., Gardner, W. R. 1972. *Soil Physics.* New York: Wiley

Bear, J. 1972. *Dynamics of Fluids in Porous Media.* New York: Elsevier

Biggar, J. W., Nielsen, D. R. 1964. Chloride-36 diffusion during stable and unstable flow through glass beads. *Soil Sci. Soc. Am. Proc.* 28:591–95

Biggar, J. W., Nielsen, D. R. 1976. Spatial variability of the leaching characteristics of a field soil. *Water Resour. Res.* 12: 78–84

Bruce, R. R., Klute, A. 1956. The measurement of soil-water diffusivity. *Soil Sci. Soc. Am. Proc.* 20:458–62

Brutsaert, W. 1974. More on an approximate solution for nonlinear diffusion. *Water Resour. Res.* 10:1251–52

Brutsaert, W. 1976. The concise formulation of diffusive sorption of water in a dry soil. *Water Resour. Res.* 12:1118–24

Cisler, J. 1974. Note on the Parlange method for the numerical solutions of horizontal infiltration of water in soil. *Soil Sci.* 117:70–73

Dixon, R. M., Linden, D. R. 1972. Soil air pressure and water infiltration under border irrigation. *Soil Sci. Soc. Am. Proc.* 36:948–53

Emerson, W. W., Bond, R. D., Dexter, A. R. 1978. *Modification of soil structure.* New York: Wiley

Gaudet, J. P., Jégat, H., Vachaud, G., Wierenga, P. J. 1977. Solute transfer, with exchange between mobile and stagnant water, through unsaturated sand. *Soil Sci. Soc. Am. J.* 41:665–71

Green, W. A., Ampt, G. A. 1911. Studies on soil physics. I. The flow of air and water through soils. *J. Agri. Sci.* 4: 1–24

Groenevelt, P. H., Kay, B. D. 1977. Water and ice potentials in frozen soils. *Water Resour. Res.* 13:445–49

Groenevelt, P. H., Parlange, J.-Y. 1974. Thermodynamic stability of swelling soils. *Soil Sci.* 118:1–5

Hayhoe, H. N. 1978. Numerical study of quasi-analytic and finite difference solutions of the soil-water transfer equation. *Soil Sci.* 125:68–74

Herkelrath, W. N., Miller, E. E., Gardner, W. R. 1977. Water uptake by plants. I. Divided root experiments. II. The root contact model. *Soil Sci. Soc. Am. J.* 41: 1033–43

Hill, D. E., Parlange, J.-Y. 1972. Wetting front instability in layered soils. *Soil Sci. Soc. Am. Proc.* 36:697–702

Jury, W. A., Gardner, W. R., Saffigna, P. G., Tanner, C. B. 1976. Model for predicting simultaneous movement of nitrate and water through a loamy sand. *Soil Sci.* 122: 36–43

Jury, W. A., Frenkel, H., Devitt, D., Stolzy, L. H. 1978. Transient changes in the soil water system from irrigation with saline water. II. Analysis of experimental data. *Soil Sci. Soc. Am. J.* 42:585–90

Krupp, H. K., Elrick, D. E. 1969. Density effects in miscible displacement experiments. *Soil Sci.* 107:372–80

Liu, P. L.-F. 1976. A perturbation solution for a nonlinear diffusion equation. *Water Resour. Res.* 12:1235–40

McLaren, A. D. 1975. Comment on kinetics of nitrification and biomass of nitrifiers in a soil column. *Soil Sci. Soc. Am. Proc.* 39: 597–98

McWhorter, D. B. 1971. Infiltration affected by the flow of air. *Hydrology Papers. Colorado State U. No. 49.* 43 pp.

McWhorter, D. B. 1975. Vertical flow of air and water with a flux boundary condition. *Ann. Meet. ASAE*, Davis, Calif. No. 75, p. 2012

Miller, R. D., Lock, J. P. G., Bresler, E. 1975. Transport of water and heat in a frozen permeater. *Soil Sci. Soc. Am. Proc.* 39:1029–38

Neuman, S. P. 1977. Theoretical derivation of Darcy's law. *Acta Mechanica* 25:153–70

Onstad, C. A., Olson, T. C., Stone, L. R. 1973. An infiltration model tested with monolith moisture measurements. *Soil Sci.* 116:13–17

Parlange, J.-Y. 1974. Water movement in soils. *Geophysical Surveys* 1:357–87

Parlange, J.-Y. 1975a. Theory of water movement in soils: 11. Conclusion and discussion of some recent development. *Soil Sci.* 119:158–61

Parlange, J.-Y. 1975b. On solving the flow equation in unsaturated soils by optimization: Horizontal infiltration. *Soil Sci. Soc. Am. Proc.* 39:415–18

Parlange, J.-Y. 1976. Capillary hysteresis and the relationship between drying and wetting curves. *Water Resour. Res.* 12: 224–28

Parlange, J.-Y., Babu, D. K. 1978. Comment

on a perturbation solution for a nonlinear diffusion equation. *Water Resour. Res.* 14:155–56

Parlange, J.-Y., Hill, D. E. 1976. Theoretical analysis of wetting front instability in soils. *Soil Sci.* 122:236–39

Parlange, J.-Y., Hill, D. E. 1979. Air and water movement in porous media—Compressibility effects. *Soil Sci.* 127:257–63

Passioura, J. B. 1976. Determining soil water diffusivities from one-step outflow experiments. *Aust. J. Soil Res.* 15:1–8

Philip, J. R. 1964. Similarity hypothesis for capillary hysteresis in porous materials. *J. Geophys. Res.* 69:1553–62

Philip, J. R. 1970. Flow in porous media. *Ann. Rev. Fluid Mech.* 2:177–204

Philip, J. R. 1975. Stability analysis of infiltration. *Soil Sci. Soc. Am. Proc.* 39:1042–49

Quisenberry, V. L., Phillips, R. E. 1976. Percolation of surface-applied water in the field. *Soil Sci. Soc. Am. J.* 40:484–89

Raats, P. A. C. 1973. Unstable wetting fronts in uniform and non-uniform soils. *Soil Sci. Soc. Am. Proc.* 37:681–85

Raats, P. A. C. 1975. Distribution of salt in the root zone. *J. Hydrol.* 27:237–48

Raats, P. A. C. 1978. Convective transport of solutes by steady flows. I. General theory. II. Specific flow problems. *Agri. Water Manage.* 1:201–32

Reeves, M., Francis, C. W., Duguid, J. O. 1977. Quantitative analysis of soil chromatography. I. Water and radio nuclide transport. *Environ. Sci. Div.* No. 1105. Natl. Tech. Inf. Serv., US Dept. Comm. 172 pp.

Ripple, C. D., James, R. V., Rubin, J. 1974. Packing-induced radial particle-size segregation: Influence on hydrodynamic dispersion and water transfer measurements. *Soil Sci. Soc. Am. J.* 38:219–22

Rose, C. W. 1976. Watershed management on range and forest land. *Proc. US–Aust. Rangelands Panel*, ed. H. F. Heady, pp. 83–94. Utah State Univ.

Rose, C. W. 1979. Field oriented models of one-dimensional transport of a sorbed solute. *Res. Coord. Meet. FAO/IAEA/GSF.* No. 1.15. 16 pp.

Saffigna. P. G., Keeney, D. R., Tanner, C. B. 1977. Lysimeter and field measurements of chloride and bromide leaching in an uncultivated loamy sand. *Soil Sci. Soc. Am. J.* 41:478–82

Sawhney, B. L., Parlange, J.-Y. 1976. Radial movement of saturated zone under constant flux: Theory and application to the determination of soil-water diffusivity. *Soil Sci. Soc. Am. J.* 40: 635–39

Scotter, D. R. 1974. Salt and water movement in relatively dry soil. *Aust. J. Soil Res.* 12:27–35

Skopp, J., Warrick, A. W. 1974. A two-phase model for the miscible displacement of reactive solutes in soils. *Soil Sci. Soc. Am. Proc.* 38:545–50

Smiles, D. E. 1978. Transient and steady flow experiments testing theory of water flow in saturated bentomite. *Soil Sci. Soc. Am. J.* 42:11–14

Smith, R. E. 1972. The infiltration envelope: Results from a theoretical infiltrometer. *J. Hydrol.* 17:1–21

Smith, R. E., Parlange, J.-Y. 1977. Optimal prediction of ponding. *Trans. ASAE* 20:493–96

Smith, R. E., Parlange, J.-Y. 1978. A parameter-efficient hydrological infiltration model. *Water Resour. Res.* 14:533–38

Sposito, G. 1973. Volume change in swelling clays. *Soil Sci.* 115:315–20

Sposito, G. 1975. On the differential equation for the equilibrium moisture profile in swelling soil. *Soil Sci. Soc. Am. Proc.* 39:1053–56

Starr, J. L., Parlange, J.-Y. 1975. Nonlinear denitrification kinetics with continous flow in soil columns. *Soil Sci. Soc. Am. J.* 39:875–80

Starr, J. L., Parlange, J.-Y. 1976a. Relation between the kinetics of nitrogen transformation and biomass distribution in a soil column during continous leaching. *Soil Sci. Soc. Am. J.* 40:458–60

Starr, J. L., Parlange, J.-Y. 1976b. Solute dispersion in saturated soil columns. *Soil Sci.* 121:364–72

Starr, J. L., Parlange, J.-Y. 1977. Plate-induced tailing in miscible displacement experiments. *Soil Sci.* 124:56–60

Starr, J. L., De Roo, H. C., Frink, C. R., Parlange, J.-Y. 1978. Leaching characteristics of a layered field soil. *Soil Sci. Soc. Am. J.* 42:386–91

Topp, G. C. 1971. Soil water hysteresis in silt loam and clay loam soils. *Water Resour. Res.* 7:111–27

Wooding, R. A., Morel-Seytoux, H. J. 1976. Multiphase fluid flow through porous media. *Ann. Rev. Fluid Mech.* 8:233–74

Ann. Rev. Fluid Mech. 1980. 12: 103–38

ANALYSIS OF TWO-DIMENSIONAL INTERACTIONS BETWEEN SHOCK WAVES AND BOUNDARY LAYERS

T. C. Adamson, Jr. and A. F. Messiter
Department of Aerospace Engineering, The University of Michigan, Ann Arbor, Michigan 48109

INTRODUCTION

If the presence of a boundary layer could be ignored, the intersection of a shock wave with a solid boundary would imply nothing more complicated than a discontinuity in surface pressure and perhaps an irregular reflection. In reality, of course, the fluid velocity decreases to zero at the wall through a laminar or a turbulent boundary layer, and an interaction occurs between the shock wave and the boundary layer. The velocity profile in the boundary layer is thereby altered, as is the wave pattern in the external flow. The changes can be local if the shock wave is very weak but are observed on a larger scale if the shock is strong enough to cause separation. The interaction can occur in the presence of an incident oblique or normal shock wave or can be caused by an irregularity in wall shape, such as a corner or a step. Some typical interactions are sketched in Figures 1*a,b,c*.

Seventy-five years have passed since Prandtl (1904) first explained how a fluid motion is influenced by small viscosity. He recognized that viscous forces are important only in thin layers and that a global kind of interaction occurs as a result of flow separation: the pressure in the boundary layer along a solid surface is determined by an external inviscid flow, which in turn is influenced by regions of separated flow bounded by thin shear layers. At supersonic speeds the corresponding alteration of a potential flow because of boundary-layer separation was observed forty years ago in wind-tunnel tests by Ferri (1939). Deflection of a trailing-edge control surface caused a weak oblique shock wave well ahead of the

0066-4189/80/0115-0103$01.00

hinge line, so that the flow field was quite different from that predicted by inviscid-flow theory. Prandtl's brief remark concerning interaction of a separated flow with an external flow at low speeds was reiterated in greater detail by Oswatitsch & Wieghardt (1942), who also showed how a local interaction can occur at supersonic speeds, noting that a local pressure increase and boundary-layer thickening can reinforce each other. This effect was demonstrated in a calculation based on an integral

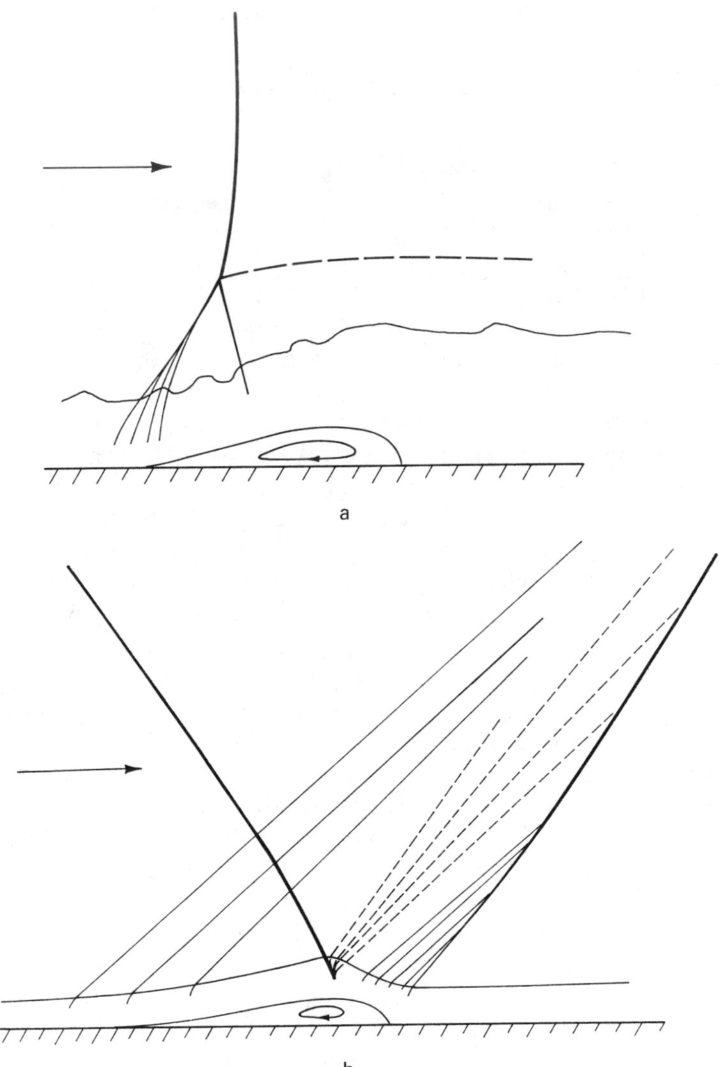

method, which included the important distinction that the local increase in displacement thickness for a turbulent boundary layer depends primarily on pressure forces, whereas for a laminar boundary layer both pressure and viscous forces play an essential role.

The first systematic investigations of the interaction between a boundary layer and a shock wave were carried out by Ackeret et al (1946) and by Liepmann (1946). Each set of experiments simulated the embedded supersonic region that appears in flow past an airfoil at high subsonic speeds, with a weak normal shock wave at a curved surface, and showed pronounced differences between laminar and turbulent interactions. Shortly afterward Fage & Sargent (1947) carried out a study of the interaction of a turbulent boundary layer with both oblique and normal shock waves. For an incident oblique shock wave and for a compression corner Liepmann et al (1949) again found marked differences between laminar and turbulent cases. Interactions resulting from a change in wall shape were emphasized in a later series of tests by Chapman et al (1957), which compared laminar, transitional, and turbulent separation caused by a ramp or a step as well as by an incident oblique shock wave. Results of the early investigations and of the many studies of the 1950s were summarized in a review article by Pearcey (1961), which dealt especially with the prevention of separation on airfoils.

As a result of these studies, certain general features of interacting flows

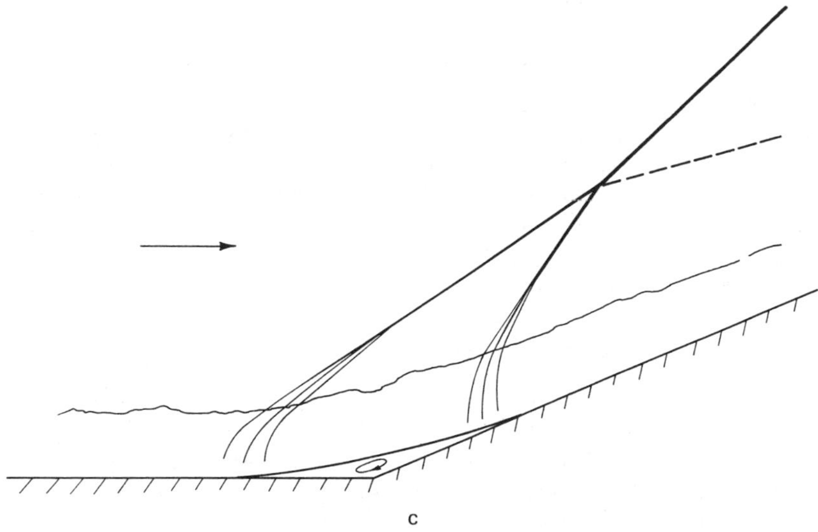

Figure 1 Typical interactions between shock waves and boundary layers: (*a*) incident normal shock wave; (*b*) incident oblique shock wave; (*c*) compression ramp or corner.

gradually became clear. A weak shock wave can extend into the boundary layer, with changing strength and inclination, but must end in the supersonic part of the flow, so that the pressure at the wall is necessarily continuous. The thickening of the subsonic part of the boundary layer extends somewhat upstream as well as downstream, and the resulting outward displacement of the streamlines causes local changes in the external flow. For laminar flows only a small pressure rise is required for separation and the separation point rapidly moves forward as the strength of the disturbance is increased. The pressure rise to separation and the constant "plateau" pressure observed downstream of the small separation region are independent of the manner in which the disturbance is introduced ("free interaction"). For turbulent flow, a larger pressure rise is required for separation and the separation point moves upstream more slowly as the disturbance becomes stronger. The pressure rise to separation is again nearly independent of the nature of the disturbance, but beyond the separation point some dependence on downstream conditions tends to persist. A careful and detailed description of many different types of interaction has been given in the comprehensive review article by Green (1970).

Early theoretical studies were undertaken in two quite different directions. Lees (1949) used a Pohlhausen integral method for calculation of pressures in the interaction of a laminar boundary layer with an incident oblique shock wave. Integral methods developed later, with different profiles and additional integral conditions, have provided satisfactory accuracy in a variety of flow problems for parameter ranges of practical interest. Much of this work is reviewed in the survey by Green (1970). Following some other early attempts at formulating approximate descriptions of these flows, Lighthill (1953) recognized that a local interaction can be described largely by inviscid-flow equations, except in a sublayer having thickness much smaller than the overall boundary-layer thickness. This important observation was the first step toward the later development of an asymptotic local-interaction theory for laminar boundary layers. Related derivations for turbulent boundary layers show some essential differences from the laminar-flow case.

Since the time of Green's (1970) review, new analytical, numerical, and experimental investigations have added substantially to our available information about these interactions. Asymptotic flow descriptions, although not always providing satisfactory numerical accuracy, have helped to identify the length scales for which different physical effects are essential, have shown the primary dependence of the flow variables on the parameters, and have allowed the calculation of some universal solutions. Advances in numerical methods and reductions in computing times have led to the possibility of obtaining accurate numerical solutions to the complete Navier-Stokes equations at reasonable cost. Experimental results have been obtained for increasingly broad ranges of important parameters; of special interest are new measurements at high Reynolds numbers. A review by Stanewsky (1973) has added to Green's (1970) survey with greater experimental detail. Sirieix (1975) has given a general review of turbulent separation, including descriptions of physical features, experimental results, and prediction methods for interactions with shock waves. Hankey & Holden (1975) have reviewed a wide variety of theoretical and experimental studies relating to boundary-layer interactions at high speed.

The present review attempts to supplement these earlier surveys, emphasizing the significant analytical and, to a lesser extent, numerical developments of the past ten years and commenting rather briefly on recent experimental work. While it has seemed appropriate to recall some of the earliest contributions, the subject has been of interest for about forty years, and the articles of Pearcey (1961) and Green (1970) are relied upon as summaries of most of the previous work. In addition, no attempt is made to consider unsteady flow or to describe applications to airfoils or to other aircraft components such as control surfaces or engine inlets; some of the important areas of application have been discussed in a survey by Korkegi (1971). Finally, the study of three-dimensional interactions has begun in earnest, but is not considered in this review.

LAMINAR BOUNDARY LAYERS

Asymptotic Theory of Free Interactions

Many aspects of laminar interactions can be discussed in terms of approximations which, following Van Dyke (1975), might be called rational. The boundary-layer equations for steady flow are, of course, derived from the steady-state Navier-Stokes equations in a particular limit as the Reynolds number Re becomes large, where Re is based on local external-flow properties and on the boundary-layer length L. At values of Re that are large but for which the flow still remains laminar, higher-order terms typically are small, and solutions to the first-order equations agree well with experiment. These solutions are, however, not uniformly valid in regions where abrupt changes in boundary conditions occur, as in the neighborhood of a shock wave. One is then led to seek different limit processes which might lead to different sets of equations capable of describing the flow accurately in regions having larger longitudinal gradients. Results derived in this way have been reviewed in broader contexts by Stewartson (1974) and by Messiter (1979).

The separation of a supersonic laminar boundary layer through a free interaction was first studied from this viewpoint independently by Stewartson & Williams (1969) and by Neiland (1969). Some of the basic ideas were also partially contained in earlier work, in particular of Lighthill (1953) and Gadd (1957). Near the wall the undisturbed velocity profile is linear, and it is here, in a sublayer where the velocity is small, that the largest relative velocity change occurs as the result of a pressure increase. Changes in viscous forces likewise are most important near the wall. Deceleration of the slow-moving fluid implies a thickening of streamtubes in the sublayer and therefore a corresponding displacement of streamlines in the external flow. Outside the boundary layer, the local pressure rise p, nondimensional with twice the undisturbed dynamic pressure, is directly proportional to the streamline inclination angle θ. These considerations lead to a self-consistent asymptotic description with, in a first approximation, $\partial p/\partial Y = 0$ everywhere and $\partial\theta/\partial Y = 0$ except in the sublayer. The interaction length is $O(\mathrm{Re}^{-\frac{3}{8}}L)$ and the sublayer thickness is $O(\mathrm{Re}^{-\frac{5}{8}}L)$, somewhat larger and somewhat smaller, respectively, than the boundary-layer thickness $O(\mathrm{Re}^{-\frac{1}{2}}L)$.

The complete asymptotic formulation involves a three-layer ("triple-deck") structure, expressed in terms of coordinates $\mathrm{Re}^{\frac{3}{8}}Y$ for the local external flow, $\mathrm{Re}^{\frac{1}{2}}Y$ for the main part of the boundary layer, and $\mathrm{Re}^{\frac{5}{8}}Y$ for the sublayer, where Y is the coordinate normal to the wall, nondimensional with L. The Navier-Stokes equations are studied in the limit as $\mathrm{Re} \to \infty$ with $\mathrm{Re}^{\frac{3}{8}}X$ and, successively, each of the three Y coordinates held fixed. Asymptotic representations of the flow variables can be written for each of these three limit processes. As an end result one obtains the incompressible boundary-layer equations in the coordinates $\mathrm{Re}^{\frac{3}{8}}X$ and $\mathrm{Re}^{\frac{5}{8}}Y$, for a temperature equal to the wall temperature. The no-slip condition is satisfied at $\mathrm{Re}^{\frac{5}{8}}Y = 0$; as $\mathrm{Re}^{\frac{5}{8}}Y \to \infty$, the velocity is linear and the pressure perturbation is proportional to the flow deflection angle. The equations might be called asymptotic local-interaction equations or, in the present context, free-interaction equations. Dependence on the parameters can be shown explicitly in the independent and dependent variables, such that the final problem formulation contains no parameters and the solution is a universal solution. In particular, the pressure change in a free interaction has the form

$$p \sim \frac{\lambda^{\frac{1}{2}}C^{\frac{1}{4}}}{(M_e^2-1)^{\frac{1}{4}}}\,\mathrm{Re}^{-\frac{1}{4}}p_1(x) \tag{1}$$

where

$$x = \mathrm{Re}^{\frac{3}{8}} \frac{\lambda^{\frac{5}{4}}(M_e^2 - 1)^{\frac{3}{8}}}{C^{\frac{3}{8}}(T_w/T_e)^{\frac{3}{2}}} X \tag{2}$$

and X is the coordinate measured along the wall with origin, say, at the separation point and made nondimensional with L. The external-flow Mach number is M_e; the ratio of wall temperature to external-flow temperature is T_w/T_e; and the slope of the undisturbed boundary-layer velocity profile at the wall is λ, equal to 0.332 for the Blasius solution. The Chapman viscosity law $\mu/\mu_e = CT/T_e$ has been assumed.

Separation through a free interaction can occur ahead of, for example, an incident oblique shock wave, a compression corner, or a forward-facing step. The separation point occurs forward of the disturbance at a distance which is large in comparison with the free-interaction length. However, the differential equations also remain the same if the strength of the disturbance is smaller, such that the pressure changes are $O(\mathrm{Re}^{-\frac{1}{4}})$ and the overall interaction length, including both separation and reattachment, is decreased to $O(\mathrm{Re}^{-\frac{3}{8}}L)$. For a still weaker disturbance with $p \ll \mathrm{Re}^{-\frac{1}{4}}$, the differential equations become linear, the length scale remaining $O(\mathrm{Re}^{-\frac{3}{8}}L)$. In any of these latter cases, the interaction is a local one. If the overall pressure change is $O(\mathrm{Re}^{-\frac{1}{4}})$, the solution for p again has the form given in Equations (1) and (2), with differences in geometry taken into account through differences in the boundary conditions and therefore different solutions for the function $p_1(x)$.

For flows with separation occurring well ahead of the imposed disturbance, numerical solution of the asymptotic free-interaction equations has been carried out with an added assumption about the backflow (Neiland 1971, Stewartson & Williams 1973, Williams 1975). Downstream of separation, as $\mathrm{Re}^{\frac{3}{8}}X \to \infty$, the separated shear layer moves away from the wall at an angle $O(\mathrm{Re}^{-\frac{1}{4}})$, with velocity profile close to the undisturbed profile except near the separation streamline $\psi = 0$. In a thin sublayer about $\psi = 0$ the flow is described in a first approximation as the merging of a uniform shear flow with fluid at rest. A low-speed backflow is considered to supply just the mass required for entrainment in the sublayer. The scaled values of plateau pressure, pressure at separation, and pressure gradient at separation have been given, respectively, as $p_1(\infty) = 1.800$, $p_1(0) = 1.026$, and $p_1'(0) = 0.263$ (Stewartson & Williams 1969, Williams 1975, Brown & Williams 1975). It would also be desirable to study the low-speed recirculating flow in a complete interaction as $\mathrm{Re} \to \infty$ in order to ascertain the manner in which downstream conditions might influence the flow near separation through terms that are presumably of higher order.

Second-order terms in expansions of the flow variables as $\mathrm{Re} \to \infty$ are smaller only by a factor $O(\mathrm{Re}^{-\frac{1}{8}})$, and may therefore be numerically important. The major effects appearing in the corrections are the variations in pressure and streamline deflection across the main part of the boundary layer. At $M_e = 2.0$ and $\mathrm{Re} \approx 10^5$, the first approximation overpredicts the separation pressure p_s and the pressure gradient $(dp/dX)_s$ at separation; a two-term approximation (Brown & Williams 1975) improves p_s but overcorrects $(dp/dX)_s$. It is clear also, from Equations (1) and (2), that (Neiland 1970a, Brown et al 1975) the hypersonic interaction parameter χ has been assumed small; for the assumed viscosity law, $\chi = C^{\frac{1}{2}} M_e^3 \mathrm{Re}^{-\frac{1}{2}}$. In a limit such that $\mathrm{Re} \to \infty$, $M_e \to \infty$, and $\chi \to 0$, analytical results show explicitly that the numerical coefficient of the second-order term is fairly small for p_s but rather large for $(dp/dX)_s$.

Neiland (1973) has noted that the interaction length can remain small at higher Mach numbers if the wall is highly cooled, but that at the same time the streamtube divergence in the main part of the boundary layer becomes important. Again, for a linear viscosity law, the results for small χ give the length as $\Delta X = O(\chi^{\frac{3}{2}} g_w^{\frac{3}{2}})$ and the first terms in the pressure as $O(\chi^{\frac{1}{2}})$ and $O(\chi^{\frac{3}{4}} g_w^{-\frac{3}{2}})$, where $g_w = T_w/T_o$ and T_o is the stagnation temperature in the external flow. The subsonic part of the boundary layer near the wall can become thin enough that a small increase in pressure causes a decrease, rather than the more typical increase, in displacement thickness. The mutual interference of increases in p and θ expected for an interaction can then no longer occur. For this "supercritical" boundary layer, Neiland (1973) has formulated a local flow description with interaction length now of the same order as the boundary-layer thickness, obtaining inviscid-flow equations with $\partial p/\partial Y \neq 0$ in the main part of the boundary layer and boundary-layer equations in a sublayer. Further elaboration of these ideas, including comparison of numerical solutions with boundary-layer and Navier-Stokes solutions, would seem desirable.

The formulation of Equations (1) and (2) likewise fails when M_e is near 1 (Messiter et al 1971). The result $\theta = O(p^{\frac{3}{2}})$ of transonic small-disturbance theory must then replace the linear pressure-angle relation for the external flow when $M_e - 1 = O(\mathrm{Re}^{-\frac{1}{5}})$. The derivations are otherwise unchanged, and it is found that the interaction length is $O(\mathrm{Re}^{-\frac{3}{10}} L)$, with the pressure rise for separation given by $p = O(\mathrm{Re}^{-\frac{1}{5}})$. Numerical solutions of the transonic free-interaction problem have been obtained by Bodonyi & Kluwick (1977).

Incident Oblique Shock Waves

As shown in Equation (1), the free-interaction theory predicts $p \sim$ (const.) $(M_e^2 - 1)^{-\frac{1}{4}} \mathrm{Re}^{-\frac{1}{4}}$ at separation. It follows that a weak incident oblique shock wave at a plane wall causes separation if $|\alpha| = (\text{const.})(M_e^2 - 1)^{\frac{1}{4}}\text{Re}^{-\frac{1}{4}}$, where α is the turning angle, and the constant factor can be found in principle by solution of the local-interaction equations. If the shock wave is somewhat stronger ($|\alpha| \gg \text{Re}^{-\frac{1}{4}}$ in the asymptotic theory), several flow regions can be distinguished. The pressure p initially increases in a free-interaction region containing the separation point, reaches a constant "plateau" value of order $\text{Re}^{-\frac{1}{4}}$, and rises to its final value through a reattachment region downstream of the shock impingement point. The separated shear layer ahead of the shock wave has thickness $O(\text{Re}^{-\frac{1}{2}}L)$ and is inclined to the wall at a constant angle of order $\text{Re}^{-\frac{1}{4}}$. Below this layer is a region of low-speed recirculating flow. The shock wave undergoes a continuous reflection within the supersonic part of the separated shear layer, over a short distance of order $\text{Re}^{-\frac{1}{2}}L$. Since no abrupt pressure change is possible below, the reflected wave observed in the external flow is an expansion. The shear layer is turned toward the wall through an angle of approximately 2α, and so the compression at reattachment occurs through an angle of about 2α. The fluid above the separation streamline continues downstream, while the small amount of fluid entrained in the shear layer from below is turned back by the pressure rise immediately ahead of the reattachment point. The outgoing compression and expansion eventually combine to form the reflected shock wave of inviscid-flow theory. Most of these features are indicated in Figure 1*b*; a sketch of this kind was first given by Liepmann et al (1949).

If it is assumed that the complete interaction can be described by boundary-layer equations, $\partial p/\partial Y$ is of course neglected, but $p(X)$ is to be determined. Typically a simple-wave pressure-angle relation is imposed at a displacement surface or at some "edge" of the boundary layer. The shock wave is usually considered to reflect as a centered expansion; thus p is continuous but θ and dp/dX have jumps, and details of an "inner" solution for $\Delta X = O(\text{Re}^{-\frac{1}{2}})$ are thereby ignored. One parameter remains undetermined after this jump, and an iteration is carried out until a suitable boundary condition is satisfied far downstream. Setting $\partial p/\partial Y = 0$ throughout the interaction is consistent with an asymptotic formulation as $\text{Re} \to \infty$, in which $\partial p/\partial Y$ would be of higher order everywhere outside the "inner" region; this feature of reattachment is noted later. A summary describing some details of various calculation methods has been given by Hankey (Hankey & Holden 1975, Part I).

The integral method of Lees & Reeves (1964) is derived from the boundary-layer equations and permits approximate calculation of the changes in wall pressure, displacement thickness, and skin friction. In the earliest formulation an adiabatic wall was considered and the total enthalpy was assumed uniform. Two ordinary differential equations are provided by integrals of the momentum equation and its product with the X-component of velocity, u. A third follows from the coupling condition with the external flow, with an expression for θ found from integration of the continuity equation. An assumed one-parameter family of velocity profiles that permit reversed flow is taken from the Stewartson and Cohen-Reshotko similarity solutions of the compressible boundary-layer equations. Calculations with heat transfer were carried out by Holden (1965) and by Klineberg & Lees (1969), by essentially the same methods but with the addition of an energy integral and a one-parameter family of enthalpy profiles.

A second widely used integral method is based on the method of integral relations (e.g. Nielsen et al 1966, 1969, Holt 1966). Here $\partial u/\partial Y$ and the total temperature are written as functions of u rather than of Y, with coefficients as functions of X to be determined. The accuracy improves as the number of parameters is increased in the chosen profiles, with additional moments of the momentum equation supplying the required number of integral conditions; in practice, however, the complexity increases rapidly.

A finite-difference solution of the boundary-layer equations has been obtained by Reyhner & Flügge-Lotz (1968). The simple-wave matching with the external flow was used at a distance from the wall large enough so that dissipative effects were very small. At this "edge" of the boundary layer the reflection of the shock wave was again represented by a jump in θ. Stability of the calculations in the region of reversed flow was achieved by omitting (or else multiplying by -0.1) the small convection terms in this region. An approximate method of including some effects contained in the boundary-layer formulation but not in the first-order asymptotic theory was proposed by Tu & Weinbaum (1976). Streamtube divergence in the main part of the boundary layer was taken into account through the integral of the continuity equation, and density changes in the sublayer were allowed by an integral method, without the requirement that the sublayer be very thin.

Except close to the shock wave, the boundary-layer equations appear to give a correct asymptotic approximation as $\mathrm{Re} \to \infty$, in the sense that equations obtained by various limit processes (i.e. for various flow regions) as $\mathrm{Re} \to \infty$ would contain no terms not already included in the boundary-layer equations. The largest of the higher-order terms in p are those associated with streamline curvature near separation and reattachment and are of order $\mathrm{Re}^{-\frac{3}{8}}$. A solution of these equations obtained by an integral method, however, can be said to provide a rational approximation (Van Dyke 1975) only if a systematic procedure is available for deriving successive approximations; this is true in principle for the method of integral relations. A comparison shown by Murphy (1969) indicates that results of the two major integral methods are in close agreement with those of the finite-difference calculations at moderate Mach numbers, say between $M_e = 2.0$ and $M_e = 4.0$. Comparisons with measured pressures given there and elsewhere suggest good agreement in this range, although the length of the interaction is somewhat over-predicted; Holden (1965) has also shown a favorable comparison at higher Mach number. One can not, of course, predict with confidence the accuracy of the integral methods outside the ranges of Reynolds number, Mach number, shock-wave strength, and wall temperature ratio for which they have been checked.

Numerical solutions of the time-dependent Navier-Stokes equations were obtained by MacCormack (1971), with a two-step finite-difference procedure having second-order accuracy. A finer mesh was used near the plate, and the difference equations were "split" into two sets of one-dimensional equations. The initial flow was uniform, with a discontinuous θ imposed along a line parallel to the wall and at a suitably large distance. More recently (MacCormack 1976) the method has been modified so as to decrease drastically the computing time required. The Y difference operator is now replaced by two operators, one for convection and one for diffusion terms, each allowing time steps far larger than were previously possible. Results obtained were in close agreement with measurements of Hakkinen et al (1959). A related method, implicit-explicit as is MacCormack's (1976) but without the splitting of convection and diffusion terms, has been developed by Shang (1978) and likewise shows substantial reduction in computing time. Still other calculations of boundary-layer interactions with an incident oblique shock wave have been carried out by Li (1976) and by Beam & Warming (1978).

If the Mach number M_e is close to one, the boundary-layer formulations have to be modified to allow coupling with an external flow described by transonic small-disturbance theory. In particular, as already noted, separation occurs for $p \sim \mathrm{Re}^{-\frac{1}{5}}F$, where F is a function of the parameter $(M_e^2 - 1)\mathrm{Re}^{\frac{1}{5}}$. An analytical solution for a weak incident oblique shock wave, such that $(M_e^2 - 1)Re^{\frac{1}{5}} \ll 1$, has been given by Brilliant & Adamson (1974).

Corners and Steps

As for an incident shock wave, it follows from Equation (1) that separation of the supersonic boundary-layer flow past a compression corner is expected to occur when $p = O(Re^{-\frac{1}{4}})$, i.e. for a corner angle $\alpha = O(\mathrm{Re}^{-\frac{1}{4}})$. If α increases further, the separation point moves forward, and a pressure plateau appears between separation and reattachment. The pressure first rises through a free interaction to a plateau value $O(\text{Re}^{-\frac{1}{4}})$ and the remaining increase, of order α, occurs in a region near reattachment. The overall pressure distribution thus resembles that for an incident oblique shock wave which turns the flow through an angle $\frac{1}{2}\alpha$ (the turning angle at reattachment is α).

Corner-flow solutions based on integral methods have been given by many authors, including Holden (e.g. Hankey & Holden 1975, Part II), Nielsen et al (1969), Klineberg & Lees (1969), Georgeff (1974), and Holt & Lu (1975). Results have been shown for supersonic and hypersonic Mach numbers and for adiabatic and cooled walls. For integral methods in which a dimensionless boundary-layer thickness δ is one of the dependent variables, the boundary layer can become "supercritical" in the sense that $d\delta/dp < 0$. In these cases Klineberg & Lees (1969) introduced a "supercritical-subcritical jump;" then separation occurs slightly downstream in a region where $d\delta/dp > 0$. Holden (1969) has removed this difficulty by adding an integral of the y-momentum equation to the set of integral conditions satisfied. Stollery & Hankey (1970) suggested that the difficulty arose when the moment of momentum equation was introduced and is removed if the kinetic-energy thickness is taken proportional to the momentum thickness. Crocco (1975) has also discussed the possibility of improved integral methods.

Werle & Vatsa (1974) have obtained numerical solutions of the boundary-layer equations with interaction by a relaxation procedure that uses an artificial time-like term introduced through the pressure gradient. The results for wall pressure in corner flows are generally in close agreement with Navier-Stokes solutions of Carter (1972) and with experimental results of Lewis et al (1968). Hung & MacCormack (1976) have carried out numerical solutions of the Navier-Stokes equations, using MacCormack's (1971) method with some modifications; Carter (1972) also solved the time-dependent equations, using another difference scheme. Rizzetta et al (1978) have obtained numerical solutions of the time-dependent form of the asymptotic local-interaction equations for corner flows. The initial flow was taken to be the undisturbed uniform shear flow and the ramp angle α was increased impulsively at time $t = 0$. Incipient separation is predicted for $\alpha = 1.57\lambda^{\frac{1}{2}}(M_e^2 - 1)^{\frac{1}{4}}(\text{Re}/C)^{-\frac{1}{4}}$. The results are found to approach the boundary-layer solutions of Werle & Vatsa (1974) downstream of the corner only as the Reynolds number is increased to unrealistically large values of about 10^8; at these values the upstream pressure still rises a little too early, so that the pressure at separation is overpredicted.

Burggraf et al (1979) have recently extended these comparisons. Their solutions of the "interacting boundary-layer equations" agree well with available experimental data and Navier-Stokes solutions at moderate Reynolds number and with asymptotic solutions at very high Reynolds number. In an asymptotic sense these equations represent a composite set of equations containing the largest of the effects that are of higher order in the asymptotic theory. Of special interest is the observation that satisfactory combination of numerical accuracy and computational efficiency was achieved only if mesh size was scaled according to the results of the asymptotic theory.

The asymptotic form of the flow at larger ramp angles, for $\mathrm{Re}^{-\frac{1}{4}} \ll \alpha \ll 1$, has been described by Burggraf (1975). Since the relative length L_{p}/L of the separated shear layer remains small, the velocity profile changes only slightly, except near the separation streamline $\psi = 0$. Here the similarity solution for the merging of a uniform shear flow with fluid at rest gives $u = O(L_{\mathrm{p}}^{\frac{1}{3}}/L^{\frac{1}{3}})$ in a sublayer of thickness $\Delta Y = O(\mathrm{Re}^{-\frac{1}{2}} L_{\mathrm{p}}^{\frac{1}{3}}/L^{\frac{1}{3}})$ just upstream of reattachment. The pressure rise along $\psi = 0$ at reattachment is, in the limit, the same as if the compression were isentropic, and is of the same order as the pressure rise in the local external flow. The two values are $O(L_{\mathrm{p}}^{\frac{2}{3}}/L^{\frac{2}{3}})$ and $O(\alpha)$, and so $L_{\mathrm{p}}/L = O(\alpha^{\frac{3}{2}})$. The same estimate of overall interaction length would be found for interaction with an incident oblique shock wave.

With the help of these results for the orders of magnitude it is possible to outline some general features of an asymptotic description of reattachment. Just ahead of the reattachment region, the sublayer about $\psi = 0$ has thickness $O(\mathrm{Re}^{-\frac{1}{2}}\alpha^{\frac{1}{2}}L)$, with fluid velocity $O(\alpha^{\frac{1}{2}})$. The continuity equation in von Mises variables then gives the reattachment interaction length directly as $O(\mathrm{Re}^{-\frac{1}{2}}\alpha^{-\frac{1}{2}}L)$. The momentum equation shows that the sublayer flow is described in a first approximation as the deceleration and turning of an inviscid rotational flow, and viscous forces are important only in a still thinner sublayer. The pressure change is found to be $O(\alpha^{\frac{3}{2}})$ across the main part of the shear layer. Since this is smaller than the overall pressure rise, $\partial p/\partial Y$ remains zero in the first approximation. A nonlinear integral equation can be derived for the pressure rise as a function of X (Messiter et al 1973). For consistency between the changes in pressure and in displacement thickness, Neiland (1970b) argued that the stagnation point occurs at infinity on the reattachment length scale, so that the pressure rise along the separation streamline equals the entire pressure rise in the external flow, consistent with the inviscid-flow description of reattachment proposed by Chapman et al (1957). This conclusion also follows from numerical (Burggraf 1973) and approximate analytical (Messiter et al 1973) solutions. These results, of course, give only the first term in an asymptotic expansion for each of the flow variables, and neglected terms may not be numerically small. A more complete study is needed to clarify the details of the approximation.

Other closely related types of local interaction occur if a boundary layer along a plane wall encounters a very shallow forward-facing or backward-facing step of height h. If h is of the same order of magnitude as the sublayer thickness $O(\mathrm{Re}^{-\frac{5}{8}}L)$ in the local-interaction theory, the asymptotic description has the same form as for a free interaction, but with modified wall boundary conditions and with an additional parameter $\mathrm{Re}^{\frac{5}{8}}h/L$. As this parameter becomes large, a pressure plateau again develops between separation and reattachment. For a forward-facing step, separation is expected to occur through a free interaction at a distance $O(h\,\mathrm{Re}^{\frac{1}{4}})$ ahead of the step, so that the plateau pressure remains $O(\mathrm{Re}^{-\frac{1}{4}})$, independent of h when $\mathrm{Re}^{\frac{5}{8}}h/L \gg 1$. For a backward-facing step, on the other hand, the nondimensional base pressure p_b is related to the pressure change at reattachment of the separated shear layer, which is described in the same manner as for flow over a compression ramp. This pressure rise is proportional to the shear-layer inclination angle h/L_p and also to $(L_p/L)^{\frac{2}{3}}$, the square of the velocity at the separation streamline. Therefore (Messiter et al 1973) $L_p/L = O\{(h/L)^{\frac{3}{5}}\}$ and $p_b = O\{(h/L)^{\frac{2}{5}}\}$, independent of Reynolds number in a first approximation; the second term in the expansion of p_b is $O(\mathrm{Re}^{-\frac{1}{4}})$. The shear-layer profile differs slightly from that for a ramp because of the small initial velocity along $\psi = 0$ resulting from the expansion through an angle $\sim h/L_p$ at the corner. The description of reattachment is consistent with Chapman's except for the estimate of the maximum velocity along $\psi = 0$. Chapman et al (1957) assumed that the change in shear-layer thickness is much larger than the initial thickness, whereas the asymptotic description shows that the opposite is true. The proper Mach-number dependence is also easily included; the derivation implies that the hypersonic interaction parameter χ should be small and that a transonic parameter $(M_e^2 - 1)(h/L)^{-\frac{1}{3}}$ should be large. Base-pressure measurements correlated according to the theory lie very close to a single curve, with values for slender wedges displaced slightly below those for steps.

Calculations of the complete base flow are quite difficult at high Reynolds number because different effects are dominant in different parts of the flow. The boundary layer begins to accelerate somewhat upstream of the step, and the flow here is described in a first approximation as an inviscid rotational flow. This was first shown by Matveeva & Nieland (1967), the expansion occurring within a distance $O(\mathrm{Re}^{-\frac{1}{2}}\alpha^{-\frac{1}{2}}L)$. Their formulation is related to Stewartson's (1970) calculation for an expansion through an angle $|\alpha| \gg \mathrm{Re}^{-\frac{1}{4}}$ and is similar to that proposed above for the description of reattachment. The separation point has been both observed (Hama 1966) and predicted (Allen & Cheng 1970) to occur on the base somewhat below the corner. At higher Mach numbers a "lip" shock wave is observed to form within the separated shear layer just downstream of the corner. The pressure downstream remains nearly constant until the shear layer approaches the wall and the reattachment pressure rise begins. Webb et al (1965) attempted an application of integral methods in a two-part calculation, one for a downstream region containing the reattachment stagnation point and one for a region closer to the base where the shear layer is thin in comparison with the step height. Burggraf (1969) carried out a numerical calculation of the complete base flow as $\mathrm{Re} \to \infty$, so that the separated shear layer is treated as thin for its entire length. The recirculating flow is assumed to have uniform vorticity (Prandtl 1904, Batchelor 1956) and the pressure is taken to be constant; the latter approximation avoids premature appearance of separation in the thin boundary layer along the base. Solutions to the time-dependent Navier-Stokes equations have been given by Allen & Cheng (1970), Roache & Mueller (1970), and Ross & Cheng (1971). In the future it might be desirable to use Navier-Stokes solutions as functions of h/L and Re (or $L\delta/h$ and Re δ) for obtaining a more complete understanding of the flow details at high Reynolds number, without the assumption of constant pressure; numerical values might be inferred for the coefficients in a two-term expansion of, say, the maximum pressure in the recirculating flow as $\mathrm{Re} \to \infty$.

TURBULENT BOUNDARY LAYERS

While laminar interactions have been studied in detail and are now fairly well understood, the more difficult description of interactions between shock waves and turbulent boundary layers is of still greater interest because of the high Reynolds numbers encountered in most practical applications. In recent studies of turbulent interactions special attention has been devoted to the use of modern methods of calculation, both analytical and numerical. The problem is complicated, of course, by the fact that averaging of the governing equations introduces new unknowns. The number of available equations is then insufficient and a closure condition must be assumed, thus far in terms of a model for the eddy viscosity. The use of modern high-speed computers and systematic analytical methods has allowed detailed investigation of the consequences of choosing a simple equilibrium model as well as comparison of results obtained using models of varying complexity.

One of the more important results from recent studies is the realization that the interaction in a turbulent boundary layer cannot be treated with methods that are simple extensions of those used for laminar interactions.

Thus, insofar as analytical methods are concerned it is necessary to do more than allow for a different mean-velocity profile in the incoming flow and a larger diffusion coefficient; this has also been noted, e.g. by Holden (Hankey & Holden 1975, Part II). For laminar flow one finds a strong interaction in the sense that the flow external to and in the main part of the boundary layer is affected by the growth of a viscous sublayer so that external-flow and sublayer solutions must be found simultaneously; this situation occurs for separated or unseparated flows, no matter how weak the shock wave may be. For turbulent flow, on the other hand, the interaction can be considered as a weak interaction as long as the flow is unseparated. That is, in a first approximation, the outer layer, comprising the external flow and the main part of the boundary layer and described by inviscid-flow equations, is unaffected by flow changes in thin sublayers, which therefore may be neglected in calculating the pressure distribution through the interaction. Calculations based on the assumption that sublayers can be ignored for attached flow, described by Roshko & Thomke (1969), Elfstrom (1972), and Carrière et al (1975) for supersonic and hypersonic flow over a ramp, show good comparison with experiment. Confirmation of this assumption, using asymptotic methods of analysis, has been given by Melnik & Grossman (1974), Adamson & Feo (1975), and Messiter (1979), for example, in studies of normal and oblique shock waves impinging on a turbulent boundary layer over a flat plate in transonic flow. These studies are described further in subsequent sections.

This fundamental difference between the two cases is evidently caused by the basic difference in structure of the laminar and turbulent boundary layers. The turbulent boundary layer has a two-layer structure described by the law of the wall and the law of the wake as given by Coles (1956) for incompressible flow and by Maise & McDonald (1968) for compressible flow, with an extension of van Driest's (1951) solution. The thickness of the layer in which the velocity is small enough that deceleration of the fluid causes a significant relative increase in streamtube size is much smaller, compared to the boundary-layer thickness, in turbulent than in laminar flow. Consequently, the actual order of magnitude of the streamline deviation induced by the slowing of the fluid near the surface is so small that it may be neglected in lower orders of approximation. Of course, when the shock wave is strong enough that separation takes place, a strong interaction does occur. It should be noted that the two-layer structure of the boundary layer is not compatible with a power-law profile.

It is evident from the above remarks that in an interaction without separation the choice of eddy-viscosity model should have little effect on the calculated pressure distribution; its major impact is confined to solutions found for the flow in a sublayer. For example, it has a significant effect on calculation of the shear stress at the wall. This general result should and does carry through to numerical solutions, as evidenced in computations made by Viegas & Horstman (1978), Figure 2, for several different models for the eddy viscosity. Indeed, it appears also to be true in those cases where a small separation bubble exists (Viegas & Coakley 1977). In general, of course, flow separation and thus a strong

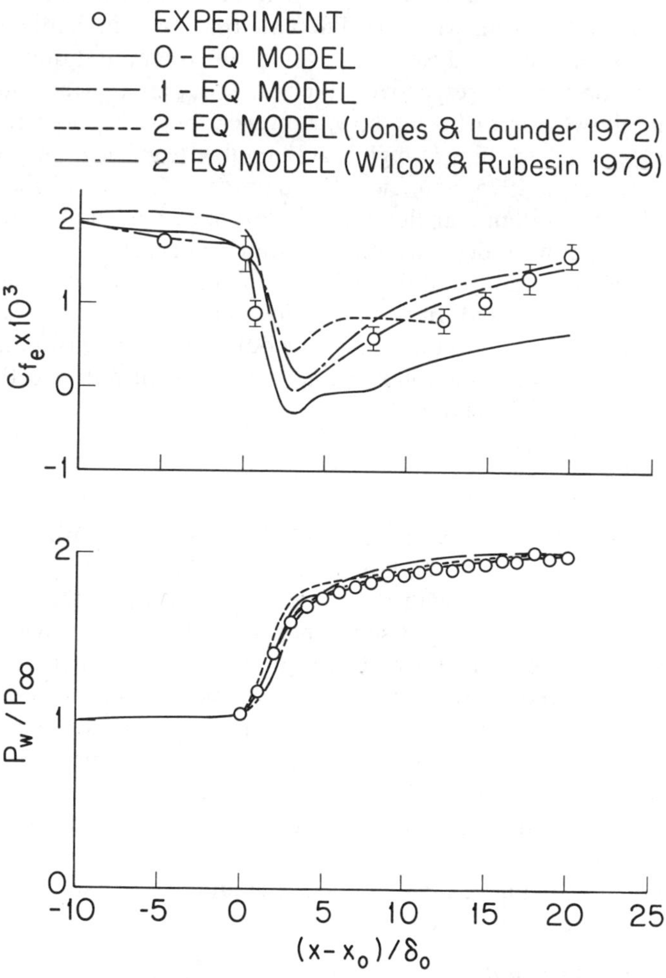

Figure 2 Comparison of skin-friction and surface-pressure measurements with computations for various eddy-viscosity models, for a normal shock wave and a turbulent boundary layer: $Re_\delta = 5.5 \times 10^5$, $M_e = 1.44$. From Viegas & Horstman (1978).

interaction implies that the turbulence model affects significantly the distribution of both the pressure and the shear stress at the wall. This is apparent in numerical computations made by Viegas & Coakley (1977) for separation induced by an incident shock wave at high Mach number; when the separation is extensive enough that a pressure plateau has developed, different models for the eddy viscosity result in quite different distributions for both variables.

The aforementioned two-layer structure in the turbulent boundary layer also leads to a Mach-number dependence for τ_w, the nondimensional shear stress at the wall, which is different from that in laminar flow. Thus, if subscripts u and d refer to positions upstream and downstream of the interaction region, respectively, then $\tau_{wd} < \tau_{wu}$ for all Mach numbers in laminar flow. For turbulent flow, however, $\tau_{wd} < \tau_{wu}$ for transonic flow, but $\tau_{wd} > \tau_{wu}$ for flow at high Mach number (e.g. Viegas & Coakley 1977, Marvin et al 1975, and Liou & Adamson 1979). The behavior of $\tau_w = (\mu\, \partial U/\partial Y)_w$ in laminar flow may be explained by noting that the viscosity at the wall depends on the wall temperature and is thus approximately invariant, and the velocity gradient decreases as the interaction region is traversed. In turbulent flow, however, the orders of the velocity gradient in the wall layer and the outer layer are vastly different and one must consider the following equation, which holds in the wall layer, to explain the behavior of τ_w:

$$-\rho\, \overline{U'V'} + \frac{\mu}{\mathrm{Re}} \frac{\partial U}{\partial Y} = \tau_w. \tag{3}$$

Here, fluid and flow properties are made dimensionless with respect to characteristic values from the flow external to the boundary layer and Y is again dimensionless with respect to the length L of the boundary layer. The notation is changed somewhat from that used previously, to conform with some of the references. Equation (3) is a statement of the fact that even in the interaction region the pressure gradient and inertia terms give higher-order corrections. As Y increases, say to a streamline somewhat outside the wall layer, the viscous stress becomes small compared with the Reynolds stress, $-\rho\, \overline{U'V'}$, which remains of the same order throughout the boundary layer. Along this streamline, as the flow is decelerated, $\overline{U'V'}$ decreases, but $\rho = PT^{-1}$, through the equation of state, increases. At high Mach numbers, the jump in pressure across the shock wave is very large and $\tau_{wd} > \tau_{wu}$.

Asymptotic Methods

Asymptotic methods of solution have been confined thus far to analyses involving shock waves impinging upon a turbulent boundary layer on a flat plate in transonic flow. Because of the important applications in airfoil theory and channel flows, the case where the shock wave is normal has received the most attention in spite of the complications introduced by the fact that the flow downstream of the shock wave is subsonic. When the flow is separated, further complications arise not only due to the presence of the separation bubble but also because the shock wave takes on the lambda configuration shown in Figure 1*a*, with significant variations in pressure in the flow external to the boundary layer (Seddon 1960, Vidal et al 1973, Kooi 1975, East 1976). Thus, the solutions obtained to date have been for those cases where the flow has not separated. In addition, the problem has been simplified further by the presumably unrestrictive assumptions of two-dimensional and steady flow; these approximations must be kept in mind when comparisons with experimental results are carried out.

A strong interaction, as in laminar flows, necessitates simultaneous consideration of mutually induced disturbances in both an outer layer and a sublayer. The use of asymptotic methods of solution permits the various layers in which different physical mechanisms are important to be clearly identified. The relationship between the pressure or pressure gradient and the fluid displacement is found in adjoining layers, and through matching the fluid displacements the inner and outer flows are coupled; the essence of the interaction is clearly demonstrated. On the other hand, when the boundary layer is turbulent and attached, use of asymptotic methods has shown that the interaction is weak, in the sense mentioned previously.

Solutions in which asymptotic methods have been employed, because they have been confined to transonic flows, can be characterized by the relative orders of two small parameters, ε and u_τ. Here $\varepsilon = U_e - 1$ is the dimensionless difference between the flow velocity and the critical speed of sound in the undisturbed flow external to the boundary layer; that is, in this section reference conditions are taken to be the critical conditions in the external flow. For transonic flow, then $\varepsilon \ll 1$. If M_e is the Mach number in the flow external to the boundary layer, then $\varepsilon = O(M_e^2 - 1)$. The parameter u_τ is defined as the dimensionless friction velocity associated with the undisturbed boundary layer at the point of impingement of the incident shock wave. As mentioned previously, the undisturbed boundary layer is taken to have the two-layer structure deduced from experiment; in asymptotic terms, then, two length scales are involved. The (outer) velocity-defect layer, where the velocity U differs from U_e by a term of order u_τ, has a dimensionless thickness $\delta = O(u_\tau)$; the thickness $\tilde{\delta} = O(\mathrm{Re}^{-1} u_\tau^{-1})$ of the (inner) wall layer, where the velocity is of order u_τ, is much smaller. Here Re is the Reynolds number written in terms of the length L and the critical conditions in the flow external to the boundary layer. For length scales intermediate to δ and $\tilde{\delta}$, $u_\tau \ll U \ll U_e$; a solution for U as a function of $u_\tau \ln Y$ in the intermediate range is consistent with these requirements, and matching with the solutions in the inner and outer layers leads to the result $u_\tau = O\{(\ln \mathrm{Re})^{-1}\}$.

The interaction of an oblique shock wave with a turbulent boundary layer on a flat plate was analyzed by Adamson & Feo (1975) for the case where the shock is weak, such that $\varepsilon \ll u_\tau$; then the sonic line in the undisturbed flow lies near the edge of the boundary layer. In independent work, Melnik & Grossman (1974) considered a stronger normal shock wave, with $\varepsilon = O(u_\tau)$; in this case, the sonic line is at an arbitrary location within the boundary layer. In recent work, Adamson & Messiter (1977), Messiter (1979), and Liou & Adamson (1979) have studied the case where the shock wave is normal with $u_\tau \ll \varepsilon \ll 1$, so that the sonic line is located deep within the boundary layer. Solutions were obtained in most of the flow field except for a small region near the foot of the shock wave having dimensions comparable with the distance from the wall to the upstream sonic line. This distance to the sonic line is an important parameter because the extent of the upstream influence is ordered by the thickness of the subsonic layer. Thus, for $\varepsilon \ll u_\tau$ the major part of the pressure increase in the interaction region takes place upstream of the impingement point of the shock wave. As ε/u_τ increases, the pressure rise downstream of the shock impingement point becomes larger until, for $\varepsilon \gg u_\tau$, it is the major part of the overall pressure rise; the upstream influence is so small that the problem may be simplified, to first order, to consideration of the undisturbed flow through a shock wave. Finally, an incident oblique shock wave, for $\varepsilon \gg u_\tau$, was considered by Adamson (1976).

An understanding of the manner in which the important mechanisms affect the flow field in an interaction may be gained from consideration of the structure found when asymptotic methods of analysis are employed. In the following, the incident shock wave is taken to be normal. The (dimensionless) horizontal length scale Δ associated with the interaction region is a function of ε and δ, where, again, $\delta = O(u_\tau)$. For weak waves, such that $\varepsilon \ll u_\tau$, $\Delta = O(\delta^2\varepsilon^{-\frac{1}{2}})$; in the remaining cases, $\Delta = O(\varepsilon^{\frac{1}{2}}\delta)$. In the direction normal to the wall, the interaction region is divided into three layers. In the outermost layer, which has a length scale of order δ and thus consists of the major part of the boundary layer and flow external to it, pressure forces are dominant over those associated with Reynolds or viscous stresses. Consequently, the flow field is described by the equations for inviscid flow. It is important to note that because Δ is large compared with the thickness of a shock wave, even as it penetrates the boundary layer and weakens to zero strength, the viscous structure of the shock wave is not an issue of importance in the analysis. Instead, as in the case of supersonic flow over a curved wall, the shock wave forms and strengthens as a result of the coalescence of weak compression waves (see Figure 3) that occur as a result of the upstream influence of the interaction. That is, deceleration of the fluid in the subsonic layer upstream of the impingement point of the shock wave causes a thickening of streamtubes there and this change in flow direction is carried out in the supersonic flow through the formation of compression waves.

The sketch in Figure 3 is representative of the results found for $\varepsilon = O(u_\tau)$ (Melnik & Grossman 1974); in this case the first-order perturbation velocities are governed by the nonlinear transonic small-disturbance equation, with prescribed vorticity, which must be solved numerically. For $\varepsilon \ll u_\tau$ (Adamson & Feo 1975) and $\varepsilon \gg u_\tau$ (Adamson & Messiter 1977, Messiter 1979) different sets of linear equations, obtained for $\varepsilon/u_\tau \to 0$ and $\varepsilon/u_\tau \to \infty$ respectively, have allowed analytical solutions to be obtained. For the case $\varepsilon \gg u_\tau$, the shock wave extends to the wall, to the scale of the outer layer, and the flow entering this shock wave is undisturbed. The solution is found in terms of a source distribution in the plane of the shock wave, within the boundary layer, symmetric about the plane of the wall. The effect on the external flow at larger distances is that of a two-dimensional source with strength proportional

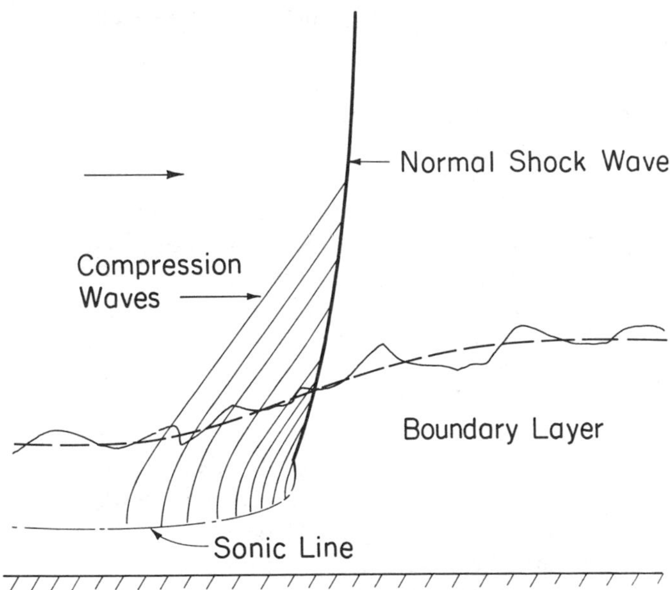

Figure 3 Incident normal shock wave with unseparated turbulent boundary layer.

to the local increase in boundary-layer displacement thickness. For axisymmetric channel flow the effect of the source distribution is that of a ring source, with significant influence on the subsonic core flow downstream of the shock wave.

As mentioned previously, lower-order solutions [to and including $O(u_\tau^2)$] in the outer inviscid layer are found without recourse to any sublayer solutions or effects. Therefore, the boundary condition used in the solutions in this layer is simply that V, the component of the velocity in the direction normal to the wall, is zero to the order considered. The pressure distribution found in the limit as the wall is approached is the pressure distribution at the wall because, to the order considered, $\partial P/\partial Y \to 0$ as $Y/\delta \to 0$. Finally, it may be noted that because, in general, $\Delta/\delta \to 0$ in the limit as $u_\tau \to 0$ and $\varepsilon \to 0$, the first term involving Reynolds stresses to be considered in the main part of the boundary layer would be $\partial(\rho\,\overline{U'U'})/\partial X$; that is, assuming both to be of the same order, a normal stress rather than a shear stress would be involved.

In the wall layer immediately adjacent to the wall, even though a relatively strong pressure gradient $\partial P/\partial X = O(u_\tau/\Delta)$ may exist, the dominant terms in the equation of motion in the flow direction are the Reynolds and viscous stress terms, i.e. one finds the same governing equation as for the undisturbed wall layer. The first integral of this equation is, therefore, Equation (3), and for $U = O(u_\tau)$ and $\overline{U'V'} = O(u_\tau^2)$, one finds, then, that the scale of the thickness of the wall layer remains $\tilde{\delta} = O(\mathrm{Re}^{-1}u_\tau^{-1})$.

From the above, it is clear that the equations of motion in the inviscid flow and wall layers have no common terms; thus solutions cannot be matched. Physically, this indicates that there is no mechanism available by which changes in momentum in the outer layer can be communicated to the wall layer. As a result, between these two layers must be inserted another layer in which momentum is transferred by turbulent means, i.e. through the Reynolds stresses. In this layer, referred to as the Reynolds-stress sublayer (Adamson & Feo 1975) or the blending layer (Melnik & Grossman 1974), the terms of importance in the equation of motion are the inertia, pressure-gradient, and Reynolds-stress terms; this layer has a thickness $\hat{\delta} = O(u_\tau\Delta)$.

The explanation for the necessity of a three-layer structure for the interaction with a turbulent boundary layer as opposed to the two layers required for laminar flow is found in the basic difference in structure of the two boundary layers. That is, in either the laminar or turbulent interaction, the major part of the flow field is governed by the equations for inviscid flow. Hence, the no-slip condition at the wall cannot be satisfied and it is necessary to consider, along the wall, a new boundary layer that is thin compared with its length, the extent of the interaction region. In laminar flow, a boundary layer has a one-layer structure and so one need consider only a viscous sublayer. On the other hand, a turbulent boundary layer has a two-layer structure and so two layers must be considered in the interaction region. In this regard, then, the Reynolds-stress sublayer is the counterpart of the velocity-defect layer in the turbulent boundary layer, and indeed, the governing equations and therefore the thickness-to-length ratios $[\hat{\delta}/\Delta = O(u_\tau) = O(\delta)]$ are the same.

Because of the difficulty in obtaining data for interactions in which the flow is truly two-dimensional, steady, and unseparated, comparison of analytical results with experiment is difficult. Thus, Melnik & Grossman (1976) extended their two-dimensional analyses to axially symmetric channel flow. They showed fairly good agreement with Gadd's (1961) experimental data for the pressure distribution at the wall. Messiter (1979) did the same and showed good agreement with the numerical solution of Melnik & Grossman downstream of the shock-wave location, these calculations being valid for the case where the upstream influence is very small ($\varepsilon \gg u_\tau$). Fairly good agreement was likewise found in a comparison of Messiter's (1979) calculations with pressure distributions measured by Ackeret et al (1946), both at and away from the wall. These solutions include a correction for wall curvature and higher-order terms, $O(u_\tau^2)$, both of which are important numerically. It was found that, for $\varepsilon \gg u_\tau$, the local curvature correction remains small in comparison with the interaction pressure already calculated; the singular inviscid-flow solution for a normal shock wave at a curved surface (Oswatitsch & Zierep 1960, Gadd 1960) can simply be added. In all of these cases, it was found that the calculated pressure gradient in the interaction region was larger than that measured. However, it should be borne in mind that if the shock wave is not stationary, measured pressures represent average values. For example, at a point in the flow that is usually but not always upstream of the oscillating shock wave, the measured pressure will be somewhat higher than that predicted by a steady-state analysis. The pressure gradient is therefore smaller than in the corresponding steady-flow distribution, with an apparently greater upstream influence. Shock-wave oscillations may have been present with amplitude sufficient to cause much of the difference mentioned above between theory and experiment.

Liou & Adamson (1979) calculated the shear-stress distribution at the wall, τ_w, using the surface-pressure distributions given by Messiter (1979), again for a normal shock wave and $\varepsilon \gg u_\tau$; the upstream influence is very small because of the strength of the shock wave. A simple mixing-length model was used for the eddy viscosity. In order to reproduce the known minimum in τ_w, it was necessary to include terms of higher order $[O(u_\tau \varepsilon)]$. Contrary to the results found for a laminar boundary layer, an asymptotic criterion for separation as $\mathrm{Re} \to \infty$ was not found. However, with the assumption that the expression derived for τ_w is a good approximation up to incipient separation, the parametric relationship between the flow Mach number and the Reynolds number at incipient separation was computed. Comparison of τ_w with an experimental distribution for unseparated flow given by Gadd (1961) resulted again in fairly good general agreement downstream of the shock wave, but in very good agreement insofar as the magnitude and position of the minimum value of τ_w were concerned. In three other cases where Gadd indicated that small separation bubbles existed, calculated results also showed separation had occurred. From computations of the Mach number at incipient separation, M_{es}, Liou & Adamson (1979) found that a large increase in Reynolds number caused only a very small increase in M_{es}, in general agreement with the experiments of Roshko & Thomke (1969) and Settles et al (1976) for a compression ramp at higher Mach numbers. A value of M_{es} computed for the case of flow over a flat plate compared favorably with a result given by Gadd (1961), but much more testing over a range of parameters is clearly necessary.

Other Approximate Methods

In an analytical model that has received considerable attention, the flow is divided into an outer inviscid flow region, scaled by the thickness of the boundary layer, and a thin viscous sublayer adjacent to the wall, as in asymptotic methods developed for laminar flow. Variations of the model include different methods of computing the solutions within and external to the boundary layer or in the sublayer, and, in the case of an incident shock wave, several approximations for the geometry and strength of the shock as it approaches the boundary layer (e.g. the shock wave remains planar, does not penetrate the boundary layer, etc). However, all are basically two-layer models and are derivatives, in some sense, of the flow model developed by Lighthill (1953) subsequent to the work of Tsien & Finston (1949) and Howarth (1948).

The crucial element of this model appears to be the treatment of the viscous sublayer, i.e. its thickness and its use in deriving the solutions. Thus, as mentioned previously, Roshko & Thomke (1969), Elfstrom (1972), and Carrière et al (1975) have shown that for attached flow over a compression ramp, good comparison with experimentally measured surface pressure distributions is obtained if the sublayer is ignored; they recognized that for a turbulent boundary layer the fluid displacement caused by the adverse pressure gradient is negligible to a good approximation. On the other hand, Rose (1970), in calculations for an incident oblique shock wave, evidently did not distinguish between interactions in laminar and turbulent boundary layers; in his solutions the displacement effect of the inner layer is a significant factor in the calculation of the pressure distribution at the wall for either case. Bohning & Zierep (1976, 1978), Inger & Mason (1976), and Inger & Zee (1978), in their solutions for a normal incident shock wave, essentially disregard these displacement effects insofar as a boundary condition for the outer layer is concerned. However, they use the characteristic thickness of the sublayer to estimate a nonzero value of the Mach number at a lower edge of the outer layer; the value obtained presumably does influence the calculated pressure distribution. Moreover, the calculated shear stress at the wall is strongly dependent upon the thickness of this assumed layer, where a linear velocity profile is used in a first approximation.

In general, boundary-layer equations for laminar flow have been used in the sublayer with the argument that this layer is in the part of the turbulent boundary layer where laminar stresses predominate. As noted previously, however, systematic methods of analysis indicate that Reynolds stresses remain important in the sublayer described by the law of the wall. Here, the proper approximate equation, given by Equation (3), has neither inertia nor pressure-gradient terms even in the interaction region, but instead shows a balance between Reynolds and viscous stresses. As a result, even though the velocity is indeed linear adjacent to the wall, the functional dependence of the characteristic thickness on the Reynolds and Mach numbers is far different from that for a viscous sublayer. Moreover, it appears that, as noted earlier, two separate sublayers are called for when the boundary layer is turbulent.

Consequently, it appears that in principle the two-layer model is suited to interactions with laminar, but not to those with turbulent, boundary layers. As long as the weak-interaction boundary condition is used at the wall (e.g. Bohning & Zierep 1976 and Inger & Mason 1976) the effect of the laminar viscous sublayer on the pressure distribution at the wall might perhaps be small, a point to be investigated by careful comparison with experiment; however, it has a major effect on skin friction. The latter effect may be illustrated, for example, in comparisons of calculations of the Reynolds-number dependence of the Mach number for incipient separation in the interaction between a normal shock wave and a turbulent boundary layer. The two-layer model of Bohning & Zierep (1978) gives a relatively large variation with Reynolds number as opposed to the small changes found by Liou & Adamson (1979). The general trend of experimental data for the related problem of flow at high Mach number over a ramp, mentioned previously (Roshko & Thomke 1969, Settles et al 1976), shows that conditions for incipient separation have a very weak dependence upon Reynolds number.

A few interaction calculations have also been carried out with integral methods; these were reviewed by Green (1970). Reshotko & Tucker (1955) used the momentum integral equation for a pressure jump across an assumed discontinuity at separation, with the contribution of skin friction neglected because the distance is very short. Comparison with experimental data, with assumptions about values of the form factor, permits a plot of pressure rise required for separation as a function of Mach number, independent of Reynolds number in this approximation. Todisco & Reeves (1969) and Hunter & Reeves (1971) have used the Lees-Reeves integral method for turbulent boundary layers. Since a turbulent boundary layer is found to be "supercritical," a "supercritical-subcritical jump" is assumed to change the boundary layer from its undisturbed form to a Stratford separation profile, and a one-parameter family of reversed-flow profiles is assumed between separation and reattachment. Holden (Hankey & Holden 1975, Part II) notes that for high Mach number this method gives plateau pressures that are larger than measured values, whereas the solution of Reshotko and Tucker continues to show the same general trend as the data. Le Balleur (1977, 1978) has proposed an altered coupling condition that replaces θ found by integration of the continuity equation with a modified value that can be thought of as obtained by extending the inviscid flow to the wall, thereby eliminating the need for a supercritical-subcritical jump. Good agreement with experiment is shown for the wall pressure distribution at a compression corner. Alber & Lees (1968) have also used an integral method in a calculation of turbulent adiabatic base flow downstream of a backward-facing step. A solution obtained for a downstream region containing the reattachment stagnation point was joined upstream to a solution for a constant-pressure free shear layer, with the initial profile found from a control-volume description of the corner expansion.

The pressure change at a backward-facing step is often estimated by methods based on the calculation of Korst (1956), which resembles that of Chapman et al (1957) for laminar flow. Korst gave an approximate description of turbulent mixing in a free shear layer and calculated the velocity at the separation streamline $\psi = 0$. He then assumed, as did Chapman, that the pressure rise along this streamline at reattachment is isentropic and equal to the pressure rise through a simple wave in the neighboring external flow. The constant-pressure mixing actually occurs primarily in a sublayer, between fluid with mean velocity close to the external-flow velocity and recirculating fluid at low speed, so that the nondimensional pressure rise at reattachment is presumably of order one. It does not seem clear whether or not the type of argument introduced in the laminar case can again be used to give a first approximation for the base pressure. In any case, neglected terms are likely to be large, and are almost certain to be important numerically for parameter values of interest. Various empirical corrections to Korst's original calculation have been proposed (e.g. Tanner 1973). Another approximate calculation has been described by Carrière et al (1975).

Numerical Methods

As might be expected, numerical methods of analysis have received a great deal of attention in recent studies of interaction between shock waves and turbulent boundary layers. Advances in computational methods, notably those due to MacCormack (e.g. MacCormack & Lomax 1979), have led to such substantial reductions in computing times that numerical investigations of quite complex interaction problems are becoming almost routine. Generally, the governing equations are taken to be the time-dependent, mass-averaged Navier-Stokes equations with the expressions for the turbulent transport of momentum and energy written in terms of eddy transport coefficients and gradients of the mean-flow properties; thus, the overall transport coefficient is the sum of the molecular and turbulent coefficients. Because a constant turbulent Prandtl number is considered, it is necessary to consider only the eddy viscosity, which is generally assumed to be isotropic. Finally, in order to close the equations, an assumption must be made concerning the dependence of the eddy viscosity on the flow properties; i.e. a model must be chosen.

As the time for computation has decreased, it has become possible to consider more complex models to describe the eddy viscosity. These models are generally classified in terms of the number of partial differential equations they add to the set of governing equations. Thus, a zero-equation model indicates that only mean-flow properties are employed in an algebraic relation; this may be either an equilibrium or a relaxation model. In the one-equation model, the eddy viscosity is written in terms of the kinetic energy of the turbulence, which defines its velocity scale, and which is governed by a partial differential equation; the length scale is written in algebraic form. For the two-equation model, additional governing equations are required for both the turbulent kinetic energy and another variable defining the length scale, both used in the relation for the eddy viscosity.

By virtue of its simplicity, the zero-equation equilibrium model has been widely used. It is generally taken to be a two-layer model, as in the form described by Cebeci & Smith (1974), with the Clauser defect law in the outer layer and the Prandtl mixing-length relation modified by the van Driest damping factor in the inner layer. Further modifications, such as intermittency correction factors or corrections to the van Driest damping factor for pressure gradient, are sometimes employed. Marvin et al (1975) used a zero-equation equilibrium model and compared numerical and experimental results for an incident oblique shock, formed by an annular shock generator, interacting with a turbulent boundary layer on a cylindrical body in hypersonic flow. It was found that the general features of the interaction region were predicted, but that significant error existed in the details of the flow field. Shea (1978) found much the same result in the comparison of his numerical solutions with the experimental data of Seddon (1960) in the case where a normal shock wave impinges on the boundary layer in transonic flow in a channel. In this investigation, however, the problem of the choice of an eddy viscosity model was overshadowed by the problems associated with the changes in the channel flow induced by the increase in the displacement thickness of the boundary layer downstream of the shock wave.

One of the criticisms of the equilibrium model is that it cannot account for any of the history of the flow field. Shang & Hankey (1975a), in an effort to overcome this deficiency, revised the model by introducing a relaxation length; Bradshaw (1973) and Rose & Johnson (1975) proposed similar formulations. This model, named the relaxation eddy-viscosity model, was used by Shang and Hankey in their computations for supersonic flow over a compression ramp. The results were compared with the experimental measurements of Law (1974) and showed close agreement and thus considerable improvement over the simple equilibrium model. Further refinements, introduced in the form of a rate equation to simulate the relaxation effect (Shang & Hankey 1975b), did not show significant improvement over the original algebraic relation. Subsequently, several numerical investigations have been made to test this model in different flow configurations and ranges of parameters. Thus, Baldwin & Rose (1975) considered an oblique shock impinging on a boundary layer on a flat surface at supersonic speeds, and found the relaxation model gave results which agreed with experimental data (Reda & Murphy 1973) better than did those calculated using several modified equilibrium models. The same problem was considered by Shang et al (1976), with similar results. Hung & MacCormack (1977) found solutions for both supersonic and hypersonic flow over a compression corner. In the supersonic case, use of the relaxation model again gave improved agreement with the experimental measurements (Law

1974) as compared with computations made using the simple equilibrium model. On the other hand, in the hypersonic case, where the experimental data of Holden (1972) were used for comparison, the solutions found using the equilibrium model agreed with experiment as well as or better than those found with the relaxation model for pressure and heat-transfer coefficient distributions along the wall; neither model was successful insofar as predictions of skin friction were concerned. Finally, Mateer et al (1976) studied the problem of a normal shock wave impinging on a boundary layer in an axisymmetric channel in transonic flow, and concluded that it was not possible to choose between the equilibrium and relaxation models employed. It thus appears that although use of the relaxation eddy-viscosity model has given good results in some problems, it has limited application. In particular, the accuracy of predictions decreases in the region downstream of reattachment in those cases where separation occurs.

A one-equation model used recently is that developed by Glushko (1965) for incompressible flow and extended for compressible flows by Rubesin (1976). Its first application was in computations made by Viegas & Coakley (1978) for comparison with two sets of experimental results already mentioned, the measurements of Mateer et al (1976) for a normal shock wave in transonic channel flow and those of Marvin et al (1975) for an oblique shock wave in hypersonic flow. In the transonic flow case, the unmodified forms of the one-equation model appeared to give significant improvement over a zero-equation model. The results in the hypersonic flow case indicated that the unmodified form of the one-equation model gave little improvement over zero-equation models, but that modifications could be made resulting in solutions that agreed better with experiment. One of these modified one-equation models was then used by Horstman et al (1977) in a numerical and experimental study of supersonic flow over a compression ramp. In addition, several zero-equation models, both equilibrium and relaxation, were considered. Although the use of the one-equation model resulted in some improvements, compared to results achieved using the other models, it did not assure accuracy of prediction of the details of the flow field.

At the present time, the two-equation models for the eddy viscosity represent the most sophisticated approach available for closure of the governing equations. Since much more information from the flow field is contained in the additional governing equations, one would expect that use of this model should result in the ability to predict more accurately the details of the flow field in the interaction region. Evidently the first numerical study involving a two-equation model, and indeed one of the first numerical studies of an interaction involving a turbulent boundary layer, was made by Wilcox (1973), who considered the problem of an oblique shock wave impinging on a flat-plate boundary layer in supersonic flow. The turbulence model equations used were those given by Saffman (1970) and modified for compressible flow by Wilcox & Alber (1972). A comparison of the calculated surface pressure distribution with the experimentally measured values given by Reda & Murphy (1973) showed good agreement. Subsequently, Saffman & Wilcox (1974) made a detailed study of the model equations and Wilcox (1975) extended his calculations for interactions to cover more cases for the oblique incident shock wave and to include supersonic flow over a compression ramp. For the incident shock wave, computed and measured surface pressure distributions and pressure profiles showed quite good agreement with no adjustment in the parameters of the model. However, such agreement was not found in the case of the flow over the compression corner. The same set of model equations was used by Baldwin & MacCormack (1975) in a numerical investigation of the interaction between an oblique incident wave and a boundary layer on a flat plate in hypersonic flow; calculated distributions of skin-friction and heat-transfer coefficients were compared with the measurements of Holden (1972). Although the solutions in which the two-equation model was used indicated some improvement over those found using a zero-equation model, the agreement with experiment was only fair.

Perhaps the most ambitious comparison of eddy-viscosity models, in their application to interactions between shock waves and boundary layers, has been performed by Viegas & Horstman (1979). In this study, a zero-equation equilibrium, a one-equation (Glushko 1965, Rubesin 1976), and two different two-equation (Jones & Launder 1972 and Wilcox & Rubesin 1977) models were used in numerical investigations of four problems: a normal shock wave in transonic flow, both in a circular duct and in a two-dimensional channel, supersonic flow over a compression ramp, and an incident oblique shock wave on a flat surface in supersonic flow. The Wilcox-Rubesin model is notable in that an effort has been made to account for anisotropic normal Reynolds stresses. Although a number of specific conclusions are stated concerning the performance of each of the models tested, perhaps the most important observations are that no matter which model was used, the results agreed qualitatively with experimental measurements, that no single model gave the best quantitative agreement with experiment, and that general improvement was obtained with more complex turbulence models.

It appears that much remains to be done insofar as turbulence modeling is concerned. Although it is generally agreed that zero-equation models give good results in those cases where the flow remains attached (e.g. MacCormack & Lomax 1979), they fail insofar as the details of the flow field are concerned when the pressure gradient is strong enough that separation occurs. The more complex one- and two-equation models do better in this regard, but apparently none has performed well enough to be considered as universally applicable.

CONCLUDING REMARKS

Insofar as laminar flow is concerned, the interaction between a shock wave and a boundary layer is, for the most part, well understood. From the viewpoint of the analyst, this means that it is possible to explain the interplay of the various physical mechanisms in terms of regions or layers, with known length scales, in which different groups of these mechanisms are important. From a practical viewpoint, this means that accurate solutions to problems with considerable geometric complexity can be found numerically, given enough time for computation.

When the boundary layer is turbulent the picture is not so clear, obviously because of continuing problems with closure of the governing equations. A simple algebraic (zero-equation) model for the eddy viscosity provides the information necessary to ascertain the gross structure of the interaction region, including the associated length scales. If it is used in numerical computations, the general features of the interaction are predicted even when the flow separates; however, details of the flow field are not given accurately. It was expected that the more complex one- and two-equation models would allow much more accurate predictions to be made because some effects of the history of the flow field could be taken into account. The use of modern high-speed computers makes introduction of these more sophisticated models feasible, albeit with longer times for computations. However, at the present time, it appears that this promise has not been fulfilled, in the sense that it is not possible to choose a given model, with a given set of constants, that gives results more accurate than those given by a simple zero-equation model, for an arbitrarily chosen interaction problem. That is, for any given problem a one- or two-equation model may be adjusted to give very accurate solutions, but this is not a universal result. It seems evident that more experiments, directed toward providing more detailed information on Reynolds stresses in flows with large adverse pressure gradients, are required.

Finally, it appears worthwhile to note the difficulty inherent in making experimental measurements in the interaction of a shock wave and a turbulent boundary layer and the attendant danger in forcing too close an agreement between experimental and numerical results. For example, it has already been noted that some unsteadiness in the position of the shock wave leads to measured properties with smaller gradients than would be found in the instantaneous distribution. Similar difficulties exist in identifying incipient separation and measuring flow properties in interactions with small separation bubbles. This point should be kept in mind when comparing experimental and calculated results and in particular when using experimental results to guide the choice of parameters in models for the eddy viscosity.

ACKNOWLEDGMENTS

The authors are grateful to C. C. Horstman, Jr., W. L. Hankey, Jr., and R. E. Melnik for their comments and suggestions during the preparation of this review.

Literature Cited

Ackeret, J., Feldmann, F., Rott, N. 1946. *Mitteil. Inst. für Aerodyn.*, ETH Zürich, Nr. 10. Engl. transl. 1947. Investigations of compression shocks and boundary layers in gases moving at high speed. *NACA TM 1113*

Adamson, T. C. Jr. 1976. The structure of shock wave–boundary layer interactions in transonic flow. *Symp. Transsonicum II*, ed. K. Oswatitsch, D. Rues, pp. 244–51. Berlin: Springer

Adamson, T. C. Jr., Feo, A. 1975. Interaction between a shock wave and a turbulent boundary layer in transonic flow. *SIAM J. Appl. Math.* 29:121–45

Adamson, T. C. Jr., Messiter, A. F. 1977. Normal shock wave–turbulent boundary layer interactions in transonic flow near separation. *Transonic Flow Problems in Turbomachinery*, ed. T. C. Adamson Jr., M. F. Platzer, pp. 392–414. Washington, DC: Hemisphere

Alber, I. E., Lees, L. 1968. Integral theory for supersonic turbulent base flows. *AIAA J.* 6:1343–51

Allen, J. S., Cheng, S. I. 1970. Numerical solutions of the compressible Navier-Stokes equations for the laminar near wake. *Phys. Fluids* 13:37–52

Baldwin, B. S., MacCormack, R. W. 1975. Interaction of strong shock wave with turbulent boundary layer. *Proc. 4th Int. Conf. Numer. Meth. Fluid Dyn., Boulder, Colo. 1974. Lect. Notes in Phys.* 35:51–56. Berlin: Springer

Baldwin, B. S., Rose, W. C. 1975. Calculation of shock-separated turbulent boundary layers. *NASA SP-347*, pp. 401–17

Batchelor, G. K. 1956. On steady laminar flow with closed streamlines at large Reynolds number. *J. Fluid Mech.* 1:177–90

Beam, R. M., Warming, R. F. 1978. An implicit factored scheme for the compressible Navier-Stokes equations. *AIAA J.* 16:393–402

Bodonyi, R. J., Kluwick, A. 1977. Freely interacting transonic boundary layers. *Phys. Fluids* 20:1432–37

Bohning, R., Zierep, J. 1976. Der senkrechte Verdichtungsstoss an der gekrümmten Wand unter Berücksichtigung der Reibung. *Z. Angew. Math. Phys.* 27:225–40. See also Adamson 1976, pp. 236–43

Bohning, R., Zierep, J. 1978. Bedingung für das Einsetzen der Ablösung der turbulenten Grenzschicht an der gekrümmten Wand mit senkrechten Verdichtungsstoss. *Z. Angew. Math. Phys.* 29:190–98

Bradshaw, P. 1973. Effects of streamline curvature on turbulent flow. *AGARDograph No. 169*

Brilliant, H. M., Adamson, T. C. Jr. 1974. Shock wave boundary-layer interactions in laminar transonic flow. *AIAA J.* 12:323–29

Brown, S. N., Stewartson, K., Williams, P. G. 1975. Hypersonic self-induced separation. *Phys. Fluids* 18:633–39

Brown, S. N., Williams, P. G. 1975. Self-induced separation III. *J. Inst. Math. Applic.* 16:175–91

Burggraf, O. R. 1969. Computation of separated flow over backward-facing steps at high Reynolds number. *Proc. ARL Symp. on Viscous Interaction*

Phenomena in Supersonic and Hypersonic Flow, pp. 463–91. Dayton, Ohio: Univ. Dayton

Burggraf, O. R. 1973. Inviscid reattachment of a separated shear layer. *Proc. 3rd Int. Conf. Numer. Meth. Fluid Dyn., Paris 1972. Lect. Notes in Phys.* 19:39–47. Berlin: Springer

Burggraf, O. R. 1975. Asymptotic theory of separation and reattachment of a laminar boundary layer on a compression ramp. *AGARD-CP-168*

Burggraf, O. R., Rizzetta, D., Werle, M. J., Vatsa, V. N. 1979. Effect of Reynolds number on laminar separation of a supersonic stream. *AIAA J.* 17:336–43

Carrière, P., Sirieix, M., Delery, J. 1975. Méthodes de calcul des écoulements turbulents décollés en supersonique. *Prog. Aerosp. Sci.* 16:385–429

Carter, J. E. 1972. Numerical solutions of the Navier-Stokes equations for the supersonic laminar flow over a two-dimensional compression corner. *NASA Rep. R-385*

Cebeci, T., Smith, A. M. O. 1974. *Analysis of Turbulent Boundary Layers.* New York: Academic

Chapman, D. R., Kuehn, D. M., Larson, H. K. 1957. Investigation of separated flows in supersonic and subsonic streams with emphasis on the effect of transition. *NACA TN 3869*. Also 1958. *NACA Rep. 1356*

Coles, D. 1956. The law of the wake in the turbulent boundary layer. *J. Fluid Mech.* 1:191–226

Crocco, L. 1975. Flow separation. *AGARD-CP-168*

East, L. F. 1976. The application of a laser anemometer to the investigation of shock-wave boundary-layer interactions. *AGARD-CP-193*

Elfstrom, G. M. 1972. Turbulent hypersonic flow at a wedge-compression corner. *J. Fluid Mech.* 53:113–27

Fage, A., Sargent, R. F. 1947. Shock-wave and boundary-layer phenomena near a flat surface. *Proc. R. Soc. London Ser. A* 190:1–20

Ferri, A. 1939. *Atti di Guidonia* No. 17. Engl. transl. 1940. Experimental results with airfoils tested in the high-speed tunnel at Guidonia. *NACA TM 946*

Gadd, G. E. 1957. A theoretical investigation of laminar separation in supersonic flow. *J. Aeronaut. Sci.* 24:759–71, 784

Gadd, G. E. 1960. The possibility of normal shock waves on a body with convex surfaces in inviscid transonic flow. *Z. Angew. Math. Phys.* 11:51–58

Gadd, G. E. 1961. Interactions between normal shock waves and turbulent boundary layers. *British ARC R&M 3262*

Georgeff, M. P. 1974. Momentum integral method for viscous-inviscid interactions with arbitrary wall cooling. *AIAA J.* 12:1393–1400

Glushko, G. S. 1965. Turbulent boundary layer on a flat plate in an incompressible fluid. *Bull. Acad. Sci. USSR, Mech. Ser. No. 4*, pp. 13–23

Green, J. E. 1970. Interactions between shock waves and turbulent boundary layers. *Prog. in Aerosp. Sci.* 11:235–340

Hakkinen, R. J., Greber, I., Trilling, L., Abarbanel, S. S. 1959. The interaction of an oblique shock wave with a laminar boundary layer. *NASA Memo 2-18-59W*

Hama, F. R. 1966. Experimental studies on the lip shock. *AIAA J.* 6:212–19

Hankey, W. L. Jr., Holden, M. S. 1975. Two-dimensional shock-wave boundary layer interactions in high speed flows. *AGARDograph No. 203*

Holden, M. S. 1965. An analytical study of separated flows induced by shock wave–boundary layer interaction. *Cornell Aeronaut. Lab. Rep. No. AI-1972-A-3*

Holden, M. S. 1969. Theoretical and experimental studies of the shock wave–boundary layer interaction on curved compression surfaces. See Burggraf 1969, pp. 213–70

Holden, M. S. 1972. Shock wave–turbulent boundary layer interaction in hypersonic flow. *AIAA Paper 72-74*

Holt, M. 1966. Separation of laminar boundary-layer flow past a concave corner. *AGARD-CP-4*

Holt, M., Lu, T. A. 1975. Supersonic laminar boundary layer separation in a concave corner. *Acta Astronautica* 2:409–29

Horstman, C. C., Settles, G. S., Vas, I. E., Bogdonoff, S. M., Hung, C. M. 1977. Reynolds number effects on shock-wave turbulent boundary-layer interactions. *AIAA J.* 15:1152–58

Howarth, L. 1948. The propagation of steady disturbances in a supersonic stream bounded on one side by a parallel subsonic stream. *Proc. Camb. Philos. Soc.* 44:380–90

Hung, C. M., MacCormack, R. W. 1976. Numerical solutions of supersonic and hypersonic laminar compression corner flows. *AIAA J.* 14:475–81

Hung, C. M., MacCormack, R. W. 1977. Numerical simulation of supersonic and hypersonic turbulent compression corner flows. *AIAA J.* 15:410–16

Hunter, L. G. Jr., Reeves, B. L. 1971. Results of a strong interaction, wake-like model of supersonic separating and reattaching turbulent flow. *AIAA J.* 9:703–12

Inger, G. R., Mason, W. H. 1976. Analytical theory of transonic normal shock–turbulent boundary-layer interaction. *AIAA J.* 14:1266–72. See also Adamson 1976, pp. 227–35

Inger, G. R., Zee, S. 1978. Transonic shock wave/turbulent-boundary-layer interaction with suction or blowing. *J. Aircraft* 15:750–54. See also *AIAA Paper 79-0005*

Jones, W. P., Launder, B. E. 1972. The prediction of laminarization with a two-equation model of turbulence. *Int. J. Heat Mass Transfer* 15:301–14

Klineberg, J. M., Lees, L. 1969. Theory of laminar viscous-inviscid interactions in supersonic flow. *AIAA J.* 7:2211–21

Kooi, J. W. 1975. Experiment on transonic shock-wave boundary-layer interaction. *AGARD-CP-168*. Also *NLR MP 75002 U*

Korkegi, R. H. 1971. Survey of viscous interactions associated with high Mach number flight. *AIAA J.* 9:771–84

Korst, H. H. 1956. A theory for base pressures in transonic and supersonic flow. *J. Appl. Mech.* 23:593–600

Law, C. H. 1974. Supersonic turbulent boundary-layer separation. *AIAA J.* 12: 794–97

LeBalleur, J. C. 1977. Couplage visqueux-non visqueux: analyse du problème incluant décollements et ondes de choc. *Rech. Aérosp.*, pp. 349–58

LeBalleur, J. C. 1978. Couplage visqueux-non visqueux: méthode numérique et applications aux écoulements bidimensionnels transsoniques et supersoniques. *Rech. Aérosp.*, pp. 65–76

Lees, L. 1949. Interaction between the laminar boundary layer over a plane surface and an incident oblique shock wave. *Princeton Univ. Aeronaut. Eng. Lab. Rep. No. 143*

Lees, L., Reeves, B. L. 1964. Supersonic separated and reattaching laminar flows: I. General theory and application to adiabatic boundary-layer/shock-wave interactions. *AIAA J.* 2:1907–20

Lewis, J. E., Kubota, T., Lees, L. 1968. Experimental investigation of supersonic laminar two-dimensional boundary-layer separation in a compression corner with and without cooling. *AIAA J.* 6:7–14

Li, C. P. 1976. A mixed explicit-implicit splitting method for the compressible Navier-Stokes equations. *Proc. 5th Int. Conf. Numer. Meth. Fluid Dyn., Enschede, The Netherlands 1976. Lect. Notes in Phys.* 59:285–92. Berlin: Springer

Liepmann, H. W. 1946. The interaction between boundary layer and shock waves in transonic flow. *J. Aeronaut. Sci.* 13: 623–37

Liepmann, H. W., Roshko, A., Dhawan, S. 1949. On reflection of shock waves from boundary layers. *Guggenheim Aeronaut. Lab. Rep.*, Calif. Inst. of Tech. Also 1951. *NACA TN 2334*

Lighthill, M. J. 1953. On boundary layers and upstream influence II. Supersonic flows without separation. *Proc. R. Soc. London Ser. A* 217:478–507

Liou, M. S., Adamson, T. C. Jr. 1979. Interaction between a normal shock wave and a turbulent boundary layer at high transonic speeds. Part II. Wall shear stress. *NASA CR* 3194. Also to appear in *Z. Angew. Math. Phys.*

MacCormack, R. W. 1971. Numerical solution of the interaction of a shock wave with a laminar boundary layer. *Proc. 2nd Int. Conf. Numer. Meth. Fluid Dyn., Berkeley, Calif. 1970. Lect. Notes in Phys.* 8:151–63. Berlin: Springer

MacCormack, R. W. 1976. A rapid solver for hyperbolic systems of equations. See Li 1976, pp. 307–17

MacCormack, R. W., Lomax, H. 1979. Numerical solution of compressible viscous flows. *Ann. Rev. Fluid Mech.* 11:289–316

Maise, G., McDonald, H. 1968. Mixing length and kinematic eddy viscosity in a compressible boundary layer. *AIAA J.* 6: 73–80

Marvin, J. G., Horstman, C. C., Rubesin, M. W., Coakley, T. J., Kussoy, M. I. 1975. An experimental and numerical investigation of shock-wave induced turbulent boundary-layer separation at hypersonic speeds, *AGARD-CP-168*. Also *NASA SP-347*, pp. 377–99. See also *AIAA Paper 75-4*

Mateer, G. G., Brosh, A., Viegas, J. R. 1976. A normal shock-wave turbulent boundary layer interaction at transonic speeds. *AIAA Paper 76-161*

Matveeva, N. S., Neiland, V. Ya. 1967. *Izv. AN SSSR, Mekh. Zhidk. i Gaza*, No. 4, pp. 64–70. Engl. transl. The laminar boundary layer near a body corner point. *Fluid Dyn.* 2: No. 4, pp. 42–46

Melnik, R. E., Grossman, B. 1974. Analysis of the interaction of a weak normal shock wave with a turbulent boundary layer. *AIAA Paper 74-598*

Melnik, R. E., Grossman, B. 1976. Interactions of normal shock waves with turbulent boundary layers at transonic speeds. See Adamson 1976, pp. 262–72. See also Adamson & Messiter 1977, pp. 415–33

Messiter, A. F. 1978. Boundary-layer separation. *Proc. 8th US Natl. Congr. Appl. Mech.*, pp. 157–79. North Hollywood, Calif.: Western Periodicals

Messiter, A. F. 1979. Interaction between a normal shock wave and a turbulent boundary layer at high transonic speeds. Part I: Pressure distribution. *NASA CR* 3194. Also to appear in *Z. Angew. Math. Phys.*

Messiter, A. F., Feo, A., Melnik, R. E. 1971. Shock-wave strength for separation of a laminar boundary layer at transonic speeds. *AIAA J.* 9:1197–98

Messiter, A. F., Hough, G. R., Feo, A. 1973. Base pressure in laminar supersonic flow. *J. Fluid Mech.* 60:605–24

Murphy, J. D. 1969. A critical evaluation of analytic methods for predicting laminar-boundary-layer shock-wave interaction. *NASA SP-228*, pp. 515–39. See also 1971. *NASA TN D-7044*

Neiland, V. Ya. 1969. *Izv. AN SSSR, Mekh. Zhidk. i Gaza*, No. 4, pp. 53–57. Engl. transl. Theory of laminar boundary layer separation in supersonic flow. *Fluid Dyn.* 4(4): 33–35

Neiland, V. Ya. 1970a. *Izv. AN SSSR, Mekh. Zhidk. i Gaza*, No. 3, pp. 22–32. Engl. transl. Asymptotic theory of plane steady supersonic flows with separation zones. *Fluid Dyn.* 5:372–81

Neiland, V. Ya. 1970b. *Izv. AN SSSR, Mekh. Zhidk. i Gaza*, No. 4, pp. 40–49. Engl. transl. Propagation of disturbances upstream with interaction between a hypersonic flow and a boundary layer. *Fluid Dyn.* 5:559–66

Neiland, V. Ya. 1971. *Izv. AN SSSR, Mekh. Zhidk. i Gaza*, No. 3, pp. 19–25. Engl. transl. Flow behind the boundary layer separation point in a supersonic stream. *Fluid Dyn.* 6:378–84

Neiland, V. Ya. 1973. *Izv. AN SSSR, Mekh. Zhidk. i Gaza*, No. 6, pp. 99–109. Engl. transl. Boundary-layer separation on a cooled body and interaction with a hypersonic flow. *Fluid Dyn.* 8:931–39

Nielsen, J. N., Goodwin, F. K., Kuhn, G. D. 1969. Review of the method of integral relations applied to viscous interaction problems including separation. See Burggraf 1969, pp. 31–82

Nielsen, J. N., Lynes, L. L., Goodwin, F. K. 1966. Theory of laminar separated flows on flared surfaces including supersonic flow with heating and cooling. *AGARD-CP-4*

Oswatitsch, K., Wieghardt, K. 1942. Lilienthal-Gesellschaft für Luftfahrtforschung Bericht S 13/1 Teil, pp. 7–24. Engl. transl. 1948. Theoretical analysis of stationary potential flows and boundary layers at high speed. *NACA TM 1189*

Oswatitsch, K., Zierep, J. 1960. Das Problem des senkrechten Stosses an einer gekrümmten Wand. *Z. Angew. Math. Mech.* 40:143–44

Pearcey, H. H. 1961. Shock-induced separation and its prevention by design and boundary layer control. In *Boundary Layer and Flow Control*, ed. G. V. Lachmann, Vol. 2, pp. 1166–1344. New York: Pergamon

Prandtl, L. 1904. *Verh. III Int. Math-Kongr. Heidelberg* 484–91. Leipzig: Teubner. Engl. transl. 1928. Motion of fluids with very little viscosity. *NACA TM 452*

Reda, D. C., Murphy, J. D. 1973. Shock wave/turbulent boundary-layer interactions in rectangular channels. *AIAA J.* 11:139–40

Reshotko, E., Tucker, M. 1955. Effect of a discontinuity on turbulent boundary-layer-thickness parameters with application to shock-induced separation. *NACA TN 3454*

Reyhner, T. A., Flügge-Lotz, I. 1968. The interaction of a shock wave with a laminar boundary layer. *Int. J. Non-Linear Mech.* 3:173–99

Rizzetta, D. P., Burggraf, O. R., Jenson, R. 1978. Triple-deck solutions for viscous supersonic and hypersonic flow past corners. *J. Fluid Mech.* 89:535–52

Roache, P. J., Mueller, T. J. 1970. Numerical solutions of laminar separated flows. *AIAA J.* 8:530–38

Rose, W. C. 1970. A method for analyzing the interaction of an oblique shock wave with a boundary layer. *NASA TN D-6083*

Rose, W. C., Johnson, D. A. 1975. Turbulence in shock-wave boundary-layer interaction. *AIAA J.* 13:884–89

Roshko, A., Thomke, G. J. 1969. Supersonic turbulent boundary-layer interaction with a compression corner at very high Reynolds number. See Burggraf 1969, pp. 109–38

Ross, B. B., Cheng, S. I. 1971. A numerical solution of the planar supersonic near wake with its error analysis. See MacCormack 1971, pp. 164–69

Rubesin, M. W. 1976. A one equation model of turbulence for use with the compressible Navier-Stokes equations. *NASA TM X-73128*

Saffman, P. G. 1970. A model for inhomogeneous turbulent flow. *Proc. R. Soc. London Ser. A* 317:417–33

Saffman, P. G., Wilcox, D. C. 1974. Turbulence-model predictions for turbulent boundary layers. *AIAA J.* 12:541–46

Seddon, J. 1960. The flow produced by interaction of a turbulent boundary layer with a normal shock wave of strength sufficient to cause separation. *British ARC R&M 3502*

Settles, G. S., Bogdonoff, S. M., Vas, I. E.

1976. Incipient separation of a supersonic turbulent boundary layer at high Reynolds numbers. *AIAA J.* 14:50–56

Shang, J. S. 1978. Implicit-explicit method for solving the Navier-Stokes equations. *AIAA J.* 16:496–502

Shang, J. S., Hankey, W. L. Jr. 1975a. Numerical solution for supersonic turbulent flow over a compression ramp. *AIAA J.* 13:1368–74

Shang, J. S., Hankey, W. L. Jr. 1957b. Supersonic turbulent separated flows utilizing the Navier-Stokes equations. *AGARD-CP-168*

Shang, J. S., Hankey, W. L. Jr., Law, C. H. 1976. Numerical simulation of shock wave–turbulent boundary-layer interaction. *AIAA J* 14:1451–57

Shea, J. R. III. 1978. A numerical study of transonic normal shock–turbulent boundary layer interactions. *AIAA Paper 78-1170*

Sirieix, M. 1975. Décollement turbulent en écoulement bidimensionnel. *AGARD-CP-168*

Stanewsky, E. 1973. Shock-boundary layer interaction in transonic and supersonic flow. *Transonic Flows in Turbomachinery, von Kármán Inst. Fluid Dyn. Lect. Ser. 59*

Stewartson, K. 1970. On supersonic laminar boundary layers near convex corners. *Proc. R. Soc. London Ser. A* 319:289–305

Stewartson, K. 1974. Multistructured boundary layers on flat plates and related bodies. *Adv. Appl. Mech.* 14:145–239. New York: Academic

Stewartson, K., Williams, P. G. 1969. Self-induced separation. *Proc. R. Soc. London Ser. A.* 312:181–206

Stewartson, K., Williams, P. G. 1973. On self-induced separation II. *Mathematika* 20:98–108

Stollery, J. L., Hankey, W. L. 1970. Subcritical and supercritical boundary layers. *AIAA J.* 8:1349–51

Tanner, M. 1973. Theoretical prediction of base pressure for steady base flow. *Prog. Aerosp. Sci.* 14:177–225

Todisco, A., Reeves, B. L. 1969. Turbulent boundary layer separation and reattachment at supersonic and hypersonic speeds. See Burggraf 1969, pp. 139–79

Tsien, H. S., Finston, M. 1949. Interaction between parallel streams of subsonic and supersonic velocities. *J. Aeronaut. Sci.* 16:515–28

Tu, K. M., Weinbaum, S. 1976. A non-asymptotic triple deck model for supersonic boundary-layer interaction. *AIAA J.* 14:767–75

van Driest, E. R. 1951. Turbulent boundary layer in compressible fluids. *J. Aeronaut. Sci.* 18:145–60

Van Dyke, M. D. 1975. *Perturbation Methods in Fluid Mechanics.* Stanford, Calif.: Parabolic

Vidal, R. J., Wittliff, C. E., Catlin, P. A., Sheen, B. H. 1973. Reynolds number effects on the shock wave–turbulent boundary layer interaction at transonic speeds. *AIAA Paper 73-661*

Viegas, J. R., Coakley, T. J. 1979. Numerical investigation of turbulence models for shock-separated boundary-layer flows. *AIAA J.* 16:293–94. See also *AIAA Paper 77-44*

Viegas, J. R., Horstman, C. C. 1979. Comparison of multiequation turbulence models for several shock boundary-layer interaction flows. *AIAA J.* 17:811–20

Webb, W. H., Golik, R. J., Vogenitz, F. W., Lees, L. 1965. A multimoment integral theory for the laminar supersonic near wake. *Proc. 1965 Heat Transfer Fluid Mech. Inst.*, pp. 168–89

Werle, M. J., Vatsa, V. N. 1974. New method for supersonic boundary-layer separations. *AIAA J.* 12:1491–97

Wilcox, D. C. 1973. Calculation of turbulent boundary-layer shock-wave interaction. *AIAA J.* 11:1592–94

Wilcox, D. C. 1975. Numerical study of separated turbulent flows. *AIAA J.* 13:555–56

Wilcox, D. C., Alber, I. E. 1972. A turbulence model for high speed flows. *Proc. 1972 Heat Transfer Fluid Mech. Inst.*, pp. 231–52

Wilcox, D. C., Rubesin, M. W. 1977. Progress in turbulence modeling for complex flow fields. *DCW Industries Inc. Rep. DCW-R-11-01*

Williams, P. G. 1975. A reverse flow computation in the theory of self-induced separation. See Baldwin & MacCormack 1975, pp. 445–51

Ann. Rev. Fluid Mech. 1980. 12: 139–58

FLUID MECHANICS OF THE DUODENUM

×8155

Enzo O. Macagno
Division of Energy and Institute of Hydraulic Research, The University of Iowa, Iowa City, Iowa 52242

James Christensen
Division of Gastroenterology-Hepatology, Department of Internal Medicine, University of Iowa Hospitals and Clinics, Iowa City, Iowa 52242

INTRODUCTION

Relevant Duodenal Anatomy and Physiology

Throughout the duodenum (the first 30 cm or so of the small intestine), the smooth muscle lies in two layers. In the outer layer, the muscle cells lie with their long axis parallel to the axis of the intestinal tube while the long axis of the cells of the inner layer lies along the circumference of the tube. The two layers are of nearly equal thickness and, together, are called the muscularis propria, or simply, the duodenal musculature. Although bound together, the longitudinal and the circular layers seem to be able to contract independently. These contractions produce quite different deformations of the tube: contractions of the outer longitudinal muscle shorten the tube locally, while contractions of the inner circular layer of muscle reduce its diameter. The muscularis propria is lined on the outside by the serosa and on the inside by the mucosa. This mucosa contains another muscle layer, the muscularis mucosae. This muscle layer is separated from the inner circular layer of the muscularis propria by a layer of loose fibrous tissue so that, presumably, considerable movement may occur between the mucosa and the muscularis propria. It is not known what kind of movement actually occurs here. The mucosa is not uniformly applied to the inner surface of the muscularis propria. Instead, it lies in folds, called the folds of Kerkring or the valvulae conniventes. The innermost lining of the duodenum, the epithelium, constitutes finger-like projections called villi. Each villus contains, at its core, a bundle of muscle connected to the muscularis mucosae. The cells of the intestinal

0066-4189/80/0115-0139$01.00

epithelium are specially adapted to the function of absorption of nutrients from the fluid in the intestinal lumen. Among these specializations is the existence of still smaller projections from the luminal surface of each cell. These projections, the microvilli, are not known to exhibit movement.

The surface area of the inside of the duodenum is thus much greater than that of a simple tube of corresponding diameter: the surface area is multiplied greatly by the presence of, first, the folds of Kerkring, second, the villi and, third, the microvilli. It will be evident from this review that, in models of intestinal flow, the complexity of the inner surface of the intestine has not yet been taken into account.

General View of Duodenal Contractions

The usual practice, in considering intestinal motility, is to describe contractions in terms of patterns that were seen decades ago by direct observation of the viscera exposed at operation or in fluoroscopic observation of the movements of radiopaque fluids along the intestinal lumen. These patterns were called *peristalsis*, *segmentation*, and *pendular movements*. Peristalsis and segmentation refer to luminal occlusions that seem to reflect contraction of the circular muscle layer: peristaltic contractions sweep along the intestine while segmenting contractions are stationary. Pendular movements are said to be contractions that sway back and forth. All these terms should probably be abandoned, for they are poorly defined at best and, in any event, are not useful.

Some terms must be introduced, however, and a simpler nomenclature can be used. Contractions both of the circular muscle layer and of the longitudinal muscle layer are rhythmic; that is, they repeat briefly, regularly, with a stereotyped time-course. They are also propagative, progressive, or sweeping along the tube. The rhythmicity and the propagative nature of the contractions of both muscle layers come from the existence of a pacemaking electrical signal in the musculature. The electrical potential across the membrane of the muscle cell fluctuates regularly. In all cells of a cross section the fluctuation is synchronous so that a gross electrode in contact with many hundreds or thousands of cells detects a unitary signal. When two electrodes are separated longitudinally by a few mm, the signals are not simultaneous: they are phase-locked with a phase lag. Thus, the signal appears to move in one direction, passing as a ring along the bowel. In the human duodenum, these omnipresent signals are generated at 12 cycles per minute, moving away from the stomach at a constant velocity. If and when a contraction occurs, it always occurs with a fixed phase relationship to the electrical signals. The constantly recurring electrical signals are called the electrical slow waves. At any point along the intestine, each electrical slow wave presents the muscle a choice to contract or not to contract. Approximately 30 percent of such cycles give rise to contractions.

KINEMATICS OF THE INTESTINAL WALL

Aspects of Wall Movements of Basic Interest in Fluid Mechanics

The duodenum is a tubular organ with muscular walls containing a liquid of complex and varied properties, the chyme. The flow, and mixing, of chyme is determined by the movements of the muscle layers. The ideal approach to the analysis of this flow should be by formulation of the equations that govern muscle movement and chyme rheology. We are far from understanding the correlations between muscle movements and properties of the intestinal content. Those correlations certainly exist, for the patterns of contractions are different during the fed and fasted states. The investigator of fluid mechanics of the intestine must begin with simplified models of the wall geometry and kinematics. In these models, one attempts to represent wall movements by means of prescribed simple functions, knowing, however, that those movements cannot in fact be uncoupled from the properties of the flow and chyme. In making decisions concerning intestinal wall movements in a mathematical model, one must take into account those facts that are known, however imperfectly, and try to represent the essential features. One of those facts is that the intestinal wall, when observed over short enough segments of the tube, shows organized movements, contractions occurring at intervals that are small multiples of the slow wave period. The individual movements of segments of the intestine are "the result of the concerted activity of an immense number of very small cells whose properties we are only beginning to understand" (Bülbring 1961). Seldom does a regular train of contraction sweep more than a few cm along the intestine; usually, little if any correlation can be detected along the intestine during the fed state. The movements of the intestine, or even of the duodenum alone, are very complex, and this has resulted in vague and conflicting descriptions in the writings on the subject. Perhaps the eclectic interpretation of Singerman (1974), who postulated simultaneous, and sequential, different modes of behavior, is valid in an organ in which both stimuli and responses are clearly varied. Until the wall motions are more completely described, the fluid mechanician must use a model of wall kinematics flexible enough to represent either stationary or propagative contractions. Highly simplified stationary and propagative contractions are represented

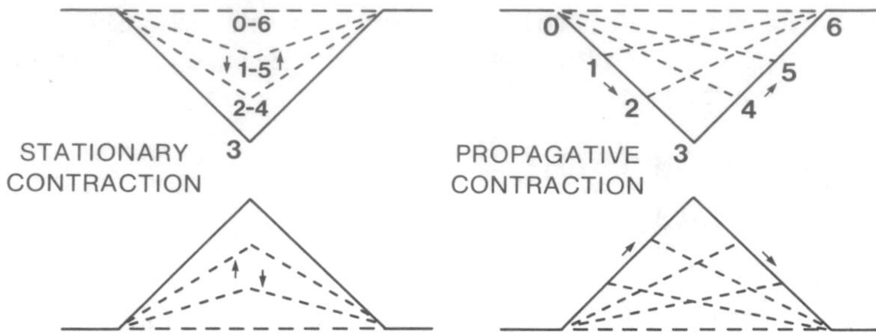

Figure 1 Stationary and propagative circular contractions. Numbers indicate successive positions of the peak of the profile. The same figures can be interpreted as diagrams, in a proper scale, of the displacements in longitudinal contractions.

in Figure 1. A stationary contraction could result from a thin annular bundle of muscle cells contracting simultaneously and dragging passive cells over a certain length on both sides. This is highly hypothetical but not without some observational basis. The propagative model is more realistic since it appears as the logical type of contraction to be determined by the propagative slow wave. One important characteristic to be included in the model is the finite length of contraction and the finite amplitude as shown in Figure 1. Longitudinal contractions can be specified by functions similar to those used for circular contractions. If the ordinates in Figure 1 are regarded as a diagram for longitudinal displacements, it is only necessary to choose a convenient scale to obtain a representation of longitudinal contractions of a certain amplitude. Specification of individual contractions is only part of the problem. Another important part is to find a way of representing the distributions of the contractions in time and space. Not much is known, however, about such distributions, and so only limited efforts are possible in this direction at present.

Ring Contractions in the Duodenum

Ring contractions in the duodenum have been studied by many different methods, but until a few years ago there were no systematic data that could be used to determine spatial and temporal correlations for these intestinal contractions. Without knowledge of such correlations, no prediction of flow due to contractions could be made. Systematic observations of ring contractions in the human duodenum were then undertaken (Christensen, Glover, Macagno et al 1971). A bundle of four polyvinyl tubes, 1 mm inside diameter and 2 mm outside diameter, were passed through the mouth into the duodenum of fasting human subjects. Each tube had a distal lateral orifice 1 mm in diameter; the distance between openings was 10 mm. Each orifice was covered with a latex rubber sleeve which was sealed against the catheter tube to make it watertight. Pressure transducers were connected at the other ends of the tubes, and the pressure variations within the closed tube were recorded on magnetic tape. Records were made for several hours after the subject had ingested a glass of skim milk. A method was developed for the computer analysis of the records (Singerman & Glover 1971). The computer program was designed so that the computer recognized practically all pressure peaks identified by eye as due to contractions, and ignored those due to causes other than intestinal contractions. The time between any two successive contractions was then recorded. The records of contractions, and of the periods of rest between contractions, were subjected to a series of statistical tests. Contractions appeared to be spatially uncorrelated. The statistical investigation was then centered on the temporal distribution of contractions and periods of rest. About 7500 events were available. From the physiological viewpoint, it was important to establish firmly the correlation between the 12/min frequency of the slow wave and that of the occurrence of contractions. In regard to fluid mechanics, the value of such a correlation and of information about the distribution of contractions and no-contractions at any point along the intestine, lies in the possibility of warranting some simplifying assumptions relating to the flow induced by contractions. The experimental distribution of intercontractile periods is shown with solid line in Figure 2 (Christensen, Glover, Macagno et al 1971). The times for the peaks and valleys of this graph were examined with a computer-aided method and found to lie at very nearly five-second intervals (Figure 3), which is the period of the slow wave. The result is really remarkable, if the different sources of uncertainty and error are taken into account. Thus the first important achievement of this study was to establish the pacing function of the electrical slow wave. From the point of view of fluid mechanics, still more important was the investigation of the statistical distribution of sequential contractions, summarized in

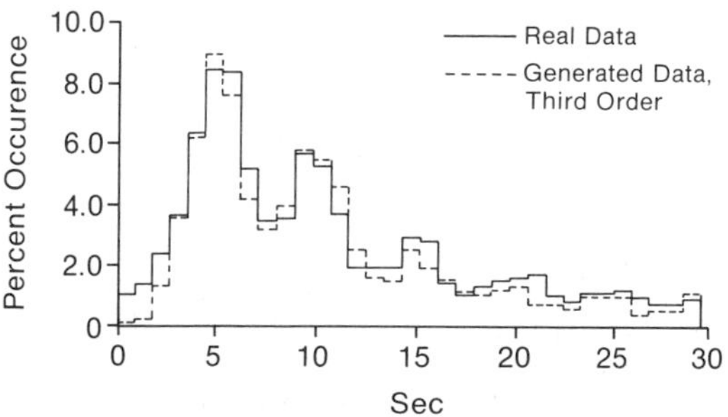

Figure 2 Frequency distribution of intercontractile times in the human duodenum.

Figure 4 (solid line). Almost 70% of the contractions are isolated events, i.e., events separated by some period of time during which the segment of duodenum is at rest. This information is necessary in attempting a description of the flow induced by contractions. A group of two successive contractions happens at a given segment only 20% of the time. More than seven contractions in a row is a rare event. The frequency distribution of periods of rest is shown in Figure 5. One can see that the two distributions shown in Figures 4 and 5 are quite different, although there is an analogy in the strong skewness of both of them. Approximately two

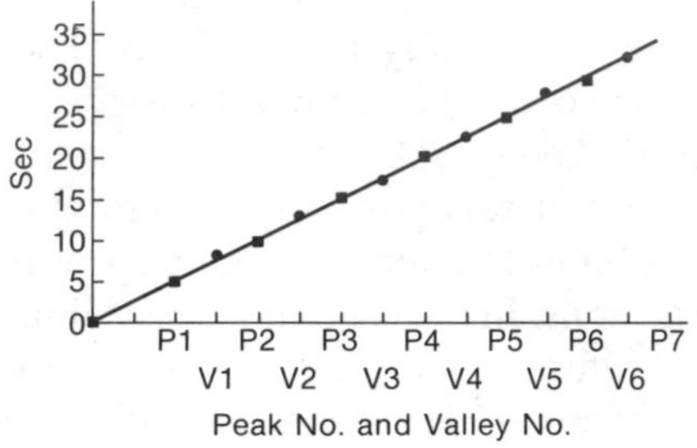

Figure 3 Correlation with time in the peaks and valleys of the frequency distribution of intercontractile periods.

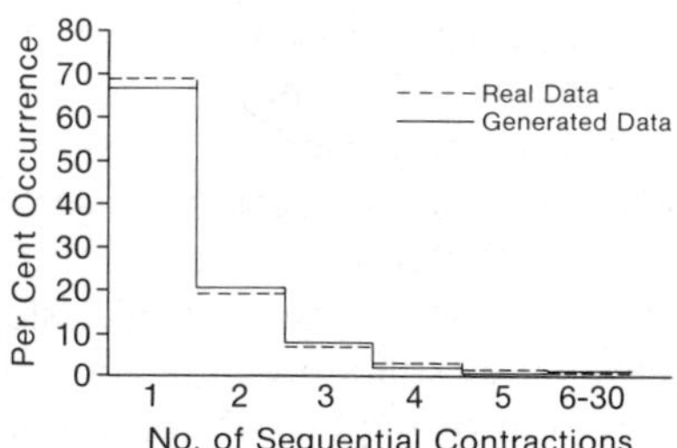

Figure 4 Frequency distribution of sequential contractions in the human duodenum.

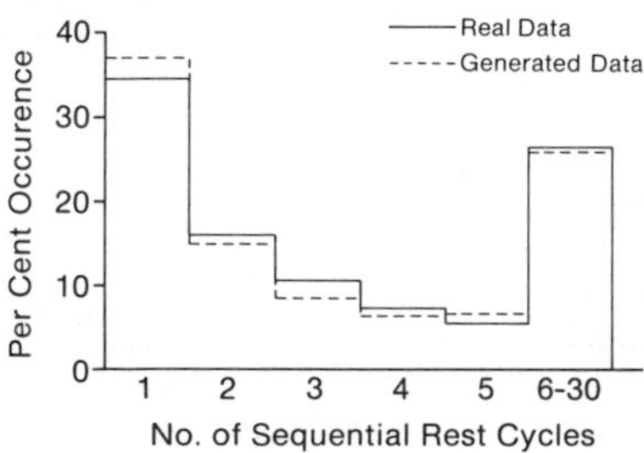

Figure 5 Frequency distribution of sequential rest cycles in the human duodenum.

out of three rest cycles are of a duration of 10 seconds or more. Such interruptions of wall motion are rather important if the flow can be assumed to be creeping flow. Then flow would cease when the wall rests. Superposition of the effects of a number of contractions may be greatly simplified.

Stochastic Model for Ring Contractions

After the statistics of the contractions were obtained, an attempt was made to model contractions at a single point of the duodenum as a simple series of independent events; the results were considered unsatisfactory, and a more refined analysis was undertaken. A Markov-type model was devised in which dependence on a given number of previous slow wave cycles was assumed and investigated (Singerman, Macagno, Glover et al 1975). Several models of different order were tested, the order being given by the number of previous events assumed to have an influence on the present event. The occurrence of duodenal contractions lends itself to an idealization in which the time is discretized in units of five seconds, the period of the slow wave in man. A decision must be made as to whether or not a contraction will occur for any of the five-second periods. To illustrate the method in a simple manner, one of the simplest generators of random numbers can be used. Assuming that the probability of a contraction is one-third, a die could be thrown; if the result were 1 or 6, a contraction would be assigned, and if it were 2, 3, 4, 5, a no-contraction would be assigned. If contraction was obtained, generation of a second random number would be used to assign the time of occurrence within the five-second period. This, of course, explains the zeroth-order model, but no essential differences appear in the others. For the actual model, two basic schemes to generate quasi-random numbers were used. Once several thousand results were obtained, generated rather than observed frequency distributions could be plotted. The third-order stochastic model appeared to give enough agreement with the real data (compare dash lines with solid lines in Figures 2, 4, and 5) to be useful in the predictions of transport in the duodenum. This model has been useful over a certain research period, but surely several improvements would be desirable if more work of this kind were to be done.

Longitudinal Movements

Longitudinal movements of suspended segments of animal intestines have been observed for a long time, and these observations are sometimes part of the laboratory work for students of physiology. Direct observation of segments of duodenum indicates that contractions of the longitudinal muscle layer can occur independently of those of the circular muscle layer. These contractions may be what Cannon (1912) and others described as pendular movements. In spite of the knowledge of these movements, no correlation with the fluid flow of the intestinal content was ever investigated until we undertook such a study. Study in human subjects was deemed too difficult, and so an investigation in vitro was undertaken (Melville, Macagno & Christensen 1975). Segments of healthy duodenum were mounted horizontally in a bath in such a way that no sagging would be visible. This introduced slight tension, and may be the reason why these rhythmic contractions were sustained for relatively long times. The serosal surface was marked with points of India ink at equal distances along the intestine. At times, the motility was purely longitudinal, and those periods were filmed with a movie camera. The results for twelve points were analyzed with a Vangard film analyzer (Figure 6). When longitudinal contractions prevailed, they occurred in an organized manner. But when circular or ring contractions were present, a random picture resulted most of the time. Until more observations are available, it is impossible to say if longitudinal contractions are preferentially more regularly distributed in time and space than circular muscle contractions.

Once in possession of records of longitudinal contractions, we attacked the problems of representing the muscle movements and of determining the induced flow. A simplified analytical model of the wall motility as controlled by the slow-wave propagation was developed. The idea was essentially to explore the possibility of modeling the contractions in a

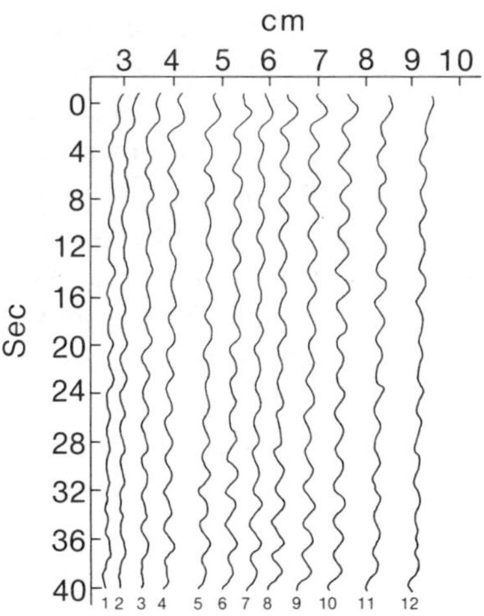

Figure 6 Experimentally determined displacements for twelve points in longitudinal contractions over a segment of duodenum.

simple case, as a first step towards much more complex modeling. The model adopted consisted of contractile and distensible elements that were paced by the propagating slow wave (Macagno, Melville & Christensen 1975). In this manner the displacements δ_0 for points of a segment of duodenum mounted with fixed ends were shown to be given by a relatively simple equation

$$\delta_0 = \frac{ax_0}{2kl}[\sin(kl-\omega t)-\sin(kx_0-\omega t)] - \frac{a(l-x_0)}{2kl}[\sin(kx_0-\omega t)+\sin\omega t]. \quad (1)$$

Herein, δ_0 is the displacement at a distance x_0 from the origin of the segment, a is a fraction defining the maximum shortening of a cell during contraction, l is the length of the intestinal specimen, k is the wave number, ω is the angular frequency of the slow wave. If the dimensionless abscissa $\xi = x_0/l$ is introduced, Equation (1) can be transformed into

$$\delta_0 = (a\lambda/4\pi)\{[\xi \sin 2\pi l/\lambda - \sin 2\pi l\xi/\lambda] \cos\omega t + [\cos 2\pi l\xi/\lambda - \xi \cos 2\pi l/\lambda - (1-\xi)] \sin\omega t\}, \quad (2)$$

which in turn can be cast into the form

$$\delta_0 = A\cos(\omega t - \alpha) \quad (3)$$

with the local amplitude A given by

$$A = (a\lambda/4\pi)\{[\xi \sin 2\pi l/\lambda - \sin 2\pi l\xi/\lambda]^2 + [\cos 2\pi l\xi/\lambda - \xi\cos 2\pi l/\lambda - (1-\xi)]^2\}^{\frac{1}{2}} \quad (4)$$

and the phase lag given by

$$\alpha = \tan^{-1}\frac{\cos 2\pi l\xi/\lambda - \xi\cos 2\pi l/\lambda - (1-\xi)}{\xi \sin 2\pi l/\lambda - \sin 2\pi l\xi/\lambda} \quad (5)$$

The dimensionless expression between brackets in Equation (4) could be expressed in terms of an amplitude function A_0 given by

$$A_0^2 = \xi^2 + 1 + (1-\xi)^2 - 2\xi\cos[b(1-\xi)] + 2(1-\xi)(\xi\cos b - \cos b\xi). \quad (6)$$

The theoretical amplitude function is given in Figure 7 in terms of the number of slow waves contained in the segment of intestine under study. As could be expected, if the segment were exactly the length λ of a slow wave, the amplitude function would be very simple. This is not easy to achieve experimentally, however. Figure 8 shows a comparison of predicted and measured displacements for a segment of length equal to 1.36 the length of the slow wave. A better agreement was obtained for the phase lag.

This analysis of longitudinal contractions must be considered to be only exploratory, and more experiments are necessary before firm conclusions can be drawn. The model probably needs more elements representing muscle behavior more realistically. Those who undertook this study were, however, much more interested in considering the fluid mechanical functions of the longitudinal contractions, which were completely unknown at the time. Work in the area of muscle behavior had to be postponed in spite of its obvious interest.

In a mathematical model of intestinal flow, Lew, Fung & Lowenstein (1971) considered very sharp ring constrictions at regular distances

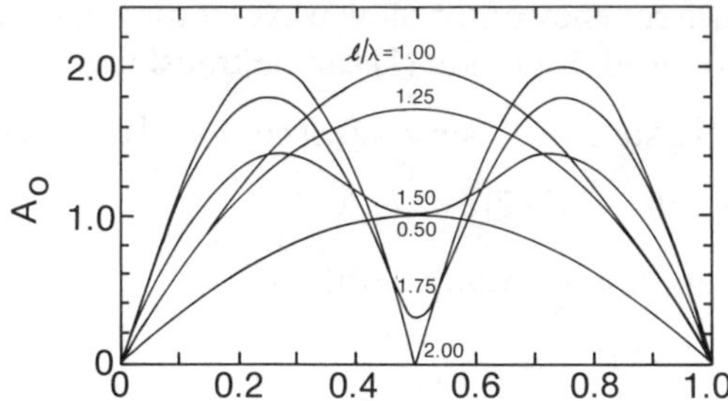

Figure 7 Amplitude function in the theoretically determined longitudinal contractions over a segment of duodenum.

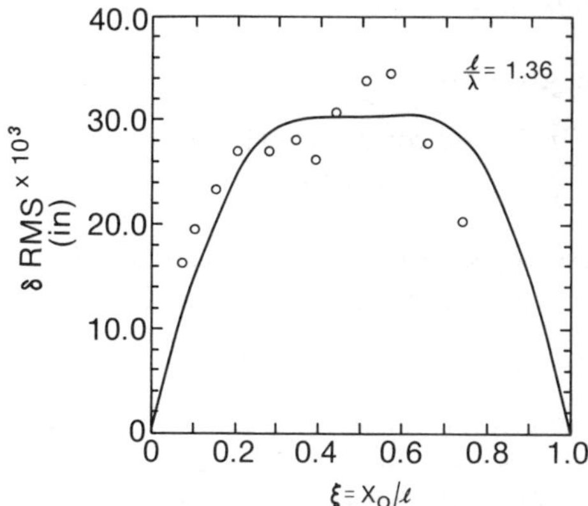

Figure 8 Best fit of theoretical to experimental rms values of amplitudes of longitudinal displacements in a segment of duodenum.

propagating along a cylindrical conduit. This assumption falls in the category of regular peristaltic waves, but it may be somewhat more realistic than the sinusoidal waves assumed by others who indicated the possibility of utilizing them for the small intestine. The use of sinusoidal waves in this connection was discussed by the authors elsewhere (Christensen & Macagno 1978).

FLOW INDUCED BY INTESTINAL MOTILITY

Only a few investigations exist on the subject of flow induced by intestinal motility. This is surely due to the uncertainties about the characteristics of wall movements in the gut. The investigations specifically related to the duodenum are based on extremely simplified assumptions concerning not only the geometry and kinematics of the walls but also the rheology of chyme. It is usual to assume that the Reynolds number and the inverse of the Strouhal number are small enough to justify use of the classic simplification of the Navier-Stokes equations that was introduced by Stokes in dealing with creeping flow. In the absence of more basic information on the intestine and its content, the simplification appears to be acceptable.

Stochastic-Deterministic Model

Singerman (1974) used a mathematical model of the duodenum to examine the classical view that stationary circular contractions (segmentation) cause mixing and propagative circular contractions (peristalsis) cause pumping. He considered only circular contractions, mainly because at that time the longitudinal contractions were little known. He assumed that each contraction generated motion of chyme only during its active period, a hypothesis consistent with creeping flow. Singerman introduced circular contractions of cylindrical shape of any desired length as the basic element of the wall motion. This length could be as short as required to produce, by a combination of elementary contractions, any spatio-temporal law for the wall kinematics, including stationary or propagative modes of any shape. This gave great flexibility to the model; only by using finite-element techniques could a more versatile and more accurate technique be developed.

In the following discussion, the cylindrical contraction will be used, although, in fact, shapes more similar to the actual ones were also used by Singerman. In Figure 9, a very simple situation is illustrated for a tube of length L divided into two inactive segments with an active one exactly in the middle. During an interval of time T, the contraction is supposed to proceed according to a specified law: $r = f(t)$. In the figure, the wall is assumed to be moving inwards and forcing fluid out to the adjacent segments. The flow is assumed to be governed by Poiseuille laws, which certainly is a crude approximation. However, a physical model was constructed to compare calculations based on this assumption with actual behavior, and discrepancies were surprisingly small. When the pressure differential $\Delta p = p_1 - p_4$ is zero, the flow is symmetric; the pressure line, also symmetric, is shown in Figure 9. If a pressure differential Δp exists, the flow becomes increasingly asymmetric as Δp increases. For a certain Δp, the flow may be arrested in one tier, as shown in the figure as the case in which the dimensionless pressure difference is $\Delta p = 6$. For larger values of Δp, the flow becomes unidirectional. Simple considerations of conservation of volume lead to expressions for the flow rate in the contracting (or expanding) element. For instance, when $\Delta p = 0$ in Figure 9 the volume flux Q is given by

$$Q = 2\pi R x V = \pi R^2 U. \tag{7}$$

In this case, Poiseuille's law for pressure drop gives then a second-degree parabola for the pressure line. In the inactive segments, the pressure drop follows a linear law. Diagrams for the volume flux Q and the pressure p

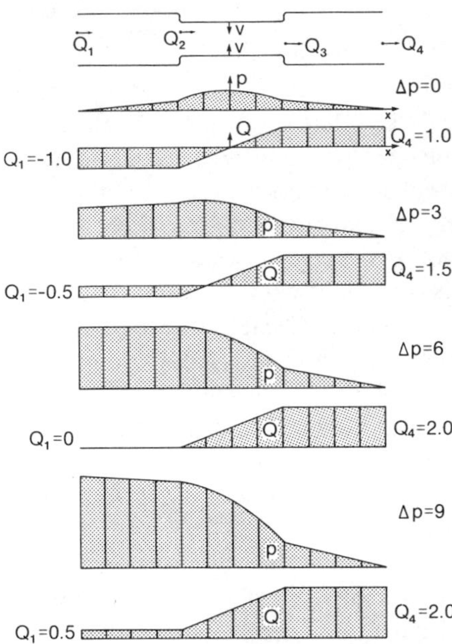

Figure 9 Elementary analysis of flow and pressure variation in a tube with a central contraction. Volume flux and pressure have been made dimensionless with the maximum values in the contractile segment when $\Delta p = 0$.

as functions of x are given in Figure 9 for four values of Δp. It should be realized that these diagrams are valid at a certain instant of time only; as the wall moves, they change. Similar calculations are possible for any number of contractions along a tube. Computer programs were developed for those calculations that could represent noncylindrical discrete contractions, and could receive as input any distribution of contractions in space and time. Thus a statistical distribution of contractions could be included if desired, but a simpler procedure was developed for such a case.

With reference to stationary symmetric contractions, at least two contractions are needed to produce pumping. Since stationary contractions are easier to model physically, a model in which such contractions could be produced was constructed to verify the assumptions made. In Figure 10 a comparison between predicted and measured values of flow rate, Q, versus phase lag of the contractions is given. In this case regular successive contractions were used.

The stochastic-deterministic model was developed for times that are long relative to the cycle of the slow wave, so that statistical averages could be used. The method can be described meaningfully for a short length of tube containing only four contractions: Singerman (1974) used the full length of the duodenum except for two inactive segments at the ends. With reference to four contractions, all possible events were considered, since the statistical study revealed that periods of rest alternate with active periods in a seemingly random fashion. It was found that 11 combinations of contractions could occur that would produce forward motion of chyme, and that another 11 could produce retrograde motion. Due to the differences in phase lag, however, a net flow in the caudad

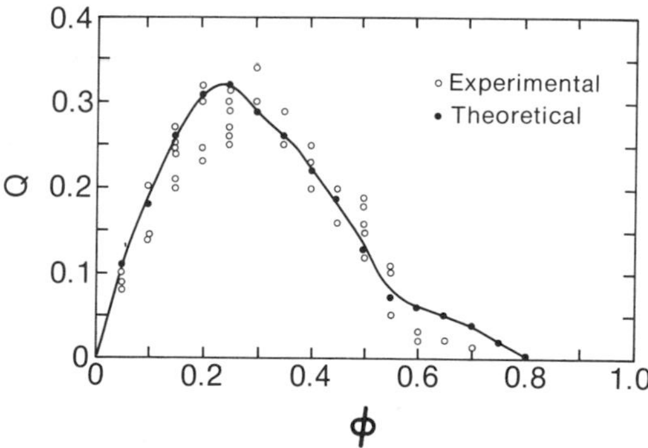

Figure 10 Theoretical curve and experimental points for flow induced in a physical model by two contractile segments. Dimensionless volume flux Q is given as a function of phase lag ϕ between contractions.

direction would result after a long number of contractions. The transport thus predicted was compared with estimates based on input from the stomach, amount of secretion and absorption, and output into the jejunum. The predicted value of flow rate was 17 cm^3/min for the human duodenum, which falls between the physiologically determined upper and lower bounds. This result is encouraging, but much more work is necessary before the method can be entirely relied upon. The many simplifications that were introduced may have to be revised, and the possibility of compensating errors should not be overlooked. One important aspect to be understood about this model is that it can easily, although clumsily, be made to include propagative contractions. Consideration of inertial effects, if it ever becomes necessary to take them into account, would be much more difficult, and probably not justified unless refinements in the hydrodynamics were introduced in this model. This would lead to a substantial change in the methodology, in fact, to another approach.

Propagative Ring Contractions

After the studies of Sancholuz (Sancholuz 1974, Sancholuz et al 1975a,b) concerning detailed spiking along and around segments of duodenum in vitro, the concept of propagative contractions of finite length was reinforced, and an investigation was undertaken to determine the flow induced by this type of contraction. The wall motion was represented by a rather simple propagative wave of triangular shape as described by Christensen & Macagno (1979) (see Figure 1). The flow was considered to be governed by the biharmonic equation, and a finite-element technique was developed utilizing a fifth-degree polynomial function. Results of these calculations are included in D. Stavitsky's doctoral dissertation (1979). Only one contractile element has been considered, and the cumulative effect of a series of contractions of this one element has been determined. The pumping effect is illustrated in Figure 11, in which a central particle initially at one side of the contractile segment is shown in its successive steps back-and-forth until it passes to the other side of the contraction.

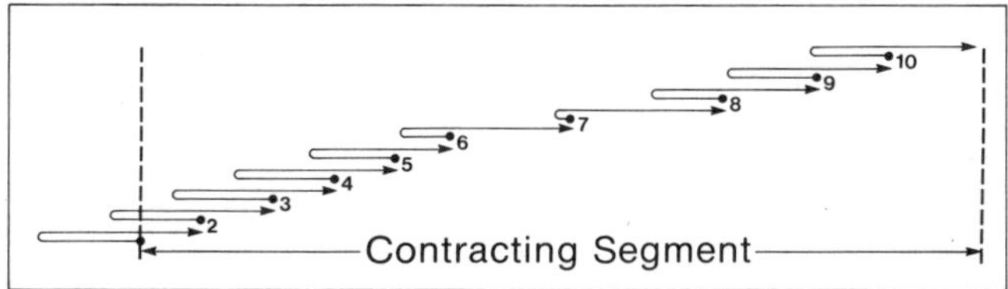

Figure 11 Displacements due to a sequence of propagative circular contractions. The segments representing displacements are shifted for the sake of clarity, but they belong all to the center line of the tube.

Study of the effects of a number of contractions becomes more difficult with this approach, not so much because of intrinsic problems as because of the large amount of computer time required. The method is now available and its future use should be determined in terms of the values attributed to this possibility as compared with others.

Traveling Sharp Ring Contractions

Deep and sharp traveling ring constrictions, uniformly distributed along a cylindrical tube of radius *a*, were assumed by Lew, Fung & Lowenstein (1971) as a model of intestinal motility. The general solution was presented as a combination of three particular solutions in terms of the dimensionless distance between constrictions, $\lambda = L/a$, and the relative amplitude of the constrictions, $\eta = c/a$. Velocity distributions were determined and patterns of flow inferred, apparently, from them. Curves representing the effectiveness of two types of peristalsis were given and are reproduced in Figures 12*a* and *b*. For the peristalsis of pure transport, the effectiveness was defined as the induced dimensionless average velocity ($u/U = u'$). For the peristalsis of pure compression, the effectiveness was defined as the induced mean average pressure gradient [$a^2(\partial p/\partial x)/\mu U = J$, where μ is the viscosity of chyme]. In connection with the usual assumption of a Newtonian fluid and a regular geometry, Lew, Fung & Lowenstein expressed the view that "an introduction of a minimum amount of irregularity in the geometry of the effect of the non-Newtonian property of chyme into the formulation of the problem results in a formidable

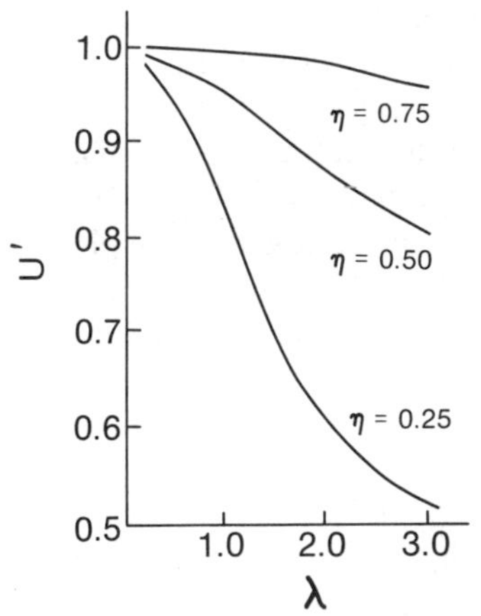

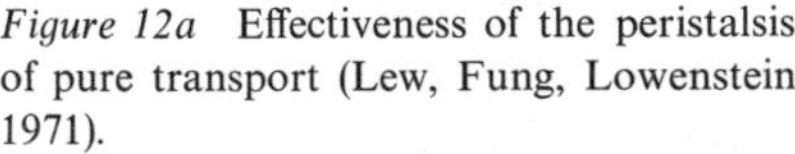

Figure 12a Effectiveness of the peristalsis of pure transport (Lew, Fung, Lowenstein 1971).

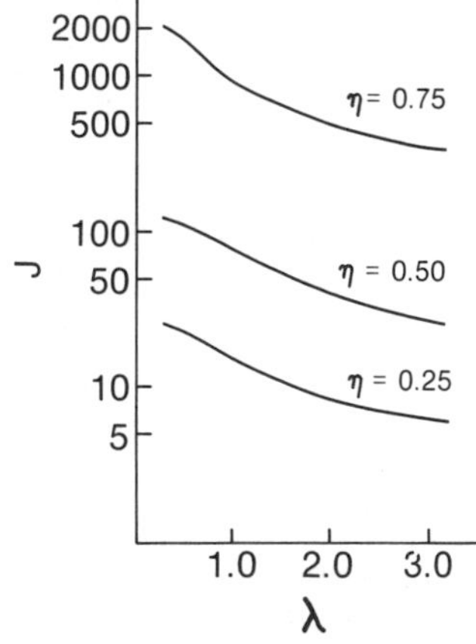

Figure 12b Effectiveness of the peristalsis of pure compression (Lew, Fung, Lowenstein 1971).

complication of the algebra in solving the problem." Unpublished measurements by R. van Winkle in 1976 seem to indicate that, at least for the fasted state, the duodenum may contain a fluid that does not behave too differently from a Newtonian liquid. At least for certain types of non-Newtonian behavior, the work of Giaquinta & Hung (1968) in flow expansions at low Reynolds numbers indicates rather small effects relative to the flow of a Newtonian fluid in the same expansion. Of course, with the ignorance of what kind of non-Newtonian behavior the chyme of different types may present, there is little that can be said in general terms. Only a systematic investigation of the rheology of chyme can solve the problem.

If the role of sweeping contractions in the duodenum were to move secretions into lower segments of the intestine, the results of Lew et al certainly provide a reasonable mechanism for such a function. As stated by these investigators, sweeping contractions may also be the mechanism by which the material in the ileum is passed to the large intestine.

TRANSPORT AND MIXING IN THE DUODENUM

In a flowing fluid, the rates of transfer processes and reactions are usually affected by the flow. This is reflected in the governing equations by the presence of convective terms together with diffusive and source-sink terms. Convective mixing of chyme has been recognized as one of the functions accomplished by intestinal motility, but besides general statements to this effect, little is known of a quantitative nature about such mixing. A first step in gaining some insight into the transport and mixing of chyme is achieved if the proper visualization of the flow induced by the wall movements is attained. Here one encounters a difficulty that is actually intrinsic to fluid mechanics. An Eulerian description of a flow is usually obtained without great difficulty, but only a Lagrangian description is really useful and meaningful to the physiologist or the physician, and even to the fluid mechanician (Denli 1975). A second difficulty of a fluid mechanical nature stems from the scanty quantitative knowledge about laminar mixing, a subject that is of crucial importance if the flow is a creeping flow or nearly so.

Mixing Induced by Circular Contractions

An isolated stationary circular contraction induces no permanent convective mixing if the flow is laminar, but propagative ring contractions produce permanent convective mixing. Successive contractions tend to mix more and more. However, from an initial evaluation of the effect of these contractions (Stavitsky 1979), it seems that they produce less mixing

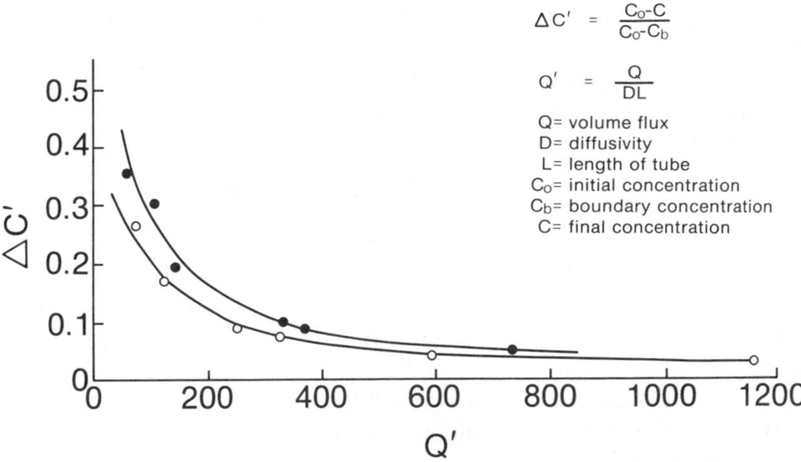

Figure 13 Relative depletion $\Delta c'$ of sodium chloride in a tube with contractions (●) and without contractions (○) in terms of dimensionless flow rate, Q'.

than would be required for a realistic increase in food uptake by the intestine. This is a very provisional conclusion, but it has gained some backing from experiments that are part of the doctoral work of Y. C. Lee (1978) at the Iowa Institute of Hydraulic Research. In these experiments, a flexible porous tube is subjected to mechanically-induced contractions. The tube contains an aqueous solution of sodium chloride, and is immersed in a bath of circulating water with a much lower sodium chloride concentration. The diffusion of salt through the walls of the tube is monitored with a conductometer for different patterns of contractions of the tube wall. The contractions definitely produce an increase in the rate of transfer (Figure 13), but it is smaller than expected (Christensen & Macagno 1979). It is quite possible that other patterns of contractions and other types of contractions could produce a more significant effect. This appears to be the case with propagative longitudinal contractions.

Mixing Induced by Longitudinal Contractions

The laminar mixing induced by longitudinal contractions of a propagative type which occur over a finite length of tube is illustrated in Figure 14 (Stavitsky 1979). The propagative contraction used has been described by Christensen & Macagno (1979). The fluid is marked assuming a certain distribution of concentrations over fluid segments. The Lagrangian description has been obtained by numerical integration after the Eulerian description of the flow was obtained. Of course, this result comes from a study under highly simplified conditions, but so do those for ring contractions. A comparison of both types of contractions from the point of view of mixing effects is warranted, and is presently underway. Figure 14 indicates the ability of a sequence of longitudinal contractions to produce effective convective laminar mixing. Much remains to be investigated in this area, including the basic aspects of laminar mixing.

Laminar Mixing: Convective and Diffusive Contributions

Given a fluid with a nonuniform concentration of a substance, the mixing resulting from purely molecular processes should, in general, be less in the fluid at rest than in fluid subject to, for example, shearing flows. One can certainly think of some nonuniform distribution that could become uniform given the proper fluid displacements, but this is most likely a very special case. It is assumed therefore that, in general, if forced convection is effected an initially nonuniform distribution would tend to become more rapidly uniform than by diffusion alone. This would be the case in Figure 15. By using a flow simpler than that induced by the longitudinal propagative contractions, but still such that a similar shearing of the fluid relative to the concentration distribution would occur, the acceleration of the mixing process can be estimated.

In Figure 15, the space between two coaxial cylinders is shown filled with two segments of fluid marked differently. It is supposed that there is a difference in concentration of any substance between the two portions of fluid. At a given moment, the outer cylinder is rotated to induce a shearing flow of the fluid. The results after one and two turns are shown

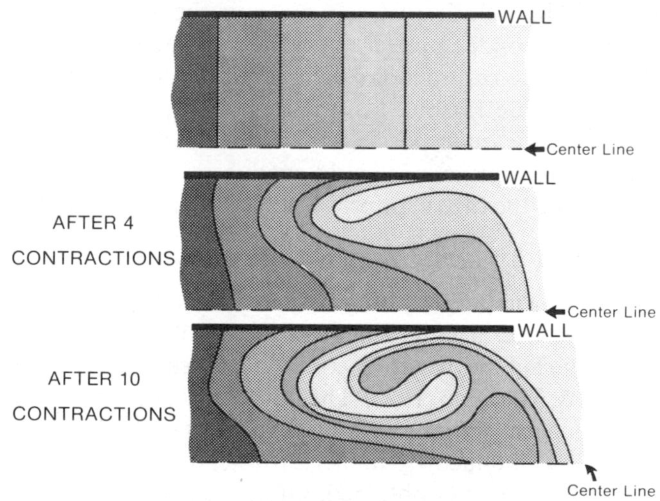

Figure 14 Lagrangian description of flow induced by longitudinal contractions.

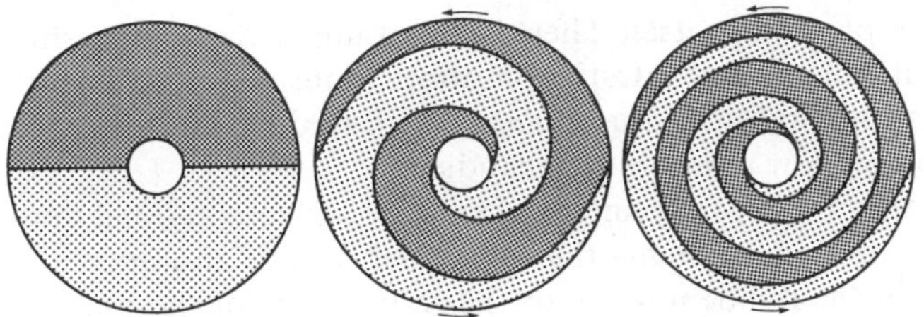

Figure 15 Lagrangian description of flow induced by circular Couette flow.

in Figure 15. It can be seen that after one rotation three layers are formed, and after two turns, there are five layers. It is easily deduced that the number of layers is $2n + 1$ if the number of turns is n. If the space between cylinders is a, the average thickness is taken as an indication of the degree of mixing; after two turns the mixing has increased in the ratio of five to three relative to one turn. If one assumes that the shearing motion occurs rather rapidly relative to the diffusional mixing, the latter can be calculated on the basis of occurring after the convective mixing. As a measure of the diffusional mixing, one can use the time necessary for a given change in concentration to occur at the centerline of a layer. This time is proportional to the square of the thickness of the layer. Therefore, increase in the number of layers greatly accelerates the mixing process. For instance, if convective mixing is increased by a factor of ten, the convective-diffusional mixing will be increased by a factor of a hundred. With this simple method, an estimate of the laminar mixing, in cases where the Lagrangian description of the layers generated is already obtained, can be made. Of course, by integration of the convective-diffusion equation more exact values could be obtained, but this would be rather expensive.

RESEARCH PROSPECTIVES

The research prospectives for the fluid mechanics of the duodenum should be contemplated within the context of the fluid mechanics of the entire alimentary canal, of which the duodenum is only one segment. The alimentary canal is obviously not uniform in its motor behavior from one place to the next, so it should be studied as a system constituted of a series of different organs with some general motor functions in common, but each having its own specialized wall motions.

The study of the fluid mechanics of the small intestine is only beginning. Nearly all the findings reported need confirmation or clarification. The matter is fraught with great difficulties not only of methodology but also of the physiology itself. There is, for example, the relationship between intestinal content and intestinal motions; intestinal movements seem to vary both in quality and in quantity with feeding, and the effect of such variation on flow remains to be studied.

In addition to the movements of circular and longitudinal muscle layers there are other movements that presumably have an important function on flow in the duodenum. Both the movements of the mucosa and the contractile activity of the villi may be important in the stirring of the boundary layer within which nutrient uptake is accomplished. These movements and the corresponding fluid mechanics remain to be investigated.

ACKNOWLEDGMENTS

This work was supported by National Institute of Health Grant AM 08901.

Literature Cited

Bülbring, E. 1961. Motility of the intestine. *Proc. R. Soc. Med.* 54:773–75

Cannon, W. B. 1912. Peristalsis, segmentation and the myenteric reflex, *Am. J. Physiol.* 30:114–18

Christensen, J., Glover, J. R., Macagno, E. O., Singerman, R. B., Weisbrodt, N. W. 1971. Statistics of contractions at a point in the human duodenum. *Am. J. Physiol.* 221:1818–23

Christensen, J., Macagno, E. O. 1978. Motility and flow in small intestine. *J. Eng. Mech. Div. ASCE* 104: Paper 13516, 11–29

Christensen, J., Macagno, E. O. 1979. Small intestinal motility: The problem of relating contractions to flow. *International Colloquium in Gastroenterology: New Frontiers in Diarrheal Disorders*, ed. H. Janowitz, D. Sacagr. Montclair, NY: Projects in Health

Denli, N. 1975. *An analytical model of flow induced by longitudinal contractions in the small intestine.* M.S. thesis. Univ. Iowa, Iowa City

Giaquinta, A. R., Hung, T. K. 1968. Slow non-Newtonian flow in a zone of separation. *J. Eng. Mech. Div. ASCE* 94:1521–38

Lew, H. S., Fung, Y. C., Lowenstein, C. B. 1971. Peristaltic carrying and mixing of chyme in the small intestine (an analysis of a mathematical model of peristalsis of the small intestine). *J. Biomech.* 4:297–315

Macagno, E., Melville, J., Christensen, J. 1975. A model for longitudinal motility of the small intestine. *Biorheology* 12: 369–76

Melville, J., Macagno, E. O., Christensen, J. 1975. Longitudinal contractions in the duodenum: their fluid-mechanical function. *Am. J. Physiol.* 228:1887–92

Sancholuz, A. G. 1974. *A statistical study of spike bursts on the slow wave in the duodenum.* M.S. thesis. Univ. Iowa, Iowa City

Sancholuz, A. G., Croley, T. E., Christensen, J., Macagno, E. O., Glover, J. R. 1975a. Phase lock of electrical slow waves and spike bursts in cat duodenum. *Am. J. Physiol.* 229:608–12

Sancholuz, A. G., Croley, T. E., Macagno, E. O., Glover, J. R., Christensen, J. 1975b. Distribution of spikes in cat duodenum. *Am. J. Physiol.* 229:925–29

Singerman, R. B., 1974. *Fluid mechanics of the human duodenum.* Ph.D. thesis. Univ. Iowa, Iowa City

Singerman, R. B., Glover, J. R. 1971. Computer recognition of contractions in the small intestine. *IIHR Rep. No. 134.* Univ. Iowa, Iowa City

Singerman, R. B., Macagno, E. O., Glover, J. R., Christensen, J. 1975. Stochastic model of contractions at a point in the duodenum. *Am. J. Physiol.* 229:613–17

Stavitsky, D. 1979. *Flow and mixing in a contracting channel with applications to the human intestine.* Ph.D. thesis. Univ. Iowa, Iowa City

Ann. Rev. Fluid Mech. 1980. 12: 159–79

DYNAMIC MATERIALS TESTING: BIOLOGICAL AND CLINICAL APPLICATIONS IN NETWORK-FORMING SYSTEMS

Larry V. McIntire
Department of Chemical Engineering, Rice University, Houston, Texas 77001

INTRODUCTION

The fluid mechanics of bodily processes is a subject of obvious importance. There are many fluids which must be transported around the various conduits of the body, including blood, lymph, and mucus. Many reviews on the subject of hemodynamics and hemorheology have been published recently, including one in this series (Goldsmith & Skalak 1975). Lymph and mucus are much less well studied. The rheological characteristics of all of these fluids are extremely important to their proper functioning. Sometimes small changes in material properties (for instance mucus elasticity) can lead to large alterations in the transport ability of parts of the body (for example ciliary transport). Because of the complex interrelationship of fluid flow and nervous control of vessel diameter in parts of the circulation, the study of the rheology of these fluids is, of course, not sufficient to predict the flow in actual in vivo experiments, but it is necessary before flow calculations can be done. Perhaps more importantly, in vitro study of the rheology of biological fluids can be used to examine the causes of pathological states, which result in grossly altered fluid mechanics. Thus knowledge of the rheology of these fluids may lead to understanding of the biochemical and biophysical basis of the disease. In addition, it may lead to a prediction of possible drug therapies that will return the fluid rheology to normal.

Since the field of biorheology and biofluid mechanics is so vast, this review will concentrate on the use of dynamic rheological testing techniques to characterize two important fluids capable of forming three-dimensional protein networks under in vivo conditions. The formation of

0066-4189/80/0115-0159$01.00

these networks enormously alters the fluid rheology, transforming a low viscosity, nearly Newtonian fluid, into a viscoelastic gel. This transition has important implications for subsequent fluid dynamic experiences of the material. The two areas to be covered are coagulation-structure formation and the viscoelastic properties of mucus. Although blood plasma and mucus appear at first glance to be quite distinct, both are capable of forming network structures when some of the proteins contained in the fluid are acted on by certain enzymes, or when the physico-chemical environment is rapidly changed. Thus, the rheological equipment necessary to investigate this structure formation is the same. Small-amplitude oscillatory testing is ideal for this purpose because the network structure is not destroyed by the measurements (as in viscosity determination). Also, the material properties obtained, an elastic and viscous modulus, can be related, using network theory, to the underlying molecular structure.

Coagulation rheology is examined first, including a brief description of the biochemistry of coagulation and the rheological instruments used for these investigations.

COAGULATION BACKGROUND

Coagulation is essentially the polymerization and crosslinking of a protein monomer fibrin, which is formed from the circulating blood protein, fibrinogen, by the action of a proteolytic enzyme. The "coagulation cascade" is a sequential activation of circulating zymogens in the blood, ultimately resulting in the conversion of fibrinogen to fibrin (Williams 1977). The enzymes of the "coagulation cascade" along with calcium and fibrinogen are termed coagulation factors and denoted by Roman numerals (Figure 1). Coagulation that occurs because of foreign surface contact (Factor XII activation) proceeds by the "intrinsic" system. If the coagulation process is initiated by Factor VII and tissue activator, the "extrinsic" system has been used. After Factor X activation, the extrinsic and intrinsic mechanism coincide (Figure 1). The term "cascade" results from the observation that enzyme activation occurs sequentially (MacFarlane 1964). In the control of the coagulation cascade, feedback effects are important. Thrombin may act in an autocatalytic role once formed and increase the rate of Factor X activation. Enzyme inhibitors also significantly affect the rate of coagulation. The principal inhibitors in the circulatory system are antithrombin III and α_2-macroglobulin. Antithrombin III is an inhibitor of the serine proteases thrombin (Abildgaard 1967), Factor X_a (Yin, Wessler & Stoll 1973), Factor XI_a (Rosenberg, McKenna & Rosenberg 1975) and Factor XII_a (Stead, Kaplan

& Rosenberg 1976), while α_2-macroglobulin is principally a thrombin inhibitor (Abildgaard 1969).

The circulating precursor of the fibrin monomer is fibrinogen, a protein of approximately 340,000 molecular weight (MW). The fibrinogen molecule is composed of three pairs of peptide chains—Aα, Bβ, and γ—having molecular weights of about 64,000, 57,000, and 48,000 respectively (McDonagh et al 1972). These peptide chains are linked by 28–29 disulfide bridges (Henschen 1964) and by electrostatic interaction and hydrogen bonding. At physiological pH, the fibrinogen molecule is globular with a diameter of 22–23 nm and a radius of gyration of 11 nm (Hudry-Clergeon et al 1975, Koppel 1970). This conformation is determined primarily by the strong electrostatic attraction between the negatively charged N-terminal disulfide knot and the positively charged C-terminal end of the Aα, Bβ, and γ chains.

Fibrinogen is converted to fibrin by the cleavage of fibrin peptides A and B (the negatively charged N-terminal ends of the Aα and Bβ chains) by thrombin (Blomback et al 1966). With the electrostatic interaction between the C- and N-terminal ends no longer present, the

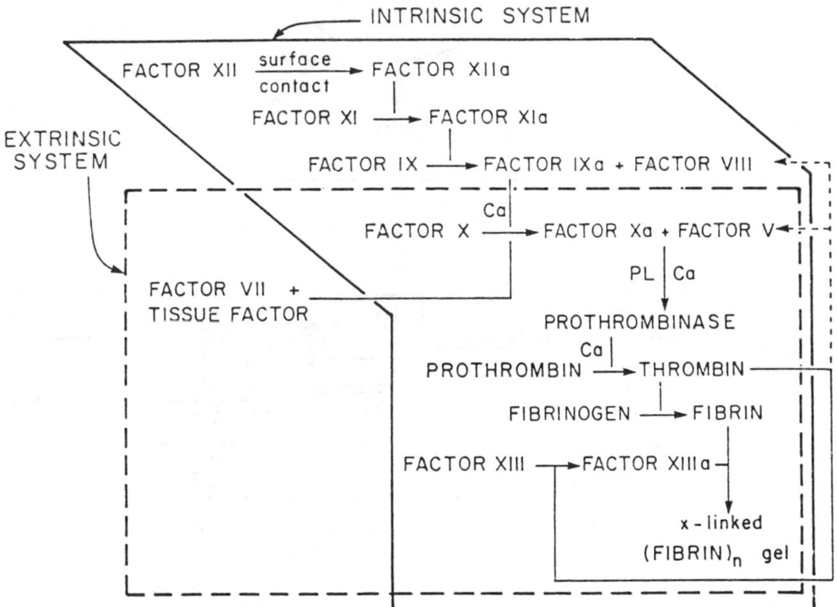

Figure 1 The cascade system for blood coagulation. Solid lines without arrowheads indicate a catalytic function. Solid arrows denote a conversion of substrate or reactant to product. Broken lines with arrows indicate the enhancement of Factors VIII and V activity by thrombin. Ca = calcium; PL = phospholipid.

molecule unfolds to yield a "squashed" T-shaped moiety (Figure 2). These fibrin monomers then associate electrostatically to yield a fibrin gel, consisting of linear arrays of the T-shaped molecules (Pouit, Hudry-Clergeon & Suscillon 1973).

Factor XIII is a thrombin-activated transamidase, which catalyzes the formation of ε-(γ-glutamyl) lysine bonds between the α and γ chains of fibrin to yield α polymers and γ dimers (Chen & Doolittle 1970, McKee, Mattock & Hill 1970, Kanaide & Shainoff 1975). A total of six covalent crosslinks are formed per fibrin monomer (Pisano et al 1972). The formation of these crosslinks yields a clot of greatly increased mechanical strength, insoluble in agents that disrupt hydrogen bonding and charge interaction.

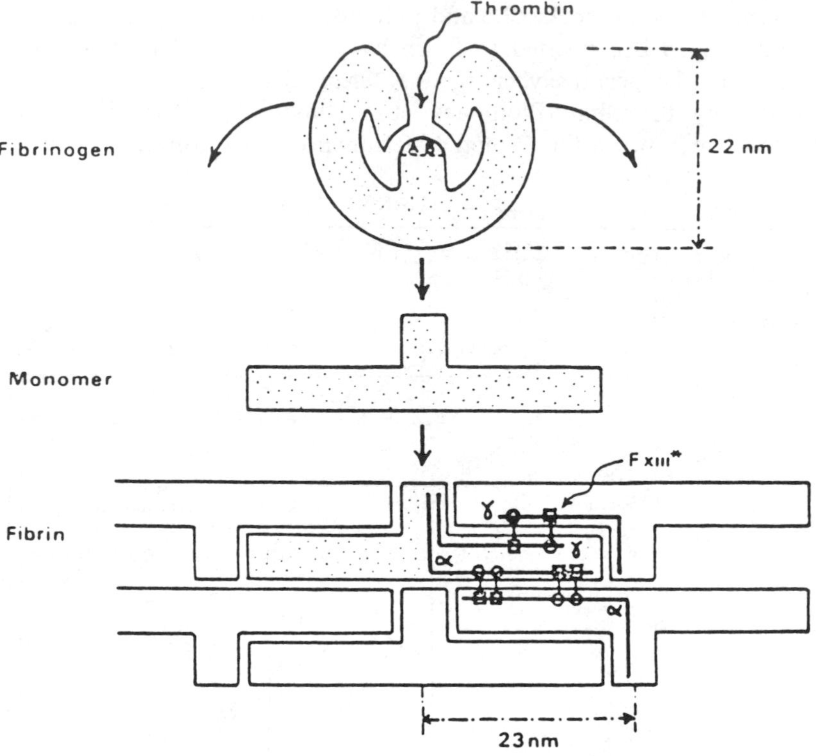

Figure 2 Enzymatic polymerization of fibrinogen and fibrin crosslinking. The cleavage of fibrinopeptides A and B reduces the C- and N-terminal interactions. The structural changes that follow expose the sites involved in the formation of the linear fibrin array (hydrogen and electrostatic bonding and hydrophobic interactions) and crosslinking (covalent bonding).

Anticoagulation is usually done in vitro by using a calcium chelator, such as citrate, EDTA, or EGTA. Because calcium is a necessary cofactor in the coagulation cascade, the chelation of this divalent cation prevents activation of the coagulation factors.

For in vivo anticoagulation therapy, heparin and dicumarol are the drugs principally administered. Heparin increases enormously the activity of antithrombin III (Stead, Kaplan & Rosenberg 1976), while coumarin anticoagulants inhibit the production of the vitamin K-dependent clotting factors by the liver (Hemker, Veltkamp & Loeliger 1968).

Platelets are disc-shaped blood cells with an average volume of 5–7.5 μm^3. In whole human blood, the normal range of platelet concentration is considered to be 1.5×10^5–4.0×10^5 platelets/mm^3. When a blood vessel is injured, subendothelial tissue (similar to collagen) is exposed. Platelets adhere to this surface due to the presence of this collagenlike material and also the released ADP from damaged cells. The platelet plug continues to grow after the initial adhesion because of the aggregation of other platelets. In the circulation, this hemostatic function is affected by blood flow, which generates wall shear stress and removes aggregating agents from the site of the injury (Baumgartner 1973).

In addition to forming hemostatic plugs (platelet aggregation) platelets also interact directly with the coagulation cascade (Niewiarowski et al 1968). The activation of Factors XI and XII may be enhanced by platelets, and Factor XI_a protected from inactivation (Schiffman, Rapaport & Chong 1973). Also, the phospholipid-rich platelet membrane possibly acts as a catalytic surface for factor activation (Silver 1965). In addition, thrombin and polymerizing fibrin induce platelet aggregation (Niewiarowski et al 1972), and thrombin's action on platelets can cause the release of platelet granules containing serotonin and other chemicals.

Platelets interact with and attach to polymerizing fibrin, and, in vitro, clot retraction occurs (Widmer & Moake 1976, Niewiarowski et al 1975). This decrease in clot size is a consequence of platelet contraction and shortened fibrin chain length. In systems in which the clot volume is not permitted to change, an increase in isometric tension or clot dynamic elastic modulus is found as a result of the tightening of the network by platelet contraction. While in the presence of flowing blood platelets are not distributed homogenously throughout the fibrin network it is likely that the retraction mechanism still plays a significant role in hemostasis. In vivo, tightening of the platelet-fibrin matrix probably serves to close a wound, and decreasing the size of a clot prevents occlusion of blood vessels.

Use of Rheology in Studying Coagulation

During the formation of fibrin clots in blood, platelet-rich, or platelet-poor plasma, enormous changes in material properties occur. The plasma is transformed from an essentially Newtonian viscous liquid into a gel-like material, with many properties of an elastic solid. These material changes have been the basis of many of the qualitative coagulation tests commonly used in clinical hematology. Recently, with the improved sensitivity of rheological techniques for measuring quantitative dynamic material properties, it has become possible to follow the changes that occur in viscous and elastic material properties during clot structure formation. The measurement of specific material properties is extremely important, because it allows the application of well-developed theories for the behavior of linear viscoelastic networks. These theories relate the

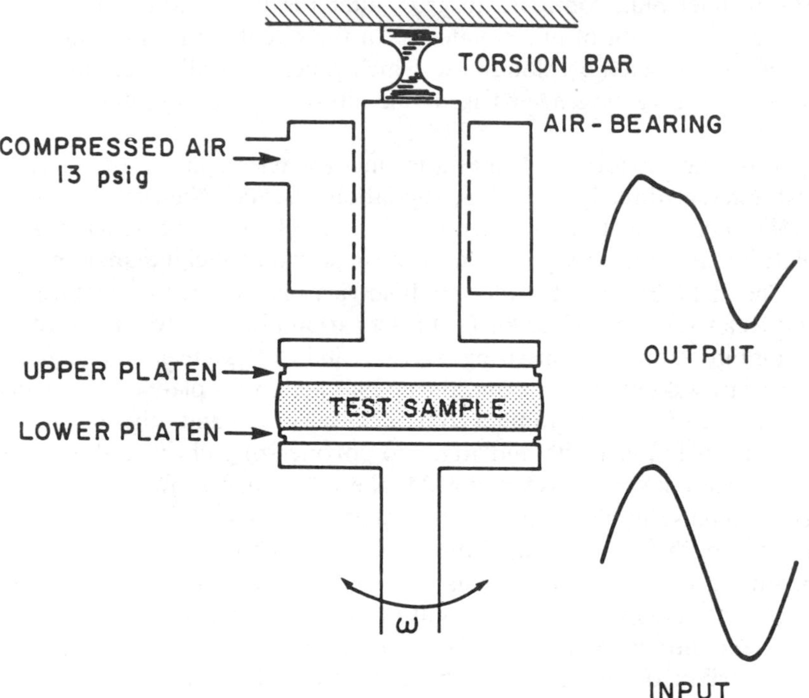

Figure 3 A schematic of typical platen geometry for the Weissenberg Rheogoniometer and the Fluids Rheometer used in dynamic testing. Data reduction can be done with software programs (Weissenberg) or with hard-wired circuit boards in a microprocessor. Digital filtration and/or cross correlation analysis is very important when working with dilute protein solutions because of the noise present in the output signal.

molecular structure of the network to measured rheological properties. Thus, one can utilize rheological data to obtain information of great biochemical importance, e.g. the number of crosslinks formed per fibrin monomer. The potential exists for using this new data to examine quantitatively alterations in the kinetics of network formation and platelet function caused by mechanical forces, clinical abnormalities, and chemotherapy.

To follow the structure formation that takes place during clotting most investigations have used dynamic materials-testing techniques. In principle, this is simple (Figure 3). The blood or plasma is put between parallel plates (or other platen geometries) and one is driven in a small-amplitude oscillation (input). The motion of the other plate is monitored (output) and a phase difference and amplitude ratio of the input and output are calculated. Using the theory of linear elasticity, a storage (or elastic) modulus G' and a loss (or viscous) modulus G'' can be calculated (Walters & Kemp 1968). The test frequency must be high enough for material property changes due to chemical reaction to be small over the data sampling period. Three machines commonly used for these measurements are the Weissenberg Rheogoniometer, the Rheometrics Fluids Rheometer, and the Viscoelastorecorder. Typical data obtained using normal recalcified plasma samples are shown in Figure 4. Zero time is the instant of recalcification. For the sample of platelet-rich plasma (PRP) the modulus rises sharply, after an initial delay that correlates well with a fibrometer time, passes through a maximum, and decreases (or relaxes) with time. The peak elastic modulus for PRP is over an order of magnitude greater than the peak modulus in platelet-free systems with the same fibrinogen concentration.

The manner in which platelets enhance clot rigidity is uncertain. Platelets contain a small amount of fibrinogen ($\sim$5 mg/100 ml for PRP) but certainly not enough to account for an order of magnitude increase in the maximum modulus. They contain phospholipid (platelet Factor III) and other components that participate in the coagulation mechanism. Platelets may act as sites of concentrated polymerization activity resulting in a final clot with optimal molecular structure (i.e. greater effective crosslink density). Perhaps the surface properties of platelets play a role in enhancing clot strength through control of fibrin polymerization (Walsh 1974).

The phenomenon of clot retraction is a manifestation of the interaction between platelets and fibrin that has been studied extensively. A possible explanation for the increase in G'_{max} shown in the Figure 4 data when platelets are present involves clot retraction, a mechanical (as far as rheological properties are concerned) rather than a chemical process in the interaction of platelets with fibrin. Evidence for this was reported by de Gaetano, Bottecchia & Vermylen (1973). In their studies changes in clot retraction during clot formation were investigated using thromboelastography. Results seemed to indicate that the greater the clot retraction, the greater the clot strength and that the two processes coincided with respect to time. Previous studies with polymers indicate that increase in modulus can be obtained as the result of elongational strain, apparently due to network reorientation (Nielsen 1962). Clot retraction occurs in the presence of platelets when the network is not strongly attached to surfaces. In the rheogoniometer the clot remains attached to the platen surfaces. Thus, instead of retraction occurring, strain (three-dimensional) is built up throughout the network. We suggest, then, that in the case of platelet-rich plasma, dynamic oscillations are deviations from a strained or "tight network" state whereas in platelet-free plasma the oscillations are deviations from an unstrained or "loose-network" state. A value for the "effective" strain in a clot that

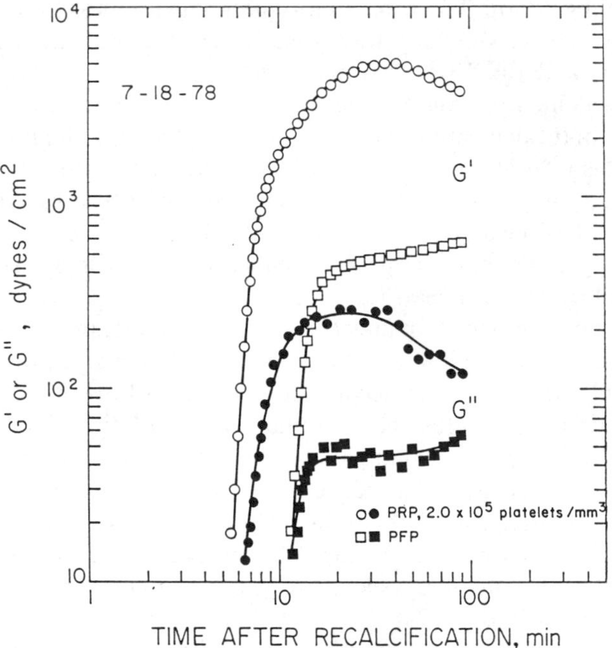

Figure 4 Typical data for the elastic modulus (G') and the viscous modulus (G'') obtained during clot formation with platelet-free plasma (PFP) or platelet-rich plasma (PRP). Time zero is the instant of recalcification of the citrated plasma sample. The delay time (t_0) is due to the cascade of reactions that must take place before fibrin monomer is formed to begin the polymerization.

contains platelets can be estimated from clot retraction data. Kuhnke & Brilla (1964) give values of clot retraction as a function of platelet concentration. They showed that the value that represented the amount of retraction was the length L (cm) that the clot shrunk. For example, if the original length was 10 cm, the remaining length was $(10-L)$. If a fiber of that length were stretched again to full length the strain would be $L/(10-L)$. Interpretation of their data would indicate that the strain or elongation ratio in a fully formed clot containing 320,000 platelets/mm^3 would be about four. In one containing 125,000 platelets/mm^3 this effective strain would be estimated to be one. Based on accepted engineering principles, a strain of one represents a doubling of the initial length of a fiber. Thus, strain during clot retraction is very large and may well be responsible for the dramatic effect of platelets on shear modulus.

The decline of the modulus in platelet-rich plasma after attaining a maximum value could also be associated with the presence of an internal strain phenomenon. It is typical of a viscoelastic material to show stress relaxation at constant strain, i.e. a time-dependent shear modulus $G(t)$,

$$\sigma(t) = \gamma G(t)$$

where $\sigma(t)$ is the time-dependent stress due to the constant strain γ. Such a relaxation could be the result of truly viscoelastic behavior of the fibrin strands, or (more likely) it could be a relaxation or fatiguing of the platelet contractile action, allowing the network to return towards its original state. It is also possible that the decline in maximum modulus is the result of fibrinolytic activity with destruction of the fibrin clots. However, the early time at which the decline occurred (approximately 30 minutes after recalcification) would indicate that fibrinolysis is not a significant factor. The ratio G''/G' gives an estimate of the importance of viscous dissipation of input energy relaxation to elastic storage of that energy in the clot.

Studies of Purified Fibrinogen-Thrombin Systems

Kaibara & Fukada (1970, 1971, 1973a,b, 1976, 1977) have carried out an extensive investigation of the rheology of fibrin gels made from purified fibrinogen. They have studied the effect of fibrinogen and thrombin concentrations, temperature, and ionic strength on the equilibrium properties of the gel and the kinetics of structure formation. Roberts, Lorand & Mockros (1973) and Mockros, Roberts & Lorand (1974) have used purified fibrinogen systems to look at the effect of Factor XIII activity on clot rheological properties, in addition to investigating ionic strength and fibrinogen concentration alterations on dynamic storage and loss moduli. Roberts et al (1974) and Gerth, Roberts & Ferry (1974) have studied the creep behavior and fluid permeation in fibrin clots, both with and without Factor XIII–induced crosslinking.

Platelet-Free Plasma

Of more clinical interest are the properties of systems of platelet-free plasma (plasma prepared from blood by high-speed centrifugation to remove all the formed elements). Here, all of the early coagulation factors are present, and alterations in proteins other than fibrinogen can be studied. Glover et al (1975a) compared in vitro Factor XIII inhibition with plasma from patients who exhibited low Factor XIII activity and found very similar crosslinking abnormalities in the clots formed (using SDS polyacrylimide gel electrophoresis) and the same very low equilibrium elastic moduli. Potential exists here for using rheological methods to examine alterations in the kinetics of structure formation caused by low factor activities, such as Factor VIII (antihemophilic factor) or Factor IX (Christmas Factor).

Platelet-Rich Plasma

The rheology of structure formation in platelet-rich plasma allows evaluation of clinically important platelet-fibrin interactions. PRP is obtained from blood by low-speed centrifugation, removing red cells, but leaving most of the platelets in the upper layer. The effect of normal platelets on clot rheology was quantified by Glover et al (1975b), who also looked at the effect of some pathological platelet abnormalities. On passing through artificial organs such as heart-lung bypass or kidney dialyzer, platelet aggregation and emboli formation is a severe clinical problem. In vitro studies examining effects of shear stress on platelet-fibrin interaction have shown that stress levels found in these devices can increase the rate of clot formation, cause the formation of platelet aggregates, and decrease the mechanical strength (G') of clots formed after stress exposure (Glover et al 1974, 1977). These in vitro studies are in general agreement with the contrasting clinical observations of increased emboli formation (platelet aggregation), and yet also severe bleeding problems from vascular sites (decreased clot mechanical strength).

Rheological studies can also be of great value in examining platelet-fibrin interaction, and thus indirectly in understanding platelet contractile biochemistry. Kuntamukkula et al (1978a) have studied the effect of internal platelet cyclic adenosine monophosphate (cAMP) on platelet contractility using dynamic materials testing, combined with biochemical assays. In addition, Kuntamukkula et al (1978b) have proposed using these rheological techniques for monitoring drug therapy (for example, cancer chemotherapy) for possible adverse platelet effects.

Cohen & DeVries (1973) and Cohen et al (1975) have employed a much different apparatus for the study of PRP clots. In this device a preformed cylindrical platelet-rich clot is held isometrically between two clamps, and the tensile force generated by the isometric tightening of the fibrin network due to platelet contraction is measured. They have estimated the contractile force generated per platelet using this method.

Studies of Anticoagulation and Clot Degradation

Heparin anticoagulation is routine for most large-scale surgical interventions. Here, there is a delicate balance required between anticoagulation to prevent emboli and clot formation and the development of internal bleeding due to excessive heparin. At present there is no really good way to monitor heparin levels in vivo, and most crude clotting tests do not give sufficient information to properly guide heparin dosage (or the amount of required antiheparin-protamine). In two very interesting papers, Overholser et al (1975, 1976) propose the use of a rheological technique for examining heparin effects on blood clotting. From both kinetic analyses and equilibrium modulus data, they conclude that this technique is probably the best currently available.

Once clots have been formed (for example in thrombophlebitis), chemical degradation of the fibrin network is a possible clinical therapy. This is done by utilizing plasmin or urokinase directly (two enzymes that attack fibrinogen or fibrin clots) or by giving a drug that will activate the circulating zymogen for plasmin (plasminogen). The most common drug used for the latter purpose is streptokinase. Again rheological techniques can be used to follow clot structure dissolution as well as formation. Kirkpatrick, McIntire & Moake (1978) have shown that streptokinase clot lysis is much slower for covalently crosslinked clots than for uncrosslinked clots. Rheological techniques have the potential to be used here also as a drug therapy monitoring technique that gives information about the kinetics of structure dissolution.

Quantitative Modeling of the Rheology of Coagulation

The coagulation cascade consists of more than thirteen factors and cofactors, whose various interactions leading to the polymerization of fibrin have not yet been fully elucidated. As a result, a detailed kinetic model relating all these individual reactions and the rheological data is not possible at present.

To circumvent this problem, most models based on measurements of the storage or loss modulus of clots lump parameters and obtain equations of a relatively simple form. These equations are generally empirical in nature and are valid for only a portion of the clotting process. It should be noted that only the coagulation of platelet-free plasma or the purified thrombin-fibrinogen system has been modeled adequately. The presence of platelets introduces a complexity that precludes the use of these simple models.

One of the first sets of equations was given by Scott-Blair & Burnett (1963, 1968), who set forth a model for the initial stages of coagulation of the form

$$G^* = G_\infty \exp(-\tau/t)$$

and for later stages

$$G^* = A\left(\frac{1}{t} - \frac{1}{t'}\right)$$

where G_∞ is the rigidity of the completely clotted sample and A, t', and τ are constants determined by fitting of experimental data—also

$$G^* = [(G'^2 + G''^2)^{\frac{1}{2}}],$$

which corresponds to the magnitude of the complex shear modulus, was obtained by U-tube gelometer and by a thromboelastograph.

In modeling the development of the storage and loss moduli, Kaibara and Fukada have proposed two models,

$$G = \alpha t^\beta$$

and

$$G = G_\infty(1 - e^{-kt}),$$

where G is either G' or G'', G_∞ is the ultimate G' or G'', and k, α, and β are empirical constants. These equations are fit to a portion of the G versus time data, and thus several sets of parameters are required to model the total clotting process.

By assuming that G' is proportional to the crosslink density of the fibrin network, Kaibara & Fukada (1973b) represent the rate of change of G' by the following expression:

$$\frac{dG'}{dt} = k(G'_\infty - G')^n$$

where k is constant and n is the order of the reaction. This approach resembles that of Mussatti & Macosko (1972), which relates the modulus to the crosslink density and the rate of crosslink formation to the number of crosslinks to be formed. Kaibara also models the formation of a fibrin network as occurring by two simultaneous first-order reactions.

Glover et al (1977) incorporate the idea of available crosslinks into their model and allow a dependence on enzyme concentration to yield the equation

$$\frac{dN}{dt} = k(N_\infty - N)^a (t - t_0)^c$$

where N is the number of crosslinks formed, N_∞ is the possible number of crosslinks, and a, c, k, and t_0 are constants. The magnitude of the complex modulus $|G^*|$ is related to N by the expression

$$|G^*| = k_1 N^m - |G_0|$$

where $|G_0|$ is the magnitude of the complex modulus for the nonclotted plasma at t_0. For the region in which only a small fraction of crosslinks have formed (small values of $t - t_0$), the modulus is described by

$$G_f = k_2(t - t_0)^{k_3}$$

where k_3 and k_2 are constants and G_f is the normalized excess modulus

$$G_f = \frac{|G^*| - |G_0|}{|G^*_\infty|}.$$

Overholser et al (1976) have developed a similar "chemorheologic" model for use in evaluating heparin anticoagulation.

Modeling of the Equilibrium Structure

The equilibrium structure of the network can be examined using models from linear viscoelasticity theories for swollen crosslinked rubber. One of the oldest network theories is that of rubber elasticity (Treloar 1967). This theory states that the equilibrium shear modulus (G_e) is directly related to the crosslink density χ by the formula

$$G_e = RT\chi$$

where

T = absolute temperature
R = gas constant
χ = crosslink density (gm moles crosslinks/cm^3).

Mooney has derived a relationship between the dynamic storage modulus (G') and G_e in crosslinked networks (see, for example, Ferry 1970).

$$G' = \chi RT\,[1+\Sigma\omega^2\tau_p^2/(1+\omega^2\tau_p)].$$

Thus the low-frequency limiting value of G' (as $\omega \to 0$) must be G_e. (The τ_p represent characteristic "relaxation times" for the viscoelastic network.) Data taken on clots in the low-frequency region ($\sim$1 hertz) have shown that the measured values of the equilibrium G' are essentially independent of frequency down to $\omega < 10^{-2}$ hertz. Thus, we feel the measured values of G' are very close to G_e. If the model on which the rubber elasticity theory is based is at least crudely applicable to fibrin networks, those measurements of G' can be used to estimate the equilibrium crosslink density in the platelet-free plasma clots.

Using data from Glover et al (1975a) and utilizing a physiological fibrinogen concentration of 300 mg/100 ml, we can easily calculate a crosslink density. If molecular weight of 330,000 is assumed for fibrinogen, the density χ can be converted to crosslinks per fibrin molecule. The calculations give 3.2 crosslinks/fibrin molecule. This is remarkably close to the value determined by Pisano et al (1971), who measured between 5 and 6 crosslinks per fibrin molecule using a biochemical analysis of normal human plasma clots. If it is assumed that the three-dimensional α polymer is the primary strength-bearing network, the rheological measurements would be in exact agreement with the biochemical assay, as it is generally believed that there are 3 to 4 crosslinks per α dimer in fibrin.

Clots formed in the absence of the Factor XIII activity (deficiency or inhibition) show decreases in G' of approximately a factor of 3, compared with normal clots with the same fibrinogen concentration. This would imply a three-fold decrease in crosslink density in the clots. Pisano et al (1972), using biochemical analyses, found a slightly greater reduction in their Factor XIII deficient patients. This is not entirely unexpected, as the uncrosslinked polymerized fibrin gel will have a nonzero, though small, elastic modulus.

With proper standardization, the method described here may provide an assay technique for crosslink density in fibrin clots that would be useful in studying Factor XIII abnormalities or inhibitors of fibrin crosslinking. The rheological techniques are well established and, with proper care, extremely reproducible. The effect of varying the amount and type of crosslink inhibitor on crosslink density can be easily determined. Using a crosslinked network theory, quantitative estimates of the actual change in the number of covalent crosslinks found between fibrin strands can be ascertained.

The theory of rubber elasticity is undoubtedly at best a crude approximation for modeling crosslinked fibrin networks. The theory requires free diffusional movement of the strands between crosslinks—certainly not possible in the fibrin structure. More realistic theories of the protein network structure are certainly needed, and protein network structure should constitute an area of active research in the coming years.

MUCUS RHEOLOGY

Mucus is a substance that serves as a lubricant and moistening agent for most of the damp surfaces of the body in contact with air. Although this fluid is normally taken for granted, alterations in mucus rheology can cause dramatic changes in the fluid mechanics of the cilia lining the bronchial tubes and in lung function. Therefore, investigation of mucus rheology and, perhaps more importantly, use of rheological techniques to investigate pathological alterations in mucus protein structure can be an extremely important area of research.

Pathological Mucus in Respiratory Disease

In many respiratory diseases copious amounts of thick gelatinous mucus are produced. By hindering the transport function of the muco-ciliary system such mucus causes obstructions, exacerbates infection, and ultimately can lead to scarring of lung tissue. Much effort has been invested to determine the chemical basis for pathological mucus in diseases such as asthma, bronchiectasis, cystic fibrosis, and chronic bronchitis. However, careful measurement of mucus viscoelasticity has shown wide variation within each of these diseases, with no uniquely characteristic values (Heggs et al 1974).

Proteins in Mucus

Mucin glycoproteins crosslinked into an extended gel network appear to account for the viscoelastic properties of mucus (Gibbons 1969, Litt & Khan 1974, 1976). Crosslinking occurs through both secondary bonds and covalent disulfide bonds (Meyer 1976, Roberts 1976, Sheffner 1963). Beginning with the isolation from sputum of a small protein that was thought to be bound to glycoprotein (Gibbons & Roberts 1963), proteins have been recognized as contributing to glycoprotein crosslinking (Gibbons 1969).

Sources of crosslinking proteins in mucus have not been quantitatively identified, although a variety of proteins have been identified in mucus gels in either soluble or noncovalently bound states. Lactoferrin, albumin, α_1-acid glycoprotein, immunoglobulin A (Masson et al 1965), and kallikrein (Gernez-Rieux et al 1967) have been identified in a soluble state in bronchial secretions uncontaminated by saliva. Furthermore, 15 nonplasma proteins have been described in sputum (Ryley 1972). Much of this protein is part of the soluble aqueous milieu surrounding the glycoproteins. It is easily removed by washing. However, some of these proteins have been found tightly bound to glycoproteins and apparently they mediate the aggregation into supramolecular structures of greater than 10^6 molecular weight. Lewis (1976) has suggested that crosslinking protein originates as mucus globule membrane protein, and that it has been difficult to measure because it forms insoluble complexes.

Glycoprotein Crosslinking

Roberts (1976) has immunologically identified lactoferrin and the serum proteins, albumin and immunoglobulin A, as forming tight noncovalent complexes with glycoproteins in sputum. Proteins bind to glycoproteins by disulfide bonds as well as secondary bonds. Sulfhydryl reagents decrease the viscoelasticity of mucus in vitro (Sheffner 1963) and decrease the molecular weight of glycoproteins isolated from mucus by high salt (Creeth et al 1977) or urea (Roberts 1976). Purified glycoproteins of $MW > 10^6$ reduced with sulfhydryl reagents decreased in molecular weight by a factor of about 6–1.8×10^5 MW, apparently became more homogenous, and released a small amount of a protein of about 10^5 MW (Creeth et al 1977). The content of sulfhydryl and acidic amino acids was higher in this protein than in the associated glycoprotein.

Roberts (1976) showed that glycoproteins in respiratory sputum are joined covalently by "naked peptide." He demonstrated that in some specimens the naked peptide forms a continuous peptide chain with the glycoproteins, and that the naked peptide consists of separated chains, linked by disulfide bridges. These aggregates were susceptible to degradation by pronase.

In one model structure that has been proposed, disulfide bonds are located on a crosslinking protein that links glycoprotein units by secondary bonds (Degand et al 1973, Gibbons 1969, Havez et al 1973). In an alternate model, intermolecular disulfide bonds link glycoprotein peptide chains, and these larger macromolecules can additionally aggregate by secondary bonding (Meyer et al 1973, Masson 1973). The data of Creeth et al (1977) support the first model in that they indicate a linking peptide. The data of Roberts (1976) suggest that both types of bonding exist to different extents in different specimens of mucus.

Correlation of Protein Content with Mucus Viscoelasticity

Noncovalently bound proteins have been found tightly associated with glycoproteins in mucus from several sources, and their amounts may regulate the physical properties of mucus. Several investigators have reported significant qualitative correlations between protein content and viscoelastic properties of mucus. However, the effect of specific protein cofactors on the viscoelastic properties of mucus has not been quantitatively evaluated.

The amount of nonmucin protein in cervical mucus varies during both the estrus and menstrual cycles and may regulate the physical properties of the mucus (Masson 1973). Excessive amounts of albumin and other serum proteins have been reported in the meconium of meconium ileus patients (Schachter & Dixon 1965) and may cause its hard rubbery consistency. The duodenal fluid of cystic fibrosis patients has a significantly higher level of albumin than normal (Knauff & Adams 1968), which may cause this fluid's higher than normal viscosity.

Heggs et al (1974) demonstrated a correlation between elasticity and viscosity at low shear rates for sputum. In asthmatic and bronchiectatic sputa this correlation was significant even at higher shear rates while in cystic fibrosis and chronic bronchitis sputa it was not. Since in cystic fibrosis and chronic bronchitis mucus gland hypertrophy is present (Reid 1954), they concluded that these sputa were higher than normal in a mucus component. Since asthma and bronchiectasis are not necessarily associated with gland hypertrophy (Reid 1960) they concluded that an increase in a serum component may confer resistance to mechanical breakdown of the gel matrix in these sputa.

Bornstein et al (1978) demonstrated a correlation between sputum viscoelasticity and its contents of disulfide bonds. This correlation was significant only in mucus with high concentration of solids, which was found during the acute exacerbation of chronic bronchitis. List et al (1978) demonstrated that addition of serum albumin in a concentration of 5–20 mg/ml to pig gastric mucin enhanced the viscosity proportional to the albumin concentration. The sensitivity of this effect to changes in pH, ionic strength, and divalent ion chelation suggested that the interactions were not due to covalent bonding, but to hydrogen and/or hydrophobic bonding.

CONCLUSIONS

Changes in fluid rheological properties can have dramatic effects on local fluid mechanics in body organs. Clots or platelet emboli can lodge in small vessels of the circulation and block off local blood flow. This can lead to local tissue necrosis. If the vessel perfuses part of a vital organ, death is a possible outcome. Alterations in mucus elasticity and viscosity can greatly modify transport processes in the lung. Understanding the biochemical and biophysical processes that lead to these rheological alterations is a crucial first step in developing chemical treatment that will prevent their occurrence. In the two systems described above, plasma coagulation and mucus rheology, the formation of a three-dimensional protein network structure causes profound changes in material elasticity and viscosity. The sequential molecular steps that occur are relatively well understood in the coagulation process, but investigations into the cause and nature of the proteins involved in mucus network formation are just beginning. Dynamic materials testing provides a vital tool for examining structure-forming systems, since the testing does not disturb the network being assembled. Force transducer technology has recently improved greatly, allowing the possibility of directly examining the kinetics of structure formation in even very dilute protein systems. This will allow investigators to work with individual purified protein fractions and thus to examine in detail the effect specific molecular alterations have on network formation. The potential exists for using these rheological techniques to guide and monitor drug therapies. Coupling this rheological information with theories relating these properties to the underlying molecular structure will allow determination of parameters of fundamental biochemical importance (such as crosslink density) that are very hard to obtain by other methods.

Finally, although the field of biorheology is a very old one, the quantitative interpretation of rheological data from network-forming biological fluids is a field that promises to provide information of interest to people both in fluid mechanics and biochemistry.

Acknowledgments

The author would like to acknowledge many useful discussions with Drs. J. P. Kirkpatrick, J. D. Hellums, and C. J. Glover. This work was supported by the National Institutes of Health through grants HL 17437 and HL 18672.

Literature Cited

Abildgaard, U. 1967. Inhibition of the thrombin-fibrinogen reaction by antithrombin III, studied by N-terminal analysis. *Scand. J. Clin. Lab. Invest.* 20:207–16

Abildgaard, U. 1969. Inhibition of the thrombin-fibrinogen reaction by α_2-macroglobulin, studied by N-terminal analysis. *Thromb. Diath. Haemorrh.* 21:173–80

Baumgartner, H. R. 1973. The role of blood flow in platelet adhesion, fibrin deposition and formation of mural thrombi. *Microvas. Res.* 5:167–79

Blomback, B., Blomback, M., Edman, P., Hessel, B. 1966. Human fibrinopeptides: isolation, characterization and structure. *Biochim. Biophys. Acta* 115:371–96

Bornstein, A. A., Chen, T. M., Dulfano, M. J. 1978. Disulfide bonds and sputum viscoelasticity. *Biorheology* 15:261–67

Chen, R., Doolittle, R. F. 1970. Isolation, characterization, and location of a donor-acceptor unit from cross-linked fibrin.

Proc. Natl. Acad. Sci. USA 66:472–79

Cohen, I., DeVries, A. 1973. Platelet contractile regulation in an isometric system. *Nature* 246:36–37

Cohen, I., Gabbay, J., Glaser, T., Oplatka, A. 1975. Fibrin-blood platelet interaction in a contracting clot. *Brit. J. Haematol.* 31:45–50

Creeth, J. M., Bhaskar, K. R., Horton, J. R., Des, I., Lopez-Vidriero, M., Reid, L. 1977. The separation and characterization of bronchial glycoproteins by density-gradient methods. *Biochem. J.* 167:557–69

de Gaetano, G., Bottecchia, D., Vermylen, J. 1973. Effect of platelets on clot structuration: a thromboelastographic study. *Thromb. Res.* 3:425–35

Degand, P., Roussel, P., Lamblin, G., Durand, G., Havez, R. 1973. Biochemical and rheological data in sputum I—the biochemical definition of mucins in sputum. *Bull. Physio-Path. Resp.* 9:199–217

Ferry, J. D. 1970. *Viscoelastic Properties of Polymers*. New York: Wiley. 671 pp.

Gernez-Rieux, C., Tacquet, A., Devulden, B., Tison, F. 1967. Avian tuburcle bacilli and their position in the etiology and epidemiology of human tuberculosis. *Arch. Inst. Pasteur Tunis* 43:257–92

Gerth, C., Roberts, W. W., Ferry, J. D. 1974. Rheology of fibrin clots II. Linear viscoelastic behavior in shear creep. *Biophys. Chem.* 2:208–17

Gibbons, R. A. 1969. The composition of mucus with special reference to its rheological properties. *Protides of Biological Fluids* (16th Colloq.), pp. 299–315

Gibbons, R. A., Roberts, G. P. 1963. Some aspects of the structure of macromolecular constituents of epithelial mucus. *Ann. NY Acad. Sci.* 106:218–32

Glover, C. J., McIntire, L. V., Leverett, L. B., Hellums, J. D., Brown, C. H., Natelson, E. A. 1974. Effect of shear stress on clot structure formation. *Trans. ASAIO* 20:463–68

Glover, C. J., McIntire, L. V., Brown, C. H., Natelson, E. A. 1975a. Rheological properties of fibrin clots. Effects of fibrinogen concentration, Factor XIII deficiency, and Factor XIII inhibition. *J. Lab. Clin. Med.* 86:644–56

Glover, C. J., McIntire, L. V., Brown, C. H., Natelson, E. A. 1975b. Dynamic coagulation studies: influence of normal and abnormal platelets on clot structure formation. *Thromb. Res.* 7:185–98

Glover, C. J., McIntire, L. V., Brown, C. H., Natelson, E. A. 1977. Mechanical trauma effect on clot structure formation. *Thromb. Res.* 10:11–25

Goldsmith, H. L., Skalak, R. 1975. Hemodynamics. *Ann. Rev. Fluid Mech.* 7:213–47

Havez, R., Laine-Bassez, A., Hayem-Levy, A., Lebas, J. 1973. Biochemical and rheological data in sputum II. Biochemical definition of proteins in sputum. *Bull. Physio-Path. Resp.* 9:219–35

Heggs, P. M., Palfrey, A. J., Reid, L. 1974. The elasticity of sputum at low shear rates. *Biorheology* 11:417–26

Hemker, H. C., Veltkamp, J. J., Loeliger, E. A. 1968. Kinetic aspects of the interaction of blood clotting enzymes III. Demonstration of an inhibitor of prothrombin conversion in vitamin K deficiency. *Thromb. Diath. Haemorrh.* 19: 346–63

Henschen, A. 1964. Number and reactivity of disulfide bonds in fibrinogen and fibrin. *Arkiv. Kemi.* 22:355–73

Hudry-Clergeon, G., Marguerie, G., Pouit, L., Suscillon, M. 1975. Models proposed for the fibrinogen molecule and for the polymerization process. *Thromb. Res.* 6: 533–41

Kaibara, M., Fukada, E. 1970. Dynamic viscoelastic study for the structure of fibrin networks in the clots of blood and plasma. *Biorheology* 6:329–39

Kaibara, M., Fukada, E. 1971. The influences of the concentration of thrombin on the dynamic viscoelasticity of clotting blood and fibrinogen-thrombin systems. *Biorheology* 8:139–47

Kaibara, M., Fukada, E. 1973a. The dynamic rigidity of fibrin gels. *Biorheology* 10:129–38

Kaibara, M., Fukada, E. 1973b. Dynamic viscoelastic study of the formation of fibrin networks in fibrinogen thrombin systems and plasma. *Biorheology* 10:61–73

Kaibara, M., Fukada, E. 1976. Dynamic viscoelasticity of fibrin gels: dependence on ionic strength. *Thromb. Res.* 8: Suppl. II, pp. 45–58

Kaibara, M., Fukada, E. 1977. Effect of temperature on dynamic viscoelasticity during the clotting reaction of fibrin. *Biochim. Biophys. Acta* 499:352–56

Kanaide, H., Shainoff, J. R. 1975. Cross-linking of fibrinogen and fibrin by fibrin-stabilizing factor (factor XIIIa). *J. Lab. Clin. Med.* 85:574–97

Kirkpatrick, J. P., McIntire, L. V., Moake, J. L. 1978. Streptokinase-induced degradation of crosslinked and uncrosslinked clots. *Thromb. Res.* 13:569–75

Knauff, R. E., Adams, J. A. 1969. Proteins and mucoproteins in the duodenal fluids of cystic fibrosis and control subjects. *Clin. Chem. Acta* 19:245–48

Koppel, G. 1970. Morphology of the fibrinogen molecule. *Thromb. Diath. Haemorrh.* 39: Suppl., pp. 71–84

Kuhnke, E., Brilla, G. 1964. Der Thrombocytenäquivalenzwert und der Retraktion als Mass für vergleichende Untersuchungen der Gerinnselretraktion. *Thromb. Diath. Haemorrh.* 11:210–21

Kuntamukkula, M. S., McIntire, L. V., Moake, J. L., Peterson, D. M., Thompson, W. J. 1978a. Rheological studies of the contractile force within platelet-fibrin clots: effects of protaglandin E_1, dibutyryl-cAMP and dibutyryl-cGMP. *Thromb. Res.* 13:957–69

Kuntamukkula, M. S., McIntire, L. V., Moake, J. L., Peterson, D. M. 1978b. The use of rheological techniques to evaluate platelet function alterations caused by drug therapy. *AIChE Symp. Ser. Biorheology* 182:74–80

Lewis, R. W. 1976. Mucus globule membrane: an hypothesis conceiving its role in determining the viscosity of mucus. *J. Theoret. Biol.* 61:21–25

List, S. J., Findlay, B. P., Forstner, G. G., Forstner, J. F. 1978. Enhancement of the viscosity of mucin by serum albumin. *Biochem. J.* 175:565–71

Litt, M., Kahn, M. A. 1974. The viscoelasticity of fractionated canine tracheal mucus. *Biorheology* 11:111–17

Litt, M., Kahn, M. A. 1976. Mucus rheology: relation to structure and function. *Biorheology* 13:37–48

MacFarlane, R. G. 1964. An enzyme cascade in the blood clotting mechanism, and its function as a biochemical amplifier. *Nature* 202:498–99

Masson, P. L. 1973. *Cervical Mucus in Human Reproduction*, ed. M. Elstein, K. S. Moghissi, R. Borth, pp. 82–92. Copenhagen: Scriptor

Masson, P. L., Heremans, J., Prignot, J. 1965. Studies on the proteins of human bronchial secretions. *Biochim. Biophys. Acta* 111:466–78

McDonagh, J., Messel, H., McDonagh, R. P., Murano, G., Blomback, B. 1972. Molecular weight analysis of fibrinogen and fibrin chains by an improved solium dodecyl sulfate gel electrophoresis method. *Biochim. Biophys. Acta* 257:135–42

McKee, P. A., Mattock, P., Hill, R. L. 1970. Subunit structure of human fibrinogen, soluble fibrin and cross-linked insoluble fibrin. *Proc. Natl. Acad. Sci. USA* 66:738–44

Meyer, F. A. 1976. Mucus structure: relation to biological transport function. *Biorheology* 13:49–58

Meyer, F. A., Eliezer, N., Silberberg, A., Vered, J., Sharon, N., Sade, J. 1973. An approach to the biochemical basis for the transport function of epithelial mucus. *Bull. Physio-Path. Resp.* 9:259–74

Mockros, L. F., Roberts, W. W., Lorand, L. 1974. Viscoelastic properties of ligation-inhibited fibrin clots. *Biophys. Chem.* 2:164–69

Mussatti, F. G., Macosko, C. W. 1972. Rheology of network forming systems. *Polym. Eng. Sci.* 13:236–40

Nielsen, L. E. 1962. *Mechanical Properties of Polymers*, p. 246. London: Reinhold. 435 pp.

Niewiarowski, S., Poplawski, A., Lipinski, B., Farbiszewski, R. 1968. The release of platelet-clotting factors during aggregation and viscous metamorphosis. *Exp. Biol. Med.* 3:121–26

Niewiarowski, S., Regoeczi, E., Stewart, G. J., Senyi, A. F., Mustard, J. F. 1972. Platelet interaction with polymerizing fibrin. *J. Clin. Invest.* 51:685–92

Niewiarowski, S., Stewart, G. J., Nath, N., Sha, A. J., Lieberman, G. E. 1975. ADP, thrombin and Bothrops atrox thrombin-like enzyme in platelet-dependent fibrin retraction. *Am. J. Physiol.* 229:737–45

Overholser, K. A., Itin, J. P., Brown, D. R., Harris, T. R. 1975. The effect of heparin on the viscoelasticity of whole blood clots. *Biorheology* 12:309–21

Overholser, K. A., Baysinger, C. G., Harris, T. R., Deveau, T. 1976. Chemorheological studies of the effect of heparin on the course of coagulation. *Thromb. Haemostasis* 35:447–59

Pisano, J. J., Finlayson, J. S., Peyton, M. P., Nagal, Y. 1971. ε-(γ-glutamyl) lysine in fibrin: lack of crosslink formation in factor XIII deficiency. *Proc. Natl. Acad. Sci. USA* 68:770–74

Pisano, J. J., Bronzert, T. J., Peyton, M. P., Finlayson, J. S. 1972. ε-(γ-glutamyl) lysine crosslinks determination in fibrin from normal and factor XIII deficient individuals. *Ann. NY Acad. Sci.* 202:98–113

Pouit, L., Hudry-Clergeon, G., Suscillon, M. 1973. Electron microscopy study of positive charges on fibrin fibers. *Biochim. Biophys. Acta* 317:99–105

Reid, L. 1954. Pathology of bronchitis. *Lancet* 1:275–78

Reid, L. 1960. Measurement of the bronchial mucus gland layer: a diagnostic yardstick in chronic bronchitis. *Thorax* 15:132–41

Roberts, G. P. 1976. Role of disulfide bonds in maintaining the gel structure of bronchial mucus. *Arch. Biochem. Biophys.* 173:528–37

Roberts, W. W., Lorand, L., Mockros, L. F. 1973. Viscoelastic properties of fibrin clots. *Biorheology* 10:29–42

Roberts, W. W., Kramer, O., Rosser, R. W., Nesther, F. H., Ferry, J. D. 1974. Rheology of fibrin clots I. Dynamic viscoelastic properties and fluid permeation. *Biophys. Chem.* 1:152–60

Rosenberg, J. S., McKenna, P. W., Rosenberg, R. D. 1975. Inhibition of human factor IXa by human antithrombin. *J. Biol. Chem.* 250:8883–88

Ryley, H. C. 1972. An immunoelectrophoretic study of the soluble secretory properties of sputum. *Biochim. Biophys. Acta* 271:300–9

Schachter, H., Dixon, G. H. 1965. A comparative study of the proteins in normal meconium and in meconium from meconium ileus patients. *Can. J. Biochem.* 43:381–98

Schiffman, S., Rapaport, S. I., Chong, M. M. Y. 1973. Platelets and the initiation of intrinsic clotting. *Brit. J. Haematol.* 24:633–42

Scott-Blair, G. W., Burnett, J. 1963. An equation to describe the rate of setting of blood and milk. *Biorheology* 1:183–91

Scott-Blair, G. W., Burnett, J. 1968. On the rates of coagulation and subsequent softening of bovine and human blood and of thrombin-fibrinogen. *Biorheology* 5:163–76

Sheffner, A. L. 1963. The reduction in vitro in viscosity of mucoprotein solutions by a new mucolytic agent, N-acetyl-L-cysteine. *Ann. NY Acad. Sci.* 106:298–310

Silver, M. J. 1965. Role of calcium ions and phospholipids in platelet aggregation and plug formation. *Am. J. Physiol.* 209:1128–36

Stead, N., Kaplan, A. P., Rosenberg, R. D. 1976. Inhibition of activated factor XII by antithrombin-heparin cofactor. *J. Biol. Chem.* 251:6481–88

Treloar, L. R. G. 1967. *The Physics of Rubber Elasticity*. London: Oxford Univ. Press. 342 pp.

Walsh, P. N. 1974. Platelet coagulant activities and hemostasis: a hypothesis. *Blood* 43:597–605

Walters, K., Kemp, R. A. 1968. On the use of a Rheogoniometer III oscillatory shear between parallel plates. *Rheologica Acta* 7:1–8

Widmer, K., Moake, J. L. 1976. Clot retraction: evaluation in dilute suspensions of platelet rich plasma and gel-separated platelets. *J. Lab. Clin. Med.* 87:49–57

Williams, W. J. 1977. In *Hematology*, ed. W. J. Williams, E. Beutler, A. J. Ersler, R. W. Rundles, pp. 1266–75. New York: McGraw-Hill. 1755 pp.

Yin, E. T., Wessler, S., Stoll, P. J. 1973. Identity of plasma-activated factor X inhibitor and antithrombin III and heparin cofactor. *J. Biol. Chem.* 240:3694–702

Ann. Rev. Fluid Mech. 1980. 12: 181–222

TRANSONIC FLOW PAST OSCILLATING AIRFOILS[1]

H. Tijdeman
Nationaal Lucht-en Ruimtevaartlaboratorium, Amsterdam, The Netherlands

R. Seebass
Departments of Aerospace and Mechanical Engineering, and Mathematics, University of Arizona, Tucson, Arizona 85721

INTRODUCTION

Under certain conditions structures like airplane wings and tail surfaces may experience vibrations of an unstable nature. This phenomenon, called "flutter," is an aeroelastic phenomenon governed by the interaction of the elastic and inertial forces of the structure with the unsteady aerodynamic forces generated by the oscillatory motion of the structure itself. In general, two or more vibration modes are involved, e.g. bending and torsional vibrations of a wing, which, under the influence of the unsteady aerodynamic forces, interact with each other in such a way that the vibrating structure extracts energy from the passing airstream. This leads to a progressive increase in the amplitude of vibration and may end with the disintegration of the structure.

For a given wing structure the aerodynamic forces increase rapidly with flight speed, while the elastic and inertial forces remain essentially unchanged. Normally there is a critical flight speed, the "flutter speed," above which flutter occurs. Because of the potentially disastrous character of this phenomenon, flutter speeds of aircraft must be well outside their flight envelope. In many cases this requirement is the determining factor in the construction of wings and tail surfaces. Because the vibration characteristics of the structure at zero airspeed can be determined accurately by current numerical methods or by ground vibration tests, the

[1] The authors thank the NLR, the AFOSR, and the ONR for their support of this review and related studies.

0066-4189/80/0115-0181$01.00

accuracy of the flutter prediction depends mainly on the knowledge of the unsteady aerodynamic forces.

In subsonic and supersonic flight unsteady aerodynamic forces can be predicted reasonably well by theoretical and numerical means. For transonic flight, with its mixed subsonic-supersonic flow patterns, prediction methods are less advanced. The current practice for wings of general planform is still rather arbitrary, with interpolations and extrapolations being made on the basis of calculated airloads for pure subsonic and supersonic flow. And, in many cases, one must resort to very expensive wind tunnel experiments.

Currently there is a renewed interest in transonic flight for both military and civil aircraft. For military aircraft this stems from the need for a new generation of air combat aircraft, like the F-16 and F-17, which require an optimal maneuverability at transonic speeds. In civil aviation there is a great need for more efficient aircraft, made possible by technological advances that include the so-called supercritical wing. Such wings make it possible to cruise at transonic speeds without the usual drag penalty associated with the presence of shock waves. This is achieved by shaping the wing geometry in such a way that the transition from local flow regions with supersonic flow to the adjacent subsonic regions does not take place with strong shock waves as it does on the conventional-type wings, but with only very weak shock waves or even without them.

In the present review we describe the nature of transonic flows past oscillating airfoils and discuss recent developments in unsteady transonic flow calculations. We place emphasis on plane flows because most of the published studies deal with this type of flow. The first section starts with a general description of the flow past airfoils. Experimental results are then reviewed and used to illustrate the interaction between the steady and unsteady flow fields, the periodic motion of shock waves, and the effects of frequency and amplitude of oscillation. In the subsequent section we discuss the inviscid equations forming the basis of the various theoretical methods and review techniques for their solution, all essentially numerical. Viscous effects and calculation methods are then described. Finally, in the last section we assess the present status of the field and the future developments expected.

The reader should be aware of four related reviews. Landahl (1976) reviews the unsteady aspects of transonic flow, while a recent review of unsteady fluid dynamics by McCroskey (1977) also has informative sections on unsteady transonic flow and unsteady boundary layers. Computational aspects of steady and unsteady transonic flows are reviewed by Ballhaus (1978) and the dissertation of Tijdeman (1977) contains a more complete discussion of many of the topics addressed here.

FLOW PAST AIRFOILS

Steady Flow

A brief survey of the behavior of steady transonic flows past airfoils provides an introduction to the discussion of the transonic flow past oscillating airfoils. When the free-stream Mach number of a purely subsonic flow past a symmetrical airfoil is increased, the flow pattern usually develops in the manner sketched in Figure 1. The so-called critical Mach number M^* is reached when the maximum local Mach number in the flow becomes unity. Beyond the critical Mach number a supersonic region appears on the airfoil, which, in general, is terminated by a nearly normal shock wave through which the flow speed is reduced from supersonic to subsonic. With a further increase of the free-stream Mach number, the shock wave moves aft and the size of the supersonic region and the shock strength both increase. After the pressure jump through the shock wave has become sufficiently large, so-called shock-induced separation

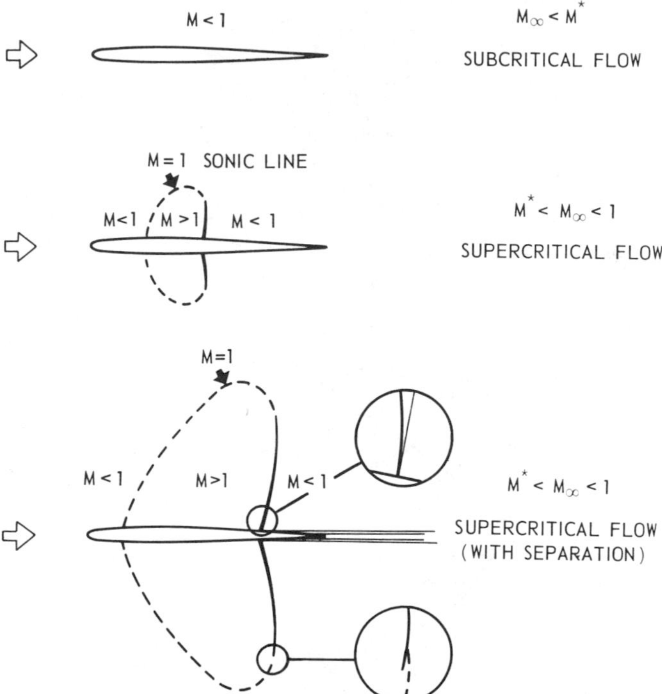

Figure 1 Influence of Mach number on flow pattern.

of the boundary layer occurs. For a turbulent boundary layer, this shock-induced separation starts when the local Mach number just upstream of the shock wave is about 1.25 to 1.3. When the boundary layer downstream of the shock wave separates completely, the flow around the airfoil is

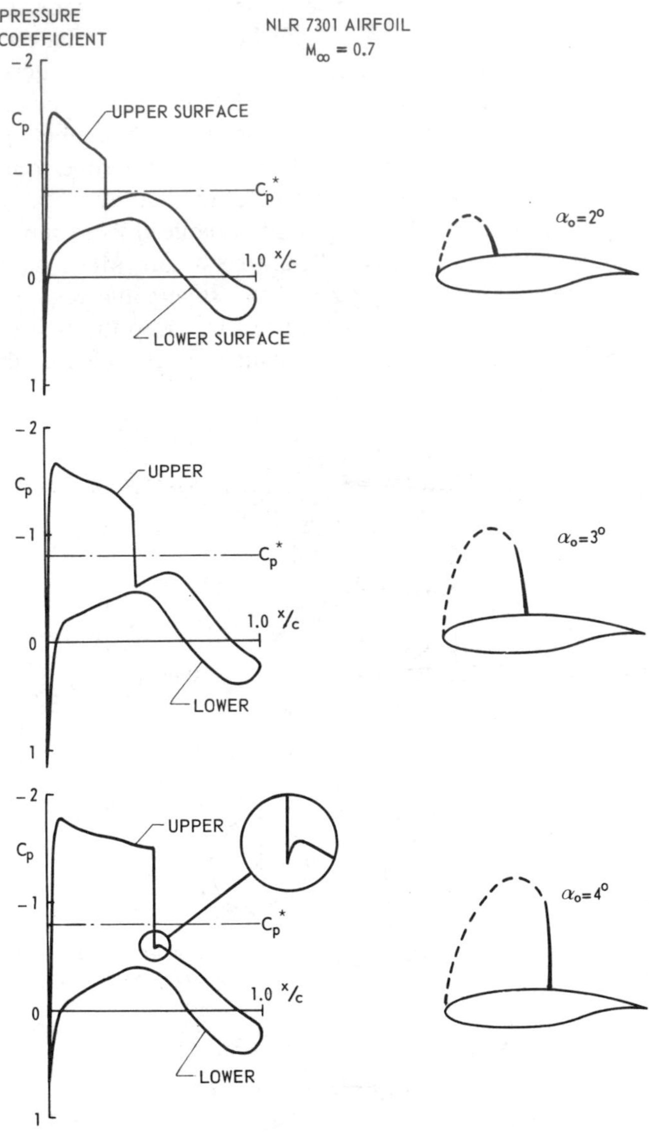

Figure 2 Influence of incidence on pressure distribution and flow pattern in transonic flow.

changed considerably, and often unsteady-flow phenomena such as "buffet" and "buzz" start to occur.

When an airfoil's incidence is increased at a supercritical Mach number, the flow changes in the manner sketched in Figure 2. Initially the airfoil carries a well-developed supersonic region on its upper surface, terminated by a shock wave. When the incidence is increased, the speed over the upper surface increases and the supersonic region and the shock wave develop in much the same way as described above for increasing free-stream Mach number. This example shows that small variations in incidence may lead to considerable changes in the pressure distribution, shock position, and shock strength.

Across the nearly normal shock wave that occurs on the airfoil, the velocity is reduced from supersonic to subsonic. In two-dimensional inviscid flow the foot of the shock wave must be normal to the airfoil. For a convex airfoil the pressure will increase and the Mach number will decrease with distance above the airfoil ahead of the shock wave. The pressures behind the shock, given by the Rankine-Hugoniot relations, must still balance the flow curvature demanded by the airfoil. This can only be accommodated if the shock wave has infinite curvature at its foot (Zierep 1966). This results in a rapid expansion there (logarithmically infinite pressure gradient); this expansion can often be noticed in surface pressure distributions of airfoils, where it manifests itself as the so-called Zierep cusp, or Oswatitsch-Zierep singularity, as sketched in Figure 2.

Usually transonic flow patterns are characterized by the presence of nearly normal shock waves on either the upper or the lower surface of the airfoil, or on both surfaces at the same time. Occasionally even two shock waves, one behind the other, occur. An exception to this rule is the flow around a so-called supercritical (shock-free) airfoil at its design condition. This type of airfoil is shaped in such a way that, for a specific combination of incidence and free-stream Mach number, the "design condition," the transition of the supersonic region to the adjacent subsonic regions takes place without a noticeable shock wave. This requires a careful tailoring of the airfoil so that a smooth recompression is obtained. Changes in the Mach number and angle of attack affect this tailoring, and away from the design condition the flow will normally have at least a weak shock wave. The flow past a supercritical airfoil in its design condition, shown in Figure 3, clearly reveals that small changes in incidence are sufficient to disturb the shock-free flow condition. An important question with respect to the practical application of supercritical airfoils is how gradually the transition from shock-free flow to the neighboring flow conditions with shock waves takes place, or, in other words, what are the margins within which the Mach number and incidence can be

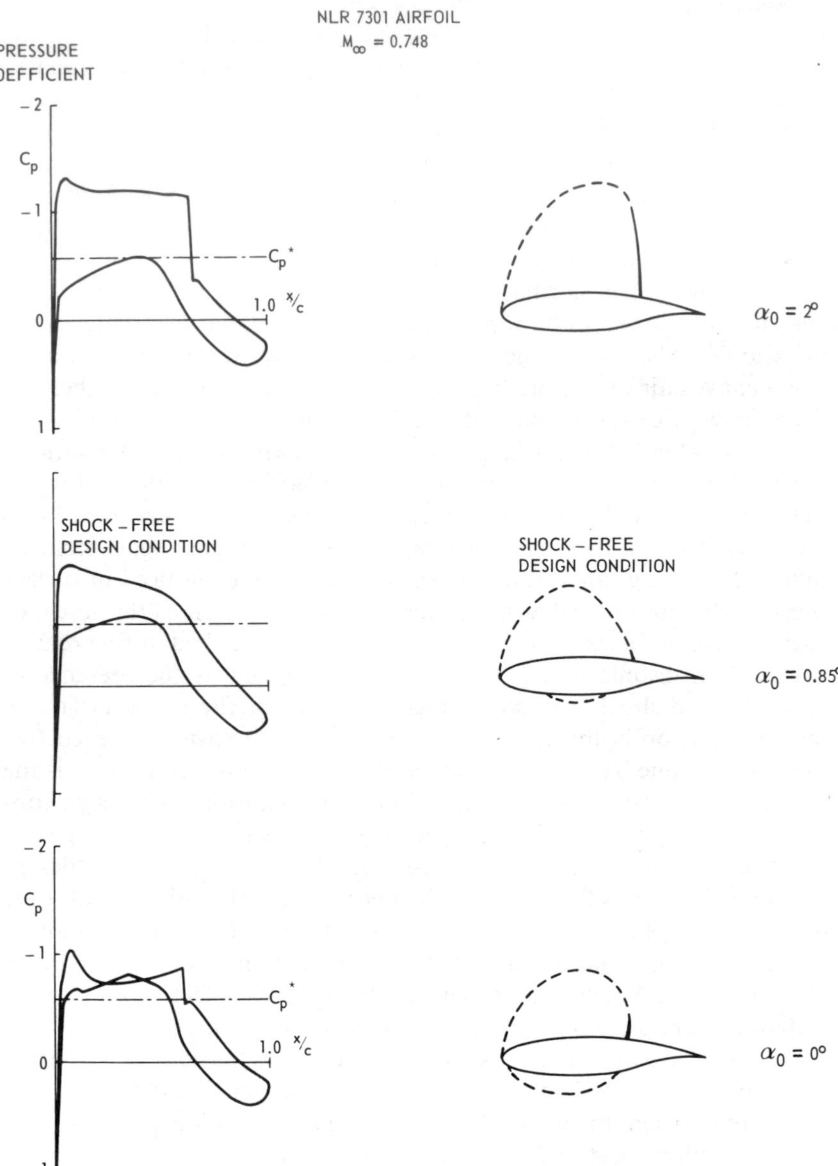

Figure 3 Influence of incidence on pressure distribution and flow pattern of a "shock-free" airfoil.

varied around the design condition without a serious deterioration of the favorably low-drag property of the shock-free flow condition.

An aspect that cannot be discarded when considering the flow past airfoils concerns the effects of viscosity. In an attached flow viscous effects are confined to a thin layer adjacent to the surface of the airfoil, the "boundary layer," and to its wake. In the boundary layer the velocity rises rapidly from zero at the surface to the local flow velocity at the outer edge. The boundary layer starts at the leading edge as a laminar boundary layer, which in cases of practical interest changes from laminar to turbulent after a small fraction of the chord (typically 5–10%). The presence of the boundary layer changes the effective contour of the airfoil and, thus, has an effect on the pressure distribution and the aerodynamic loading. The magnitude of this effect depends, among other things, on the Reynolds number, which is an important parameter for the growth of the boundary-layer thickness and the location of the transition point. The behavior of the boundary layer is of even more importance in transonic flow than in subsonic flow, since here it has a considerable influence on the position and strength of the shock wave.

Unsteady Flow

When an airfoil performs sinusoidal oscillations around a given mean position, the circulation and, hence, the lift force and local pressures show periodic variations. In order to keep the total vorticity constant (according to Helmholtz's theorem), each time-dependent change in circulation around the airfoil is compensated by the shedding of free vorticity from the trailing edge. This vorticity, which has the same strength as the change in circulation but is of opposite sign, is carried downstream by the flow as sketched in Figure 4. Due to the velocities induced by the free vortices around the airfoil, the instantaneous incidence of the airfoil is changed in such a way that the oscillating part of the lift lags behind the motion of the airfoil.

The main parameter governing the unsteady flow is the so-called reduced frequency k, defined as $k = \omega l/U_\infty$, which is proportional to the ratio of the chord length $2l$ and the wavelength L (see Figure 4). The reduced frequency is a measure of the unsteadiness of the flow; for

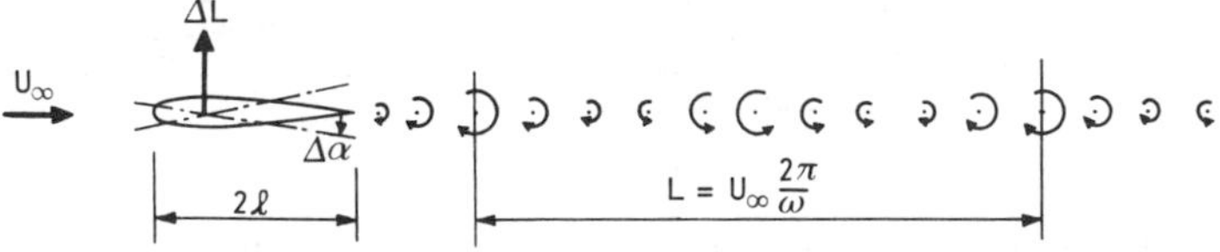

Figure 4 Flow around an oscillating airfoil.

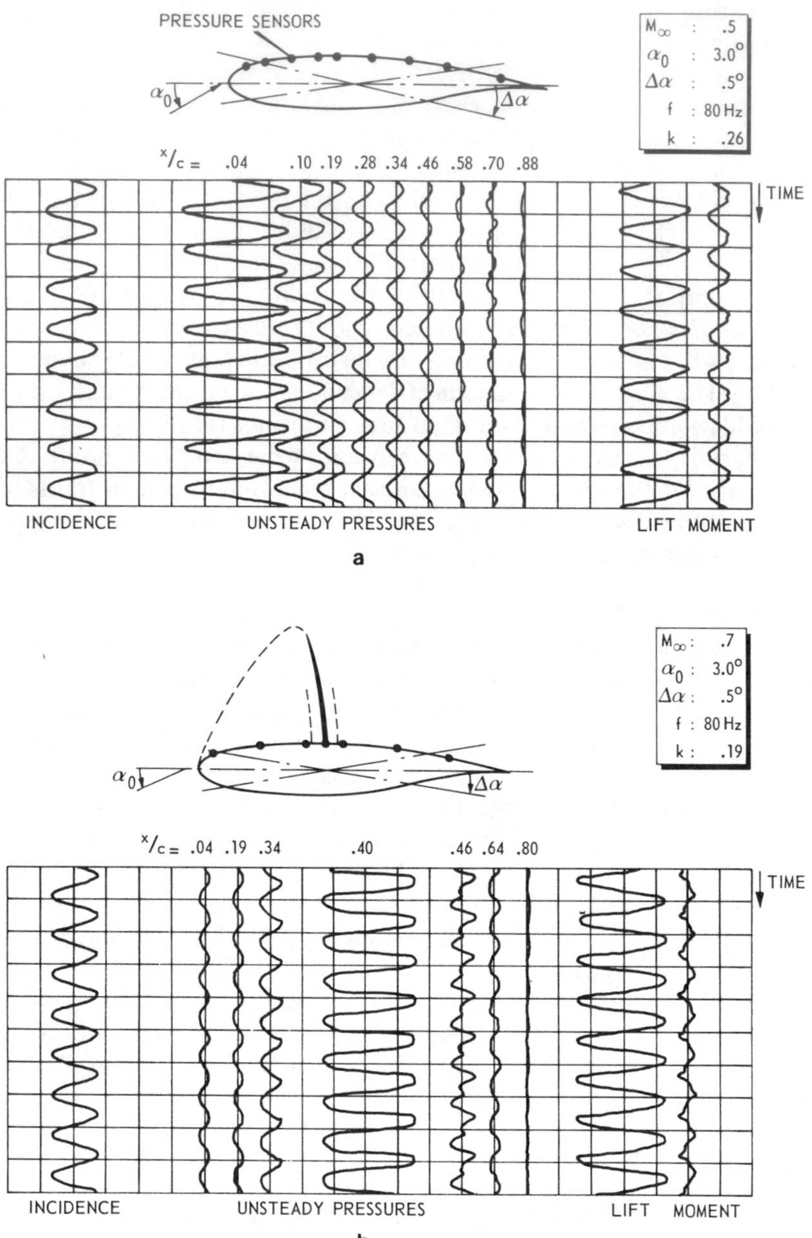

Figure 5 Unsteady pressure signals and overall loads on an oscillating airfoil: (*a*) subsonic flow; (*b*) transonic flow.

similarity of the flow past an oscillating full-scale airfoil and its wind-tunnel model representation it is required that the reduced frequencies be the same.

Figure 5*a* shows the time histories of the local pressures and the resulting lift and moment on an airfoil performing oscillations in pitch in a subsonic flow. Both the pressures and the overall loads show almost sinusoidal variations around their mean values. In this case the pressures and loads may be described by the first harmonic of a Fourier series, viz.,

$$p = p_s + \Delta p' \cos \omega t + \Delta p'' \sin \omega t = p_s + |\Delta p| \cos (\omega t - \varphi)$$

where p and p_s denote the local and mean pressures, while $\Delta p'$ and $\Delta p''$ are the components of the fundamental. The coefficient $\Delta p'$ can be interpreted as the actual pressure perturbation at the instant the oscillating airfoil reachcs its maximum deflection, while $\Delta p''$ represents the pressure perturbation at the instant the airfoil passes its midposition.

This way of describing unsteady pressures or loads is valid only if the aerodynamic quantities vary sinusoidally in time, or, in other words, as long as a linear relationship exists between the displacement of the airfoil and the unsteady pressures. This is usually the case for moderately subsonic and supersonic flow, at least as long as the flow remains attached. For transonic flow, however, this is no longer true, particularly in the region of a shock wave, as illustrated in Figure 5*b*. In such cases one has to give the complete time history of the signals or to add higher harmonics to the Fourier series.

Nonlinear Character of Unsteady Transonic Flow

The combined influence of airfoil thickness, incidence, and amplitude of oscillation is different for moderately subsonic and supersonic flow than for transonic flow. For subsonic and supersonic flow both the equations and the corresponding boundary conditions can usually be linearized. This implies that the problem of an oscillating airfoil can be decomposed into a steady problem (thickness + incidence) and the unsteady problem of an infinitely thin plate oscillating in a uniform flow. The unsteady-flow problem can be treated independently of the steady-flow problem. The main parameters for the unsteady flow then are the reduced frequency k, the free-stream Mach number M_∞, and the mode of vibration.

For transonic flows at low to moderate reduced frequencies the equations governing the motion cannot be linearized. This implies that the unsteady flow field can no longer be treated independently of the steady flow field. For the aeroelastician this means a considerable complication. In addition to the aforementioned parameters for subsonic and supersonic flow, he has to consider also the mean steady flow field around the airfoil, which is determined by the geometry of the airfoil and its mean incidence with respect to the oncoming flow. For a normal flutter investigation in subsonic and supersonic flow, unsteady airloads already have to be computed for 50 to 100 combinations of reduced frequency, free-stream Mach number, and vibration mode. For the transonic flow regime this number increases considerably because then the computations have to be performed for different values of the incidence. The complication becomes even worse if it is not possible to linearize the unsteady transonic-flow problem by assuming the unsteady flow to be a small perturbation superimposed upon a given mean steady flow field. Then the unsteady airloads are no longer linear functions of the amplitude of motion, which implies that in aeroelastic calculations, where the unsteady aerodynamic forces have to be combined with inertial, stiffness, and damping forces of the aircraft structure, linear systems of equations no longer apply. So it is quite evident that from the practical point of view there is a strong demand for some sort of linearization. Of course, in theory, this linearization can always be enforced by making the amplitude of oscillation small enough, but the question arises whether the amplitudes that occur in practice will be that small.

OBSERVATIONS FROM EXPERIMENTS

Results Available

The first transonic-flutter accidents occurred during World War II with aircraft of advanced design at that time that were able to penetrate the transonic regime during a diving flight. These accidents gave the transonic regime its veil of mysticism and contributed to the many myths about the difficulties associated with crossing the "sound barrier." At that time it was impossible to get aerodynamic data for the transonic speed range because there were no transonic wind tunnels and there was little or no support from theory. During the first fifteen years after the war, the knowledge of transonic flows improved considerably through experience gained with research aircraft, like the Bell X-1, and the development of transonic wind tunnels with slotted and porous walls. The latter greatly enlarged the possibilities for obtaining aerodynamic data under controlled conditions. From that period stem a number of unsteady aerodynamic load measurements on oscillating wind tunnel models. The majority of these experiments had an ad hoc character and were directly related to problems encountered in flight.

The first measurements of local unsteady pressures on an oscillating wind tunnel model in transonic flow were made by Erickson & Robinson (1948). Their method, in which electrical pressure cells installed flush with the model surface are used, was applied successfully by Wyss & Sorenson (1951) and their colleagues at the then NACA. Although they actually measured the pressures on oscillating control surfaces and on airfoils and wings oscillating in pitch, only overall aerodynamic coefficients were published, along with some typical oscillograph records of local pressure fluctuations. The first detailed unsteady pressure distributions in the transonic regime were reported by Lessing, Troutman & Meness (1960) and by Leadbetter, Clevenson & Igoe (1960).

A series of exploratory wind tunnel investigations on some characteristic airfoil sections was initiated by the NLR in the late sixties. With the aid of a special technique in which scanning valves and pressure tubes were used (Bergh 1965), steady and unsteady pressures were measured on the conventional NACA 64A006 airfoil with oscillating trailing-edge flap, on the shock-free NLR 7301 airfoil oscillating in pitch (Tijdeman & Bergh 1967, Tijdeman & Schippers 1973, Tijdeman 1977), and on the same airfoil with an oscillating control surface (Schippers 1978). Parallel with this basic research program, unsteady pressure distributions were measured on a variety of three-dimensional wings, under contract with aircraft industries (see, for example, Bergh, Tijdeman & Zwaan 1970, Tijdeman 1976). About the same time Triebstein (1969, 1972) performed experiments on a rectangular wing with and without control surface. In his tests he also applied the measuring technique with tubes and scanning valves. In France, successful tests on a supercritical airfoil section with control surface were reported by Grenon & Thers (1977), who used a large number of miniature pressure sensors placed inside the model. Recently, a similar technique was applied by Davis & Malcolm (1979), who performed exploratory tests on the conventional NACA 64A010 airfoil and the NLR 7301 airfoil of Figure 3. In the NASA experiments a sophisticated test rig that makes it possible to drive the models into pitch as well as plunge motions was used; moreover, the tests could be performed at large Reynolds numbers. Finally, a preliminary series of measurements on an NACA 64A010 airfoil oscillating in pitch has been reported (Davis 1979).

Interaction Between the Steady and Unsteady Flow Fields

Some of the results for an airfoil with flap (Tijdeman 1977) will be recalled here to demonstrate the mechanism of the interaction between the steady and unsteady flow fields. This example is chosen because it lends itself well to physical interpretation.

Low-speed steady and unsteady pressure distributions on the symmetrical NACA 64A006 airfoil with flap are shown in Figure 6-1. Figure 6-1*a* shows the steady pressure distributions along the upper surface of the airfoil for three flap angles, viz., -1.5, 0, and 1.5 degrees, respectively.

From these steady distributions, the "quasi-steady" pressure differences,

$$\Delta C_p = \frac{C_p(\delta_0 - \Delta\delta) - C_p(\delta_0 + \Delta\delta)}{2\Delta\delta},$$

are determined, and the resulting chordwise distribution is shown in Figure 6-1*b*. This quasi-steady pressure distribution can be interpreted as the "unsteady" pressure distribution when the oscillations are infinitely slow. Figure 6-1*c* depicts the first harmonic of the unsteady pressure distribution for a frequency of 120 Hz and an amplitude of about one degree. The unsteady pressure distribution very much resembles the quasi-steady distribution; both show the characteristic peaks at the leading edge of the airfoil and at the hinge axis at 75% chord. For the unsteady example the results of "thin-airfoil theory," assuming an infinitely thin

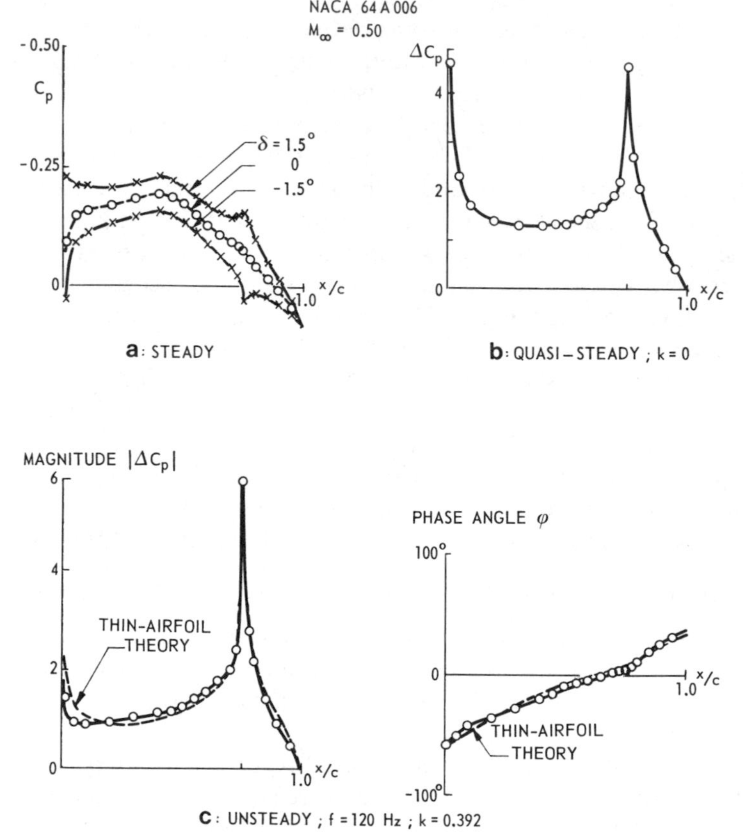

Figure 6-1 Steady, quasi-steady, and unsteady pressure distributions in low subsonic flow.

wing in a uniform main flow, are also given. At this low speed, the agreement with the experimental amplitude and phase distributions is satisfactory, which indicates that the unsteady-flow problem can be treated independently of the steady flow pattern around the airfoil.

Results for the same configuration in high subsonic flow are shown in Figure 6-2. Around the 50% chord point, where the flow is almost critical, a bulge occurs in the magnitude of the measured distribution of both the quasi-steady and unsteady pressures. This bulge and the phase variation are not predicted by the thin-airfoil theory. Another characteristic feature is that the phase lag on the front part of the airfoil is consistently larger than that given by the theory.

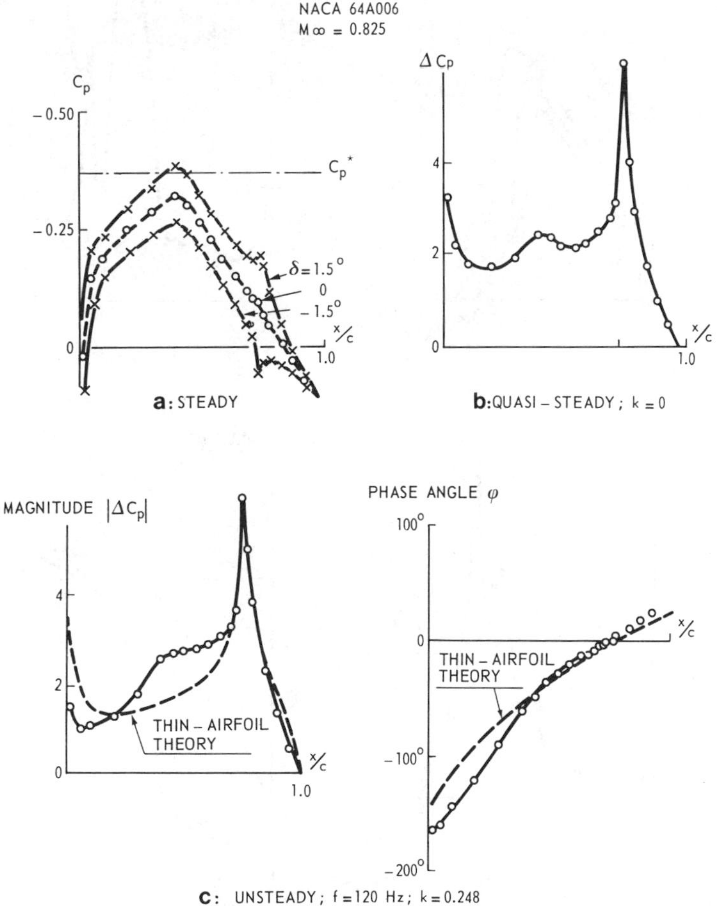

Figure 6-2 Steady, quasi-steady, and unsteady pressure distributions in high subsonic flow.

Results that are typical for a transonic flow with a nearly normal shock wave are given in Figure 6-3. It is clear from this pressure distribution that a change in flap angle is followed by a shift in shock position, and this leads to a peak in the magnitude of the quasi-steady and unsteady pressures in the vicinity of the shock. This peak, which is a significant contribution to the overall unsteady lift and moment, cannot, of course, be predicted by a fully linear theory. Note that the pressure perturbations in front of the shock wave are smaller than those predicted by theory,

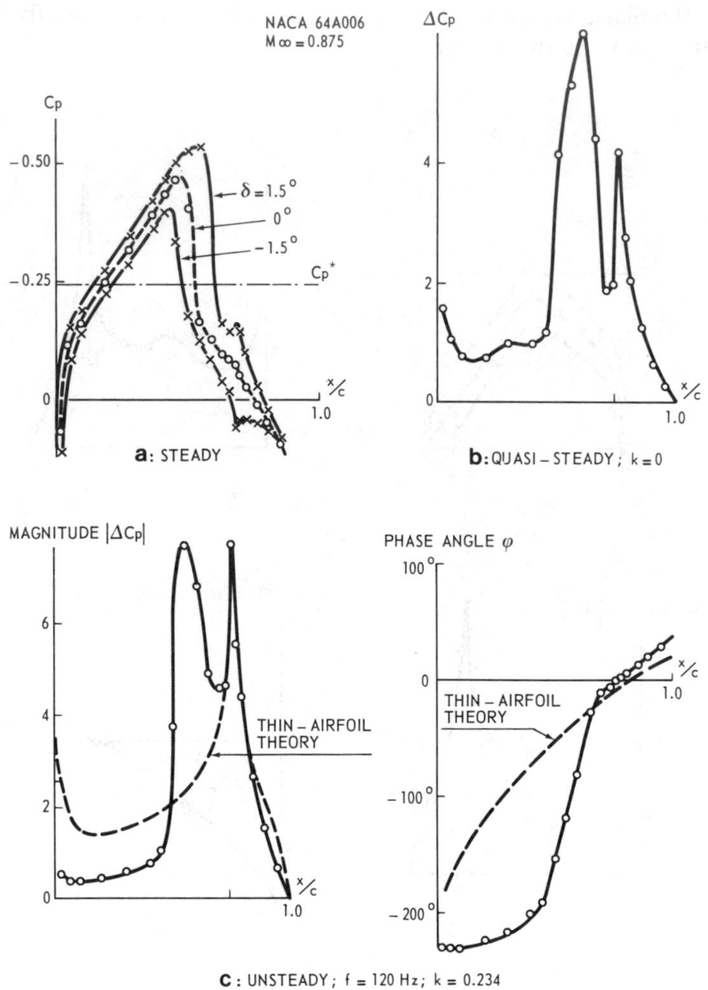

Figure 6-3 Steady, quasi-steady, and unsteady pressure distributions in slightly supercritical flow.

and that the measured phase variation shows a sharp change in gradient in the region of the shock wave.

To illustrate that the observed effects in the preceding examples for high subsonic and transonic flow are caused by the interaction of the steady and unsteady flow field, a graphical experiment has been performed. A pulsating pressure disturbance is assumed to be located at the airfoil's hinge axis. Acoustic waves propagate from this disturbance into the surrounding nonuniform flow. The acoustic wave patterns, as obtained with the well-known construction of Huygens for the airfoil under consideration, are shown in Figure 7. This figure displays the position of the wavefronts after equal time intervals Δt for two different Mach numbers. The part of the figure above the airfoil depicts the time histories of the wavefronts in the actual flow field. Below the airfoil the same wavefronts are shown, but now for a steady uniform flow field in which the local Mach number everywhere is equal to the free-stream Mach number. The corresponding travel times (time lags) are given in the diagrams at the bottom of Figure 7. At $M = 0.8$ the flow is subcritical and the upstream-moving wavefronts encounter more "head wind" in the actual flow than in the

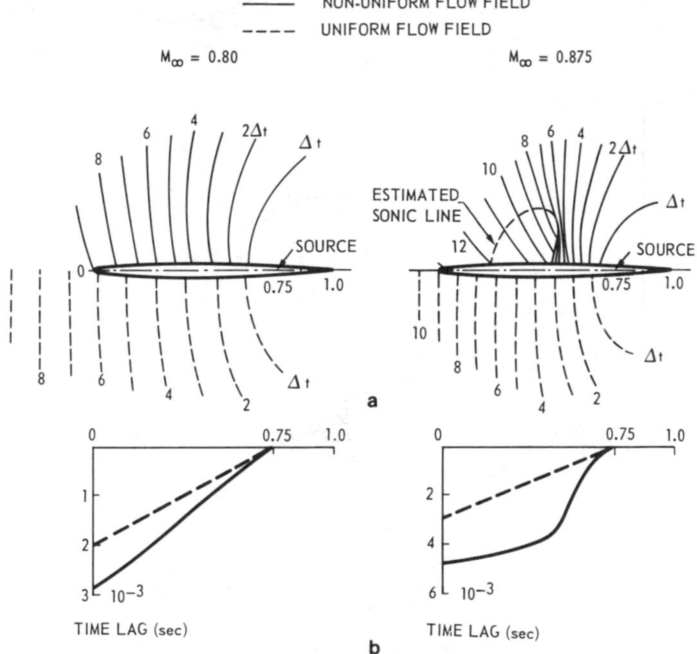

Figure 7 Upstream propagation of wavefronts generated by a source at the hinge axis: (*a*) wave propagation; (*b*) time lag derived from wave pattern.

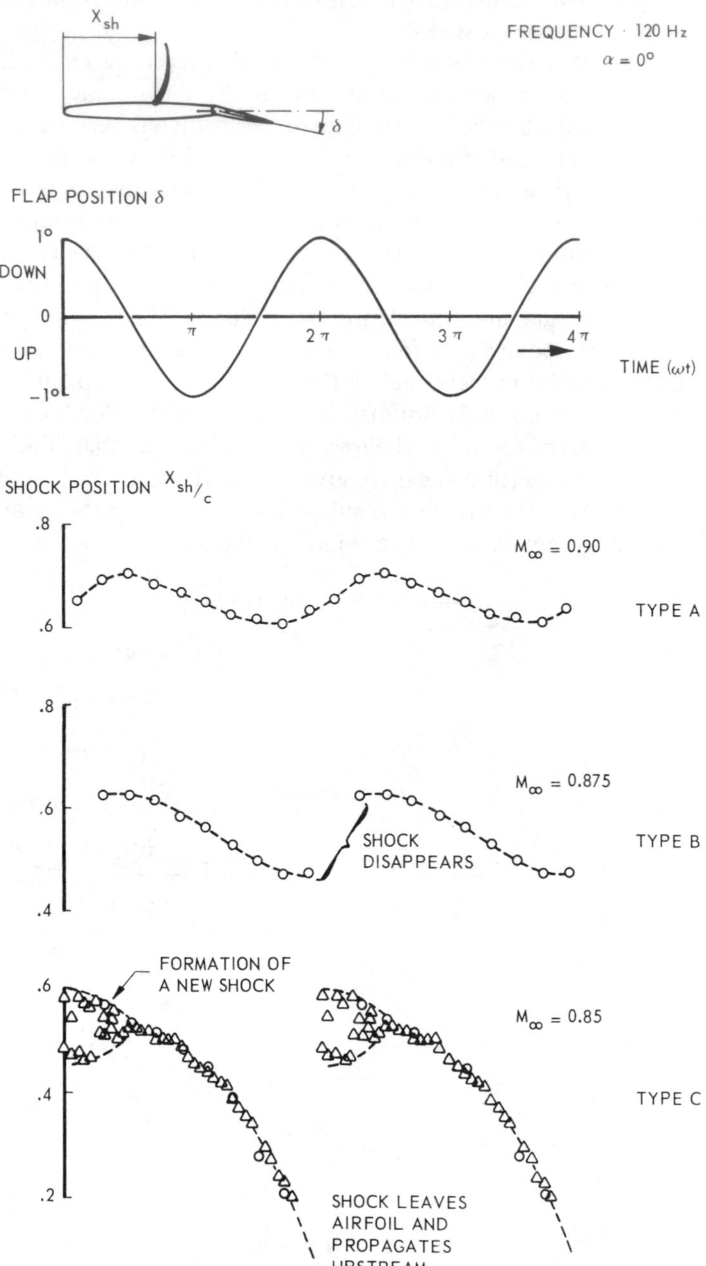

Figure 8 Observed types of periodic shock-wave motion.

uniform flow, as can be seen from the closer spacing of the fronts and from the time-lag curves. Moreover, the velocity gradients normal to the chord in the actual flow cause a forward inclination of the wavefronts. When it is recognized that the spacing of the wavefronts is a measure of the intensity of the local pressure perturbation gradient, while the time lag is a measure for the phase shift, it becomes clear that the high subsonic effects observed in Figure 6-2 can be attributed mainly to the influence of the nonuniform steady flow field. At $M = 0.875$, when a supersonic region is present and terminated by a shock wave, the inclination of the wavefronts is essential to enable the waves to penetrate the region of supersonic flow. Some portions of the upstream-moving wavefronts close to the airfoil surface merge into the shock while other portions bend around the top of the shock and penetrate the supersonic region. This is reflected in the time-lag curve. Since the energy content of the wavefronts penetrating the supersonic region has decreased, due to geometric dilatation, only small pressure changes occur in front of the shock wave. These findings correlate very well with the effects observed in the wind tunnel results presented in Figures 6-2 and 6-3. Note that the main contribution to the peak at the shock position is due to the oscillatory displacement of the shock waves, which is not included in this simple graphical result.

Periodic Motion of Shock Waves

From the preceding discussions it should be clear that the periodic motion of shock waves makes an important contribution to the overall unsteady airloads. Optical flow studies on an airfoil with flap (Tijdeman 1976) have shown that in the oscillating case three different types of periodic shock-wave motions can be distinguished. They have the following main characteristics, which are depicted in Figure 8.

SINUSOIDAL SHOCK-WAVE MOTION (TYPE A) This type resembles, more or less, the shock-wave motions discussed by Lambourne (1958) and Nakamura (1968). The shock moves almost sinusoidally and remains present during the complete cycle of oscillation, although its strength varies. Due to the dynamic effect, phase shifts exist between the model motion and shock position and between shock strength and shock position. The maximum shock strength is not reached during the maximum downstream position of the shock, as in quasi-steady flow, but during its upstream motion.

INTERRUPTED SHOCK-WAVE MOTION (TYPE B) This motion is similar to Type A, but now the magnitude of the periodic change in shock strength becomes larger than the mean steady shock strength and, as a consequence, the shock wave disappears during a part of its backward motion.

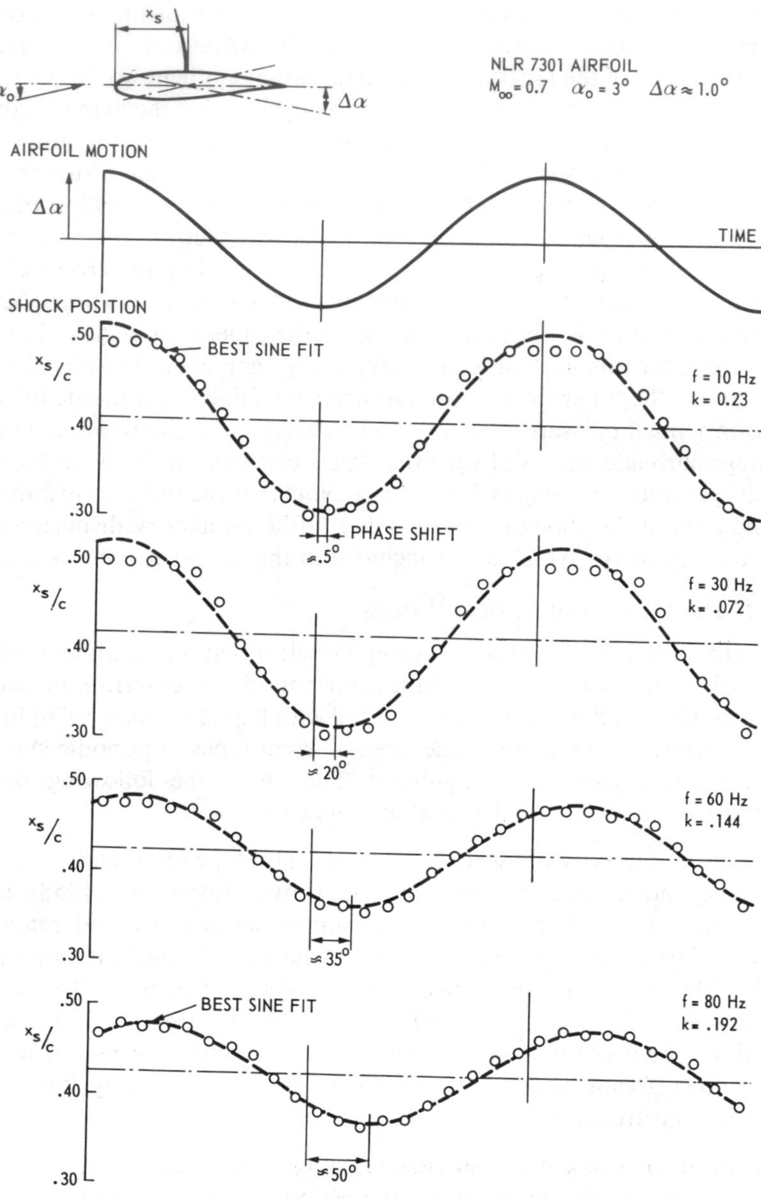

Figure 9 Periodic shock-wave motion for various frequencies.

UPSTREAM-PROPAGATED SHOCK WAVES (TYPE C) At slightly supercritical Mach number a third type of periodic shock-wave motion is observed, which differs completely from the preceding types. Periodically a shock wave is formed on the upper surface of the airfoil. This shock moves upstream while increasing its strength. The shock wave weakens again, but continues its upstream motion, leaves the airfoil from the leading edge, and propagates upstream into the oncoming flow as a (weak) free shock wave. This phenomenon is repeated periodically and alternates between upper and lower surface.

The type of shock-wave motion that occurs in a given situation depends on the Mach number, its chordwise distribution ahead of the shock wave, and on the amplitude and frequency of the motion of the airfoil. The types of shock-wave motions mentioned above are observed not only on oscillating airfoils, but also on steady airfoils with severe flow separation downstream of the shock waves (see, for example, Finke 1975, McDevitt 1979). Under these conditions the shocks do not always remain normal, but cyclically become lambda-type shocks.

Optical flow studies on an airfoil oscillating in pitch (Tijdeman 1977) reveal that the amplitude of the shock-wave motion decreases with frequency, as shown in Figure 9. This is consistent with the observation made in the next section dealing with the theoretical developments. As a consequence, the large pressure peaks generated by the periodic shock motion become smaller with increasing frequency, or, in other words, the contribution of the shock waves to the overall unsteady airloads, which forms one of the dominant effects in transonic flow, will be largest at low to moderate frequencies.

Another interesting feature, observed in the mentioned experiments and in those of Grenon & Thers (1977), is that an almost linear relationship is found between the frequency of oscillation and the phase shift between the motion of the airfoil and the motion of the shock wave for low to moderate reduced frequencies. This means that there is a constant time lag between the motion of the airfoil and the shock-wave motion. This corresponds with the findings of Erickson & Stephenson (1947), who observed a fixed relation between the phase lag of the shock motion and the time required by a pressure impulse to travel from the trailing edge to the shock wave.

In a flow pattern with a well-developed shock wave, the shock motion takes place nearly sinusoidally, with an amplitude of the shock motion that is proportional to the amplitude of the motion of the airfoil. Further, in spite of the presence of the oscillating shocks, the overall lift also varies almost sinusoidally while the moment sometimes shows irregularities (Figure 5*b*). Locally, underneath the foot of the moving shock wave, the pressure signal is highly nonlinear. However, as illustrated in the data of Figure 10 from Schippers (1978), the distributions of the higher harmonics of the pressure signals along the chord show the characteristic that they do not contribute noticeably to the overall lift and only slightly to the moment. This behavior can be explained very well by considering the pressure changes at such a point to be entirely due to a sinusoidal oscillation of a shock that spends part of its time on each side of the point (Tijdeman 1977). Neglecting the local pressure gradient, then, gives a field whose higher harmonics do not contribute to the net lift and only to second order in shock amplitude to the net moment.

Effect of Frequency and Amplitude of Oscillation

As mentioned earlier, it is important for the aeroelastician to know to what extent a linear relationship holds between unsteady airloads and

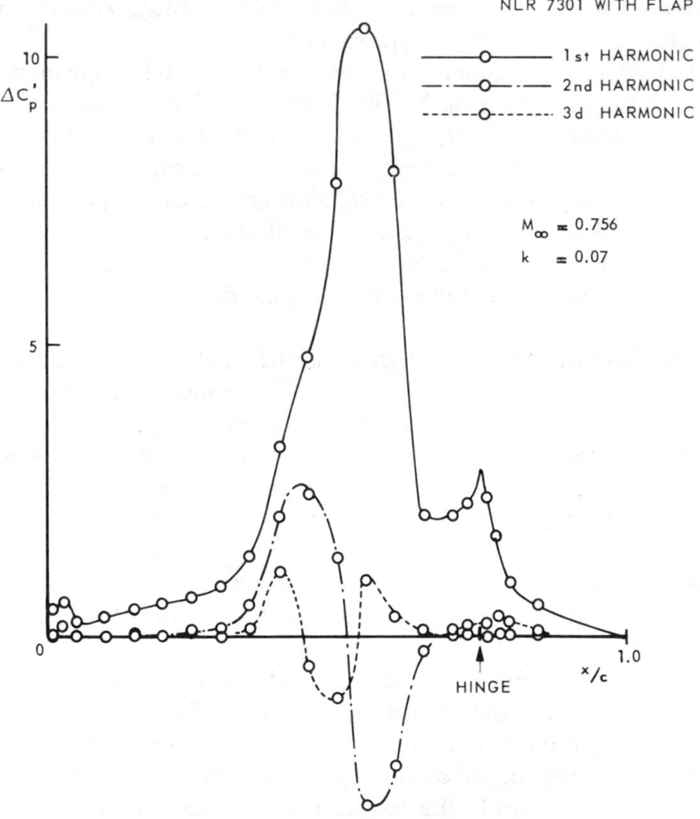

Figure 10 Unsteady pressure distribution showing the first three harmonics.

the amplitude of motion of the airfoil. For his purpose, linearization of the unsteady airloads makes sense only as long as small, but still practical changes in incidence or flap angle (say of the order of 0.5 and 1 degree, respectively) give rise to linear changes in the aerodynamic loads. Also, he must have a procedure for selecting a minimum number of suitable mean steady flow conditions for which the unsteady airloads should be determined. At this moment the experimental and theoretical evidence is not yet conclusive in this respect, but there are a number of observations that certainly throw some light on the problem.

Theoretical analysis (see the next section) shows that the unsteady transonic flow problem becomes a linear one for sufficiently high frequencies. From this, coupled with the observation that the amplitudes of periodic shock-wave motions are largest at low frequency, we can expect that the nonlinear characteristics of unsteady transonic flows will manifest themselves mainly at low to moderate frequencies and in quasi-steady flow. Therefore, we can expect that the behavior of the slopes of the steady lift and moment curves versus incidence or flap angle may serve as an effective guide to detect possible nonlinear regions and to select the mean incidences or flap angles around which linearization is possible.

As noted above, we may also expect that at sufficiently high frequency the measured results will approach the results of thin-airfoil theory. Unfortunately, the experiments show that this does not happen within the frequency range of interest for flutter investigations ($k \lesssim 0.5$). Figure 11 compares some of the results of Davis & Malcolm (1979), whose studies cover a considerable range of frequencies, with a thin-airfoil theory. It is clear that the unsteady airloads cannot be calculated with a linear theory that does not account for the mean steady flow field. However, if the

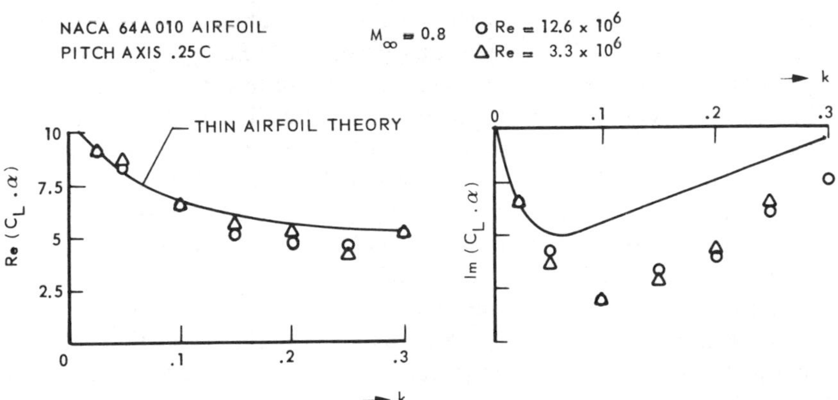

Figure 11 Effect of frequency on unsteady airloads.

amplitude of oscillation is small, we can expect that in most cases linearization for a practical range of amplitudes of oscillation should be possible as long as the flow remains attached. This expectation is confirmed very well by the tests Davis and Malcolm performed on the NACA 64A010 airfoil in a flow condition with a well-developed supersonic region terminated by a relatively strong shock wave. They found a linear relationship between the real and imaginary parts of the lift coefficient with amplitude for 0.25, 0.50, and 1.0 degrees, and further demonstrated that the unsteady airloads for a motion that was a linear combination of two other modes were a linear combination of the airloads for those modes.

THEORETICAL AND NUMERICAL METHODS

In this section we discuss the theoretical and numerical methods developed in recent years to calculate the inviscid transonic flow past oscillating airfoils, stressing the conclusions we may deduce from them. The discussion proceeds from the so-called Euler equations, which can be considered as the most complete set of equations describing inviscid flow problems, through increasingly less complete models to the small-perturbation equation.

Euler Equations

Because the flows of interest here occur at high Reynolds numbers, viscous effects are, for the most part, confined to the boundary layers and wakes. For flows in which the boundary layer remains attached over most of the airfoil, the inviscid flow is the correct first approximation for most of the flow field.

For inviscid flows, the conservations of mass, vector momentum, and energy give a system of four first-order partial differential equations in five scalar unknowns. These equations are usually referred to as the Euler equations. An additional equation relating the state variables is required. The system of equations is hyperbolic and quasilinear. Weak solutions to a hyperbolic system, viz., solutions with discontinuous behavior, may be found numerically if the difference equations are deduced from the conservative form of the equations, that is, from the equations in the form of a space-time divergence of a vector unknown. The physics governing the structure of shock waves is one of a balance between viscous and inertial terms involving viscous dissipation in the wave. Thus, the difference schemes are usually constructed so that the truncation error is predominately dissipative rather than dispersive. In such calculations the mesh size must be small in order that the viscosity implied by the truncation error be small compared with the length scale typical of flow in the vicinity of "captured" shock waves.

Numerical solutions of the Euler equations have been carried out by a number of investigators. Warming, Kutler & Lomax (1973) describe an effective third-order and consequently dissipative, finite-difference scheme with minimal dispersion. Beam & Warming (1976) report comparable results for a dispersive-dominant algorithm, with a switch from central to one-sided differencing where appropriate. The desire to compute steady flow fields originally motivated such computations. Some of the first methods used were explicit in time, time-accurate solutions being computed until a steady flow was achieved. A series of publications by Magnus & Yoshihara (1975, 1976, 1977) and Magnus (1977a, 1977b, 1978) provide many useful results for oscillating airfoils and include studies of the influence of various approximations on the accuracy of the solution. Other interesting results were reported by Beam & Warming (1974), who made a computation for small unsteady perturbations to an already established steady flow, by Laval (1975), and by Lerat & Sidès (1977).

Magnus & Yoshihara (1975) gave detailed unsteady pressure distributions for an airfoil oscillating in pitch; subsequently (1976) they gave analogous results for an airfoil with an oscillating flap, including here an ad hoc procedure to account for the shock–boundary-layer interaction. More recent studies were for the NLR 7301 supercritical airfoil (Magnus 1978). The results, among other things, reproduce the three types of shock motion observed experimentally. To illustrate this, Figure 12 depicts the instantaneous surface pressure distributions for pitch oscillations of an NACA 64A410 airfoil displaying a Type A shock motion. An interesting feature that appears in all numerical results is that the lift and moment vary nearly sinusoidally, despite the presence of strong shock waves. Figure 13 illustrates this behavior for an NACA 64A006 airfoil with an oscillating flap.

The paper of Magnus (1977a), which summarizes his previous studies using the Euler equations and comments on errors introduced by the boundary conditions used, is of considerable interest. For the low to moderate reduced frequencies, the finite computational domain often leads to the reflection of disturbances from the far-field boundary and their interference with the flow field near the airfoil before a harmonic motion is established. In Magnus' studies the far-field data were determined from vortex and doublet singularities located somewhere near the airfoil. Another approximation invoked in these studies was that the unsteady boundary conditions were applied at the steady-state location of the oscillating airfoil. Substantial differences are found when the correct boundary conditions are imposed, with the incorrect boundary conditions causing larger disturbances in the flow. Other studies, e.g. those of Lerat & Sidès (1977) and Steger (1978), avoid this approximation at the computational expense of time-dependent mappings. It is clear from these studies that it is essential to impose boundary conditions on the airfoil unless the amplitude of the unsteady motion is small, and to use a computational region that applies boundary conditions that either allow

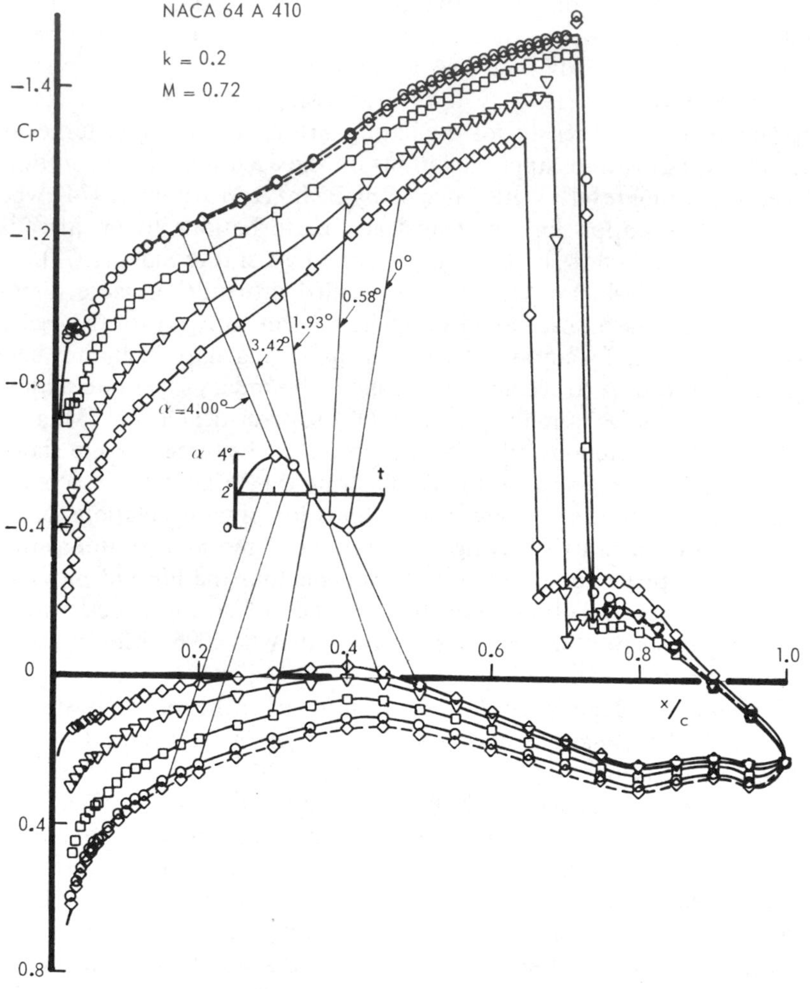

Figure 12 Instantaneous pressure distributions for an NACA 64A410 airfoil pitching around midchord calculated using the Euler equations.

waves to pass through it, or that is sufficiently large that incorrect boundary conditions on its extremities do not cause unacceptable errors.

Explicit calculations such as those discussed above require substantial central processing (CPU) time. Three or four cycles of a harmonic motion may require a comparable number of hours on a CDC 7600. Their main inefficiency occurs at the low to moderate reduced frequencies of practical interest, where improved computational times can be achieved using implicit methods. Beam & Ballhaus (1975) report that the numerical effort per time step for the implicit solution of the Euler equations is four times that required for their explicit solution. For reduced frequencies below about 0.2, however, the larger time step allowed by an implicit calculation results in reduced computational times.

The Potential Approximation

The flows of primary interest here are, for the most part, irrotational, with vorticity introduced by viscous effects in the boundary layers and

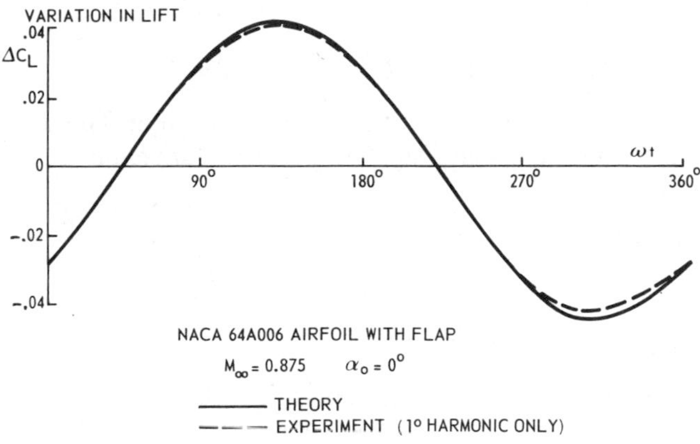

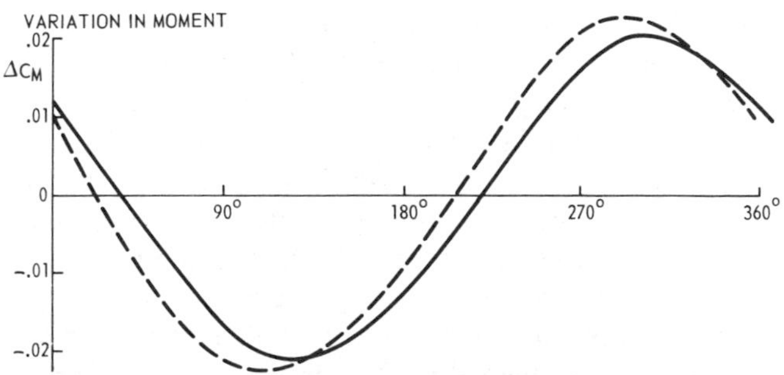

Figure 13 Time histories of the unsteady airloads calculated using the Euler equations.

shock waves. In the inviscid approximation shock-free flows that are originally irrotational will remain so; flows with shock waves will have a rotational component downstream of the shock wave. Crocco's theorem implies that in steady flow the vorticity behind the shock wave will be proportional to the cube of the change in the pressure coefficient across the shock wave multiplied by the speed of sound and divided by the vertical extent of the shock wave. For many practical situations this means that the flow behind the shock wave, and hence the overall inviscid flow, will be affected to a nonnegligible extent by the vorticity introduced by the shock wave. However, the assumption of irrotational flow makes it possible to simplify the problem to one with only a single unknown, the velocity potential. This leads to a considerable reduction in CPU time and storage requirements.

Under the assumptions mentioned above, the following equation for the velocity potential, Φ, can be derived:

$$\Phi_{tt}+(\nabla\Phi)_t^2+\tfrac{1}{2}\nabla\Phi\cdot\nabla(\nabla\Phi)^2-a^2\nabla^2\Phi=0, \tag{1}$$

where

$$a=a_\infty^2-(\gamma-1)[\Phi_t+\tfrac{1}{2}(\nabla\Phi)^2-\tfrac{1}{2}U^2],$$

and the symbols have their usual dimensional meaning. This is subject to the boundary condition that the flow remains tangential to the airfoil surface:

$$F_t+(\nabla\Phi)\cdot\nabla F=0$$

on

$$F=y-\delta Y(x)-\tilde{\delta}\tilde{Y}(x,t)=0.$$

Here δ and $\tilde{\delta}$ are amplitude parameters for the steady and unsteady motions.

Numerical algorithms for the steady potential equations, due principally to Jameson (1974, 1978), are now highly developed and provide reliable results for shock-free flows. Isogai (1977) has solved (1) using a non-conservative time-marching algorithm with unsteady boundary conditions applied at the airfoil's mean surface, a mean steady-state far-field, and an approximate vortex-wake condition. Results for a supercritical airfoil at its design condition oscillating in pitch (Isogai 1978) are shown in Figure 14. The more nearly linear behavior of the unsteady pressure distribution with an increase in frequency is clearly evident.

More recently, Chipman & Jameson (1979) have developed a conservative alternating-direction algorithm that uses a time-varying coordinate system to satisfy the exact boundary conditions. A conservative calculation will capture "shock" waves that conserve mass but add momentum to the flow to balance the wave drag on the body. These will be stronger discontinuities than Rankine-Hugoniot shock waves for a given Mach number ahead of the "shock" wave (van der Vooren & Slooff 1973). More definite comparisons of numerical solutions to (1) with those for the Euler equations are needed to determine quantitatively the adequacy of the potential approximation when shock waves appear in the flow. Isogai's computations indicate, as expected, that the results are unreliable if the cube of the pressure jump across the shock wave is not small.

The calculation of one cycle of harmonic motion requires the equivalent of a minute or so of CDC 7600 CPU time; the number of cycles required to achieve harmonic results will depend on the reduced frequency, with even the lower frequencies requiring at least three cycles.

The Small-Disturbance Approximation

The most basic approximation in inviscid aerodynamics is that of a small disturbance. With the assumption of small disturbances, the vorticity

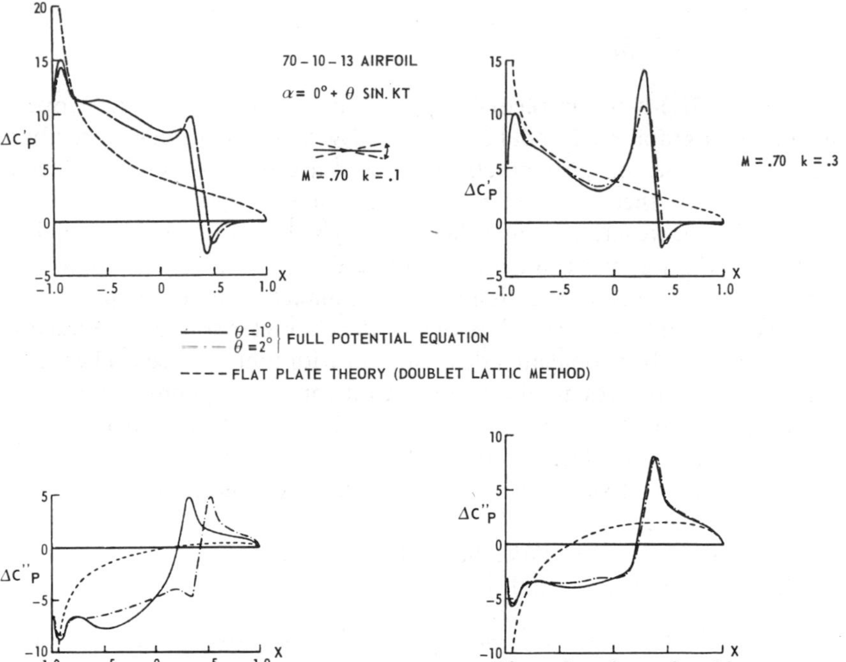

Figure 14 Unsteady pressure distributions on a 70-10-13 supercritical airfoil oscillating in pitch around the design condition calculated using the potential equations.

introduced by a shock wave will be small and the flow may be assumed irrotational. The least trivial balance of terms then leads to the equation

$$\frac{k^2 M_\infty^2}{4\delta_0}\phi_{tt}+\frac{k}{\delta_0}M_\infty^2\phi_{xt}-\left\{\frac{1-M_\infty^2}{\delta_0}-(\gamma+1)M_\infty^2\left[\phi_x+\frac{1}{2}\frac{\gamma-1}{\gamma+1}k\phi_t\right]\right\}\phi_{xx}-\phi_{yy}=0, \qquad (2)$$

where the perturbation velocity potential ϕ is defined by

$$\Phi = Uc[x+\delta_0\phi],$$

where Φ is the velocity potential. Here the time has been nondimensionalized by the circular frequency and the coordinates by the airfoil chord, and the vertical coordinate scaled by $\delta_0^{\frac{1}{2}}$. Application of the mean-surface boundary conditions requires that

$$\delta_0^{\frac{3}{2}}\phi_y(t,x,0) = -\delta\,\frac{Y'(x)}{c}-\tilde{\delta}\left[\frac{\tilde{Y}'(x,t)}{c}+\frac{k}{2c}\tilde{Y}_t(x,t)\right], \qquad (3)$$

and hence that

$$\delta_0^{\frac{3}{2}} = \max\,(\delta,\tilde{\delta},k\tilde{\delta}).$$

For $k = O(1)$ a linear theory applies. This theory is fully developed in the monograph by Landahl (1961) and will not be discussed further here as the cases of most practical concern are those of small reduced frequency. The theoretical limit of interest here is that with $k = O(\delta_0)$. In practice, however, k is often large enough that contributions of this size, viz., $O(k)$, are important, and retaining them provides a bridge to the linear theory. Neglecting the first term of (2) makes the equation parabolic and is equivalent to disregarding one of the characteristics and requiring disturbances to be propagated downstream with infinite speed. There are numerical advantages to doing this. Couston & Angélini (1978) and Houwink & van der Vooren (1979) have shown that marked improvements are obtained at larger values of k when terms of order k are retained in the transport of shed vorticity, the boundary condition (3), and the pressure coefficient.

In the strict limit $k = O(\delta_0)$, that is, for low reduced frequencies, a fully nonlinear theory applies and the conservation of mass takes the form

$$-(k/\delta_0)M_\infty^2\phi_{xt}+\{(1-M_\infty^2)/\delta_0-(\gamma+1)M_\infty^2\phi_x\}\phi_{xx}+\phi_{yy}=0, \qquad (4)$$

a result first given by Lin, Reissner & Tsien (1948). The corresponding "shock" jump relations are

$$\frac{-kM_\infty^2}{\delta_0}[\![\phi_x]\!]^2\left(\frac{dx}{dt}\right)_s-\left\{\frac{1-M_\infty^2}{\delta_0}-(\gamma+1)M_\infty^2\bar{\phi}_x\right\}[\![\phi_x]\!]^2+[\![\phi_y]\!]^2=0,$$

$$\left(\frac{dy}{dx}\right)_s=-[\![\phi_x]\!]/[\![\phi_y]\!] \tag{5}$$

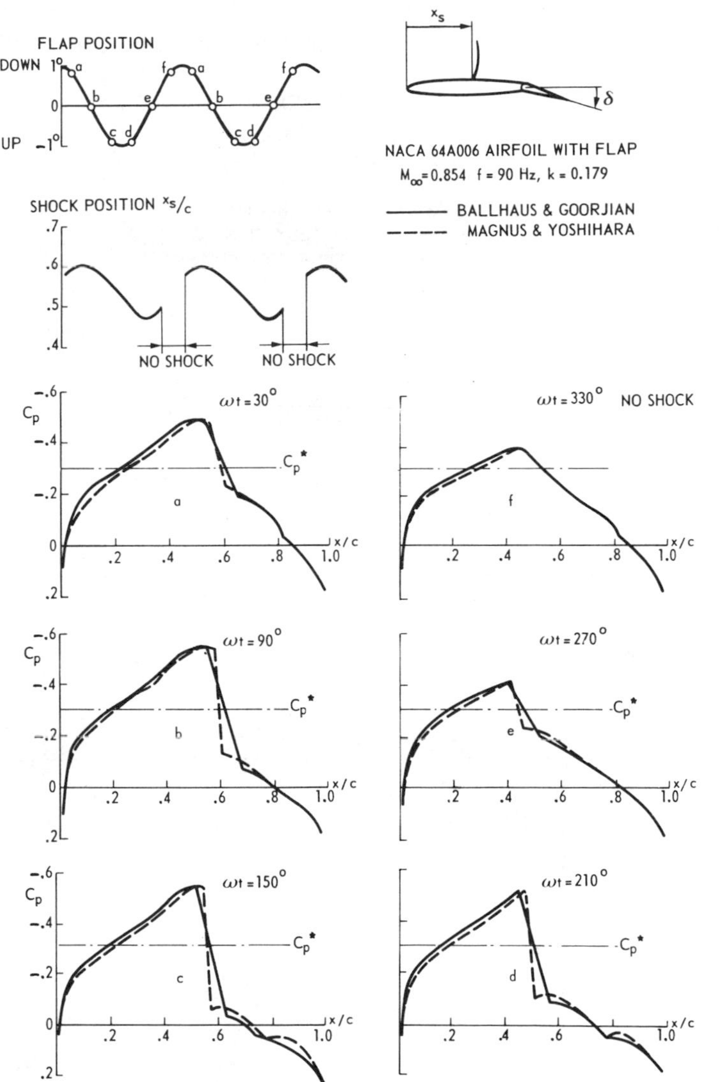

Figure 15 Instantaneous pressure distributions on the upper surface, showing type-B shock motion, calculated using the Euler and the small-disturbance equations.

with $\bar{\phi}_x$ the average value of ϕ_x, and $[\![\phi_x]\!]$ the jump in ϕ_x, across the discontinuity.

Explicit procedures for the solution of (5) are unstable unless the time step is much smaller than that required to resolve low-frequency motions. A fully implicit alternating-direction algorithm with a time-step limitation consistent with that required for accuracy has been developed by Ballhaus & Steger (1975). The far-field boundary conditions used are mean steady-state values on all but the downstream boundary where $\phi_x = 0$. This

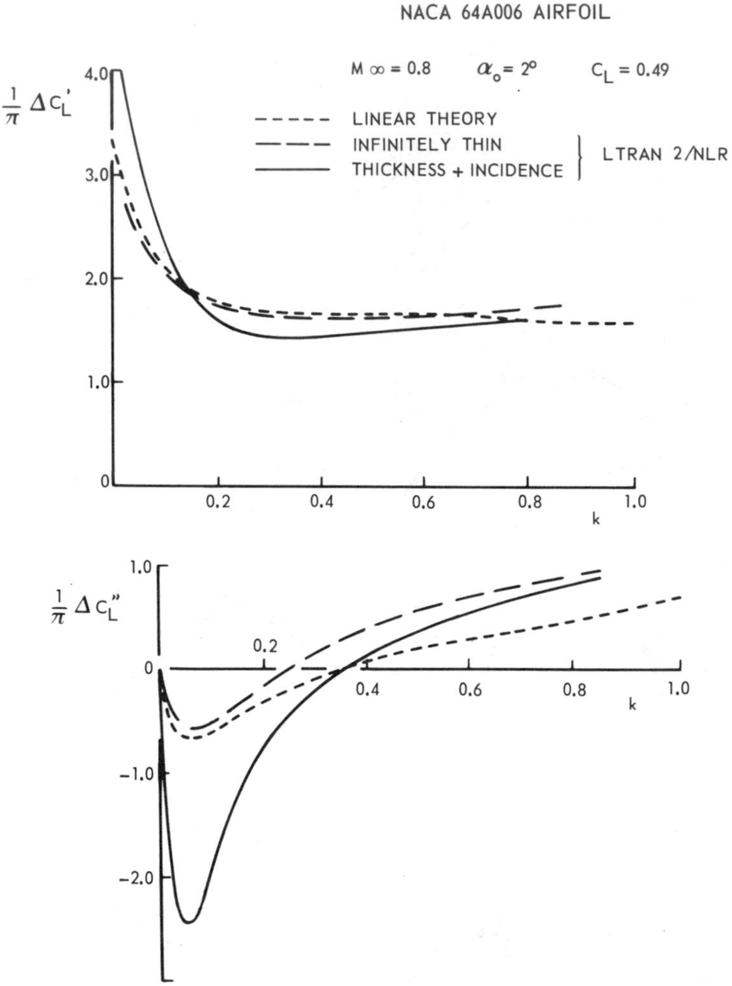

Figure 16 Unsteady lift coefficients for a pitching airfoil showing the effect of reduced frequency.

algorithm resolves shock waves and their motion well, provided the flow changes from supersonic to subsonic across the shock. Using this algorithm Ballhaus & Goorjian (1977) were able to reproduce observed features of experimental studies. To illustrate this, Figure 15 gives the time history of pressure distributions, which reveal a Type B shock-wave motion. Another interesting feature, shown in Figure 15, is that good agreement with results obtained with the Euler equations can be achieved. This is accomplished by introducing arbitrary powers of the Mach number into the equations and boundary conditions to "tune" the results to provide this agreement.

As another example of small-disturbance theory, Figure 16 shows the two components of the lift coefficient for a pitching airfoil as a function of reduced frequency. These computations, by Houwink & van der Vooren (1979), include wake vorticity transport in the Ballhaus-Steger algorithm. They agree reasonably well with the linear theory that applies for $k = O(1)$, but which also retains the ϕ_{tt} term. Figure 16 clearly shows that results for the airfoil with thickness and incidence approach the results for the infinitely thin plate as the frequency increases. Calculations of this type may be useful to determine, for a given airfoil, the frequency range within which a transonic computation method should be applied.

Time-Linearization

The small-disturbance equation (2) can be linearized by assuming the unsteady flow field to be a small perturbation superimposed upon a given mean steady flow field, or, in other words,

$$\phi(x,t) = \phi_0(x,y) + (\tilde{\delta}/\delta)\tilde{\phi}(x,y,t) + o(\tilde{\delta}/\delta)$$

where

$$\tilde{\delta}/\delta = o(1).$$

The linearized version of (5) must be retained to account for shock motion. The steady flow field may be defined either by experimental or numerical means, providing an accurate description of the shock wave's geometry and strength. Because of the ease of the practical implementation in aeroelastic computations, time-linearization is attractive for the flutter specialists. Their main interests are the magnitude and phase of the lift and moment perturbations for the relevant modes of motion. Since the unsteady loads are supposed to be linear, they can be solved in two ways, namely, in the frequency domain or in the time domain. In the frequency domain a steady equation with the frequency as parameter has to be solved for each frequency of interest. In the time domain the time history of the aerodynamic response to a step input (indicial response) has to be calculated. The results for a harmonic motion of the required frequencies then can be obtained by linear superposition, using Duhamel's integral. Solutions in the time domain have been obtained by Beam & Warming (1974), who considered small unsteady disturbances to a basic steady state defined by the Euler equations. The results compare favorably with the linear results obtainable from the theory of Heaslet, Lomax & Spreiter (1948). Noting the advantage of the indicial approach, Beam & Warming further observe that a ramp change (i.e. linear growth for an appropriate time) can be used in place of a step change, avoiding some difficulties in the numerical computations.

Ballhaus & Goorjian (1978) have used the nonlinear algorithm of Ballhaus & Steger (LTRAN2) to calculate indicial responses. They must use an amplitude that is small enough to produce linear results and yet large enough for the response to be computed correctly, a minor difficulty that is avoided by a strict time-linearization. Fung, Yu & Seebass (1978) have given a time-linearized version of the LTRAN2 algorithm that includes explicitly the effects of shock-wave motion through the time-linearized analog of (5). Steady-state shock jumps are obtained from another modification of Ballhaus' algorithm that uses shock fitting (Yu, Seebass & Ballhaus 1978). A detailed comparison of nonlinear and time-linearized calculations (Seebass, Yu & Fung 1978) verifies that the latter is accurate when $\tilde{\delta}/\delta$ is $\leqq 10^{-1}$. If shock-wave motion is not included, the lift and moment variations are incorrect. It also appears that most indicial responses are approximately exponential for the frequency range of most interest and behave as $\exp(-t/\tau)$, where τ is usually large, say 15. This implies that the amplitude of any response is the asymptotic change times $[1+(2k\tau)^2]^{-\frac{1}{2}}$, while the phase angle is $2k\tau[1+(2k\tau)^2]^{-\frac{1}{2}}$. This behavior is consistent with the experimental observation noted earlier, viz., that the amplitude of shock-wave excursions is proportional to k^{-1} for moderate reduced frequencies, and the phase shift is proportional to k for low reduced frequencies (K.-Y. Fung, private communication).

Time-linearized algorithms for the frequency domain have been developed by Ehlers (1974), with subsequent studies by Weatherill, Sebastian & Ehlers (1977, 1978), by Traci, Farr & Albano (1975), and by Fritz (1978). They solve for $\tilde{\phi}$ using a relaxation procedure. Computations in the frequency domain have an inherent limitation on $kM_\infty^2/(1-M_\infty^2)$ that depends on the mesh size. This is a consequence of the generation of standing-wave solutions, which Weatherill, Sebastian & Ehlers (1978) have unsuccessfully tried to eliminate by various means. This restriction is a serious one. More importantly, none of these studies allows for shock-wave motion, although in principle the procedure of Fung, Yu & Seebass (1978) can be carried over to the frequency domain.

To conclude this section on theoretical methods, a few words have to be said on the integral-equation method, since this type of approach has been highly successful for unsteady flow problems governed by linear equations. Such methods were probably the first to provide results for transonic flows with shock waves. The integral method is more suited for the time-linearized equations and, as described by Nixon (1978) for the low-frequency small-disturbance equation, shock-wave motions can be included. A disadvantage is that time-linearized calculations require an effective definition of the steady-state solution. Because finite-difference type time-accurate methods are frequently competitive with other methods of finding the required steady-state solution, one most probably will select a time-accurate procedure to determine the steady-state solution. However, in that case it is more convenient to continue the study of the unsteady response with the same method, instead of switching over to the integral-method approach.

Before the recent advances in computational methods, local linearization proved to be a useful but limited tool (for a review see Spreiter & Stahara 1975; for some recent results see Dowell 1977). Better tools, namely finite-difference algorithms, are now available.

Remarks on the Kutta Condition

With the approximation of inviscid flow, we must impose a Kutta condition at the trailing edge in order to obtain a unique solution. The imposition of the Kutta condition in the form of continuity of the pressure at the trailing edge requires that neither the average velocity nor the jump in velocity be zero if the circulation is to change with time. Consequently, in the strict inviscid limit the flow must follow either the upper or the lower surface of the airfoil at the trailing edge. Which surface it follows depends upon the past history of the motion. More specifically, the rate of change of circulation, Γ (measured counterclockwise), is given by

$$\frac{d\Gamma}{dt} = \frac{d}{dt}\oint \mathbf{q}\cdot d\mathbf{r} = \oint \frac{d\mathbf{q}}{dt}\cdot d\mathbf{r} + \oint \frac{dq^2}{2} = -\oint\left(\frac{dp}{\rho} - \frac{dq^2}{2}\right)$$

$$= -\frac{(q_u + q_l)}{2}(q_u - q_l), \tag{6}$$

where the right-hand side is to be evaluated at the trailing edge and the velocities are those sketched in Figure 17*a*.

Though this view is satisfactory from a computational standpoint, it is rather too narrow. The Kutta condition is an idealization of the behavior for infinite Reynolds numbers. The correct picture is given by Sears (1976), who notes that for viscous flow, Equation (6) applies provided u and l are

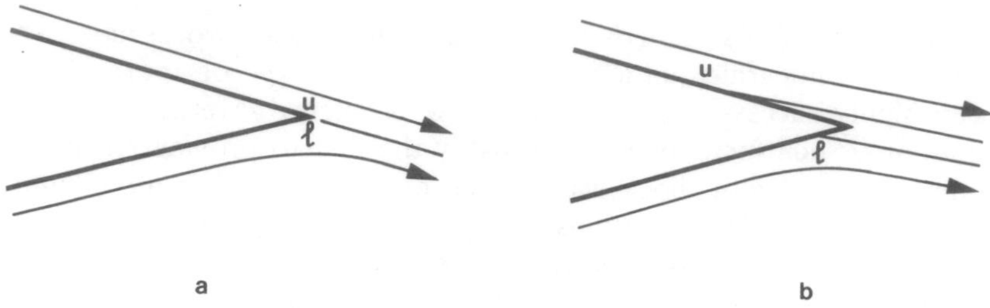

Figure 17 Flow patterns at the trailing edge for inviscid and viscous unsteady flow.

the instantaneous locations of the upper and lower points of boundary-layer separation adjusted for their motion, as shown in Figure 17*b*. Unsteady computations using the Euler equations normally impose some requirement equivalent to the (steady) Kutta condition, e.g. that the flow leaving the trailing edge bisect the trailing-edge angle. While this is inconsistent with results given above (Basu & Hancock 1978), the error involved in doing so is usually inconsequential, nor are time lags of consequence at the frequencies of interest here (see, for example, McCroskey 1977).

Viscous Flows

Viscosity determines, through the Kutta condition and the presence of a boundary layer and a wake, the basic structure of the flow past airfoils. The important parameter here is the Reynolds number, which has a significant influence on the thickness of the boundary layer, on the location of transition and separation points, and on the way in which the boundary layer interacts with a shock wave. A relatively simple method to determine the viscous flow past airfoils is the use of a combination of an algorithm to compute the inviscid flow field with an algorithm to compute the boundary layer. For steady, attached flows such methods are available: first, the inviscid flow field is determined; next the boundary layer is computed and the displacement thickness of the boundary layer is added to the airfoil contour. For this new airfoil the inviscid flow field is calculated again, followed by a new calculation of the boundary layer and so on. An illustration of a result obtained in this way with the method of Bauer et al (1975) is given in Figure 18. This figure reveals that both the steady and the associated quasi-steady pressure distributions are significantly altered by the presence of the boundary layer and that a considerably improved agreement with experiment is obtained. For quasi-steady flow (Figure 18*b*) the effect of the boundary layer is even of the same order of magnitude as the effect of wing thickness. This indicates that reliable predictions of the unsteady airloads on actual airfoils can be obtained

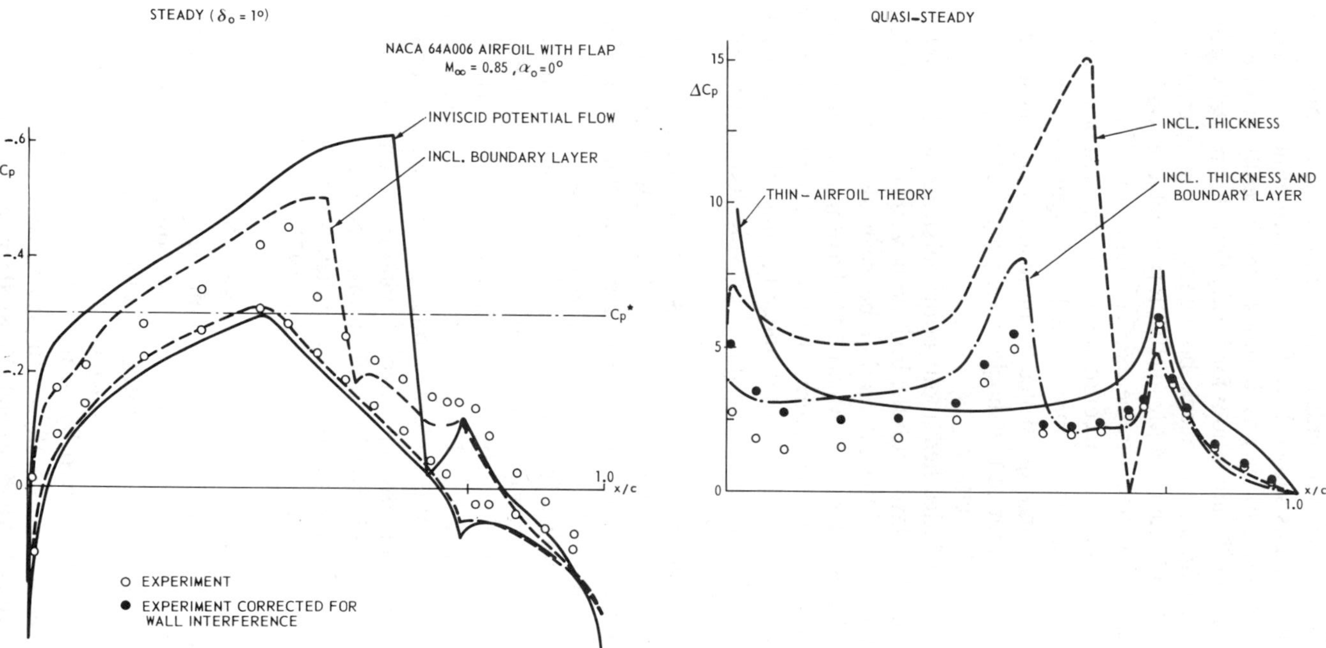

Figure 18 Effect of thickness and boundary-layer displacement on steady and quasi-steady pressure distribution in transonic flow.

only if the boundary layer is included. Provided the flow remains unseparated, computations as shown here for quasi-steady flow are satisfactory also for unsteady flow, as was demonstrated recently by Grenon, Desoper & Sidès (1979). Computations for steady flow that model viscous effects at the trailing edge and the shock–boundary-layer interaction have been reported by Melnik, Chow & Mead (1977); they achieve good agreement with experimental measurements. To take the shock-wave–boundary-layer interaction into account the unsteady methods are still limited to the simple ad hoc procedure devised by Magnus & Yoshihara (1976). They incorporate, in a quasi-steady manner, a wedge-nosed displacement ramp at the foot of the shock wave, providing a qualitative improvement in inviscid results.

There is much yet to be learned about both the steady and unsteady coupling of the boundary layer with the inviscid flow. The main problem areas are the modeling of the interaction between shock waves and boundary layers (Melnik, Chow & Mead 1977 have made considerable progress in this respect) and the accurate treatment of the flow past the trailing edge. Further, a better physical understanding of turbulence is essential if the more complex models are to provide acceptable results for modeling separated flows.

The equations that govern the complete viscous flow are the Navier-Stokes equations for a compressible medium. Questions regarding the existence and uniqueness of the solutions to these equations, even when the medium is incompressible, are generally unanswered. From an engineering point of view, flows of practical interest have turbulent boundary layers, and the Reynolds-averaged form of the Navier-Stokes equations, in conjunction with suitable turbulence models, is an appropriate and necessary basic approximation. The main difficulties that arise are in determining the adequacy of turbulence models, delineating their ability to produce reliable results over a range of conditions, and providing the central processing time needed, and storage required, to effect a numerical solution of the equations. Comprehensive reviews of such calculations may be found in Peyret & Viviand (1975) and MacCormack & Lomax (1979).

Algorithmic advances for the Reynolds-averaged Navier-Stokes equations have recently improved computational efficiency by more than an order of magnitude. One marked improvement was the time-split method due to MacCormack (1976) that separates the equations into a hyperbolic part, which is treated explicitly with a local characteristic method, and a parabolic part, which is solved by an implicit method. The efficiency of this algorithm is sufficient to allow complex three-dimensional flows to be calculated (Hung & MacCormack 1978). Beam & Warming (1978) have extended an earlier algorithm for the Euler equations to the Reynolds-averaged equations, and Steger (1978) has implemented this algorithm with unsteady grid generation and the "thin-layer" version of these equations. The "thin-layer" approximation is essentially the boundary-layer approximation, except that the normal momentum equation is retained, obviating difficulties in matching viscous and inviscid calculations (see Baldwin & Lomax 1978).

In the framework of the Reynolds-averaged Navier-Stokes equations, Seegmiller, Marvin & Levy (1978) studied the flow past an 18% thick circular-arc airfoil in a channel at Mach numbers of 0.76 and 0.79 for a Reynolds number based on a chord of 10^7. This study is part of a continuing investigation to determine the adequacy of numerical algorithms. They use the time-split algorithm of MacCormack. Separate turbulence models are used for the boundary layer ahead of the shock, the separation bubble following the shock, the wake of the separation bubble, and the outer boundary layer and wake. Each is modeled with a simple scalar

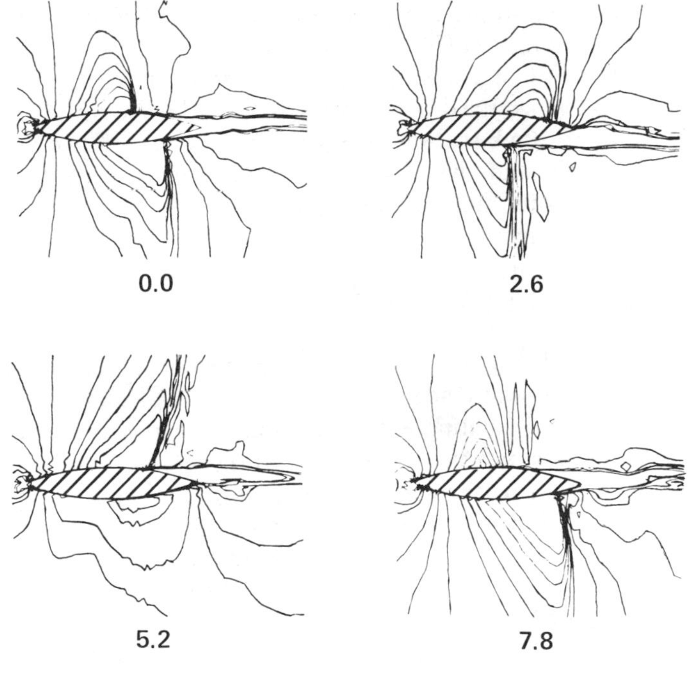

Figure 19 Computed Mach contours for flow past an 18% thick circular-arc airfoil at $M_\infty = 0.754$, Re $= 10^7$, from Levy (1978).

eddy diffusivity. This numerical simulation yielded an unsteady solution when the Mach number was 0.76, but a steady solution was found again when the Mach number was increased to 0.79. This unsteady motion, an alternate fore-aft motion of the shock wave with shock-induced separation on each side of the airfoil, was also observed experimentally over a narrow range of Mach numbers for the Reynolds number used in the calculations. The numerical results reproduce well (within 20%) the frequency of this oscillation at the lower Mach number. Figure 19, from Levy (1978), depicts the Mach contours exhibiting this behavior for a free-stream Mach number of 0.754. Velocity profiles at various chordwise stations were found to be in qualitative agreement with those measured. The main difficulties encountered in this numerical simulation were the inadequacy of the turbulence modeling near the trailing edge and a tendency of the numerical algorithm to capture strong (supersonic to subsonic) shocks where weak (supersonic to supersonic) shocks were observed. The time to carry out a cycle of the unsteady computation on a CDC 7600 was 1.8 hours.

In an exploratory study, using the "thin-layer" algorithm mentioned above, Steger & Bailey (1979) simulated the aileron buzz observed in flight tests of the P-80 and subsequently documented in the Ames 16-foot wind tunnel. These coupled aeroelastic computations were able to reproduce the Mach number of the observed onset of buzz at two angles of attack. This result, and that discussed above, demonstrate that contemporary algorithms and computer hardware are able to simulate complex flow phenomena.

CONCLUDING REMARKS

During the past five years sufficient experimental observations and measurements have been made to provide a good understanding of the transonic flow past oscillating airfoils. Furthermore, recent studies have provided results essential for the design of transonic aircraft. The main limitations of these experiments are their failure, for the most part, to duplicate full-scale Reynolds numbers and an inability to duplicate free-flight conditions due to wind tunnel wall interference. Experimental studies, both in progress and planned for the future, will be more nearly at full-scale Reynolds number, and eventually these Reynolds numbers will be obtained with minimum wall interference in new facilities now under development.

Paralleling this progress has been a rapid development of reliable, and in the small-perturbation approximation, efficient numerical algorithms for the computation of inviscid flows. Numerical results from these methods are in qualitative agreement with the experimental observations, with the main discrepancies in quantitative prediction as a consequence of the inviscid approximation. For steady flows coupled inviscid–boundary-layer calculations of unseparated flows obtain quantitative agreement with experimental measurements. We can expect this to be true for unsteady flows in the near future. The numerical simulation of unsteady separated flows is demonstrably possible, but the two orders of magnitude improvement in computer speed that is projected for a special-purpose aerodynamic computer will be essential for this simulation to have practical consequences.

It is the authors' opinion that the satisfactory prediction of unsteady airloads for aeroelastic applications is within reach. This can be accomplished by "tuning" inviscid boundary conditions to model an experimentally determined steady flow and then computing its unsteady response using an inviscid small-perturbation algorithm. Thus, the time is ripe to start with the incorporation of the new methods in aeroelastic practice as recently demonstrated by Ashley (1979). Of course, the use of two-dimensional methods is justified only for large aspect-ratio wings. To treat the low aspect-ratio configuration the next, and not difficult, step has to be made, namely, the development of prediction methods for three-dimensional flows.

Literature Cited

Ashley, H. 1979. On the role of shocks in the "sub-transonic" flutter phenomenon. *AIAA Pap. 79-0765*

Baldwin, B. S., Lomax, H. 1978. Thin layer approximation and algebraic model for separated turbulent flows. *AIAA Pap. 78–257*

Ballhaus, W. F. 1978. Some recent progress in transonic flow computations. *Numerical Methods in Fluid Dynamics*, ed. H. J. Wirz, J. J. Smolderen, pp. 155–235. New York: McGraw-Hill

Ballhaus, W. F., Goorjian, P. M. 1977. Implicit finite-difference computations of unsteady transonic flows about airfoils. *AIAA J.* 15:1728–35

Ballhaus, W. F., Goorjian, P. M. 1978. Computation of unsteady transonic flows by the indicial method. *AIAA J.* 16:117–24

Ballhaus, W. F., Steger, J. L. 1975. Implicit approximate-factorization schemes for the low-frequency transonic equation. *NASA Tech. Memo X-73,082*

Basu, B. C., Hancock, G. J. 1978. The unsteady motion of a two-dimensional aerofoil in incompressible inviscid flow. *J. Fluid Mech.* 87:159–78

Bauer, F., Garabedian, P., Korn, D., Jameson, A. 1975. Supercritical wing sections II. *Lect. Not. Econ. Math. Syst. No. 108*. Berlin: Springer

Beam, R. M., Ballhaus, W. F. 1975. Numerical integration of the small-disturbance potential and Euler equations for unsteady transonic flow. *NASA SP-347*, Part II, pp. 789–809

Beam, R. M., Warming, R. F. 1974. Numerical calculations of two-dimensional, unsteady transonic flows with circulation. *NASA Tech. Note D-7605*

Beam, R. M., Warming, R. F. 1976. An implicit finite-difference algorithm for hyperbolic systems in conservation-law form. *J. Comput. Phys.* 22:87–110

Beam, R. M., Warming, R. F. 1978. An implicit factored scheme for the compressible Navier-Stokes equations. *AIAA J.* 16:393–402

Bergh, H. 1965. A new method for measuring the pressure distribution on harmonically oscillating wings. *Proc. 4th ICAS-Congress, Paris, 1964*, ed. R. Dexter. London: MacMillan. See also *NLR MP.224*, 1964

Bergh, H., Tijdeman, H., Zwaan, R. J. 1970. High subsonic and transonic effects on pressure distributions measured for a swept wing with oscillating control surface. *Z. Flugwiss.* 18(9/10): 339–46

Chipman, R., Jameson, A. 1979. Fully conservative numerical solutions for unsteady irrotational transonic flow about airfoils. *AIAA Pap. 79-1555*

Couston, M., Angélini, J. J. 1978. Solution of nonsteady two-dimensional transonic small disturbances potential flow equation. In *Nonsteady Fluid Dynamics: Proc. ASME Winter Annual Mtg.*, Dec. 10–15, 1978, San Francisco, pp. 233–44. See also *ONERA TP-No. 1978-69*

Davis, J. A. 1979. Unsteady pressures on an NACA 64A410 airfoil: experimental and theoretical results. *AIAA Pap. 79-0330*

Davis, S., Malcolm, G. 1979. Experiments in unsteady transonic flow. *AIAA Pap. 79-0769*

Dowell, E. H. 1977. A simplified theory of oscillating airfoils in transonic flow: Review and extension. *Proc. AIAA Dynamics Specialist Conf., San Diego, 1977*, pp. 209–24

Ehlers, F. E. 1974. A finite difference method for the solution of the transonic flow around harmonically oscillating wings. *NASA Contract Rep. 2257*

Erickson, A. L., Robinson, R. C. 1948. Some preliminary results in the determination of aerodynamic derivatives of control surfaces in the transonic speed range by means of a flush type electrical pressure cell. *NACA RM A8H03*

Erickson, A. L., Stephenson, J. D. 1947. A suggested method of analyzing for transonic flutter of control surfaces based on available experimental evidence. *NACA RM A7F30*

Finke, K. 1975. Shock oscillations in transonic flows and their prevention. *Symp. Transsonicum II*, ed. K. Oswatitsch, D. Rues, pp. 57–65. Berlin: Springer

Fritz, W. 1978. Transsonische Strömung um harmonisch schwingende Profile. *Dornier GMBH Rep. 78/16 B*

Fung, K-Y., Yu, N. J., Seebass, R. 1978. Small unsteady perturbations in transonic flows. *AIAA J.* 16: 815–22

Grenon, R., Desoper, A., Sidès, J. 1979. Effets instationnaires d'une gouverne en écoulement bidimensionnel subsonique et transsonique. *AGARD CP-262*

Grenon, R., Thers, J. 1977. Étude d'un profil supercritique avec gouverne oscillante en écoulement subsonique et transsonique. *AGARD CP-227*

Heaslet, M. A., Lomax, H., Spreiter, J. R. 1948. Linearized compressible-flow theory for sonic flight speeds. *NACA Rep. 956*

Houwink, R., van der Vooren, J. 1979. Results of an improved version of LTRAN-2 for computing unsteady airloads on airfoils oscillating in transonic flow. *AIAA Pap. 79-1553*

Hung, C. M., MacCormack, R. W. 1978. Numerical solution of three-dimensional shock wave and turbulent boundary-layer interaction. *AIAA J.* 16: 1090–96

Isogai, K. 1977. Calculation of unsteady transonic flow over oscillating airfoils using the full potential equation. *Proc. AIAA Dynamics Specialist Conf., San Diego, 1977*

Isogai, K. 1978. Numerical study of transonic flow over oscillating airfoils using the full potential equation. *NASA Tech. Pap. 1120*

Jameson, A. 1974. Iterative solution of transonic flows over airfoils and wings, including flows at Mach 1. *Comm. Pure Appl. Math.* 27: 283–309

Jameson, A. 1978. Transonic flow calculations. In *Numerical Methods in Fluid Dynamics*, ed. H. J. Wirz, J. J. Smolderen, pp. 1–87. New York: McGraw-Hill

Lambourne, N. C. 1958. Some instabilities arising from the interaction between shock waves and boundary layers. *AGARD Rep. 182*

Landahl, M. T. 1961. *Unsteady Transonic Flow*. New York/London: Pergamon

Landahl, M. T. 1976. Some developments in unsteady transonic flow research. *Symp. Transsonicum II*, ed. K. Oswatitsch, D. Rues. Berlin: Springer

Laval, P. 1975. Calcul de l'écoulement instationnaire transsonique autour d'un profil oscillant par une méthode à pas fractionnaires. *ONERA TP No. 1975-115*

Leadbetter, S. A., Clevenson, S. A., Igoe, W. B. 1960. Experimental investigation of oscillatory aerodynamic forces, moments and pressures acting on a tapered wing oscillating in pitch at Mach numbers from 0.40 to 1.07. *NASA TN-D1236*

Lerat, A., Sidès, J. 1977. Calcul numérique d'écoulements transsoniques instationnaires. *AGARD Spec. Mtg. on Unsteady Airloads in Separated and Transonic Flow, Lisbon, April 1977*. See also *ONERA TP No. 1977-19E*

Lessing, H. C., Troutman, J. L., Meness, G. P. 1960. Experimental determination of the pressure distribution on a rectangular wing oscillating in the first bending mode for Mach numbers from 0.24 to 1.30. *NASA TN-D344*

Levy, L. L., Jr. 1978. Experimental and computational steady and unsteady transonic flows about a thick airfoil. *AIAA J.* 16: 564–72

Lin, C. C., Reissner, E., Tsien, H. S. 1948.

On two-dimensional non-steady motion of a slender body in a compressible fluid. *J. Math. Phys.* 3:220–31

MacCormack, R. W. 1976. An efficient numerical method for solving the time-dependent compressible Navier-Stokes equations at high Reynolds number. *NASA Tech. Memo X-73, 129*

MacCormack, R. W., Lomax, H. 1979. Numerical solution of compressible viscous flows. *Ann. Rev. Fluid Mech.* 11: 289–316

McCroskey, W. J. 1977. Some current research in unsteady fluid dynamics. *Trans. ASME, March 1977, J. Fluids Eng.*, pp. 8–39

McDevitt, J. B. 1979. Supercritical flow about a thick circular-arc airfoil. *NASA Tech. Memo 78549*

Magnus, R. J. 1977a. Computational research on inviscid, unsteady, transonic flow over airfoils. *ONR CASD/LVP 77-010*. 68 pp.

Magnus, R. J. 1977b. Calculations of some unsteady transonic flows about the NACA 64A006 and 64A010 airfoils. *AFFDL Tech. Rep. 77-46*

Magnus, R. J. 1978. Some numerical solutions of inviscid, unsteady, transonic flows over the NLR 7301 airfoil. *ONR CASD/LVP*. 37 pp.

Magnus, R. J., Yoshihara, H. 1975. Unsteady transonic flows over an airfoil. *AIAA J.* 13:1622–28

Magnus, R. J., Yoshihara, H. 1976. Calculation of the transonic oscillating flap with "viscous" displacement effects. *AIAA Pap. 76-327*

Magnus, R. J., Yoshihara, H. 1977. The transonic oscillating flap. *AGARD-CP-226*

Melnik, R. E., Chow, R., Mead, H. R. 1977. Theory of viscous transonic flow over airfoils at high Reynolds number. *AIAA Pap. 77-680*

Nakamura, Y. 1968. Some contributions on a control surface buzz at high subsonic speeds. *J. Aircraft* 5:118–25

Nixon, D. 1978. Calculation of unsteady transonic flows using the integral equation method. *AIAA J.* 16:976–83

Peyret, R., Viviand, H. 1975. Computation of viscous compressible flows based on the Navier-Stokes equations. *AGARDograph No. 212*

Schippers, P. 1978. Results of unsteady pressure measurements on the NLR 7301 airfoil with oscillating control surface. *NLR TR 78124 C*

Sears, W. R. 1976. Unsteady motion of airfoils with boundary layer separation. *AIAA J.* 14:216–20

Seebass, A. R., Yu, N. J., Fung, K-Y. 1978. Unsteady transonic flow computations. *AGARD-CP-227*

Seegmiller, H. L., Marvin, J. G., Levy, L. L. Jr. 1978. Steady and unsteady transonic flow. *AIAA J.* 16:1262–70

Spreiter, J. R., Stahara, S. S. 1975. Unsteady transonic aerodynamics—An aeronautics challenge. *Proc. Unsteady Aerodynamics Symp.*, Univ. of Arizona, Vol. II, pp. 553–82

Steger, J. L. 1978. Implicit finite-difference simulation of flow about arbitrary two-dimensional geometries. *AIAA J.* 16:679–86

Steger, J. L., Bailey, H. E. 1979. Calculation of transonic aileron buzz. *AIAA Pap. 79-0134*

Tijdeman, H. 1976. High subsonic and transonic effects in unsteady aerodynamics. *Part of AGARD Rep. 636*; see also On the motion of shock waves on an airfoil with oscillating flap. *Symp. Transsonicum II*, ed. K. Oswatitsch, D. Rues, pp. 43–56. Berlin: Springer

Tijdeman, H. 1977. *Investigations of the Transonic Flow Around Oscillating Airfoils*. Doctoral thesis. Technische Hogeschool Delft, The Netherlands. 148 pp.

Tijdeman, H., Bergh, H. 1967. Analysis of pressure distributions measured on a wing with oscillating control surface in two-dimensional high subsonic and transonic flow. *NLR-TR F.253*

Tijdeman, H., Schippers, P. 1973. Results of pressure measurements on an airfoil with oscillating flap in two-dimensional high subsonic and transonic flow (zero incidence and zero mean flap position). *NLR TR 73078 U*

Traci, R. M., Farr, J. L., Albano, E. 1975. Perturbation method for transonic flows about oscillating airfoils. *AIAA Pap. 75-877*

Triebstein, H. 1969. Instationäre Druckverteilungsmessungen an einem schwingenden Tragflügel im subsonischen und transsonischen Geschwindigkeitsbereich. *DLR FB 69-49*

Triebstein, H. 1972. Instationäre Druckverteilungsmessungen an einem harmonisch schwingenden Flügelmodell in drei-dimensionaler, kompressibler Strömung. *Z. Flugwiss.* 21(11):400–12; see also *DLR FB 72-55*

van der Vooren, J., Slooff, J. W. 1973. On inviscid isentropic flow models used for finite difference calculations of two-dimensional transonic flows with embedded shocks about airfoils. *NLR MP 73024 U*

Warming, R. F., Kutler, P., Lomax, H. 1973. Second- and third-order noncentered

difference schemes for nonlinear hyperbolic equations. *AIAA J.* 2:189–96
Weatherill, W. H., Sebastian, J. D., Ehlers, F. E. 1977. Application of a finite difference method to the analysis of transonic flow over oscillating airfoils and wings. *AGARD-CP-226*
Weatherill, W. H., Sebastian, J. D., Ehlers, F. E. 1978. The practical application of a finite difference method for analyzing transonic flow over oscillating airfoils and wings. *NASA Contract Rep. 2933*
Wyss, J. A., Sorenson, R. M. 1951. An investigation of the control surface flutter derivatives of an NACA 65-213 airfoil in the Ames 16-foot high speed wind tunnel. *NACA RM A51J10*
Yu, N. J., Seebass, A. R., Ballhaus, W. F. 1978. Implicit shock-fitting scheme for unsteady transonic flow computations. *AIAA J.* 16:673–78
Zierep, J. 1966. Theorie der Schallnahen und der Hyperschallströmungen. Karlsruhe: G. Braun

Ann. Rev. Fluid Mech. 1980. 12: 223–36

SCIENTIFIC PROGRESS ON FIRE

Howard W. Emmons
Division of Applied Sciences, Harvard University, Cambridge, Massachusetts 02138

INTRODUCTION

A fire is the combustion—usually unwanted—of some oxidizable structure or material in air. This review will be limited to fires in buildings. In this case the need to know is the desire to design buildings and their contents such that 1. persons in the building will be warned of a fire in time to get out, and 2. the building itself has a low probability of ignition and/or fire spread so as to keep life and property losses low.

Many of the present fire safety procedures—as embodied in city fire codes for example—are purely empirical and contain obvious contradictions as disclosed, for example, by some round-robin flammability tests (Emmons 1967, 1974). Such failures of a practical safety system can best and most rapidly be corrected by developing an understanding of the fundamental phenomena, in this case the "Science of Fire."

THE PROBLEM

When we look at fire in a burning building our first impression is one of great complexity resulting from the intimate interaction of many chemical and physical processes. The scientific approach to such a problem is as always to study it closely, to discern all the individual processes that are occurring, and to break the total process down into a group of somewhat separable phenomena and the interactions between them. The science of fire has advanced to the point that this breakdown into component phenomena is complete and the frontier is now composed of two parts: 1. the development of the scientific understanding of the component phenomena and their interactions, and 2. the development of the techniques by which the component knowledge can be used to understand the fire as a whole.

0066-4189/80/0115-0223$01.00

THE COMPONENT PHENOMENA

Consider first the burning piece of wood, cloth, cellular plastic, or solid plastic of, for example, a chair. The solid is hot and being heated. It is pyrolyzing by issuing more or less flammable gases through a more or less thick layer of char. The pyrolysis process inside the solid produces gaseous products which flow and diffuse both inward and outward. As the gases leave the surface they mix with air and burn more or less completely in a turbulent plume. The plume exists because of the buoyancy of the hot reacted gas that moves upward to form a hot layer below the room ceiling. As the hot layer deepens, the gas flows out of the room of origin into adjacent rooms or out of doors thus spreading soot and hot and toxic gas. Often there is insufficient oxygen in the room of origin of the gaseous fuel to burn it all and the flames move on to spread the fire throughout the structure.

Back in the original fire room, all objects are being heated in small part by actual contact with hot gases and in major part by radiation from the chemically reacting flame gases and incandescent soot and in part by hot objects. Thus initially unignited objects begin to pyrolyze and eventually to burn. Since all unignited objects heat at about the same rate, all ignite at about the same time and produce an often rapid increase in the rate of gas production, a rapid growth in flame size, and a rapid growth in the fire. Thus occurs what the firefighter calls a "flashover," an ill-defined but sometimes spectacular growth from a "fire in a room" to "full room involvement."

Included among the component problems are

1. Solid-phase response to heat,
2. Flow of gases (both fire-produced gases and "fresh" air),
3. Heat and mass transfer processes, especially radiative transfer,
4. Movement of particulates (both fire-produced soot and the droplets of a fire extinguishing agent, generally water).

Complex chemical activity is of course fundamental to the whole fire process but is not explicitly mentioned because it is an inherent aspect of all four of the above named parts.

Fire presents one additional problem. When we try to solve a system of complex partial differential equations we try to estimate the magnitude of the various terms and retain only those of comparable magnitude in each of various regions. The same process must be carried out for the fire as a whole. However, the various pertinent areas of science have advanced to greatly different levels, and so the equivalent of choosing terms of comparable magnitude is the choosing of comparable levels of precision of description of various phenomena.

Given the temperature and composition of a hot gas mixture and appropriate geometry, we can, *in principle*, make rather precise radiant-flux estimates, highly accurate view-factor integrations, and thus good radiative heat-transfer predictions. In fact, the attainment of much of that possible accuracy is at present a pleasant technical exercise in view of the impossibility of knowing the gas composition with adequate precision and the wildly fluctuating currently unpredictable shapes of the flames.

THE GAS PHASE PHENOMENA

The energy feedback mechanisms interacting with temperature-dependent chemical-reaction rates have been known for a long time (Mallard & Le Chatalier 1883), but it was not until the work of Hirshfelder et al (1953) and von Kármán (1955) that a really sound theory was formulated for the relatively simple problem of flame speed in a premixed mixture of gases. The precise prediction of laminar-flame spread through a premixed mixture of fuels is now held up by a lack of knowledge of the exact chemical mechanisms and their kinetic rates. The chemical complexity is shown by the flame in a mixture of hydrogen, oxygen, and inert gases which requires twenty kinetic equations between eight specie to get flame-speed predictions close to experimental accuracy (Stephensen & Taylor 1973). Experimental work by Wilson et al (1969) shows the rise and fall of various specie through a flame but little has been done to make careful comparisons of these results with the available flame theory. In fires, most gaseous reactions occur in turbulent-diffusion flames (gases not premixed), which radiate strongly. Thus the limited knowledge of chemical kinetics adds but little to the difficulties of a problem already beyond our current capabilities.

The laminar-diffusion flames in forced- and free-convection boundary layers over fuel surfaces have been solved (Spalding 1960, Emmons 1956, Kim et al 1971). Fernandez-Pello & Williams (1977) have developed an approximate theory of flame spread over a solid surface that shows fair agreement with experiments, while Carrier & Fendell (1979) have developed an exact theory for downwind flame spread under simplified conditions. There have been no exact solutions of boundary-layer burning including radiation-heat transfer, and only limited empirical studies (Fernandez-Pello 1977, Williams 1976, Hirano & Tazawa 1978).

When combustible gases leave a pyrolyzing fuel and burn in a plume flame the resulting flame heights as measured have been correlated through dimensional arguments (Thomas 1963, Becker & Liang 1978). Direct theoretical predictions have been extensions of the classical plume theory of Morton, Taylor & Turner (1956). The theory by Stewart (1970) makes a first attempt at solution and while the result serves well to provide correlating dimensionless groups, it falls short of adequate. The first attempt to use recent turbulent-simulation equations by Lockwood & Naguib (1975) and by Tamanini (1977) shows some promise but involves very complex turbulent simulations and a lot of computation time. The development of new numerical approaches to turbulent-flow calculations is very active at present (Launder & Spalding 1972, Rubesin 1977) but is in its infancy as disclosed by the large number of different sets of complex ad hoc assumptions to be found in current literature.

Present fire-plume theories leave a number of essential problems untouched. The original pyrolysis gases mix and burn in a very nonsteady flame. The flame is made luminous by the content of incandescent soot—mostly elemental carbon. The chemistry of soot formation is not adequate to predict the amount of soot produced from a given fuel. Furthermore, there is no theory to predict how much of the soot and fuel gases will remain unburned in the gas above the flames. Why doesn't all of it burn as oxygen becomes available higher in the flames? Presumably this is because the particles radiate energy faster than they receive energy from their surroundings or produce it by reaction. If this scenario is correct, soot and smoke are the result of the interaction of radiative heat transfers and the turbulent mixing supply of oxygen.

In recent work Pagni & Shih (1977) compute the fraction of fuel burned at various levels in a laminar-boundary-layer flame but do not tackle the more difficult problem of why the flame stops with its job only partially done. In a small-scale apparatus Tewarson (1979) finds that flame combustion burns only limited fractions of the fuel. Polystyrene foamed plastic, for example, burns only about 50% of the mass pyrolyzed; the remainder appears as a dense cloud of soot and some unburned fuel gases—partially burned pyrolysis gases of polystyrene. A start in understanding these processes has been made by Magnussen & Hjertager (1977). They use soot formation and burning rates based upon mean quantities in a turbulent flow and omit the soot radiation that this author believes to be critical to the process.

As plume gases rise by their buoyancy to the ceiling, they turn and flow outward as a ceiling jet (Alpert 1975). At the start of a fire the buoyancy is small and the plume sets up a general three-dimensional circulation in the room. However, as the buoyancy becomes greater the hot gas at the ceiling is unable to descend and so it merely turns at the walls and flows back under the top layer at the ceiling. Essentially a hot layer is formed. The ceiling layer loses heat by convection to the ceiling and by radiation to all objects below. Therefore the size of the room is important in determining how much room circulation (ceiling to floor) occurs relative to layer formation. What criterion determines just what buoyant flow occurs? This question is easily posed but hard to answer. To date there are no criteria from which to judge the importance of general circulation given the size of the fire and the size of the room. Experience suggests that under most fire conditions, layer formation is most important and occurs very soon after ignition.

For ideal gases of a constant specific heat, the energy per *unit volume* is independent of the temperature at a given pressure. In a fire the pressure is atmospheric except for relatively small buoyancy differences. Hence as a fire adds energy (and a little mass) to the gas in a room other gas must move out. In sealed rooms the pressure rises dangerously (the Apollo capsule burst), while in all ordinary rooms there is considerable gas outflow.

As the hot layer of gas at the ceiling deepens to below the top of a vent (door or window), the buoyancy moves hot gas with higher velocity. In fact when the layer becomes deep enough the buoyancy-induced flow carries out more gas than the energy-release rate requires and hence gas (air) from outside flows in at the bottom of the vent. Finally, the flow does not change with the depth of the hot layer when the hot-cold interface falls below the sill (of a window).

These flows in principle could be solved as laminar three-dimensional flows on a large computer. The real flows are turbulent and include cooled boundary layers and flow separation. For fire gases just as for orifices and nozzles generally, it is more practical to derive hydraulic-type equations and then apply empirical flow coefficients. This is the approach developed by Prahl & Emmons (1975).

While the two-layer hydraulic-flow theory with flow coefficients is sufficient for most fire analysis, it can be seriously wrong under two circumstances.

1. Near the start of the fire the three-dimensional room circulation will alter the vent flow. While scientifically interesting, this regime is not of much fire importance.
2. A rapidly burning object has its plume so near the vent that the interactions are all important. There has been no work on this problem to date.

When the flow of fire gases throughout a large building is considered, we find vents of various sizes connecting rooms with various hot-layer depths and densities. If all possible combinations of vent, sill, and soffit locations are considered with all possible two-layer density stratifications in each of the two rooms, there are a total of seventy different flow conditions requiring identification and calculation. One can either derive special flow equations for each of these circumstances or use the fact that each flow layer in the vent (of which there may be as many as six) has a linear vertical pressure drop during the flow. The special equation approach has been used by Prahl & Emmons (1975), Tanaka (1978), and Kubota & Zukoski (1978) (only a small fraction of the seventy required cases have been developed). The second approach is currently under development by the author.

Regardless of how each vent is handled (and it is not yet clear which method will be more practical), there remains the problem of combining the vent performance with the flow effect of connecting rooms in a building. This is done by writing room mass-conservation equations in terms of the pressure drop between rooms (say at floor level) and then adjusting the pressures until conservation is satisfied. That this method will work in principle is most easily seen by considering a building of one story containing identical square rooms with an identical vent in every wall. If we assume the (unreal) condition that each vent flow is proportional to the pressure difference, mass conservation leads to

$$ {}_0\Delta_1 p + {}_0\Delta_2 p + {}_0\Delta_3 p + {}_0\Delta_4 p = 0 $$

or

$$ p_1 + p_2 + p_3 + p_4 - 4p_0 = 0 \tag{1} $$

where room 0 is surrounded by rooms 1, 2, 3, and 4.

Equation (1) will be recognized as a finite-difference form of the Laplace equation. For such a building, the solution to the Laplace equation would be an approximation to the desired solution of the finite-difference system rather than the more usual situation in which the Laplace solution is desired and the finite-difference equations are the numerical approximation.

The first step in converting a technical art into a science is the derivation of an adequate mathematical model. The appropriate gas-phase multicomponent equations can be found in the work of Hirshfelder et al (1954) and are presented by Williams (1965). However, the balance of inaccuracies in fire science suggests many simplifications for fire purposes. Generally, common values of specific heat, thermal conductivity, viscosity, and diffusion coefficients for all gases and gas mixtures are sufficiently accurate. Thus the most useful forms of the equations are

$$ \text{specie continuity} \quad \rho \frac{dY_s}{dt} + (\rho D Y_{s,i})_i = W_s $$

momentum $\rho \frac{dv_i}{dt} = -p_{,i} + \tau_{ji,j} - \rho g_i$

with $\tau_{ij} = \mu(v_{i,j} + v_{j,i}) - \frac{2}{3}\mu v_{k,k}\delta_{ij}$ (2)

energy $\rho \frac{dh}{dt} + \left(\frac{\lambda}{C_p} h_{,i}\right)_{j,i} + Q_r + Q_{rad} = 0$

reaction rate $W_s = A Y_F^{\nu_F} Y_O^{\nu_O} \exp(-E/RT)$,

where subscript r = reaction, F = fuel, O = oxygen.

While many interesting scientific questions can be asked involving, for example, thermal diffusion, diffusion stresses, dissipation function, gravity work, and nonsteady pressure changes, they are not now significant fire questions. If a proper treatment of radiation and turbulence is not included in a fire calculation many other effects are quantitatively irrelevant for the prediction of the course of a fire with presently attainable accuracy.

RADIATIVE HEAT TRANSFER

As noted earlier, a fire burns at a rate dependent upon the energy feedback from the fire to the fuel. Although conductive and convective heat transfers are important, the controlling heat transfer in essentially all unwanted fires is radiation. Radiant energy transfer is under active study at present. De Ris (1978) has just prepared a general review of radiation work. Current progress is being made on three fronts.

1. The development of approximate methods by which overall radiative emission and absorption properties can be practically computed from the very extensive knowledge of atomic and molecular radiative spectral data (Modak 1975, Yuen & Tien 1977, Taylor & Foster 1974).
2. The measurement and correlation of flame radiative properties (Markstein 1978, Hägglund & Persson 1974).
3. The study of flame size and shape (Modak 1977).

These studies are listed in the order of increasing importance to fire science. The obviously important problem of the radiative properties of fuel surfaces is not critical at present since with most solid fuels some char is formed and this is close to black although the emissivity of liquid melt surfaces may be somewhat lower.

THE SOLID-PHASE PHENOMENA

The most important solid-phase phenomena are ignition, fire spread, and burning rate. For all three, the pyrolysis process of the fuel is critical. For a few solid fuels like polymethyl methacrylate (plexiglass) the pyrolysis is nearly an unzipping surface reaction, i.e. the polymer breaks up as one monomer molecule after another comes off the end with very little char formation. For most solid fuels, however, the process is chemically far more complex (Parker & Lipska 1969, Broido 1976, Bankston et al 1979, Handa et al 1978). No attempt has yet been made to calculate the gases and char produced by pyrolysis in a chemically correct way. For fire purposes, an overall Arrhenius reaction rate is generally assumed with the constants adjusted to fit some particular set of experiments.

As most solids are heated in a fire and begin to pyrolyze, a pyrolysis front begins at the surface and moves slowly into the solid leaving a cracked charcoal layer behind. The combustible gases flow out through the charcoal layer (especially the cracks) and depending upon the porosity may also flow back into the virgin fuel (Min & Emmons 1972) and partially condense. As the front moves deeper, any condensed products are further pyrolyzed and may catalytically modify the virgin-fuel pyrolysis process.

Because of the chemical and physical complexities created by the cracked charcoal layer, only a few solid fuel analyses have been performed (Min & Emmons 1972, Kung 1972, Kansa et al 1977). These same complexities are probably the reason that accurate theories for smoldering and flaming combustion have not yet appeared. The existing theories of smoldering (Kinbara et al 1966, Summerfield et al 1978) are of ad hoc character incapable of general extrapolation of data. There are no criteria for transitions from smoldering to flaming or vice versa although some empirical data exist (Smith 1972, Toong et al 1978, Summerfield et al 1978).

With the scientific basis for pyrolysis in disarray, it is no surprise that while heat loss and heat feedback theories of ignition exist (Shivadev & Emmons 1974, Kindelan & Williams 1977), nothing of general quantitative, predictive value is available. While many experimental studies of cellulose ignition have been made in the arc imaging furnace to simulate an atomic-bomb radiant pulse (Butler et al 1956), no generally applicable criteria have emerged. Much work is needed for fire purposes in elucidating the mechanisms of pyrolysis, smoldering, flaming, and all the transitions possible between them. Simms (1960) has made a particularly interesting observation—if eventually confirmed—that flaming ignition occurs most easily if the combustible gas movements become turbulent. Thus a low-velocity convective (ambient temperature) air flow over a surface may lead to earlier ignition if the flow becomes turbulent in spite of the fact that it convectively cools the system.

The production of fuel gases at the surface of an evaporating solid fuel surface is an endothermic process—by the latent heat of vaporization. A solid that chars at high temperature also requires heat input to continue pyrolyzing. However, existing interpretations of data on the pyrolysis of cellulose (Roberts 1970, Shivadev & Emmons 1974, Murty & Blackshear 1967) show a range for wood from -80 to $+500$ cal/gm with variations dependent upon the size, shape, and past history of the sample. This "nonsense" is almost certainly caused by an inadequate theory of the whole complex dynamic pyrolysis process used in data interpretation.

As would be expected in view of all the above, not enough is known about the magnitude of all the complex chemical and dynamic pyrolysis effects to write a set of (differential or otherwise) equations whose solution would assure us of the proper quantitative (or even qualitative) understanding of most cases.

Using some level of approximation of pyrolysis, energy feedback, and ignition, the phenomena of flame spread over solid (or liquid) fuels have been calculated and compared with limited ranges of fire spread data. This area of work has been reviewed recently by Williams (1976).

The pyrolysis products, gases and particulates (smoke and soot), present challenging scientific problems not only in their production and burning but also in their transport and other properties. As particulates move away from a fire they agglomerate into larger particles. Baum & Mulholland (1979) have made recent progress in its quantitative description. This process is of major practical importance since various types of smoke detectors are variously sensitive to particle size.

Both the gaseous and particulate components of smoke are important in fire safety since they determine the toxicological properties. While this is a most challenging field it will not be reviewed here both because it is largely a physiological problem and because its empirical development is in its infancy.

THE SCIENCE OF FIRE EXTINGUISHMENT

So what's the problem? Just put water on it and it goes out. This has been the almost universal attitude for too long. How do you "put water on it"? Exactly what does the water do when it gets to the fire? There are many hand-waving "explanations" and empirical—mostly qualitative—studies using a fire hose. Such work may be "evolution" but it is not "science."

A recent study by Bhagat in the author's laboratory has shown that a small amount of water in the form of very small drops (not vapor) can *increase* the burning rate of charcoal by 30%. Thus while extinguishing a fire in one room by a hose stream, the fine drop splash carried with air may significantly increase the burning rate in an adjacent room. With this one reference the survey is complete. It is clear that what a water drop does in extinguishing a fire is a fertile field for future research.

Somewhat more progress has been made with the problem of how water is applied. Droplets from a sprinkler or a hose stream not only move forward in the general direction in which they were ejected, but by drag forces with the surrounding gases change their motion and in turn change the motion of the gas. (Most sprinklers fail to put any significant water on a vigorous fire because the droplets are so small that they are driven aside by the strong buoyant plume. They control fires excellently by soaking everything nearby and thus the fire cannot spread and burns itself out.)

Various equations describing the interaction of a spray of droplets and the surrounding gas have been written. Most of these have prescribed a gas-velocity field and computed drop trajectories (Beyler 1977). Only limited work with interaction between droplets and the air flow, with inclusion of the modification of flow of each, has been done (Crowe et al 1977).

A careful derivation of the equations of motion of the gas and the particle stream by kinetic-theory methods has been carried out by M. Krook (1978, personal communication). If $f = f(t,r,v,a,T)$ is the distribution function for droplets in a differential range of size at radius a, temperature T, velocity v, radius r, and at a time t, f must satisfy a kinetic equation

$$f_{,t}+v_i f_{,i}+(\dot{v}_i f)_{,v_i}+(\dot{a}f)_{,a}+(\dot{T}f)_{,T}=0$$

when it is assumed that each droplet interacts with the surrounding moist air but not at all with the other droplets. If we choose to consider at $t = 0$ drops of one size, one velocity (vector), one temperature, and one point of origin (called a beam), f can be expressed as a sum of beams

$$f = \sum_{j=0}^{N} n^{(j)}\delta(v_i - v_i^{(j)})\,\delta(a - A^{(j)})\,\delta(T_d - T_d^{(j)})$$

where the sum is over all beams (j) and n, v_i, A, T_d are current values of the beam variable at r, t. For each beam the kinetic equation reduces to

$$n_{,t}+(nU)_{,i}=0,$$

$$U_{,t}+U_k U_{i,k}=(\dot{v})=(j)\text{ droplet acceleration by gas drag,}$$

$$A_{,t}+U_k A_{,k}=(\dot{a})=(j)\text{ droplet radius change by evaporation,}$$

$$T_{d,t}+U_k T_{,k}=(\dot{T})=(j)\text{ droplet temperature change by energy balance.}$$

The motion of the gas is determined by the Equations (2)—except that each equation needs an added term on the right-hand side; the continuity equation needs the rate of mass evaporated from the drops, the momentum equation needs the (negative of the) drag on the drops, and the energy equation needs the heat transfer to the evaporating drops.

No analytical two- or three-dimensional solutions to these equations have yet been found. A considerable number of instructive one-dimensional solutions are known and will be the subject of a future paper.

Again the field is just now opening and offers many problems for the inspired and the diligent.

THE STRUCTURAL PROBLEM

As a fire grows, it may consume a structural member and cause collapse or it may heat and thus weaken a noncombustible structural member with the same final effect. The heating of insulated structural members can be computed by the relatively well-developed techniques of heat transfer as soon as the fire environment is known (Bresler 1976). The weakening and failure of combustible members requires adding a stress analysis to the theory of pyrolysis of solid fuels (Bresler et al 1976). Little has been done, except some work on spalling. Almost nothing scientific has been done with the problem of the thermal-stress failure of window glass. I leave these problems for the future.

THE FIRE

Once a quantitative description of each of the various components that make up a fire is available, it is possible to assemble the appropriate equations and basic data into a program to compute a fire from beginning to end. Fire science has just reached the start of this possibility and several attempts at several levels of sophistication are being made to build such a computer program. Nearly all of these programs use a modular approach by which I mean that equations are written for each of the components as finite entities and the interactions are computed. Thus it is possible to step forward in time as the fire develops. This approach, in a very real but unconventional sense, is a finite-element method of treatment of a problem that otherwise would be a hopelessly complex three-dimensional time-dependent problem. Quintiere (1976) has a model that treats a fire as a succession of steady states. This properly describes some periods of some fires. However, it encounters some large transitions that the real fire and its transient description pass through smoothly. Pape & Waterman (1977) started to develop a probabilistic model of fire spread but changed to a deterministic treatment to the degree that it is possible. It is at present a partial model since some essential input is taken from the experiment it is in the process of predicting. Reeves & MacArthur (1976) have a code that uses certain burning-rate data from a conventionalized test (Smith 1972) to compute the spread of fire over the seats and up the walls of an aircraft cabin. Emmons et al (1977) are developing a Computer Fire Code that is designed ultimately to compute the progress of a fire in a large building from the plan of the building and a description of its furnishings. At present a fire in a single room can be computed using theory and basic experimental data with nothing adjusted to fit the experiment being predicted. This program has attained fair precision (Emmons 1978, 1979). The extension of this program to include flow of fire gases through a large building is in progress as briefly described in the section on gas-phase phenomena.

CONCLUSIONS

The phenomenon of fire brings together chemistry, thermodynamics, heat transfer, and aerodynamics in a geometry unique to each building. Many of these phenomena have been studied intensively for many years in their respective fields. However, many new questions are raised and old questions are presented from a new direction so that numerous scientific studies will have to be completed before fire can take its place beside other great areas as an engineering field based upon a sound understanding of what will happen and why.

The fact that present-day building codes are written in very specific terms should not be interpreted to mean that either fire safety is thus assured or that the scientific understanding of the phenomenon is anything like adequate. Similarly, the fact that computer programs are being constructed to compute the growth of fire in a building is not an indication that component phenomena are clear. Guesses and poor empirical information are currently being used for want of something better and only the growth of fire science can make our world fire-safe in the future.

Literature Cited

Alpert, R. 1975. Turbulent ceiling-jet induced by large scale fire. *Combust. Sci. Tech.* 11:197

Bankston, C. P., Powell, E. A., Cassanova, R. A., Zinn, B. T. 1979. Detailed measurements of the physical characteristics of smoke particulates generated by flaming materials. *Fire Flam.* In press

Baum, H. R., Mulholland, G. W. 1979. Coagulation of smoke aerosol in a buoyant plume. *J. Coll. Interface Sci.* In press

Becker, H. A., Liang, D. 1978. Visible length of vertical free turbulent diffusion flames. *Combust. Flame* 32:115–37

Beyler, C. L. 1977. The interaction of fire and sprinklers. *Dept. Fire Prot. Eng., Univ. Maryland. Proj. Rep.*

Bresler, B. 1976. Response of reinforced concrete frames to fire. *Rep. No. UCB FRG 76-12.* Univ. Calif., Berkeley

Bresler, B., Thielen, G., Nigamuddin, Z., Iding, R. 1976. Limit state behavior of reinforced concrete frames in fire environments. *Rep. No. UCB FRG 76-12.* Univ. Calif., Berkeley

Broido, A. 1976. Kinetics of solid-phase cellulose pyrolysis. In *Thermal Uses and Properties of Carbohydrates and Lignins*, ed. F. Shafizadah, pp. 19–36. New York: Academic

Butler, C. P., Lai, W., Martin, S. 1956. Thermal radiation damage to cellulosic materials. Part II. Ignition of α cellulose. Berkeley, Calif.: US Forest Service, Div. Fire Res.

Carrier, G. F., Fendell, F. E. 1979. Downwind fire spread. *J. Fluid Mech.* Submitted for publication

Crowe, C. T., Sharma, M. P., Stock, D. E. 1977. The particle-source-in-cell (PS1-Cell) model for gas-droplet flows. *J. Fluid Eng.* 99(2): 325–32

De Ris, J. 1978. Fire radiation—A review. *17th Symp. (Int.) Combust.* 1003–16

Emmons, H. W. 1956. The film combustion of liquid fuel. *Z. A. Math. Mech.* 36: 60–71

Emmons, H. W. 1967. Fire research abroad. *Fire Tech.* 3: 225–31

Emmons, H. W. 1974. Fire and fire protection. *Sci. Am.* 231: 21–7

Emmons, H. W., Mitler, H. E., Trefethen, L. 1977. Computer fire code III. *Home Fire Proj. Rep. No. 25.* Harvard Univ.

Emmons, H. W. 1978. The prediction of fires in buildings. *17th Symp. (Int.) Combust.*, 1101–12

Emmons, H. W. 1979. Fire. *8th US Cong. Appl. Mech.*

Fernandez-Pello, A. C. 1977. Downward flame spread under the influence of externally applied thermal radiation. *Combust. Sci. Tech.* 17: 1–9

Fernandez-Pello, A. C., Williams, F. A. 1977. A theory of laminar flame spread over flat surfaces of solid combustibles. *Combust. Flame* 28: 251–77

Hägglund, B., Persson, L.-E. 1974. An experimental study of the radiation from wood flames. *FoU-brand (Fire Res. & Dev. News, Swedish Fire Protect. Assoc.)* 1974, no. 1: 2–6

Handa, T., Sada, T., Nagashima, T., Kaneka, K., Yamamura, T., Takahashi, Y., Suzuki, H. 1978. Size determination of submicron particulates by optical counter using laser and characteristics of smoke from polymerized materials. *Fire Res.* 1: 255–63

Hirano, T., Tazawa, K. 1978. A further study of the effects of external radiation on flame spread over paper. *Combust. Flame* 32: 95–105

Hirshfelder, J. O., Curtiss, C. F., Campbell, D. E. 1953. The theory of flames and detonations. *4th Symp. (Int.) Combust.*, pp. 190–211

Hirshfelder, J. O., Curtiss, C. F., Bird, R. B. 1954. *Molecular Theory of Gases and Liquids*, pp, 694–756. New York: Wiley. 1219 pp.

Kansa, E. J., Perlee, H. E., Chaiken, R. F. 1977. Mathematical model of wood pyrolysis including internal forced convection. *Combust. Flame* 29: 311–24

Kim, J. S., De Ris, J., Kroesser, F. W. 1971. Laminar free-convection burning of fuel surfaces. *13th Symp. (Int.) Combust.*, pp. 949–61

Kinbara, T., Endo, H., Sega, S. 1966. Downward propagation of smoldering combustion through solid materials. *11th Symp. (Int.) Combust.*, pp. 525–31

Kindelan, M., Williams, F. A. 1977. Gas phase ignition of a solid with in-depth absorption of radiation. *Combust. Sci. Tech.* 16: 47–58

Kubota, T., Zukoski, E. E. 1978. Computer model for fluid dynamic aspects of a transient fire in two room structures. *Cal. Inst. Tech. Rep.* prepared for Natl. Bur. Stand., Ctr. Fire Res.

Kung, H. C. 1972. A mathematical model of wood pyrolysis. *Combust. Flame* 18: 185–95

Launder, B. E., Spalding, D. B. 1972. *Mathematical Models of Turbulence.* New York: Academic

Lockwood, F. C., Naguib, A. S. 1975. The prediction of the fluctuations in the properties of free, round-jet, turbulent diffusion flames. *Combust. Flame* 24: 109–24

Magnussen, B. F., Hjertager, B. H. 1977. On mathematical modeling of turbulent combustion with special emphasis on soot formation and combustion. *16th Symp. (Int.) Combust.*, pp. 719–29

Mallard, F. E., Le Chatelier, H. L. 1883. Combustion des mélanges gazeux explosifs. *Annal. Mines* sér. 8, 3: 274–378

Markstein, G. H. 1978. Radiative properties of plastics fires. *17th Symp. (Int.) Combust.*, 1053–62

Min, K., Emmons, H. W. 1972. The drying of porous media. *Proc. 1972 Heat Transfer Fluid Mech. Inst.*, pp. 1–18

Modak, A. T. 1975. Non-luminous radiation from hydrocarbon-air diffusion flames. *Combust. Sci. Tech.* 10: 245–59

Modak, A. T. 1977. Thermal radiation from pool fires. *Combust. Flame* 29: 177–92

Morton, B., Taylor, G. T., Turner, J. 1956. Turbulent gravitational convection from maintained and instantaneous sources. *Proc. R. Soc. London Ser. A* 234: 1–23

Murty, K. A., Blackshear, P. L. 1967. Pyrolysis effects in the transfer of heat and mass in thermally decomposing organic solids. *11th Symp. (Int.) Combust.*, pp. 517–23

Pagni, P. J., Shih, T. M. 1977. Excess pyrolyzate. *16th Symp. (Int.) Combust.*, pp. 1329–43

Pape, R., Waterman, T. 1977. Modifications to the RFIRES preflashover room fire computer model. *IITRI Rep. J6400* prepared for Natl. Bur. Stand.

Parker, W. J., Lipska, A. E. 1969. A proposed model for the decomposition of cellulose and the effect of flame retardants. *OCD Work Unit 2531C-NRDL-TR-69*

Prahl, J., Emmons, H. W. 1975. Fire induced flow through an opening. *Combust. Flame* 25:369–85

Quintiere, J. 1976. The growth of fire in building compartments. *ASTM-NBS Symp. Fire Standards and Safety*

Reeves, J. B., MacArthur, C. D. 1976. Dayton aircraft cabin fire model. *FAA-RD-76-120*

Roberts, A. F. 1970. A review of kinetics data for the pyrolysis of wood and related substances. *Combust. Flame* 14: 261–72

Rubesin, M. W. 1977. Numerical turbulence modeling. *AGARD-LS-86*

Shivadev, U., Emmons, H. W. 1974. Thermal degradation and spontaneous ignition of paper sheets in air by irradiation. *Combust. Flame* 22:223–36

Simms, D. L. 1960. Ignition of cellulosic materials by radiation. *Combust. Flame* 4:293–300

Smith, E. E. 1972. Measuring rate of heat, smoke, and toxic gas release. *Fire Tech.* 8:237

Spalding, D. B. 1960. A standard formulation of the steady convective mass transfer problem. *Int. J. Heat Mass Transfer* 1:192–207

Stephensen, P. L., Taylor, R. G. 1973. Laminar flame propagation in hydrogen, oxygen, nitrogen mixture. *Combust. Flame* 20:231–44

Stewart, F. R. 1970. Prediction of the height of turbulent buoyant flames. *Combust. Sci. Tech.* 2:203–12

Summerfield, M., Ohlemiller, T. J., Sandusky, H. W. 1978. A thermophysical mathematical model of steady-draw smoking and predictions of overall cigarette behavior. *Combust. Flame* 33:263–79

Tamanini, F. 1977. Reaction rates, air entrainment and radiation in turbulent fire plumes. *Combust. Flame* 30:85–101

Tanaka, T. 1978. *3rd US-Japan Panel on Fire Research & Safety*. NBS

Taylor, P. B., Foster, P. J. 1974. The total emissivity of luminous and non-luminous flames. *Int. J. Heat Mass Transfer* 17: 1591–1605

Tewarson, A. 1979. Experimental evaluation of flammability parameters of polymeric materials. In *Flame Retardant Polymeric Materials*. Lewin, M., Atlas, S., Pearce, E. Vol. 3: New York: Plenum. In press

Thomas, P. H. 1963. The size of flames from natural fires. *9th Symp. (Int.) Combust.*, pp. 844–59

Toong, T. Y. et al. 1978. Smoldering combustion of cellular plastics and its transition to flaming or extinguishment. *Products Research Committee NBS*

von Kármán, T. 1955. The present status of the theory of laminar flame propagation. *6th Symp. (Int.) Combust.*, pp. 1–11

Williams, F. 1965. *Combustion Theory*, pp. 1–17. Reading, Mass.: Addison-Wesley. 447 pp.

Williams, F. 1976. Mechanisms of fire spread. *16th Symp. (Int.) Combust.*, pp. 1281–94

Wilson, W. E., O'Donovan, J. T., Fristrom, R. M. 1969. Flame inhibition by halogen compounds. *12th Symp. (Int.) Combust.*, pp. 929–42

Yuen, W. W., Tien, C. L. 1977. A simple calculation for the luminous-flame emissivity. *Combust. Sci. Tech.* 9:41–7

Ann. Rev. Fluid Mech. 1980. 12: 237–69

TOWARD A STATISTICAL THEORY OF SUSPENSION ×8159

Richard Herczyński and Izabela Pieńkowska
Department of Fluid Mechanics, Institute of Fundamental Technological Research, Polish Academy of Sciences, Warsaw, Świetokrzyska 21, Poland

1 INTRODUCTION

This review deals with attempts to describe the flow of suspensions of rigid spherical particles embedded in an incompressible Newtonian liquid. The reason for re-examining this classical subject is that some important new steps have recently been made to bridge the gap between the microscopic description or, to use a newly coined term, microhydrodynamics, and the macroscopic behavior of such suspensions. In fact, such an undertaking is one of the main tasks of the physical sciences.

The field of microhydrodynamics, according to the inventor of this term, G. K. Batchelor (1976a), is very broad and covers all flows of particles in the size range from 0.1 μm to 10 μm, where Brownian effects usually become negligible. Particles may be differently shaped, either rigid or deformable, and can be subjected to breakage or coagulation. Inertia effects are in most cases negligible. Interaction between the particles can be of different origin (i.e. electric or van der Waals forces) and can be differently coupled with hydrodynamic forces.

Some of these problems were recently reviewed, for instance in Batchelor's review mentioned above. Dispersions in which the particle-particle interaction is mainly of nonhydrodynamic origin were described by Mewis & Spaull (1976), with an up-to-date review of experimental findings and comparison of theories with experimental data. The rheological properties of rigid particle suspensions, and methods of description and verification of these properties, were reviewed by Jeffrey & Acrivos (1976). A different outlook on rheological properties of dilute suspensions with greater emphasis on the role of Brownian motion was presented in a review by Leal & Hinch (1973). A comprehensive review of dilute suspensions with particles of different shapes treated in the framework of an ingenious formalism can be found in Brenner (1974).

0066-4189/80/0115-0237$01.00

The common feature of these reviews is the joint treatment of various problems that previously were considered separately. The common limitation is also obvious: all the rigorous approaches are confined (in essence, though not in formulation) to dilute suspensions, and methods effectively treating dense suspensions have not yet been established.

The only working theories for dense suspensions have a semiempirical character. They give some insight into the role of interparticle interactions, but their predictive value is limited and so far they are not well anchored to physical principles. For a review of these models see Happel & Brenner's book (1965).

In contrast to other classical works now of interest almost exclusively to historians of science, the Einstein paper (1905) from which the theory of suspensions started is still read, interpreted, and discussed. In that paper the formula for the effective viscosity of suspensions

$$\eta = \eta_0(1+\tfrac{5}{2}\Phi_0) \tag{1.1}$$

was found. Here η_0 and η are, respectively, the viscosity of the ambient fluid and the effective viscosity of the suspension; Φ_0 is a constant volume fraction of particles. The formula was derived by calculating the dissipation of kinetic energy in a homogeneous suspension of noninteracting spherical particles (i.e. in a dilute suspension). It is the method of derivation which causes continuing controversy; because the perturbation of fluid velocity due to the presence of a sphere vanishes as r^{-2}, r being the distance from the center of the particle, the integral expressing additional energy dissipation is not absolutely convergent. Its value depends on the shape of the region enveloping the test particle.

A new method of derivation of the Einstein formula has been suggested in a tiny section on suspensions in Landau & Lifshitz's *Fluid Mechanics* (1959). On the microscopic scale the flow of a suspension is always inhomogeneous and unsteady. Thus, to obtain any macroscopic characteristic of a suspension it is necessary to employ some kind of averaging, and this can be conveniently done if the system of equations used to describe the composed medium is valid at each point of the space occupied by the suspension.

Landau & Lifshitz employed volume averaging of the stress tensor in the suspension, which seems at first glance a physical heresy. They showed that such averages can be found without a knowledge of the stress inside the particles by calculating the average stress in a large spherical volume of fluid encompassing the test particle. By relating the averaged stress tensor to the averaged rate of strain tensor, the constitutive equation of the suspension was formulated. It should be noted that in this analysis the problem of non–absolutely converging integrals has not been overcome.

Peterson & Fixman (1963) used the proposed method as a starting point for the calculation of the Huggins coefficient K_H appearing in the expression for effective viscosity if the generalization of (1.1) is sought in the form

$$\eta = \eta_0[1+\tfrac{5}{2}\Phi_0+K_H(\tfrac{5}{2})^2\Phi_0^2+\ldots]. \tag{1.2}$$

At the time of publication of the Peterson-Fixman paper, the problem was controversial because the method of calculating K_H was obscure. Thus the importance of that paper lies not so much in determining K_H as in applying concepts not exploited previously in the theory of suspensions, such as response functions and multipole expansions, and in mentioning the possible use of a grand canonical ensemble. A few recent papers can be treated as a direct continuation of Peterson & Fixman's work; they are reviewed in some detail in Sections 4 and 6.

The ideas of Landau and Lifshitz inspired Batchelor, who realized that the original method, if refined, could provide a powerful tool in the theory of suspension. In the problem of suspensions of spherical particles these refinements were connected with a detailed analysis of interaction of pairs of spheres, with relating the probability density function of separation of spheres to the flow field, and with introducing a certain method to bypass the non–absolute convergence of the integrals appearing in the theory (Batchelor 1970, Batchelor & Green 1972a).

Batchelor's approach proved successful not only in the determination of K_H but also in a number of other problems (Batchelor 1974); these include sedimentation (Batchelor 1972), bulk properties of suspensions of highly elongated particles in pure straining motion (Batchelor 1971), and, recently, hydrodynamical interaction of Brownian particles (see Section 8).

The problem of interaction of two spheres in the flow field, which forms a basis of these theories, recently received further attention (Arp & Mason 1977). In particular the possibility of collisions between particles, and the formation and dispersion of pair aggregates, was studied. In the analysis the influence of Brownian motion is not taken into account. Because it is of considerable theoretical importance to understand the influence of Brownian motion on spherical particle suspensions, this problem is briefly discussed in Section 8.

The most controversial point of Batchelor's papers is the device to bypass non–absolutely convergent integrals. This problem received considerable attention. Jeffrey (1978) sees it as a problem of interpretation, which has already been solved. Hinch (1977), on the other hand, views it as a kind of normalization procedure often used in different branches of physics.

The problem seems to be of a mathematical, rather than physical, significance. The asymptotic behavior of velocity follows from the linearization of the Navier-Stokes equation and from the assumption of incompressibility. If these assumptions are relaxed, the asymptotic behavior can be different. Unfortunately, the asymptotic behavior for the full Navier-Stokes equation is not known but there are reasons to believe that in such a case the velocity perturbation is of the order of $r^{-2-\varepsilon}$, $\varepsilon > 0$. If this is so, the whole problem can be treated as a mathematical artifact with little relevance to the physical picture.

Still another line of research should be mentioned, that represented by Buyevich and co-workers. This differs from all approaches mentioned above in one important point, namely that each phase of a suspension is considered separately. The main, far-reaching aim of this approach is to find a firm basis for the theory of dense suspensions. However, as it stands at present, the difficulties in solving any specific problem involving the complicated system of equations derived by Buyevich are very great if not insurmountable. Recently a complete review of these works was published (Buyevich & Shchelchkova 1978) and the interested reader can find there a firsthand account.

The present review starts with the description of ensemble averaging, which is frequently used in the theory of suspensions. The equations of flow and the hierarchy of equations that follow from the averaging are discussed in Section 3. Here we discuss also a problem of closing the hierarchy.

The point-particle method used recently for solving the equation of flow by Felderhof (1976a,b) is analyzed in Section 4.

In Section 5 we return to ensemble averaging, this time in the sense of a grand canonical ensemble. This method has been used in the paper by Bedeaux, Kapral & Mazur (1977); their method is presented in detail in Section 6.

The two sections that follow deal with wall effects (Section 7) and Brownian motion (Section 8). A short conclusion of a general nature ends the review.

2 ENSEMBLE AVERAGING

The positions of particles in space can be described by the characteristic function, which assumes the value 1 at points in the fluid and 0 at points in the particles. For spherical particles the analytical expression for the characteristic function can be easily written as

$$H(\mathbf{r},t;\mathbf{R}_1,\mathbf{R}_2,\ldots,\mathbf{R}_N) = 1 - \sum_{i=1}^{N} \mathscr{H}[a_i - |\mathbf{r} - \mathbf{R}_i(t)|] \tag{2.1}$$

where $\mathscr{H}$ is the Heaviside function

$$\mathscr{H}(x) = \begin{cases} 0 \text{ for } x < 0 \\ 1 \text{ for } x \geqq 0 \end{cases} \tag{2.2}$$

and $\mathbf{R}_i(t)$ $i = 1, \ldots, N$ are instantaneous positions of the centers of spheres in a given domain D, a_i being the radius of the ith sphere.

For suspensions built up of spheres of the same radius a the characteristic function (2.1) carries information equal to that of the number density function, defined as

$$n(\mathbf{r},t) \equiv n(\mathbf{r},t;\mathbf{R}_1,\mathbf{R}_2,\ldots,\mathbf{R}_N) = \sum_{i=1}^{N} \delta[\mathbf{r} - \mathbf{R}_i(t)]. \tag{2.3}$$

Both these descriptions are of little value because they contain much redundant information, and thus, to obtain any consistent suspension theory, the use of some averaging procedure is inevitable.

For this purpose a distribution function $\Psi_N(t;\mathbf{R}_1,\mathbf{R}_2,\ldots,\mathbf{R}_N)$ is introduced. The function is defined in D^N, where D^N is the N-time Cartesian product of D. The spheres are assumed to be indistinguishable, and thus the distribution function Ψ_N is symmetric with respect to $\mathbf{R}_i$, $\mathbf{R}_j$. The normalization condition for all t reads

$$\int_{D^N} \Psi_N(t;\mathbf{R}_1,\ldots,\mathbf{R}_N)\, d\mathbf{R}_1\, d\mathbf{R}_2 \cdots d\mathbf{R}_N = 1 \tag{2.4}$$

and implies that in the case of an unbounded domain D the distribution function Ψ_N decreases rapidly for large values of $|\mathbf{R}_i|$. Hence, in the present form, the distribution function Ψ_N does not allow one to treat a homogeneous distribution of particles in a domain D with infinite volume.

In addition, the information contained in Ψ_N is too detailed and one has to use reduced distribution functions of lower orders. The first of these is the one-particle distribution function $\Psi_1(t,\mathbf{R}_1)$, which gives the probability that at the instant t the center of a sphere is placed at $\mathbf{R}_1$

$$\Psi_1(t;\mathbf{R}_1) = \int_{D^{N-1}} \Psi_N(t;\mathbf{R}_1,\mathbf{R}_2,\ldots,\mathbf{R}_N)\, d\mathbf{R}_2\, d\mathbf{R}_3 \cdots d\mathbf{R}_N. \tag{2.5}$$

Similarly the two-particle distribution function $\Psi_2(t;\mathbf{R}_1,\mathbf{R}_2)$ gives the probability that simultaneously the centers of two spheres are, respectively, at $\mathbf{R}_1$ and $\mathbf{R}_2$

$$\Psi_2(t;\mathbf{R}_1,\mathbf{R}_2)=\int_{D^{N-2}}\Psi_N(t;\mathbf{R}_1,\mathbf{R}_2,\ldots,\mathbf{R}_N)\,d\mathbf{R}_3\,d\mathbf{R}_4\cdots d\mathbf{R}_N. \tag{2.6}$$

It follows from these definitions and from (2.4) that all Ψ_N are normalized.

The ensemble average of any arbitrary function $f(\mathbf{r},t;\mathbf{R}_1,\mathbf{R}_2,\ldots,\mathbf{R}_N)$ is now by definition[1]

$$\langle f(\mathbf{r},t)\rangle=\int_{D^N}f(\mathbf{r},t;\mathbf{R}_1,\mathbf{R}_2,\ldots,\mathbf{R}_N)\,\Psi_N(t;\mathbf{R}_1,\ldots,\mathbf{R}_N)\,d\mathbf{R}_1\cdots d\mathbf{R}_N. \tag{2.7}$$

Using (2.7) one obtains from (2.3) and (2.5) the averaged number density in the expected form

$$\langle n(\mathbf{r},t)\rangle=N\Psi_1(t;\mathbf{r}). \tag{2.8}$$

The averaged volume fraction

$$\langle\Phi(\mathbf{r},t)\rangle=\langle n(\mathbf{r},t)\rangle\tfrac{4}{3}\pi a^3 \tag{2.9}$$

follows either from (2.8) or directly from the definition of the characteristic function (2.1).

In the theory of suspensions one needs to calculate forces acting on a chosen particle (a test, or reference particle). One then encounters the problem of the distribution of spheres when the position of the test particle is fixed. This calls for the use of the conditional probability describing the above situation. For one fixed sphere the conditional distribution function is

$$\Psi^c_{N-1}(t;\mathbf{R}_2,\mathbf{R}_3,\ldots,\mathbf{R}_N\,|\,\mathbf{R}_1)=\Psi_N(t;\mathbf{R}_1,\ldots,\mathbf{R}_N)/\Psi_1(t;\mathbf{R}_1). \tag{2.10}$$

Similarly, if the positions of two spheres are fixed, the conditional distribution function for the remaining $N-2$ spheres is given by the formula

$$\Psi^c_{N-2}(t;\mathbf{R}_3,\mathbf{R}_4,\ldots,\mathbf{R}_N\,|\,\mathbf{R}_1,\mathbf{R}_2)=\Psi^c_{N-1}(t;\mathbf{R}_2,\ldots,\mathbf{R}_N\,|\,\mathbf{R}_1)/\Psi_2(t;\mathbf{R}_1,\mathbf{R}_2). \tag{2.11}$$

The conditional average of a function f with the use of the distribution (2.10) results in form

[1] The notation used here differs slightly from that used elsewhere, but we adopt it for this review because of its simplicity and convenience. Note that the variables with respect to which the averaging is *not* performed are indicated within the averaging signs $\langle\ \rangle$.

$$\langle f(\mathbf{r},t\,|\,\mathbf{R}_1)\rangle = \int_{D^{N-1}} f(\mathbf{r},t;\mathbf{R}_1,\ldots,\mathbf{R}_N)\,\Psi^c_{N-1}(t;\mathbf{R}_2,\ldots,\mathbf{R}_N\,|\,\mathbf{R}_1)\,d\mathbf{R}_2\,d\mathbf{R}_3\cdots d\mathbf{R}_N. \tag{2.12}$$

Similarly for the distribution (2.11)

$$\langle f(\mathbf{r},t\,|\,\mathbf{R}_1,\mathbf{R}_2)\rangle = \int_{D^{N-2}} f(\mathbf{r},t;\mathbf{R}_1,\ldots,\mathbf{R}_N)\,\Psi^c_{N-2}(t;\mathbf{R}_3,\ldots,\mathbf{R}_N\,|\,\mathbf{R}_1,\mathbf{R}_2)\,d\mathbf{R}_3\cdots d\mathbf{R}_N. \tag{2.13}$$

From the above expressions it follows that

$$\begin{aligned}\langle f(\mathbf{r},t)\rangle &= \int_D \Psi_1(t;\mathbf{R}_1)\langle f(\mathbf{r},t\,|\,\mathbf{R}_1)\rangle\,d\mathbf{R}_1\\ &= \int_{D^2} \Psi_2(t;\mathbf{R}_1,\mathbf{R}_2)\langle f(\mathbf{r},t\,|\,\mathbf{R}_1,\mathbf{R}_2)\rangle\,d\mathbf{R}_1\,d\mathbf{R}_2.\end{aligned} \tag{2.14}$$

The first conditional average of number density is

$$\langle n(\mathbf{r},t\,|\,\mathbf{R}_1)\rangle = N\Psi^c_2(t,\mathbf{r}\,|\,\mathbf{R}_1) = N\Psi_2(t;\mathbf{r},\mathbf{R}_1)/\Psi_1(t;\mathbf{R}_1). \tag{2.15}$$

The main advantage of ensemble averaging (in contrast to volume averaging) is that the operations of differentiation and ensemble averaging commute. This follows from the appropriate theorems about differentiation of integral expressions. And yet in many papers volume averaging serves as the main working tool. The explanation of this fact is two-fold.

First, as already mentioned, the ensemble average formalism does not allow consideration of flows of homogeneous suspensions in domains of infinite volume. Such problems attract most attention because of their fundamental significance and because of the considerable difficulties encountered in solving problems with "boundary effects."

Second, the elegant formulation of ensemble averages is not very efficient as an analytical tool because very little is known about the distribution of spheres (not to mention particles of more complicated forms). This difficulty is of a fundamental, rather than technical, character and stems from the condition of nonoverlapping of particles.

Finally, it should be noted that the equivalence of ensemble and volume averages has been discussed in a number of papers and plausible arguments in favor of ergodicity were formulated. This question is, however, a very subtle one, especially in the case of objects with finite volume and, as far as we know, no rigorous proof of ergodicity is available in the class of problems considered here.

3 EQUATIONS GOVERNING SUSPENSIONS

In the linear approximation, the Navier-Stokes equations governing the flow of a liquid and the continuity equation for an incompressible fluid are

$$\rho \frac{\partial \mathbf{v}}{\partial t} = -\nabla \cdot \mathbf{P} + \mathbf{F}, \qquad \nabla \cdot \mathbf{v} = 0. \tag{3.1}$$

Here ρ is fluid density, assumed to be constant, $\mathbf{F}$ is an external force, $\mathbf{P}$ is the stress tensor defined by the relation

$$\mathbf{P}(\mathbf{r},t) = p(\mathbf{r},t)\mathbf{1} - 2\eta_0 \overline{\nabla \mathbf{v}(\mathbf{r},t)} \tag{3.2}$$

where $\overline{\quad}$ denotes the symmetric traceless part of a tensor, i.e.

$$\overline{A_{ij}} = \tfrac{1}{2}(A_{ij} + A_{ji}) - \tfrac{1}{3} A_{kk} \delta_{ij}, \tag{3.3}$$

p is the pressure and $\mathbf{1}$ denotes the unit tensor. A summation convention is assumed. Equations (3.1) are valid outside of the spheres, in the region where $|\mathbf{r} - \mathbf{R}_i(t)| > a$.

The motion of rigid spheres is given by

$$m \frac{d\mathbf{u}_i}{dt} = -\int_{S_i} \mathbf{P}(\mathbf{r},t) \cdot \mathbf{n} \, dS + \mathbf{F}_p, \qquad i = 1, 2, \ldots, N \tag{3.4}$$

where the first term on the right-hand side represents force exerted by the fluid on the surface $S_i = S(\mathbf{R}_i)$ of the spherical particle centered at $\mathbf{R}_i$; m is the mass of the particle, $\mathbf{F}_p$ is an external force acting on the sphere; $\mathbf{n}$ is the outward normal.

The above equations are coupled by the condition on the surface of a sphere setting (in the absence of slip) the velocity of the particle equal to that of the fluid,

$$\mathbf{v}(\mathbf{r},t) = \mathbf{u}_i(t) + \mathbf{\Omega}_i(t) \times [\mathbf{r} - \mathbf{R}_i(t)] \quad \text{for} \quad |\mathbf{r} - \mathbf{R}_i(t)| = a \tag{3.5}$$

where $\mathbf{\Omega}_i$ is the angular velocity of the ith sphere.

These equations are nonlinear in the velocities due to the dependence of $\mathbf{R}_i$, and hence also of $S(\mathbf{R}_i)$, on time. In the framework of linear theory consistent with Equation (3.1) this time dependence should be neglected. The linearization is equivalent to omitting the term of order $(\mathbf{u} \cdot \nabla)\mathbf{v}$ (Berker 1963).

As already stated, it is convenient to introduce one equation of motion valid in the whole space occupied by a suspension. This can be done not only on the level of average values but also on the microscopic level. The principal idea of this procedure is to replace a particle by a fictitious force which should be added to the right-hand side of (3.1). In the case of force-free and torque-free particles, this force will be called hereafter (Bedeaux et al 1977) the induced force and denoted by $\mathbf{F}_{\text{ind}}$. The velocity which enters Equation (3.1) should be extended to the whole space and the natural choice is to put

$$\mathbf{v}(\mathbf{r},t) = \mathbf{u}_i(t) + \boldsymbol{\Omega}_i(t) \times [\mathbf{r} - \mathbf{R}_i] \quad \text{for} \quad |\mathbf{r} - \mathbf{R}_i| \leqq a \tag{3.6}$$

inside the particles. We assume that the pressure inside the particles vanishes, so the definition of stress (or pressure) tensor $\mathbf{P}$ given in (3.2) remains valid. If the density of particles is the same as that of fluid, the equations for both phases read

$$\rho \frac{\partial \mathbf{v}}{\partial t} = -\nabla \cdot \mathbf{P} + \mathbf{F}_{\text{ind}} + \mathbf{F} = -\nabla \cdot \mathbf{P} - \nabla \cdot \boldsymbol{\sigma} + \mathbf{F};$$

$$\nabla \cdot \mathbf{v} = 0; \qquad \nabla \cdot \boldsymbol{\sigma} = -\mathbf{F}_{\text{ind}}. \tag{3.7}$$

The tensor $\boldsymbol{\sigma}$ is called the induced pressure tensor. The induced force has the form

$$\mathbf{F}_{\text{ind}} = \begin{cases} 0 \quad \text{for} \quad |\mathbf{r} - \mathbf{R}_i| > a, \\ \rho \dfrac{d}{dt} [\mathbf{u}_i(t) + \boldsymbol{\Omega}_i(t) \times (\mathbf{r} - \mathbf{R}_i)] \quad \text{for} \quad |\mathbf{r} - \mathbf{R}_i| < a \end{cases} \tag{3.8}$$

and it should be emphasized that $\mathbf{F}_{\text{ind}}$ vanishes outside the particles.

The use of Equation (3.7) instead of the set of equations written separately for each phase proves effective in the derivation of the generalized Stokes formula as well as of the unsteady version of the Faxen law (Mazur & Bedeaux 1974).

The price paid for the above simplification is the necessity of introducing the generalized functions into the equations because both $\mathbf{F}_{\text{ind}}$ and $\boldsymbol{\sigma}$ are singular at least at the surface of spheres. By completing $\mathbf{P}$ from (3.2) with the induced pressure tensor $\boldsymbol{\sigma}$ we obtain the "submacroscopic" stress tensor for the suspension $\mathbf{P}_s$,

$$\mathbf{P}_s = p\mathbf{1} - 2\eta_0 \overline{\nabla \mathbf{v}} + \boldsymbol{\sigma}. \tag{3.9}$$

It is instructive to approach the same problem from a different point (Lundgren 1972). To separate singularities present in (3.7) the characteristic function H defined in (2.1) is employed. Because $\nabla \mathbf{v}$ vanishes inside the particles [see (3.5)]

$$\langle \overline{\nabla \mathbf{v}} \rangle = \langle (1-H) \overline{\nabla \mathbf{v}} \rangle + \langle H \overline{\nabla \mathbf{v}} \rangle = \langle H \overline{\nabla \mathbf{v}} \rangle \tag{3.10}$$

and, consequently, the average stress is

$$\langle H(\mathbf{r},t)\mathbf{P}(\mathbf{r},t)\rangle = \langle p\rangle\mathbf{1} - 2\eta_0\langle H\overline{\nabla\mathbf{v}}\rangle = \langle p\rangle\mathbf{1} - 2\eta_0\langle\overline{\nabla\mathbf{v}}\rangle. \tag{3.11}$$

To obtain the desired result we confine ourselves to the steady case, i.e. to the Stokes equation

$$\nabla\cdot\mathbf{P} = 0, \tag{3.12}$$

from which it follows that

$$\nabla\cdot\langle H\mathbf{P}\rangle = \langle\mathbf{P}\cdot\nabla H\rangle \tag{3.13}$$

and

$$\langle\mathbf{P}\cdot\nabla H\rangle = \langle p\rangle\mathbf{1} - 2\eta_0\langle\overline{\nabla\mathbf{v}}\rangle. \tag{3.14}$$

The average $\langle\mathbf{P}\cdot\nabla H\rangle$ can be calculated with the use of the definition of H [see (2.1)] which leads to

$$\nabla H = \sum_{i=1}^{N} \frac{\mathbf{r}-\mathbf{R}_i}{|\mathbf{r}-\mathbf{R}_i|}\,\delta(|\mathbf{r}-\mathbf{R}_i| - a). \tag{3.15}$$

Now we obtain

$$\begin{aligned}\langle\mathbf{P}\cdot\nabla H\rangle &= \sum_{i=1}^{N}\int_{D^N}\mathbf{P}\cdot\nabla H\,\Psi_N\,d\mathbf{R}_1\cdots d\mathbf{R}_N\\ &= N\int_D \delta(|\mathbf{r}-\mathbf{R}_1| - a)\cdot\frac{\mathbf{r}-\mathbf{R}_1}{|\mathbf{r}-\mathbf{R}_1|}\Psi_1(\mathbf{R}_1)\\ &\quad\times\left\{\int_{D^{N-1}}\mathbf{P}\Psi_{N-1}^{c}\,d\mathbf{R}_2\ldots d\mathbf{R}_N\right\}d\mathbf{R}_1\\ &= \int_D\langle n(\mathbf{R}_1)\rangle\langle\mathbf{P}(\mathbf{r}|\mathbf{R}_1)\rangle\cdot\frac{\mathbf{r}-\mathbf{R}_1}{|\mathbf{r}-\mathbf{R}_1|}\,\delta(|\mathbf{r}-\mathbf{R}_1| - a)\,d\mathbf{R}_1\\ &= \langle n(\mathbf{r})\rangle\int_D\langle\mathbf{P}(\mathbf{r}|\mathbf{R}_1)\rangle\,d\mathbf{R}_1\end{aligned} \tag{3.16}$$

where (2.8) has been used.

By converting the last integral into a surface integral taken over the surface of a sphere centered at $\mathbf{R}_1$, $S_1 = S(\mathbf{R}_1)$, the resulting equation for $\langle v(\mathbf{r})\rangle$ reads

$$\nabla\langle p(\mathbf{r})\rangle - \eta_0\nabla^2\langle\mathbf{v}(\mathbf{r})\rangle = \langle n(\mathbf{r})\rangle\int_{S_1}\langle\mathbf{P}(\mathbf{r}|\mathbf{R}_1)\rangle\cdot\mathbf{n}_1\,dS \tag{3.17}$$

and the integral gives the conditionally averaged stress on the surface of the test sphere. According to (3.2), the integrand is

$$\langle \mathbf{P}(\mathbf{r}\,|\,\mathbf{R}_1)\rangle = \langle p(\mathbf{r}\,|\,\mathbf{R}_1)\rangle \mathbf{1} - 2\eta_0 \langle \overline{\nabla \mathbf{v}(\mathbf{r}\,|\,\mathbf{R}_1)} \rangle \tag{3.18}$$

and thus in (3.16) both $\langle \mathbf{v}(\mathbf{r})\rangle$ and $\langle \mathbf{v}(\mathbf{r}\,|\,\mathbf{R}_1)\rangle$ appear.

To find the equation for $\langle \mathbf{v}(\mathbf{r}\,|\,\mathbf{R}_1)\rangle$ the whole procedure starting from (3.10) should be repeated using the conditional average [$\mathbf{R}_1$ fixed, see (2.12)] instead of the unconditional one. The resulting equation

$$\begin{aligned}
&\nabla\langle p(\mathbf{r}\,|\,\mathbf{R}_1)\rangle - \eta_0 \nabla^2 \langle \mathbf{v}(\mathbf{r}\,|\,\mathbf{R}_1)\rangle \\
&= N \int \delta(|\mathbf{r}-\mathbf{R}_2|-a)\frac{\mathbf{r}-\mathbf{R}_2}{|\mathbf{r}-\mathbf{R}_2|}\Psi_1^c(\mathbf{R}_2\,|\,\mathbf{R}_1) \\
&\quad \times \left\{ \int_{D^{N-2}} \mathbf{P}\Psi_{N-2}^c \, d\mathbf{R}_3 \ldots d\mathbf{R}_N \right\} d\mathbf{R}_2 \\
&= \langle n(\mathbf{r}\,|\,\mathbf{R}_1)\rangle \int_{S_2} \langle \mathbf{P}(\mathbf{r}\,|\,\mathbf{R}_1,\mathbf{R}_2)\rangle \cdot \mathbf{n}_2 \, dS
\end{aligned} \tag{3.19}$$

is of a form similar to (3.16). Another form often used can be obtained with an additional assumption that

$$\Psi_2(\mathbf{R}_1,\mathbf{R}_2) = \begin{cases} \Psi_1(\mathbf{R}_1)\Psi_1(\mathbf{R}_2) & \text{for } |\mathbf{R}_1-\mathbf{R}_2| > 2a \\ 0 & \text{for } |\mathbf{R}_1-\mathbf{R}_2| \leqq 2a \end{cases} \tag{3.20}$$

which means that positions of spheres are uncorrelated and the spheres can not overlap. This additional assumption leads [see (2.10)] to

$$\begin{aligned}
&\nabla\langle p(\mathbf{r}\,|\,\mathbf{R}_1)\rangle - \eta_0 \nabla^2 \langle \mathbf{v}(\mathbf{r}\,|\,\mathbf{R}_1)\rangle \\
&= N \int_{D\cap\{|\mathbf{R}_1-\mathbf{R}_2|>2a\}} \delta(|\mathbf{r}-\mathbf{R}_2|-a)\frac{\mathbf{r}-\mathbf{R}_2}{|\mathbf{r}-\mathbf{R}_2|}\Psi_1(\mathbf{R}_1) \\
&\quad \times \left\{ \int_{D^{N-2}} \mathbf{P}\Psi_{N-2}^c \, d\mathbf{R}_3 \cdots d\mathbf{R}_N \right\} d\mathbf{R}_2
\end{aligned} \tag{3.21}$$

$$\begin{aligned}
&= \langle n(\mathbf{r})\rangle \int_D g(|\mathbf{R}_1-\mathbf{R}_2|)\langle \mathbf{P}(\mathbf{r}\,|\,\mathbf{R}_1,\mathbf{R}_2)\rangle \\
&\quad \times \frac{\mathbf{r}-\mathbf{R}_2}{|\mathbf{r}-\mathbf{R}_2|}\delta(|\mathbf{r}-\mathbf{R}_2|-a)\, d\mathbf{R}_2
\end{aligned} \tag{3.22}$$

where the function g in the last integral is the simplest form of the pair distribution function for hard spheres:

$$g(r) = \begin{cases} 0 & \text{if} \quad r < 2a \\ 1 & \text{if} \quad r > 2a, \qquad r = |\mathbf{r}|. \end{cases} \tag{3.23}$$

This function is obtained from $g(r) = \exp(-W_2(r)/kT)$ when the two-particle interaction potential $W_2(r)$ is infinite for $r < 2a$ and zero elsewhere (see Section 5).

By repeating the procedure which leads to (3.16) and (3.22), one obtains the whole hierarchy of equations which on each level should be supplemented by an appropriate continuity equation. Such equations for the first and the second level read, respectively, $\nabla \cdot \langle \mathbf{v}(\mathbf{r}) \rangle = 0$ and $\nabla \cdot \langle \mathbf{v}(\mathbf{r}|\mathbf{R}) \rangle = 0$.

As is usual when a hierarchy of equations appears, it is necessary to enforce closing at some level, and then to solve the truncated hierarchy starting from the highest order equation, that on which the closure has been made. In the case in question the closing is always bound to a particular hydrodynamical problem and always involves some additional statistical assumptions.

On the lowest level, the hydrodynamical problem is to find the flow field around an isolated test sphere fixed at $\mathbf{R}_1$ ignoring the presence of all other particles. The corresponding statistical assumption is that particles are noninteracting and therefore independent. Thus the average stress on the surface of the sphere, S_1, is equal to that calculated for an isolated sphere. Of course the above procedure coincides with the derivation of the bulk stress to the order $O(\Phi_0)$.

On the second level of hierarchy, pairs of particles are involved and the hydrodynamical problem is that of the interaction of two spheres moving freely in different velocity fields. This problem has been solved by Batchelor & Green with noslip boundary conditions (1972b) and, independently, in a point-particle approximation, with more general boundary conditions by Felderhof (1977). These solutions play an important role in considering the interaction of Brownian particles (see Section 8). The statistical assumption involves the pair distribution function. The analysis, by no means simple, leads to the determination of the bulk stress to the order $O(\Phi_0^2)$.

As far as we are aware, no one has solved the problem of the interaction of three buoyant spheres, and thus the higher level equations of the hierarchy have never been used. Nevertheless the problem of closure is of great theoretical interest and has attracted considerable attention.

Brinkman (1947), and thereafter Lundgren (1972), used a formal closure equation [see (3.16)]

$$\langle \mathbf{P} \cdot \nabla H \rangle \doteq \mathscr{F}(\langle \mathbf{v} \rangle) \tag{3.24}$$

where $\mathscr{F}$ is a linear functional acting on $\langle \mathbf{v} \rangle$. As mentioned by Lundgren, this assumption has a hereditary character, which implies that a relation similar to (3.24) and with the same $\mathscr{F}$ holds for $\langle \mathbf{v}(\mathbf{r} \mid \mathbf{R}_1) \rangle$, $\langle \mathbf{v}(\mathbf{r} \mid \mathbf{R}_1, \mathbf{R}_2) \rangle$, etc. This type of closure leads to

$$\mathscr{F}(\langle \mathbf{v} \rangle) \sim \nabla^2 \langle \mathbf{v} \rangle \tag{3.25}$$

and the resulting expression for the effective viscosity

$$\eta = \eta_0 (1 - \tfrac{5}{2}\Phi_0)^{-1} \tag{3.26}$$

differs slightly from the Einstein formula (1.1).

A more systematic approach to the problem of closure on any level of hierarchy has been proposed by Howells (1974) in connection with the problem of permeability. Howells considers a given number of particles chosen from an infinite array of particles (all fixed) and proposes a kind of self-consistent scheme.

With reference to the closure problem, we should mention the work of Jeffrey (1974), although the problem of closure and truncation is not explicitly considered there. Jeffrey introduced group expansions that describe contributions to a given physical magnitude (say, bulk stress) coming from groups of 1, 2, etc. particles. The similarity between this paper and that of Howells is that both use expansions resembling the so-called cluster expansions exploited in statistical mechanics. However, in neither of these papers is the connection with well-established methods of statistical mechanics fully realized.

The problem of closing a hierarchy of equations is often encountered in different physical theories. The most penetrating analysis of an analogous problem has been carried out for the so-called BBGKY hierarchy in the theory of gases, and perhaps some useful lessons can be learned from this work (Ferziger & Kaper 1972).

The first lesson, a positive one, is that the full meaning of the ad hoc closure that led Boltzmann to his equation can be appreciated only by analyzing the whole structure of the hierarchy. Different derivations based on arguments, which looked at first very plausible on physical grounds, showed themselves more often than not to be misleading. It should be noted that rigorous derivation of the hierarchy can be based on the cluster expansion.

The second, less optimistic lesson is that the integrals responsible for triple, quadruple, and higher order collisions are not absolutely convergent, even though the interparticle forces encountered in gases are short-range. This divergence is quite distinct in origin from divergences discussed above (Section 1). It is therefore possible that in the theory of suspensions similar difficulties will arise when the calculation of the effects of interactions of three or more particles is attempted.

There is, however, one basic difference between the hierarchy of equations in the theory of suspensions and that in the kinetic theory of gases. In the latter case, the hierarchy describes distribution functions of different orders whereas in the former the unknown functions are velocities, and the distribution functions involved are assumed to be known independently. It is for this reason that in considering the closure problem it is necessary to add some statistical assumptions. Thus the ideal type of closure would be such that the hierarchy of equations of motion be considered together with equations determining flow-controlled particle distribution functions. The closure procedure would involve both types of equations on the same level of hierarchy. The principal difficulty in accomplishing such a scheme is in obtaining equations for flow-related distribution functions.

One final remark: The difficulty in using the ensemble averages introduced in Section 2 for a homogeneous distribution of spheres has a mathematical character; in practice the additional assumptions, made on the first and the second level of the hierarchy, allow one to consider $\langle n(\mathbf{r})\rangle$ in (3.16) and (3.22) as a constant independent of $\mathbf{r}$. A more fundamental solution to this problem is discussed in Section 5.

4 POINT-PARTICLE APPROXIMATION

Some of the questions arising in solving the equations of motion

$$-\nabla p+\eta_0\nabla^2\mathbf{v} = -\mathbf{F}(\mathbf{r}), \quad \nabla\cdot\mathbf{v}=0 \tag{4.1}$$

can be illustrated in a relatively simple case, when the forces appearing on the right-hand side are assumed to act at isolated points. In the recent papers by Felderhof (1976a,b) this assumption has been exploited in connection with polymer suspensions. The distinctive feature of these papers is the explicit confrontation of the local and the ensemble-averaged velocities. It should be noted that our exposition at some points differs from that of Felderhof.

The presence of particles is modelled by multipoles at the points $\mathbf{R}_1, \mathbf{R}_2, \ldots, \mathbf{R}_N$ and thus the force $\mathbf{F}(\mathbf{r})$ takes the form

$$\mathbf{F}(\mathbf{r}) = \sum_{k=1}^{N}\sum_{n=0}^{\infty}\frac{(-1)^n}{n!}\mathbf{f}_k^n(\mathbf{R}_k)\odot\nabla^n\delta(r-\mathbf{R}_k) \tag{4.2}$$

where $\mathbf{f}_k^n$ is the tensor of rank n dependent on $\mathbf{R}_k$, and the symbol $\odot$ denotes the appropriate contraction.

The line of reasoning in Felderhof's paper can be presented in the following steps:

1. The solution of an averaged equation

$$-\nabla\langle p\rangle+\eta_0\nabla^2\langle \mathbf{v}\rangle = -\langle \mathbf{F}\rangle, \quad \nabla\cdot\langle \mathbf{v}\rangle = 0 \tag{4.3}$$

is

$$\langle \mathbf{v}(\mathbf{r})\rangle = \mathbf{v}^0(\mathbf{r})+\int \mathbf{T}(\mathbf{r}-\mathbf{r}')\langle \mathbf{F}(\mathbf{r}')\rangle\, d\mathbf{r}' \tag{4.4}$$

where $\mathbf{T}$ is the Oseen operator

$$\mathbf{T}(\mathbf{r}) = \frac{1}{8\pi\eta_0}\left(\frac{\mathbf{1}}{r}+\frac{\mathbf{rr}}{r^3}\right), \quad r = |\mathbf{r}| \tag{4.5}$$

and $\mathbf{v}^0(\mathbf{r})$ is the velocity field corresponding to the homogeneous equation, i.e. the velocity field due to the unperturbed flow, when the particles are absent.

The averaged force can be expressed using the function

$$\boldsymbol{\phi}^n(\mathbf{r}) = \left\langle \sum_{k=1}^{N} \mathbf{f}_k^n(\mathbf{R}_k)\delta(\mathbf{r}-\mathbf{R}_k)\right\rangle. \tag{4.6}$$

which leads to

$$\langle \mathbf{F}(\mathbf{r})\rangle = \sum_{n=0}^{\infty} \frac{(-1)^n}{n!}\nabla^n \odot \boldsymbol{\phi}^n(\mathbf{r}). \tag{4.7}$$

2. Due to the assumed form of the force (4.2) the solution of the original Equation (4.1) is

$$\begin{aligned}\mathbf{v}(\mathbf{r}) &= \mathbf{v}^0(\mathbf{r})+\int \mathbf{T}(\mathbf{r}-\mathbf{r}')\mathbf{F}(\mathbf{r}')\, d\mathbf{r}' \\ &= \mathbf{v}^0(\mathbf{r})+\sum_{n=0}^{\infty}\sum_{k=1}^{\infty}\frac{(-1)^n}{n!}\nabla^n\mathbf{T}(\mathbf{r}-\mathbf{R}_k)\odot \mathbf{f}_k^n(\mathbf{R}_k).\end{aligned} \tag{4.8}$$

This solution is valid at all points, also at $\mathbf{r} = \mathbf{R}_j$ provided that the sum in (4.8) is taken over all $\mathbf{R}_k$ except $\mathbf{R}_j$.

Local velocity is defined as

$$\mathbf{v}_{\text{loc}}(\mathbf{r}) = \mathbf{v}^0(\mathbf{r})+\lim_{\varepsilon\to 0}\sum_{n=0}^{\infty}\int_{D\setminus K_\varepsilon(\mathbf{r})} \nabla^n\mathbf{T}(\mathbf{r}-\mathbf{r}')\odot \boldsymbol{\phi}^n(\mathbf{r}')\, d\mathbf{r}' \tag{4.9}$$

and it differs from $\mathbf{v}(\mathbf{R}_j)$ in that $\mathbf{f}_k^n$ is replaced by an average, $\boldsymbol{\phi}^n$ [see (4.6)], and that an infinitesimal sphere around $\mathbf{R}_j(=\mathbf{r})$, $K_\varepsilon(\mathbf{R}_j)$ is excluded.

3. The difference between $\mathbf{v}_{\text{loc}}$ and $\langle \mathbf{v}\rangle$

$$\mathbf{v}_{\text{loc}}-\langle \mathbf{v}\rangle = \lim_{\varepsilon\to 0}\sum_{n=0}^{\infty}\frac{(-1)^n}{n!}\int_{D\setminus K_\varepsilon(\mathbf{r})}[\nabla^n\mathbf{T}(\mathbf{r}-\mathbf{r}')-\mathbf{T}(\mathbf{r}-\mathbf{r}')\nabla^n] \odot \boldsymbol{\phi}^n(\mathbf{r}')\, d\mathbf{r}' \tag{4.10}$$

is calculated using surface integrals over the surface of $K_\varepsilon(\mathbf{r})$, ∂K_ε, and over ∂D, where D is the region containing all N particles.

There is a delicate point in the integration over ∂K_ε because generalized functions are involved and thus $\nabla\langle\mathbf{v}\rangle$ is not equal to $\langle\nabla\mathbf{v}\rangle$ at $\mathbf{r}=\mathbf{R}_j$.

The integral on the outer boundary, ∂D, is neglected. It is at this point that the question of non–absolutely convergent integrals enters the problem. It is true that $\mathbf{v}_{\text{loc}}$ and $\langle\mathbf{v}\rangle$ exhibit the same asymptotic behavior and their difference vanishes at infinity. However, in the framework of the formalism used one can not expand D to infinity because it is involved in the definition of the particle distribution function.

4. In order to take into account the finite size of the spheres the effective velocity is introduced

$$\mathbf{v}_{\text{eff}}(\mathbf{r})=\mathbf{v}_{\text{loc}}(\mathbf{r})-\sum_{n=0}^{\infty}\frac{(-1)^n}{n!}\int_D[1-g(|\mathbf{r}-\mathbf{r}'|)]\nabla^n\mathbf{T}(\mathbf{r}-\mathbf{r}')\odot\boldsymbol{\phi}^n(\mathbf{r}')\,d\mathbf{r}' \tag{4.11}$$

with g defined by (3.23). (Felderhof allows a wider class of radial distribution functions.) The calculations of $\mathbf{v}_{\text{eff}}-\langle\mathbf{v}\rangle$ are based on previous calculations of $\mathbf{v}_{\text{loc}}-\langle\mathbf{v}\rangle$.

Note the use of the averages $\boldsymbol{\phi}^n$ in (4.9) and (4.11) instead of conditional averages. This, and the limiting process $\varepsilon\to 0$, are in fact equivalent to a closure procedure.

5. Up to now the coefficients of the multipole expansion (4.2) have been unrelated to the flow field. In order to find the flow-controlled solutions, a linear flow field

$$\mathbf{v}(\mathbf{r})=\mathbf{v}_c+\boldsymbol{\Omega}\times\mathbf{r}+\mathbf{B}\cdot\mathbf{r};\quad \mathbf{B}=\overline{\mathbf{B}} \tag{4.12}$$

with constants $\mathbf{v}_c$, $\boldsymbol{\Omega}$, $\mathbf{B}$ is considered.

Assuming a sphere embedded in this field, a solution of the problem is sought at large distances from the sphere. The multipole coefficients obtained are linear functions of $\mathbf{v}_c$, $\boldsymbol{\Omega}$, and $\mathbf{B}$. Only four lowest order multipole coefficients $\mathbf{f}^0$, $\mathbf{f}^1$, $\mathbf{f}^2$, and $\mathbf{f}^3$ do not vanish.

To derive $\boldsymbol{\phi}^n$ using these coefficients it is assumed that the number density, n_0, is constant throughout the space, and thus [see (4.6)]

$$\boldsymbol{\phi}^n=n_0\mathbf{f}^n,\quad n=0,1,2,3. \tag{4.13}$$

6. There is still one degree of freedom at our disposal, namely the relative velocity $\mathbf{u}$ of the sphere with respect to the fluid.

If this relative velocity $\mathbf{u}$ is constant (sedimentation problem) the resulting friction coefficient is

$$c_d=-6\pi\eta_0 a(1+\tfrac{11}{2}\Phi_0) \tag{4.14}$$

and the numerical coefficient of Φ_0 differs from the value 6.55 obtained by Batchelor (1972) who considers spheres rather than point particles.

If the torque and force acting on the particle vanish (suspension) all velocities are equal: $\langle \mathbf{v} \rangle = \mathbf{v}_{loc} = \mathbf{v}_{eff}$ and also $\mathbf{\Omega} = \nabla \times \langle \mathbf{v} \rangle$. In this case, as expected, the Einstein formula for the effective viscosity is obtained.

The main advantage of the point-particle method is that, in spite of ignoring the shielding effects, it yields reliable results. Of course, for higher concentrations these effects become important but none of the existing theories is well suited to dealing with dense systems.

The main deficiency of the method described is related to the use of the averages defined in Section 2. Thus the formula (4.13) can not be rigorously derived from the assumptions used in the Felderhof paper.

5 ENSEMBLE AVERAGING—OTHER APPROACHES

The spatial distribution of different geometrical objects is a problem analyzed in the field of geometrical probability. The spatial distribution of spheres appears to be a simple question to solve but, in spite of being of considerable interest to physics, chemistry, and astronomy, it is not well understood (Kendall & Moran 1963). The results obtained are preliminary and often can be considered as useful approximations rather than mathematical theorems. In particular, some knowledge of this kind was gained with respect to the nearest-neighbor distribution function (Buyevich 1968, Herczyński 1975); this can be used as a tool in the closure procedure discussed in Section 3.

The situation is much better understood if one considers a random distribution of points related to the point-particle approximation. The problem closest to the one considered here is the so-called Holtsmark problem (Chandrasekhar 1954) in which the distribution of gravitational forces at some point in space is sought, when stars (treated as points) are randomly distributed (with a constant density) in the whole space. The similarity consists in that in both cases the difficulty has the same origin: the r^{-2} asymptotic behavior. Thus in both cases the problem of non–absolute convergence appears.

The Holtsmark method of finding the force distribution function w_N at a given test point consists in finding first the characteristic function, i.e. the Fourier transformation of w_N. This can be represented as a product of the characteristic functions of N point-forces contained in a large sphere of radius R surrounding the test point. The thermodynamic limit (i.e. $R \to \infty$, $N \to \infty$ with $3N/4\pi R^3 = n_0 = \text{const.}$) is now taken, and it is shown that this limit does exist.

For the theory of suspensions the most important point, however, is the possibility of taking into account interactions between particles; this can not be done in the framework of purely geometrical considerations.

Another method of averaging is related to the use of the grand canonical ensemble rooted in statistical mechanics. Some results will be given here without derivation. We will start with a few words about the canonical ensemble.

When a classical (i.e. nonquantum) subsystem S of a given volume V and a given number of particles N interacts exchanging energy with its environment, the canonical distribution function should be employed and it reads

$$\Psi_N^{ce}(\mathbf{q},\mathbf{p}) = \frac{\hbar^{-sN}}{N!Z} e^{-\beta H(\mathbf{q},\mathbf{p})}; \tag{5.1}$$

$$\mathbf{q} = \{\mathbf{q}_1, \mathbf{q}_2, \ldots, \mathbf{q}_N\}, \quad \mathbf{p} = \{\mathbf{p}_1, \mathbf{p}_2, \ldots, \mathbf{p}_N\},$$

where H is the Hamiltonian of the system depending on generalized coordinates: positions, $\mathbf{q}$, momenta, $\mathbf{p}$. The factor $\hbar^{-sN}/N!$ with Planck's constant, $\hbar$, and the number of degrees of freedom, s, exhibited by each particle, is introduced to provide a proper link with quantum statistics (Landau & Lifschitz 1958, Balescu 1975). The canonical partition function Z,

$$Z = \frac{\hbar^{-sN}}{N!} \int e^{-\beta H(\mathbf{q},\mathbf{p})}\, d^N\mathbf{q}\, d^N\mathbf{p}, \tag{5.2}$$

can be treated as a normalizing function. Its central role in thermodynamic theory (in equilibrium conditions) stems from the fact that it is directly related to the free energy F by

$$-\beta F(T,V,N) = \ln Z, \quad \beta = 1/kT \tag{5.3}$$

where T is absolute temperature, and k is the Boltzmann constant. The free energy of the system is a thermodynamic potential and because all thermodynamic functions can be derived from it, F (and therefore Z) gives a full thermodynamic description of the system.

The canonical average of any function f of $\mathbf{q}$ and $\mathbf{p}$ is

$$\langle f \rangle^{ce} = \frac{1}{Z} \int f(\mathbf{q},\mathbf{p})\, e^{-\beta H(\mathbf{q},\mathbf{p})}\, d^N\mathbf{q}\, d^N\mathbf{p} \tag{5.4}$$

and here as in (5.2) the integration should be performed over the entire accessible phase space.

If we allow also exchange of particles between the chosen subsystem

S and its environment (i.e. if the number of particles in S varies) the probability of finding N particles in a given position of phase space is determined by the grand canonical distribution function Ψ_N^{gc}

$$\Psi_N^{gc}(\mathbf{q},\mathbf{p}) = \frac{\hbar^{-sN}}{N!\Xi} \exp\{-\beta H_N(\mathbf{q},\mathbf{p}) + \beta\mu N\} \tag{5.5}$$

where μ is chemical potential and Ξ is the grand canonical partition function

$$\Xi = \sum_{N=1}^{\infty} \frac{z^N}{N!} \int e^{-\beta H_N(\mathbf{q},\mathbf{p})}\, d^N\mathbf{q}\, d^N\mathbf{p} \tag{5.6}$$

with the activity $z = \hbar^{-s} e^{\beta\mu}$.

The significance of Ξ is analogous to that of the canonical partition function Z, and it is important to note that Ξ can be represented in a series form

$$\Xi(T,V,\zeta) = \sum_{N=1}^{\infty} \zeta^N Z(T,V,N) \tag{5.7}$$

where $\zeta = \hbar^s z = e^{\beta\mu}$, showing how Ξ depends on the canonical partition functions of "clusters" with $1, 2, \ldots,$ particles.

Assuming that the Hamiltonian has its usual form of a sum of kinetic and potential energy, $W_N = W_N(\mathbf{q}_1, \mathbf{q}_2, \ldots, \mathbf{q}_N)$ and that the former does not play any important role in the theory of suspensions, the grand canonical average of a function $f = f[n(\mathbf{r})]$ [see (5.4)] is (we return to previous notation, Bedeaux et al 1977)

$$\langle f \rangle^{gc} = \Xi^{-1} \sum_{N=1}^{\infty} \frac{z^N}{N!} \int e^{-\beta W_N(\mathbf{R}_1,\mathbf{R}_2,\ldots,\mathbf{R}_N)}\, d\mathbf{R}_1 \ldots d\mathbf{R}_N. \tag{5.8}$$

In the form of an infinite series the grand canonical average is of little practical use, and the commonly employed method is virial expansion. For activity, z, and the partition function Ξ, this yields

$$z = n_0 - 2b_2 n_0^2 + O(n_0^3)$$

$$\Xi = 1 + n_0 \int d\mathbf{R}_1 + O(n_0^2) \tag{5.9}$$

where b_2 is the second virial coefficient

$$b_2 = \tfrac{1}{2} \int (e^{-\beta W_2(\mathbf{r})} - 1)\, d\mathbf{r}, \quad \mathbf{r} = \mathbf{R}_1 - \mathbf{R}_2. \tag{5.10}$$

The grand canonical average $\langle f \rangle^{gc}$ up to the second power in density is

$$\langle f \rangle^{gc} = n_0 \int f(\mathbf{R}_1)\, d\mathbf{R}_1 + \tfrac{1}{2} n_0^2 \int e^{-\beta W_2(\mathbf{r})} [f(\mathbf{R}_1,\mathbf{R}_2) - f(\mathbf{R}_1) - f(\mathbf{R}_2)]\, d\mathbf{R}_1\, d\mathbf{R}_2. \tag{5.11}$$

Modern derivation of the above formulae is based on an expansion in the powers of density, corresponding to the number of particles at a given level. This type of expansion is called a cluster expansion and is presented in many textbooks and monographs on statistical physics (see Hill 1956, Balescu 1975). A short review of different types of cluster expansions is given in the book by Croxton (1974).

In view of what has been said, the averages defined in Section 2 can be considered as averages at a given order of density. The methods of group expansion used by Howells (1974) and Jeffrey (1974) and mentioned earlier in Section 3 are in fact particular variants of such an expansion.

The use of the grand canonical ensemble removes some logical difficulties related to the use of unbounded domains with an infinite number of particles. However, it is fair to say that the use of grand canonical ensemble for suspensions needs some additional consideration and does not resolve automatically all difficulties.

The most important is that the virial expansion can be employed only for dilute suspensions because it converges only up to some critical value of density and this is the reason why in the statistical theory of fluids other methods are used. Incidentally, these methods lead again to a hierarchy of equations. And yet, the great importance of the use of the grand canonical ensemble formalism in the theory of suspensions lies in the fact that problems, originally studied using their own specific methods, have been transplanted to the well-cultivated ground of statistical physics.

A more specific question is related to the form of the interaction potential W_N that should be used in the theory of suspensions. For hard spheres, W_2 can be assumed in the form

$$W_2(\mathbf{R}_1,\mathbf{R}_2) = W_2(\mathbf{R}_1 - \mathbf{R}_2) = \begin{cases} 0 & \text{for } |\mathbf{R}_1 - \mathbf{R}_2| > 2a \\ \infty & \text{for } |\mathbf{R}_1 - \mathbf{R}_2| < 2a \end{cases} \tag{5.12}$$

which leads to the pair distribution function $g(r)$ defined by (3.23). It remains unclear, however, which distribution should be assumed for W_3 and for higher orders of virial expansion.

6 VISCOSITY AS AN OPERATOR

In this section we intend to present the line of reasoning that has been used in the paper by Bedeaux, Kapral & Mazur (1977). The most striking result of this paper is that the viscosity of a suspension is an operator, which in the limiting case reduces to the scalar magnitude.

The distinctive features of the proposed method can be described as follows:

1. Ensemble averages are used in the sense described in the preceding section. Thus, instead of a hierarchy of equations only one equation should be considered and the closure procedure is replaced by truncation of the series (5.8). No additional statistical assumptions are required.

2. Instead of solving a separate hydrodynamic problem on each level of the hierarchy, a one-sphere response function is used in the multipole approximation. The multipole coefficients of higher order than that obtained by Felderhof [see (4.13)] are obtained by finding the response function describing the perturbation of a quadratic flow field due to the presence of a sphere.

3. The governing equation is a linearized, unsteady Navier-Stokes equation (3.1). Thus inertial effects are partly taken into account.

4. Fourier transformation is used consistently. The presence of generalized functions in Equation (3.7) calls for the use of Fourier transformation as the most appropriate mathematical tool. It is important to note that, although for generalized functions the operations of differentiation and averaging do not commute, they do when the Fourier transformation is taken.

5. The resulting effective viscosity is an operator. In the $(\mathbf{k},\omega)$ representation this operator in the limiting case $\mathbf{k} = 0$, $\omega = 0$ reduces to a scalar depending on Φ_0. In the order $O(\Phi_0)$, the Einstein formula (1.1) is obtained. The calculation of the limiting viscosity up to the order $O(\Phi_0^2)$ shows the effectiveness of the proposed approach.

A similar method has been used also by Kapral & Bedeaux (1978) to find shear viscosity of a regular array of suspended spheres.

Rigorous Theory

It is convenient to present the derivation of the effective viscosity of suspensions in three steps.

In the first step the generalized Oseen operators are used to find the relation between $\boldsymbol{\sigma}$ and $\ulcorner\nabla\mathbf{v}\urcorner$. From the Equations (3.7)

$$\rho\frac{\partial\mathbf{v}}{\partial t} - \eta_0\nabla^2\mathbf{v} = -\nabla p - \nabla\cdot\boldsymbol{\sigma}, \quad \nabla\cdot\mathbf{v} = 0 \tag{6.1}$$

it follows that

$$\nabla^2 p = -\nabla\nabla : \boldsymbol{\sigma}. \tag{6.2}$$

The solution of (6.1) and (6.2) is

$$\mathbf{v} = \mathbf{v}^0 - \mathscr{G}[\nabla(p-p^0) + \nabla\cdot\boldsymbol{\sigma}], \quad p = p^0 + \mathscr{G}_0\nabla\nabla : \boldsymbol{\sigma} \tag{6.3}$$

where $\mathscr{G}$ and $\mathscr{G}_0$ are Green operators [called propagators in quantum theory, Schiff (1968)]. As previously, $\mathbf{v}^0$ and p^0 describe the unperturbed flow field. One of the following Fourier transformations is employed: $f(\mathbf{r},t) \to f(\mathbf{r},\omega)$, or $f(\mathbf{r},t) \to f(\mathbf{k},\omega)$. The operators $\mathscr{G}$ and $\mathscr{G}_0$ assume in the $(\mathbf{k},\omega)$ representation the form

$$\mathscr{G}(\mathbf{k},\omega) = (-i\rho\omega + \eta_0 k^2)^{-1}, \quad \mathscr{G}_0(\mathbf{k}) = k^{-2}, \tag{6.4}$$

whereas in the $(\mathbf{r},\omega)$ representation

$$\mathscr{G}(\mathbf{r},\omega) = (4\pi\eta_0 r)^{-1} e^{-\alpha r}, \quad \mathscr{G}_0(\mathbf{r}) = (4\pi r)^{-1} \tag{6.5}$$

here $\alpha = (-i\omega\rho/\eta_0)^{\frac{1}{2}}$.

In terms of these operators

$$\mathbf{v} = \mathbf{v}^0 - \mathscr{G}(\mathbf{1} + \mathscr{G}_0 \nabla\nabla)\nabla : \boldsymbol{\sigma} \tag{6.6}$$

where $\mathbf{1}$ denotes the appropriate unit tensor.

The operator $\mathscr{G}_0$ is the Oseen operator for pressure and $\mathscr{G}(\mathbf{1} + \mathscr{G}_0 \nabla\nabla)$ is the Oseen operator for velocity [see (4.5)], generalized for the unsteady case.

Taking the gradient of both sides of Equation (6.6) we arrive at the desired result

$$\overline{\nabla\mathbf{v}} = \overline{\nabla\mathbf{v}^0} - \mathbf{G} : \overline{\boldsymbol{\sigma}} \tag{6.7}$$

where $\mathbf{G}$ is a fourth-rank tensor.

In the second step a one-sphere response function is used to find the relation between $\boldsymbol{\sigma}$ and $\overline{\nabla\mathbf{v}}$ complementary to (6.7).

Let the center of the ith sphere be placed at $\mathbf{R}_i(t)$ and let $\mathbf{v}^i_{\text{eff}}$ be the velocity at that point. Clearly $\mathbf{v}^i_{\text{eff}}$ depends on the imposed velocity field $\mathbf{v}^0$ and on the presence of all particles.

Due to the linearity of the equations used the contribution of the ith sphere to the stress has the form of a linear operator[2]

$$\overline{\boldsymbol{\sigma}^i(\mathbf{r},t)} = -2 \int \chi[\mathbf{r} - \mathbf{R}_i(t), t \,|\, \mathbf{r}' - \mathbf{R}_i(t'), t'] \, \overline{\nabla' \mathbf{v}^i_{\text{eff}}(\mathbf{r}',t')} \, d\mathbf{r}' \, dt' \tag{6.8}$$

where $\boldsymbol{\chi}$ is the response function. In a more compact form (6.8) is

$$\overline{\boldsymbol{\sigma}^i} = -2\chi^i : \overline{\nabla\mathbf{v}^i_{\text{eff}}}. \tag{6.9}$$

The effective velocity $\mathbf{v}^i_{\text{eff}}$ can be calculated using (6.7) to give

[2] There is no summation convention for superscripts.

$$\overline{\nabla \mathbf{v}^i_{\text{eff}}} = \overline{\nabla \mathbf{v}^0} - \sum_{j \neq i} \int \mathbf{G}(\mathbf{r}-\mathbf{r}', t-t') : \overline{\boldsymbol{\sigma}^j(\mathbf{r}', t')}\, d\mathbf{r}'\, dt'$$

$$= \overline{\nabla \mathbf{v}^0} - \int \mathbf{G}(\mathbf{r}-\mathbf{r}', t-t')\, \Theta(\mathbf{r}' - \mathbf{R}_i(t')) : \overline{\boldsymbol{\sigma}^j(r', t')}\, d\mathbf{r}'\, dt' \tag{6.10}$$

or in an abbreviated form

$$\overline{\nabla \mathbf{v}^i_{\text{eff}}} = \overline{\nabla \mathbf{v}^0} - \mathbf{G}^i : \overline{\boldsymbol{\sigma}}. \tag{6.11}$$

In (6.10) the total stress

$$\boldsymbol{\sigma}(\mathbf{r}, t) = \sum_i \boldsymbol{\sigma}^i(\mathbf{r}, t) \tag{6.12}$$

has been used. The function Θ in (6.10) is defined by

$$\Theta(r) = \begin{cases} 0 & \text{for} \quad r \leqq a + \varepsilon \\ 1 & \text{for} \quad r > a + \varepsilon, \quad \varepsilon > 0, \end{cases} \tag{6.13}$$

and it prohibits the spheres not only from overlapping, but also from touching each other.

Substitution of (6.11) into (6.9) with the aid of (6.12) gives

$$\overline{\boldsymbol{\sigma}} = -2 \sum_i \boldsymbol{\chi}^i : \overline{\nabla \mathbf{v}^0} + 2 \sum_i (\boldsymbol{\chi}^i : \mathbf{G}^i) : \overline{\boldsymbol{\sigma}} \tag{6.14}$$

hence, with the use of (6.7)

$$\overline{\boldsymbol{\sigma}} = -2 \boldsymbol{\eta}_b : \overline{\nabla \mathbf{v}} \tag{6.14a}$$

where

$$\boldsymbol{\eta}_b = (\mathbf{1} - 2\mathbf{H}_\gamma + 2\boldsymbol{\gamma}_b : \mathbf{G})^{-1} : \boldsymbol{\gamma}_b \tag{6.15}$$

$$\mathbf{H}_\gamma = \sum_i \boldsymbol{\chi}^i : \mathbf{G}^i \tag{6.16}$$

$$\boldsymbol{\gamma}_b = \sum_i \boldsymbol{\chi}^i. \tag{6.17}$$

The expression (6.14) is the desired complementary relation to (6.7). The operator $\boldsymbol{\eta}_b$ defined by (6.15) is called the submacroscopic contribution to viscosity.

In the final step the constitutive equations for the suspension and the effective viscosity are obtained.

Using (6.7) and (6.14) $\overline{\nabla \mathbf{v}}$ and $\overline{\boldsymbol{\sigma}}$ can be expressed in terms of the gradient of the imposed velocity field, $\overline{\nabla \mathbf{v}^0}$, namely

$$\overline{\nabla \mathbf{v}} = (\mathbf{1} - 2\mathbf{G} : \boldsymbol{\eta}_b)^{-1} : \overline{\nabla \mathbf{v}^0} \tag{6.18}$$

$$\overline{\boldsymbol{\sigma}} = -2\boldsymbol{\eta}_b : (\mathbf{1} - 2\mathbf{G} : \boldsymbol{\eta}_b)^{-1} : \overline{\nabla \mathbf{v}^0}. \tag{6.19}$$

This is a convenient place to introduce averaging since the averaging leaves $\overline{\nabla \mathbf{v}^0}$ invariant. Thus

$$\begin{aligned} \langle \overline{\nabla \mathbf{v}} \rangle &= \langle (\mathbf{1} - 2\mathbf{G} : \boldsymbol{\eta}_b)^{-1} \rangle : \overline{\nabla \mathbf{v}^0} \\ \langle \overline{\boldsymbol{\sigma}} \rangle &= -2 \langle \boldsymbol{\eta}_b : (\mathbf{1} - 2\mathbf{G} : \boldsymbol{\eta}_b)^{-1} \rangle : \overline{\nabla \mathbf{v}^0} \end{aligned} \tag{6.20}$$

and by eliminating $\overline{\nabla \mathbf{v}^0}$

$$\langle \overline{\boldsymbol{\sigma}} \rangle = -2 \boldsymbol{\Delta\eta} : \langle \overline{\nabla \mathbf{v}} \rangle \tag{6.21}$$

where the tensor operator $\boldsymbol{\Delta\eta}$ is defined by

$$\boldsymbol{\Delta\eta} = \langle \boldsymbol{\eta}_b : (\mathbf{1} - 2\mathbf{G} : \boldsymbol{\eta}_b)^{-1} \rangle : \langle (\mathbf{1} - 2\mathbf{G} : \boldsymbol{\eta}_b)^{-1} \rangle^{-1}. \tag{6.22}$$

Assuming that the distribution of spheres is translationally invariant and stationary and using the symmetry properties of $\boldsymbol{\Delta\eta}$, it can be shown that in the $(\mathbf{k},\omega)$ representation $\overline{\boldsymbol{\sigma}}$ can be expressed as a diadic

$$\langle \overline{\boldsymbol{\sigma}(\mathbf{k},\omega)} \rangle = -2i \Delta\eta(\mathbf{k},\omega) \langle \overline{\mathbf{k}\mathbf{v}(\mathbf{k},\omega)} \rangle \tag{6.23}$$

with a scalar operator $\Delta\eta(\mathbf{k},\omega)$.

From this equation and from the averaged Equation (6.1) it follows that

$$-i\omega\rho \langle \mathbf{v}(\mathbf{k},\omega) \rangle = -[\eta_0 + \Delta\eta(\mathbf{k},\omega)] k^2 \langle \mathbf{v}(\mathbf{k},\omega) \rangle \tag{6.24}$$

and thus

$$\eta(\mathbf{k},\omega) = \eta_0 + \Delta\eta(\mathbf{k},\omega) \tag{6.25}$$

plays the role of viscosity. It follows from the derivation of $\Delta\eta$ that it is, as it should be, independent of the imposed flow (i.e. $\mathbf{v}^0$). However, $\Delta\eta$ depends on $\mathbf{k}$ and ω or, in the $(\mathbf{r},t)$ representation, on the field gradient and time derivative. Thus, in the present approach, the viscosity of a suspension is an operator. In the isotropic case $\mathbf{k} = 0$; for steady cases $\omega = 0$.

Approximations

In order to get any definite results some approximations are unavoidable. The first one we will present here is related to the calculation of the response function $\boldsymbol{\chi}$ defined in (6.8). The method of multipole expansion is employed and only the first and the second terms are used. The result leads to

$$\gamma_b(\mathbf{r},t \mid \mathbf{r}',t') = \tfrac{5}{2} \eta_0 v_s \boldsymbol{\Delta} M n(\mathbf{r},t) \delta(\mathbf{r} - \mathbf{r}') \delta(t - t') \tag{6.26}$$

with the operator M given by

$$M = 1 + \frac{7\rho a^2}{20\eta_0}\frac{\partial}{\partial t} + \frac{1}{10}a^2(\nabla^2 + \nabla'^2 + \tfrac{7}{4}\nabla\cdot\nabla'); \quad \nabla = \frac{\partial}{\partial \mathbf{r}}, \nabla' = \frac{\partial}{\partial \mathbf{r}'}$$

and tensor $\boldsymbol{\Delta}$ defined as

$$\Delta_{ijkl} = \tfrac{1}{2}\delta_{ik}\delta_{jl} + \tfrac{1}{2}\delta_{il}\delta_{jk} - \tfrac{1}{3}\delta_{ij}\delta_{kl}; \tag{6.27}$$

v_s is the volume of a sphere of radius a. The number density in (6.26) is

$$n(\mathbf{r},t) = \sum_i \delta(\mathbf{r} - \mathbf{R}_i(t)). \tag{6.28}$$

The second approximation in the calculation of $\boldsymbol{\chi}$ [see (6.8)] concerns identification of the position of the ith sphere at the instant t, $\mathbf{R}_i(t)$ and t', $\mathbf{R}_i(t')$. This can be considered as a kind of linearization [see comment after (3.5)].

For the calculation of higher coefficients of the multipole expansion an elegant method is devised. It is a generalization of the method used by Landau & Lifshitz (1959) who consider the flow around a sphere in the linear velocity field, $\mathbf{v} = \mathbf{B}\cdot\mathbf{r}$, $\mathbf{B} = \overline{\mathbf{B}}$, while Bedeaux et al consider also a bilinear field, $\mathbf{v} = \mathbf{B}'\!:\mathbf{rr}$, $\mathbf{B}'$ being a third-rank tensor symmetric and traceless with respect to all pairs of indices.

The above approximation allows us to write the tensor $\boldsymbol{\Delta\eta}$ defined in (6.22) in the form

$$\boldsymbol{\Delta\eta} = [\mathbf{1} + 2\gamma\!:\!(\mathbf{G} - \gamma_b^{-1}\!:\!\mathbf{H}_\gamma)]^{-1}\!:\gamma \tag{6.29}$$

with

$$\gamma = \langle\mathbf{F}\rangle\!:\!(\mathbf{1} + 2\gamma_b^{-1}\!:\!\mathbf{H}_\gamma\!:\!\langle\mathbf{F}\rangle)^{-1} \tag{6.30}$$

$$\mathbf{F} = \gamma_b\!:\!(\mathbf{1} - 2\gamma_b^{-1}\!:\!\mathbf{H}_\gamma\!:\gamma_b)^{-1}. \tag{6.31}$$

Finally, the use of only two terms in the virial expansion (5.8) can be considered as an additional approximation. To calculate $\langle\mathbf{F}\rangle$ we must have at our disposal the partial sums appearing in (5.8). These are obtained by replacing $n(\mathbf{r},t)$ (6.28) by

$$n_N(\mathbf{r},t) = \sum_{i=1}^{N} \delta(\mathbf{r} - \mathbf{R}_i(t)) \tag{6.32}$$

which consequently leads to definitions of γ_b^N and $\mathbf{H}_\gamma^N$ in place of γ_b and $\mathbf{H}_\gamma$ in (6.17) and (6.16).

By applying $\langle\mathbf{F}\rangle$, the tensor γ (6.30) can be presented in the form

$$\gamma = n_0\gamma_1 + n_0^2\gamma_2 + O(n_0^3) \tag{6.33}$$

where n_0 is the number density,

$$\gamma_1 = \int \mathbf{F}(\mathbf{R}_1)\, d\mathbf{R}_1,$$

and γ_2 depends on both $\mathbf{F}_1(\mathbf{R}_1)$ and $\mathbf{F}_2(\mathbf{R}_1,\mathbf{R}_2)$.

Results and Comments

The resulting formula for the effective viscosity up to the first power of the volume fraction Φ_0 in the $(\mathbf{k},\omega)$ representation is

$$\eta(\mathbf{k},\omega) = \eta_0[1+\tfrac{5}{2}\Phi_0(1+\tfrac{7}{20}\alpha^2 a^2 - \tfrac{1}{40}k^2 a^2)] \tag{6.34}$$

and in the $(\mathbf{r},t)$ representation

$$\eta(\mathbf{r},t) = \eta_0\left[1+\frac{5}{2}\Phi_0\left(1+\frac{7}{20}\frac{a^2\rho}{\eta_0}\frac{\partial}{\partial t}+\frac{a^2}{40}\nabla^2\right)\right] \tag{6.35}$$

$\eta(\mathbf{k},\omega)$ reduces to the Einstein formula at $\omega = 0$ and $\mathbf{k} = 0$.

Much more complicated calculations are needed to find the Huggins coefficient K_{H} [see (1.2)]

$$\eta(\mathbf{k} = 0,\ \omega = 0) = \eta_0[1+\tfrac{5}{2}\Phi_0+(\tfrac{5}{2})^2 K_{\mathrm{H}}\Phi_0^2+O(\Phi_0^3)]. \tag{6.36}$$

The expression for $\eta(\mathbf{k},\omega)$ for nonvanishing ω or for the nonisotropic case ($\mathbf{k} \neq 0$) is not available. To get (6.36) it is necessary to calculate the tensor γ_2 in (6.33). It has been shown that the use of the lowest order multipoles only leads to $K_{\mathrm{H}} = 0.90$. The use of two lowest order multipoles and some ad hoc estimation of the higher order terms gives $K_{\mathrm{H}} = 0.77$. This result is close to that found by Batchelor & Green (1972a), $K_{\mathrm{H}} = 0.832$. The difference between these values can be attributed to the fact that the multipole expansion is not an accurate one when particles are separated by a distance comparable to their radii. This remark is made by Batchelor & Green (1972a) with respect to the results of Peterson and Fixman.

The most striking result of the Bedeaux et al paper (1977) is that the effective viscosity ceases to be a scalar (or tensor) characteristic of the suspension and becomes an operator [see (6.21), (6.22), (6.29)].

The presence of the term proportional to $\omega(\sim \alpha^2)$, but with a different coefficient, has been found already by Buyevich & Markov (1973) who considered it as responsible for the frequency dispersion of rheological characteristics of suspensions. This phenomenon has to be taken into account when the frequency is very large. Assuming spheres with radius $a = 1\ \mu$m suspended in water, the contribution of the frequency-dependent term is comparable with unity for frequencies of the order of 10^8 c/s.

Similarly the significance of the term proportional to k^2 seems to be confined to velocity fields with large and strongly nonlinear velocity gradients.

The use of the effective viscosity in the form of a differential operator will change the order of the equation of motion of the suspension and additional boundary conditions will be needed. It is difficult to imagine how these additional conditions can be derived.

7 THE EFFECT OF WALLS

One of the most important questions in the theory of suspensions is the influence of walls and the related question of the boundary conditions that should supplement the equation of motion of the suspension. This problem is crucial because, first, in the Stokes approximation the walls affect flow at large distances; second, all experiments involve wall effects and, as a rule, it is not possible to compare any apparent viscosity measured experimentally with theoretical results without precise knowledge about the wall-particle interactions.

In spite of the importance of the above problem very few papers deal with it. The simplest attempts are based on purely geometrical considerations; effective viscosity is a function of volume fraction, and that is modified by the presence of walls because some part of the space is inaccessible for spheres. A more sophisticated approach, still based on geometrical considerations, is suggested by Buyevich & Shchelchkova (1978).

The important point in the analysis, however, is connected with the fact that the perturbation of flow introduced by the presence of particles is strongly affected by the proximity of walls. The first successful attempt to take this fact into account is by Cox & Brenner (1971) who find that there is an apparent slip of the suspension on the wall. This slip is also found experimentally.

Recently, wall-particle interactions have been reexamined by Tözeren & Skalak (1977). These authors consider a rigid sphere placed in the vicinity of a plane wall, and neglect interparticle interactions. Both these assumptions seem to be justified in the proximity of the wall. The interaction of a particle with the wall is obtained using the method of O'Neill (1964) and Dean & O'Neill (1963). The way of averaging was the main question to be answered. Volume averaging could not be exploited because the distribution of particles near the wall differs considerably from that in the bulk. The authors used area averaging over a plane parallel to the wall which, undoubtedly, better describes the actual particle distribution. However, the relation between the density and the distance from the wall is not included in the Tözeren-Skalak analysis, and the authors have to assume some forms of this relation. Thus the results have only a qualitative character, though they confirm the main conclusion of Cox & Brenner (1971) about the slip on the wall.

Near the wall determination of the flow should be coupled with determination of the distribution function. This difficult problem, as far as we know, has never been successfully attacked.

8 BROWNIAN MOTION

In recent years numerous papers have been published on hydrodynamic interactions of Brownian particles. The growing interest in the subject stems from the fact that in many cases, especially in solutions of macromolecules, the volume fraction of Brownian particles is large enough to produce significant deviation from what can be expected from treating each Brownian particle separately. In a number of experiments (e.g. light scattering) the results depend on the collective behavior of Brownian particles. All theories describing these effects have their roots, of course, in hydrodynamics. Perhaps this subject requires a special review. Nevertheless, in this article we cannot ignore Brownian effects and some results will be briefly presented.

From the hydrodynamic point of view the main problem is the influence of Brownian motion on the bulk behavior of suspensions. For nonspherical particles the mechanism is well understood. Consider first a very dilute suspension with no interactions between particles. The orientation of particles (rods, for example) depends on the kind of flow field in which they are embedded. The imposed flow implies ordering of nonspherical particles, which is counteracted by Brownian motion. The relative influence of these effects—ordering by flow and randomizing by Brownian motion—is described by the Fokker-Planck equation for the orientation distribution function. The effect of Brownian motion on nonspherical particles in dilute suspensions is evident, e.g. in the non-Newtonian behavior of such systems. In particular, the coefficient α in the formula for the effective viscosity $\eta = \eta_0(1+\alpha\Phi_0)$ depends on the shape of particles and on the imposed flow.

The situation with a suspension built up of spherical particles is different. As there is no distinctive orientation, the Einstein formula remains valid for dilute suspensions, even in the presence of Brownian motion. Thus the main problem is whether Brownian motion can influence the higher order terms in the expression for bulk stress and viscosity. Batchelor (1977) gives a positive answer to this question, and finds how Brownian motion affects the Huggins coefficient.

Now, let us consider hydrodynamic interactions between Brownian particles in the fluid at rest. The probability function $\Psi(t,\mathbf{R}_1,\ldots,\mathbf{R}_N)$ obeys the Smoluchowski equation

$$\frac{\partial\Psi}{\partial t} = -\partial_i \mathscr{I}_i \tag{8.1}$$

where

$$\mathscr{I}_i = -D_{ij}\partial_j\Psi$$

are components of a flux, and D_{ij} components of the diffusion tensor.

The time scale t_d, in which the distribution function (or the spatial configuration of spheres) varies, is of a different order of magnitude than the time scale t_m of momentum relaxation, i.e. the time scale of velocity changes of Brownian particles, $t_m \ll t_d$. This allows us to write for the components of the diffusion tensor:

$$D_{ij} = D_{ij}(\mathbf{R}_1,\ldots,\mathbf{R}_N) = \int_0^\infty \langle \mathbf{v}_i(0)\mathbf{v}_j(t)\rangle_0 \, dt \tag{8.2}$$

where the averaged value $\langle\ \rangle_0$ is obtained with the use of the distribution function at time $t = 0$.

The distribution function Ψ and thus also Equation (8.1) can be generalized to take into account rotational Brownian motion, coupling between translational and rotational motion, as well as a mean force potential.

The hydrodynamical problem consists in finding $\mathbf{v}$ and the velocity correlations $\langle \mathbf{v}_i(0)\mathbf{v}_j(t)\rangle_0$ which depend on the positions of all particles.

There are several ways of attacking this problem. One is based on using the Langevin equation for the fluctuating velocity of the Brownian particle. The relation between drag coefficients entering this equation and the diffusion tensor forms the generalized Einstein relation (Wolynes & Deutch 1977).

Another way is employed by Felderhof (1979) who uses the modified Smoluchowski Equation (8.1) for one-particle, and two-particle distribution functions. The diffusion tensor is represented as a sum of contributions due to isolated particles and pairs of particles. The results of the previous work of Felderhof (1977) concerning hydrodynamic interaction of two spheres in the point-particle approximation allow the author to find the diffusion tensor as a function of volume fraction up to the order Φ_0. It should be noted that results are obtained not only for noslip boundary conditions but also when slip is allowed. In spite of the use of the point-particle approximation the results obtained are extremely close (for the noslip case) to those of Batchelor (1976b).

The method used by Batchelor (1976b) is a generalization of Einstein's method. The diffusive flux due to Brownian motion is identified with the flux due to the thermodynamic force. This, in turn, is related to non-uniformity of the function $\Psi(\mathbf{R}_1,\ldots,\mathbf{R}_N)$ (assumed to be time-independent) for a group of N particles which at some instant can be treated as separated from other Brownian particles. Detailed hydrodynamic analysis has been performed for a group of two spherical particles. The diffusion tensor follows from this analysis.

In a complementary paper (Batchelor 1977) stress due to the presence of Brownian motion of rigid spherical particles is calculated up to the order Φ^2 using the method employed previously on the Huggins coefficient. Two contributions to the local stress are distinguished, a mechnical one, $\boldsymbol{\sigma}^{\mathrm{B}}$, and a thermodynamical one, $\boldsymbol{\tau}$, both of which are stationary random functions. Since the suspension as a whole is at rest, these partial stresses are related one to the other

$$\nabla\cdot\boldsymbol{\sigma}^{\mathrm{B}} = -\nabla\cdot\boldsymbol{\tau} \tag{8.3}$$

and the bulk stress is

$$\mathbf{P}^{\mathrm{B}} = \langle\boldsymbol{\sigma}^{\mathrm{B}}\rangle + \langle\boldsymbol{\tau}\rangle. \tag{8.4}$$

From the Einstein approach it follows that the thermodynamic force acting on the ith sphere is

$$\mathbf{F}_i \sim -kT\frac{\partial}{\partial\mathbf{R}_i}\log\Psi(\mathbf{R}_1,\ldots,\mathbf{R}_N) \tag{8.5}$$

and the average thermodynamic stress can be represented in the form

$$\langle\boldsymbol{\tau}\rangle = -\frac{1}{V}\sum_{i=1}^{N}\mathbf{R}_i\mathbf{F}_i, \tag{8.6}$$

where V is the volume containing the N particles. The relation between the force $\mathbf{F}_i$ and the mechanical stress on the surface S_i of the ith particle is

$$\mathbf{F}_i = -\int_{S_i}\boldsymbol{\sigma}^{\mathrm{B}}\cdot\mathbf{n}\,dS. \tag{8.7}$$

In the above analysis of Batchelor it is assumed that there is no flow of the suspension, but the method works also in the case when flow is present. In that case, due to the linearity of the theory, the bulk stress can be presented as a sum of stresses resulting from the imposed flow and from Brownian motion. The important point is that the flow-induced stress depends implicitly on Brownian motion through the pair distribution function (the diffusion tensor appears in the differential equation for the pair distribution function). When the Brownian motion is strong enough to overcome the hydrodynamic effects, the effective viscosity calculated up to the term Φ^2 reduces to

$$\eta = \eta_0(1+2.5\Phi_0+6.2\Phi_0^2). \tag{8.8}$$

As follows from the above remarks and in contrast to the situation encountered only a few years ago the knowledge of hydrodynamic interactions between Brownian spherical particles is now well advanced. There are a number of methods allowing successful treatment of this subject.

9 CONCLUSIONS

In this review the theory of a suspension of rigid spherical, equal-sized particles has been considered. This classical subject still attracts attention because it is the most convenient model for discussing fundamental problems arising in the theory of suspensions.

We conclude the review with a few general comments.

First, we think that the use of the grand canonical ensemble is an important step in building a statistical theory of homogeneous suspensions. It allows one to see the particular difficulties encountered in that theory in a wider context of fundamental problems of statistical mechanics.

Second, hydrodynamics now provides a firm basis for understanding the collective interactions of Brownian particles. These interactions should be taken into account in a number of problems encountered in physics and chemistry, and also in purely mechanical phenomena, such as bulk stress in suspensions.

On the other hand, only limited progress can be reported on problems of most interest in hydrodynamics, i.e. on the flow of suspensions (especially dense suspensions) in the presence of walls. The apparently simple question of the flow through a circular tube still awaits solution.

ACKNOWLEDGMENT

We are grateful to Professor Andrzej Ziabicki for his numerous comments and thoughtful criticism.

Literature Cited

Arp, P. A., Mason, S. G. 1977. The kinetics of flowing dispersions. VIII. Doublets of rigid spheres (theoretical). *J. Colloid Interface Sci.* 61:21–43

Balescu, R. 1975. *Equilibrium and Non-Equilibrium Statistical Mechanics.* New York: Wiley

Batchelor, G. K. 1970. The stress system in a suspension of force free particles. *J. Fluid Mech.* 41:545–70

Batchelor, G. K. 1971. The stress generated in a non-dilute suspension of elongated particles by pure straining motion *J. Fluid Mech.* 46:813–29

Batchelor, G. K. 1972. Sedimentation in a dilute dispersion of spheres. *J. Fluid Mech.* 52:245–68

Batchelor, G. K. 1974. Transport properties of two-phase materials with random structure. *Ann. Rev. Fluid Mech.* 6:227–55

Batchelor, G. K. 1976a. In *Theoretical and Applied Mechanics*, ed. W. T. Koiter, pp. 33–55. Preprints of the IUTAM Congress Delft, The Netherlands, 30 August–4 September 1976. Amsterdam: North-Holland

Batchelor, G. K. 1976b. Brownian diffusion of particles with hydrodynamic interaction. *J. Fluid Mech.* 74:1–29

Batchelor, G. K. 1977. The effect of Brownian motion on the bulk stress in a suspension of spherical particles. *J. Fluid Mech.* 83:97–117

Batchelor, G. K., Green, J. T. 1972a. The determination of the bulk stress in a suspension of spherical particles to order c^2. *J. Fluid Mech.* 56:401–27

Batchelor, G. K., Green, J. T. 1972b. The hydrodynamic interaction of two small freely-moving spheres in a linear flow field. *J. Fluid Mech.* 56:375–400

Bedeaux, D., Kapral, R., Mazur, P. 1977. The effective shear viscosity of a uniform suspension of spheres. *Physica Ser. A* 88:88–121

Berker, R. 1963. In *Handbuch der Physik*, Vol. VIII/2, ed. S. Flügge. Berlin: Springer

Brenner, H. 1974. Rheology of a dilute suspension of axisymmetric Brownian particles. *Int. J. Multiphase Flow* 1:195–341

Brinkman, H. C. 1947. A calculation of the viscous force exerted by a flowing fluid on a dense swarm of particles. *Appl. Sci. Res. Ser. A* 1:27

Buyevich, Yu. A. 1968. Fluctuation of the number of particles in dense dispersive systems. *Inzh. Fiz. Zh.* 14:454–59

Buyevich, Yu. A., Markov, V. G. 1973. Continual mechanics of mono-disperse suspensions. Rheological equations of state for suspensions of moderate concentration. *Prikl. Mat. Mekh.* 37:1059–77

Buyevich, Yu. A., Shchelchkova, I. N. 1978. Flow of dense suspensions. *Prog. Aerospace Sci.* 18:121–50

Chandrasekhar, S. 1954. In *Selected Papers on Noise and Stochastic Processes*, ed. N. Wax. New York: Dover

Cox, R. G., Brenner, H. 1971. The rheology of a suspension of particles in a Newtonian fluid. *Chem. Eng. Sci.* 26:65–93

Croxton, C. A. 1974. *Liquid State Physics—A Statistical Mechanical Introduction.* Cambridge Univ. Press

Dean, W. R., O'Neill, M. E. 1963. A slow motion of viscous liquid caused by the rotation of a solid sphere. *Mathematika* 10:13–24

Einstein, A. 1905. *Eine neue Bestimmung der Moleküldimensionen.* Inaugural Dissertation. Bern: K. J. Wyss (see also: 1906. *Ann. Physik.* 19:298–306)

Felderhof, B. U. 1976a. Rheology of polymer suspensions. I. Local flow effects. *Physica Ser. A* 82:596–610

Felderhof, B. U. 1976b. Rheology of polymer suspensions. II. Translation, rotation and viscosity. *Physica Ser. A* 82:611–22

Felderhof, B. U. 1977. Hydrodynamic interaction between two spheres. *Physica Ser. A* 89:373–84

Felderhof, B. U. 1979. Diffusion of interacting Brownian particles. *Physica Ser. A* In press

Ferziger, J. H., Kaper, H. G. 1972. *Mathematical Theory of Transport Processes in Gases.* Amsterdam: North-Holland

Happel, J., Brenner, H. 1965. *Low Reynolds Number Hydrodynamics with Special Applications to Particulate Media.* Englewood Cliffs, N.J.: Prentice-Hall

Herczyński, R. 1975. The distribution function for random distribution of spheres. *Nature* 255:540–41

Hill, T. L. 1956. *Statistical Mechanics. Principles and Selected Applications.* New York: McGraw-Hill

Hinch, E. J. 1977. An averaged-equation approach to particle interactions in a fluid suspension. *J. Fluid Mech.* 83:695–720

Howells, I. D. 1974. Drag due to the motion of a Newtonian fluid through a sparse random array of small fixed rigid objects. *J. Fluid Mech.* 64:449–75

Jeffrey, D. J. 1974. Group expansions for the bulk properties of a statistically homogeneous random suspension. *Proc. R. Soc. London Ser. A* 338:503–16

Jeffrey, D. J. 1978. In *Continuum Models of Discrete Systems*, ed. J. W. Provan. *Proc. Second Int. Symp. on Continuum Models of Discrete Systems*, Mont Gabriel, Quebec, Canada, June 26–July 2, 1977. Waterloo, Ontario: Univ. Waterloo Press

Jeffrey, D. J., Acrivos, A. 1976. The rheological properties of suspension of rigid particles. *AIChE J.* 22:417–32

Kapral, R., Bedeaux, D. 1978. The effective shear viscosity of a regular array of suspended particles. *Physica Ser. A* 91:590–602

Kendall, M. G., Moran, P. A. P. 1963. *Geometrical Probability.* London: Charles Griffin

Landau, L. D., Lifshitz, E. M. 1958. *Statistical physics.* London: Pergamon

Landau, L. D., Lifshitz, E. M. 1959. *Fluid Mechanics*. London: Pergamon

Leal, L. G. Hinch, E. J. 1973. Theoretical studies of a suspension of rigid particles affected by Brownian couples. *Rheol. Acta* 12: 127–32

Lundgren, T. S. 1972. Slow flow through stationary random beds and suspensions of spheres. *J. Fluid Mech.* 51: 273–99

Mazur, P., Bedeaux, D. 1974. A generalization of Faxen's theorem to nonsteady motion of a sphere through an incompressible fluid in arbitrary flow. *Physica* 76: 235–46

Mewis, J., Spaull, A. J. B. 1976. Rheology of concentrated dispersions. *Adv. Colloid Interface Sci.* 6: 173–200

O'Neill, M. E. 1964. A slow motion of viscous liquid caused by a slowly moving sphere. *Mathematika* 11: 67–74

Peterson, J. M., Fixman, M. 1963. Viscosity of polymer solutions. *J. Chem. Phys.* 39: 2516–23

Schiff, L. I. 1968. *Quantum Mechanics*. New York: McGraw-Hill. 3rd ed.

Tözeren, A., Skalak, R. 1977. Stress in a suspension near rigid boundaries. *J. Fluid Mech.* 82: 289–307

Wolynes, P. G., Deutch, J. M. 1977. Dynamical orientation correlations in solution. *J. Chem. Phys.* 67: 733–41

Ann. Rev. Fluid Mech. 1980. 12: 271–301

COASTAL CIRCULATION AND WIND-INDUCED CURRENTS

×8160

Clinton D. Winant
Scripps Institution of Oceanography, La Jolla, California 92093

INTRODUCTION

Coastal circulation has been a subject for inquiry by marine scientists since the earliest days of fishing and shipping. The two most easily distinguishable causes for coastal motion are winds and tides, but the response of coastal waters to these forces varies widely, influenced by climate, geomorphology, and stratification. The last decade has seen a veritable explosion of field studies of coastal circulation, made possible by the development of recording current meters which, deployed in arrays over the shelf, provide long time series of horizontal currents, temperature, and, in some instances, salinity. Results from some of these studies are reviewed here, with emphasis on the wind-driven circulation, which until now has been the primary focus of attention.

1 TEMPORAL AND SPATIAL VARIABILITY

As a result of current meter observations, good descriptions of the frequency spectrum of currents have become available, but unfortunately the spatial variability is not as well documented. Nonetheless, comparison of spectra obtained in different locations highlights differences that may be induced by geographical factors.

A spectrum of summertime currents off Del Mar, California, near the surface in 60 m of water, is reproduced in Figure 1. The rotary component representation (Gonnella 1972) is used, since it represents best the characteristics of a two-component vector field. The clockwise spectrum is associated with currents that rotate in time in a clockwise sense. A measure of the ellipticity of the motion is given by the difference between clockwise and anticlockwise spectral estimates. If they are equal,

0066-4189/80/0115-0271$01.00

the motion is linear. The orientation of the major axis of the motion is shown in the frame beneath the spectra. In the coordinate system used here, an orientation of 0° is across the shelf, while an orientation of 90° corresponds to longshore motion. Spectra of kinetic energy obtained on the New England shelf and off Oregon are reproduced in Figure 2. These spectra illustrate several fundamental features of the variability of coastal currents. The spectrum is inherently red (as are spectra of wind speed), with significant peaks at well-defined frequencies. Note that the highest frequency resolved in Figure 1, 360 cycles per day (cpd), corresponds to

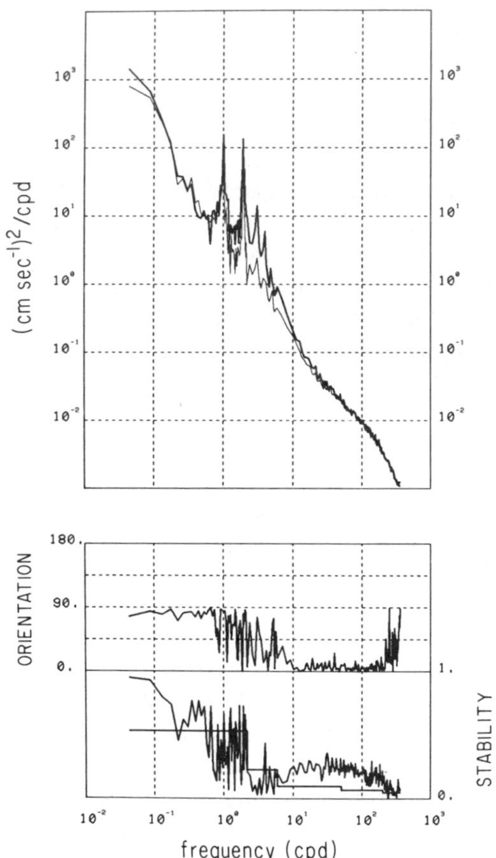

Figure 1 Rotary spectra of currents off Del Mar, California. The observations are from instruments 10 m beneath the surface in 60 m of water, at the shelf break. The degrees of freedom increase with frequency, with 16 degrees of freedom for the lowest band resolved here. In the top frame, the heavy line represents clockwise spectral estimates, and the lighter line represents counterclockwise estimates.

a period of 4 minutes, so that motions induced by surface waves are not included.

In the high frequency part of the spectrum ($f > 10$ cpd) motion is predominantly in the cross shore direction. In this band, currents may be due to either baroclinic or barotropic processes. The energy is several orders of magnitude lower than that associated with neighboring bands (tides at lower frequencies, surface waves at higher frequencies), but these motions are important with regard to dispersion in coastal waters. Because they occur at periods substantially shorter than the inertial period, their dynamics are independent of rotation.

High frequency baroclinic motions on the shelf are ubiquitous when the stratification is high. At these frequencies, the wavelength of internal waves is small (<1 km) and these motions are observed as progressive waves traveling toward shore. Winant & Olson (1976) report mean square velocities on order 1 $cm^2 s^{-2}$. Individual waves can have amplitudes that are a substantial fraction of the total water depth, but are intermittent. C. Y. Lee & R. C. Beardsley (1974) explain the high frequency fluctuations observed by Halpern (1971) and Haury, Briscoe & Orr (1979) in terms of nonlinear internal waves. Maxworthy (1978) demonstrates mechanisms by which long internal waves can be generated by the flow over an isolated topographic feature.

Atmospheric events can generate high frequency barotropic edge waves. Munk et al (1956) showed how such a wave may be forced by traveling atmospheric perturbations. Beardsley et al (1977b) note that

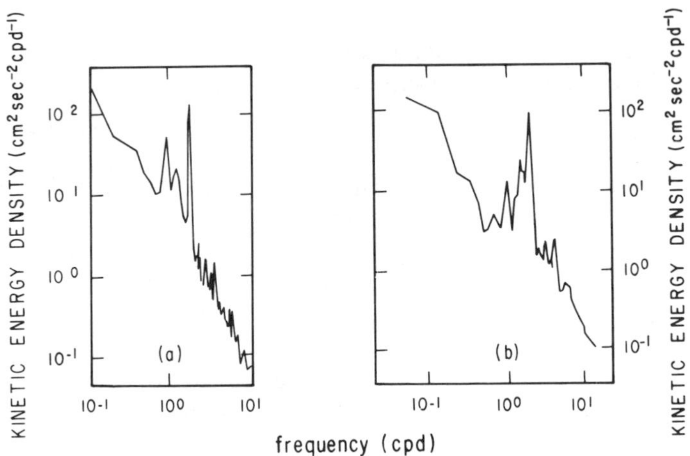

Figure 2 Kinetic energy spectra from (*a*) the New England shelf (courtesy R. C. Beardsley) and (*b*) the Oregon shelf (courtesy D. Halpern).

these storm-driven fluctuations may have amplitudes on order 1 m, and determine that their observed dynamics are at least in qualitative agreement with edge wave theory. In an experiment designed to determine spatial and temporal scales of the background of edge wave noise on the shelf, Munk et al (1964) verified the edge wave dispersion relation.

The tidal band (1 cpd $> \sigma >$ 10 cpd) is characterized by large peaks at the tidal frequencies and, in some instances (Figure 1), at their harmonics. In current spectra these peaks are broader than in spectra of sea surface elevation, in which the harmonic energy is notably less. These differences are explained by the presence of both barotropic and baroclinic tides. Sea level is a measure of only the barotropic motions, but both barotropic and baroclinic currents may exist. Observations of tidal currents on the shelf are reviewed by Winant (1979). Although there exist shelf geometries that can support Kelvin, Poincaré, or other waves at the tidal frequency, the wavelengths of the offshore barotropic tides are so much larger than typical shelf widths, that tidal motion on the shelf is expected to occur principally as a readjustment of sea level to offshore tidal elevation, the simplest case corresponding to barotropic cross shelf motions, spilling enough water onto the shelf to adjust sea level there to that offshore. Barotropic tides are described by their ellipticity ε (the ratio of cross shore to longshore currents), and impedance I (the ratio of sea surface elevation to longshore current). Over flat bottom topography, for waves trapped to the coast, Munk et al (1970) show

$$\varepsilon^2 + g\,I^2/H = 1$$

(g is the acceleration of gravity and H the shelf depth). Over a shelf with bottom slope β, Petrie (1975) shows

$$I/\varepsilon = \beta/\sigma$$

where σ is the frequency of the tidal motion. The ratio I/ε is the ratio of sea surface elevation and cross shore velocity, and this relation, which results directly from continuity considerations, implies that the tidal elevations are for the most part due to cross shelf motions. Over wide shelves such as the eastern seaboard of the United States, $I/\varepsilon \sim 5$ s. On narrower shelves I/ε can range up to 100 s, and the currents that are responsible for the tidal elevation are undetectable from noise.

The relative amplitude of barotropic and baroclinic tides over the shelf clearly depends on geometry. Over the Scotian shelf, Petrie (1975) shows that the barotropic component has much more energy than the baroclinic component. On the narrow shelf of Southern California, the observations of Winant & Olson (1976) imply roughly the same energy in both components, but whereas the barotropic motions are primarily directed alongshore, the baroclinic motions occur in the cross shelf plane. In models that idealize the stratification as a two-layer fluid (e.g. Rattray 1960) and ignore dissipation, the baroclinic mode is standing on the shelf. This is confirmed by Winant (1979). If the phase speed of internal waves at these frequencies and in these depths is of the same order as the fluid velocities, the motion is expected to be highly nonlinear, with strong harmonics such as those shown in Figure 1. Note that the amplitude of tidal harmonics varies locally. The harmonics in Figure 1 do not appear as explicitly in either of the spectra shown in Figure 2. The baroclinic tides are, as in the deeper ocean (Munk et al 1970, Wunsch 1975), characterized by a high degree of intermittency. Although this is clearly related to low frequency changes in the overall stratification, the relationship remains obscure, as does the question of generation, or coupling with the barotropic tide.

Motions occurring at longer periods than the tidal periods but less than a month are generally related to atmospheric forcing. Recent observations are discussed in the following section.

Current systems that fluctuate on time scales of the order of a few weeks to a year have been observed since the earliest days of shipping. Comparisons of classical observations (based on hydrographic measurements, drifters, estimates of ship drift) with recent direct current measurements are to be found in the reviews of Beardsley et al (1976) for the currents on the eastern seaboard of the United States, and Hickey (1979) for the California current system. Very generally these low frequency or seasonal currents can be categorized into three classes depending on the forcing.

1. Direct meteorological forcing, which imparts both momentum (wind stress) and vorticity (curl of the wind stress) to the water. As Munk (1950) originally suggested, the California current system is an example of this (Hickey 1979).
2. Forcing as a result of adjoining deep ocean systems (steady circulation as well as waves) for which the shelf acts as a boundary layer. The best example of this is the northeastern seaboard of America, on which steady southwestward flows of order 10 cm s^{-1} are observed (Beardsley et al 1976). Numerical models of oceanic circulation (Semtner & Mintz 1977) result in sea surface slopes along the western boundary of the basin of the same order (10^{-7}) as required to maintain the observed flow in the presence of bottom friction. Beardsley & Winant (1979) examine the possible dynamical causes, and conclude that this large scale steady feature is in all likelihood forced by the large scale oceanic circulation, as a boundary layer feature of the subpolar gyre.

3. Forcing from the runoff of large river systems, which can generate shelf motions by causing large scale perturbations in the density field. These perturbations result in misalignment of constant density lines with isobaths, which results in a source of vorticity (the topographic torque). Hendershott & Rizzoli (1976) conclude that the seasonal circulation in the Adriatic Sea is strongly influenced by outflow from the Po River and drainage from Italy and Yugoslavia.

Although not a steady feature, the encroachment of deep sea structures (eddies and topographic Rossby waves) onto the shelf is observed worldwide. Kroll & Niiler (1976) examine the transmission of topographic Rossby waves onto the shelf. These waves can propagate into shallow water, but because of friction, the energy must substantially decay before water depths of order 25 m are reached. Garrett (1979) interprets the fluctuations in currents off east Australia, deduced from ship drift observations (Hamon et al 1975), as components of topographic Rossby waves. Observation of spinoff eddies from the Florida current are reported by T. N. Lee (1975) and T. N. Lee & D. A. Mayer (1977). These cyclonic eddies are characterized by advection of warm saline water onto the narrow continental shelf that separates Florida from the Florida current. They also play an important dynamic role, transferring energy between the offshore current systems and shelf water.

2 WIND-DRIVEN CIRCULATION

Longshore wind stress drives a complex coastal circulation; the dynamics involve the effects of stratification, rotation, friction, and topography. Experimental programs have been devised with the specific objective of documenting such motions. The results are reviewed in this section.

Long Waves

In all observation of wind-driven coastal circulation, currents and sea level respond simultaneously and rapidly to the wind, and the high coherence between longshore currents and wind stress observed on the shelf decays with distance offshore. Niiler (1975) points out that these are the elements that suggest that the shelf acts in some ways as a wave-guide for wind-induced motions. Theoretical considerations suggest that different wave systems can propagate alongshore, trapped to the coast either by topography or by rotation. These effects have been reviewed by Clarke (1977) and further comprehensive reviews are included in this volume. A good deal of attention has been devoted to providing experimental evidence of longshore wave motion on the shelf. Since this evidence is also reviewed elsewhere in this volume, only two examples are mentioned here, the first dealing with forced wave motion on the shelf, the second dealing with free waves.

Wang & Mooers (1977) examine two month-long time series of sea level, surface atmospheric pressure, winds, and longshore currents, the latter obtained during the CUE-2 program (Pillsbury et al 1974). Sea level and atmospheric pressure records at stations along the west coast of the United States covering longshore distances over 1500 km are included. Energetic fluctuations in sea level at a ten-day period are found to be coherent at all the stations. The fluctuations in atmospheric pressure at a ten-day period are dominant, and high coherence between stations is observed for periods between 3 and 20 days. The winds are not found to be coherent between stations, probably as a result of local topographical effects on the wind. Using methods of empirical orthogonal function (E.O.F.) analysis, coherence, phase, and amplitude distributions are determined at the various stations (Figure 3). It is shown that both sea level and atmospheric pressure fluctuations are spatially coherent over alongshore distances in excess of 1500 km and that both sea level and atmospheric pressure fluctuations have northward phase propagation. In addition, where the phase speed of the atmospheric pressure and that of the sea level are comparable to the phase speed of long, coastal trapped waves, the response is in near resonance, and the amplitude of adjusted sea level variations (adjusted sea level = sea level + atmospheric pressure) builds up rapidly.

An extensive field study of coastal currents off the coast of Peru is reported by R. L. Smith (1978). Observations of longshore winds and sea level from two coastal stations, San Juan and Callao, separated by 400 km, are reported as well as current measurements in water depths of 80 m offshore from the stations. The wind observations at San Juan and Callao are found to be coherent. The sea level observations are coherent with each other, as are the currents, and the latter two pairs of observations are coherent, although they are not significantly coherent with the winds. These results imply that the atmospheric field on one hand and the sea level and current field on the other are spatially coherent over distances greater than those separating the observations, but do not interact strongly. Analysis of phase relationships in the sea level and current fields suggests low frequency (<0.25 cycles per day) fluctuations travel poleward with a phase speed (200 km/day) independent of frequency. These motions are shown to be consistent with the nondispersive propagation of free internal Kelvin waves.

Local Wind Forcing

The local forcing of coastal waters by wind is simply defined to be the motion resulting on the shelf from local wind systems. Reports of such

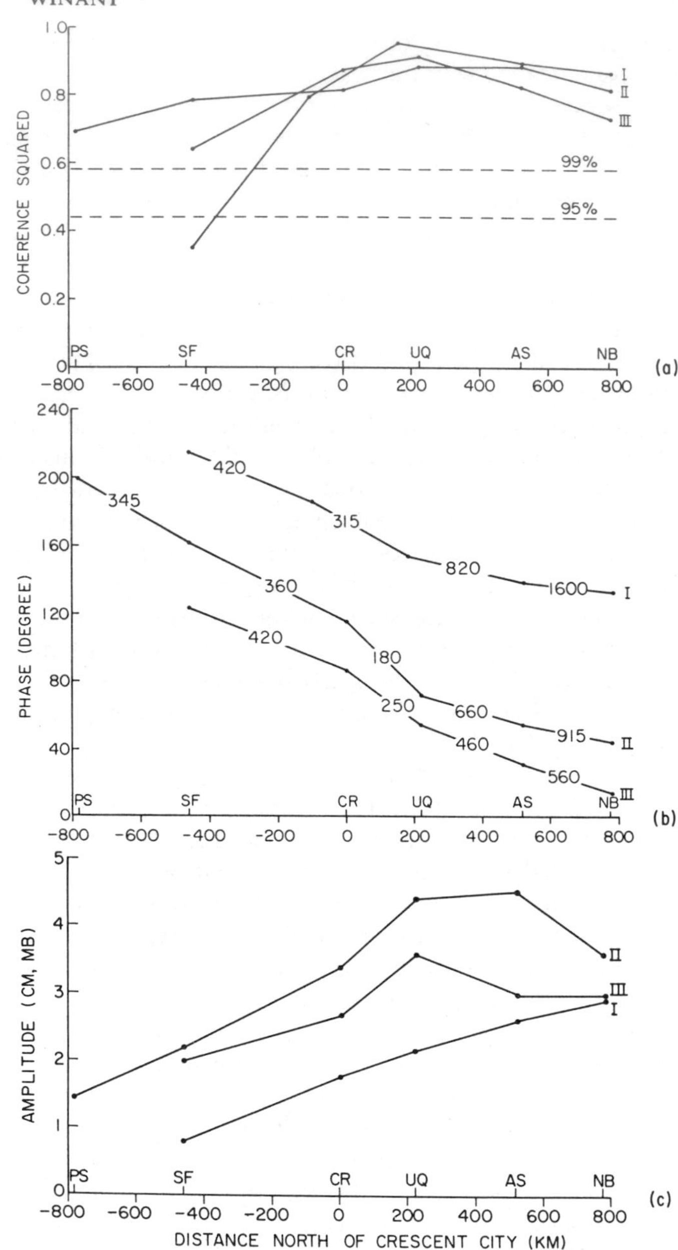

Figure 3 Alongshore distribution of the first empirical modes of 10-day fluctuations for surface atmospheric pressure (I), sea level (II), and adjusted sea level (III): (*a*) the coherence squared (95% and 99% significance levels are marked), (*b*) the phase (numbers are the northward phase speed in km day^{-1}), and (*c*) the rms amplitude. From Wang & Mooers (1977).

motions now abound in the literature. In the US experiments have been carried out in the Gulf of Maine (Brown & Beardsley 1978, Vermersch et al 1979, Noble & Butman 1979, Butman et al 1979), on the New England continental shelf and off Long Island (Beardsley & Butman 1974, Scott & Csanady 1976, Beardsley et al 1977b), off the coast of New Jersey (Bennett & Magnell 1979), on the shelf between Cape May and Cape Hatteras (Boicourt & Hacker 1976), on the shelf south of Cape Hatteras, on the eastern Florida shelf (Brooks & Mooers 1977), on the western Florida shelf (Weatherly & Van Leer 1977), on the Louisiana shelf (Murray 1975, Daddio et al 1978), off the coast of Texas (Forristal et al 1977, N. P. Smith 1978), on the southern California shelf (Winant & Olson 1976, Winant 1979, Winant 1980), on the shelf of Washington and Oregon (Collins & Pattulo 1970, D. L. Cutchin & R. L. Smith 1973, R. L. Smith 1974, Huyer et al 1975, Halpern 1976, Huyer et al 1978, Allen & Kundu 1978), and off Alaska (Hayes & Schumacher 1976, Hayes 1979). In addition, local wind-induced currents have been observed off Baja California, Mexico, as well as off British Columbia (Huyer et al 1976) and Nova Scotia (B. Petrie & P. C. Smith 1977, P. C. Smith 1978). The number of studies involved on the shelf of North America alone underlines the differences in observations. To say that the only common denominator between these reports is that when the wind blows alongshore, the water moves in the same direction is an oversimplification, yet clearly the wide variety of climatological and geomorphological settings results in different shelf response to wind forcing. An estimate of variation in the response may be obtained from Winant & Beardsley (1979) in which wind-induced currents off Long Island, Chesapeake Bay, and southern California in 30 m depth are compared. If the response is defined to be the ratio of bottom stress to applied wind stress in the longshore direction, then the response varies by about a factor of two. It is important to note, however, that such a simple definition of response begs some important questions, in particular the role played by longshore pressure gradients.

THEORETICAL FRAMEWORK In order to compare the various observations it is best to begin by setting the fluid mechanical framework in which they will be viewed, the linearized equations of motion in a rotating frame. The coordinate system chosen is illustrated in Figure 4; u is the cross shore velocity taken positive onshore, v is the longshore velocity, and w the vertical velocity, positive upwards. The continuity and momentum equations are

$$\nabla \cdot \mathbf{u} = 0 \tag{1}$$

$$\mathbf{u}_t + 2\mathbf{\Omega} \times \mathbf{u} = -\nabla p/\rho + \nu\nabla^2\mathbf{u}. \tag{2}$$

Beyond neglecting the convection terms, certain other assumptions are made that simplify the equations, but retain the important dynamical features. The motion is assumed to take place on an f-plane so that the rotation of the coordinate system is taken to be constant, aligned with the z axis and with magnitude $f = 2\Omega \sin \phi$ where Ω is the rotation rate of the earth and ϕ is the latitude of the observations. Vertical stresses are neglected, as are horizontal gradients of the stress vector

$$\boldsymbol{\tau}(z,t) = \mathbf{i}\tau^x(z,t)+\mathbf{j}\tau^y(z,t).$$

The adoption of this form for stress precludes any further consideration of the curl of the wind stress as a motive force for coastal motion. In component form, the momentum equation then reduces to

$$u_t - fv = -p_x/\rho + \tau_z^x/\rho \qquad (3)$$

$$v_t + fu = -p_y/\rho + \tau_z^y/\rho \qquad (4)$$

$$w_t = -p_z/\rho - g. \qquad (5)$$

In addition, taking the curl of these equations yields a vorticity equation

$$\omega_t = f\mathbf{u}_z - (\nabla p \times \nabla \rho)/\rho^2 + \nabla \times (\boldsymbol{\tau}_z/\rho) \qquad (6)$$

in which changes in the vorticity vector (ω) are seen to be due to the tilting of planetary vorticity, misalignment of lines of constant pressure and density, or diffusion. If vertical velocities are small (observations suggest that they are one to two orders of magnitude less than horizontal velocities in the frequency band of interest) the pressure is nearly hydro-

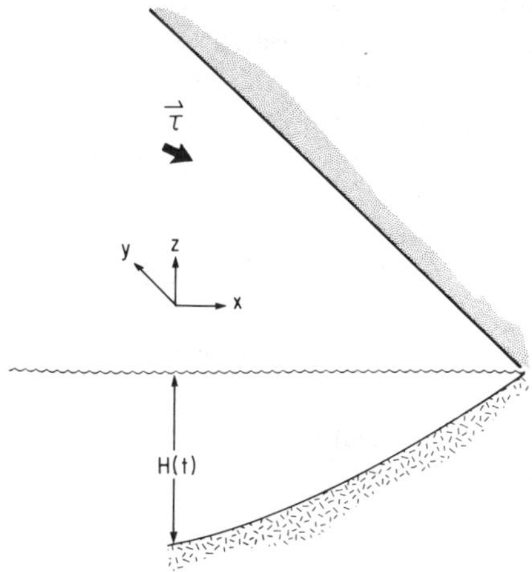

Figure 4 Coordinate system.

static, and lines of constant pressure are nearly horizontal, so that the vorticity equation further reduces to

$$\omega_t = f\mathbf{u}_z + (\mathbf{g} \times \nabla\rho)/\rho + \nabla \times (\boldsymbol{\tau}_z/\rho), \tag{7}$$

which becomes the thermal wind equation if the rate of change of vorticity and the curl of the vertical stress gradients are small.

Overall momentum balances are easier to verify with vertically integrated equations. If the continuity and horizontal momentum equations are integrated from the bottom ($z = 0$) to the free surface [$z = H(x,y,t)$] the following equations are obtained:

$$H(U_x + V_y) + H_t = 0 \tag{8}$$

$$U_t - fV = -\frac{1}{H}\int_0^H p_x/\rho dz + (\tau_s^x - \tau_b^x)/\rho H \tag{9}$$

$$V_t + fU = -\frac{1}{H}\int_0^H p_y/\rho dz + (\tau_s^y - \tau_b^y)/\rho H \tag{10}$$

where U, V denote vertically averaged properties,

$$U = \int_0^H u dz/H; \qquad V = \int_0^H v dz/H,$$

and τ_s and τ_b respectively represent surface and bottom stresses.

In the absence of cross shore wind stress ($\tau_s^x = 0$) either a longshore wind stress or a longshore pressure gradient will accelerate the water column. In this regard it is important to note in Equation (10) that these two terms play an equivalent role in terms of forcing; it is the combined effect of p_y and τ_s^y that drives the motion. Models of coastal circulation (e.g. Allen 1973, Pedlosky 1974, Csanady 1978) determine the sea surface elevation field and in particular the longshore pressure gradient, but Welander (1961) demonstrates that considerations of the longshore variability of the forcing and the topography, as well as the motion, are required. Unfortunately few current observations include reports of the sea surface elevation field and most of those that do simply consider coastal measurements from tide gauges, which yield no information concerning the offshore variability. A complete survey of sea surface elevation field in the Gulf of Maine reported by Vermersch, Beardsley & Brown (1979) concludes that the velocity field there is incoherent with the pressure field, probably as a result of complex topography. In view of these difficulties it is preferable to adopt a local longshore view of the motion, i.e. to simply consider a cross shore slice, in which case the longshore pressure gradient must be viewed as a forcing term.

A notable example of the importance of the longshore pressure gradient is provided in Beardsley & Butman (1974), in which large wind stresses ($|\tau_2| > 2$ dy cm^2) acting in opposite direction are observed to produce remarkably different currents. These observations are reproduced in Figure 5. The wind forcing is shown in the top two panels while sea

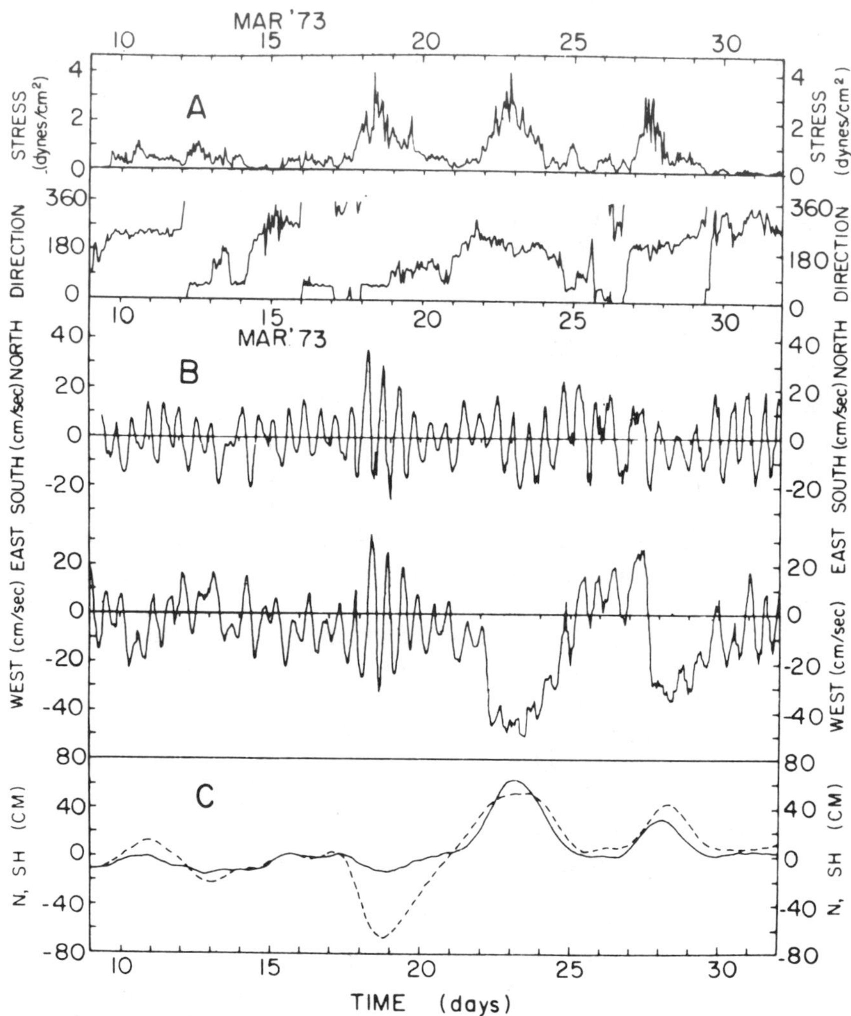

Figure 5 Meteorological and oceanographic variables monitored on the continental shelf of southern New England from March 8 to April 1, 1973. Different graphs, from top to bottom are (*A*) magnitude and direction of surface wind stress at Block island, (*B*) ocean currents at 42 m, and (*C*) corrected sea level at Nantucket Island (solid line) and Sandy Hook (dashed line). From Beardsley & Butman (1974).

level observations from Sandy Hook and Nantucket Island are reproduced in the bottom frame. While the two later wind stress events with winds blowing from the northeast produced large longshore currents, the first event, with winds blowing from the west, produced small longshore currents, but large differences in sea surface elevation. The longshore sea surface gradient may be estimated as the sea level difference (60 cm) divided by the distance between sites (300 km) resulting in a longshore pressure gradient estimate of 2×10^{-3} cm^{-2} in the proper direction to partly balance the surface wind stress.

As soon as the longshore motion develops, cross shore accelerations occur to balance the Coriolis term in the cross shore momentum equation, but near the coast these motions must result in sea surface elevation changes, because of continuity and the restriction of no flow into the coast. The cross shore pressure gradients thus established may balance the Coriolis term in the cross shore momentum equation. The time (t_s) required for such sea surface adjustments to be made is that required for long surface waves to cross the shelf or $t_s \sim L(gH)^{-\frac{1}{2}}$ if L represents the shelf width. Since the time required for sea level adjustment is much smaller [by the factor $(\Delta\rho/\rho)^{\frac{1}{2}}$] than that required for baroclinic adjustments, the primary contribution to the cross shore pressure gradient is expected to be barotropic. For a 10 km wide shelf, with average depth of 100 m, $t_s \sim 5$ minutes; for a 100 km shelf width $t_s \sim 1$ hour. In both cases the time required for geostrophic balance to be established on the shelf is small compared to time scales (t_w) typical of the forcing (t_w is variable but usually ranges between one and five days). In order to examine whether such a balance is observed, it is necessary to compare longshore velocities with cross shelf pressure gradients. While the former are relatively simple to determine, the latter are not, although in some cases two-point measurements of bottom pressure may be used to infer pressure gradient. Observations of longshore current and bottom pressure off Del Mar, California, (Winant 1980) are presented in Figure 6. The top curve represents the alongshore quasi-barotropic flow as determined from the E.O.F. analysis of a vertical array of five current meters in 60 m depth. On 5 September 1978, tropical storm Norman swept by Del Mar traveling to the north along the coast of California. Peak wind stress reached 2 dyn cm^{-2}, and the average longshore current exceeded 50 cm s^{-1}. The time dependence of bottom pressure as determined from two sensors in 30 m and 12 m of water along a common shelf transect with the current meter mooring are also shown. It is clear that the sea level rises synchronously with the current, and the cross shelf pressure gradient is in the proper direction to balance the Coriolis force. The pressure difference between the two sensors separated by one kilometer is of order 1 cm. In this instance $p_y/\rho \sim 10^{-2}$ cm s^{-2}, which is of the same order as the cross shelf Coriolis term with $f = 10^{-4}$ s^{-1} and $V = 50$ cm s^{-1}.

Huyer et al (1978) present further evidence in support of the geostrophic cross shelf balance, in the long term measurements of alongshore currents off the Oregon shelf. Six hourly values of currents from vertical arrays on the inner shelf (depth $\sim$50 m), the midshelf ($\sim$100 m), and the outer shelf ($\sim$200 m) are subjected to E.O.F. analysis. The mode that accounts for the greatest variability is quasi-barotropic. The correlation of this mode with sea level is evaluated at various lags. Results for the midshelf array are shown in Table 1, which shows that the maximum correlation occurs for zero lag during winter and spring. The maximum correlation during summer occurs at a six-hour lag, but the correlation coefficient is not significantly higher than the zero lag correlation. If the cross shore pressure gradient is estimated as the coastal sea level divided by the shelf width L ($L \sim 30$ km off Oregon), geostrophic balance would imply

$$R_1 = v/\eta = g/fL \sim 3 \text{ s}^{-1}.$$

This value is compared with regression estimates between the first mode

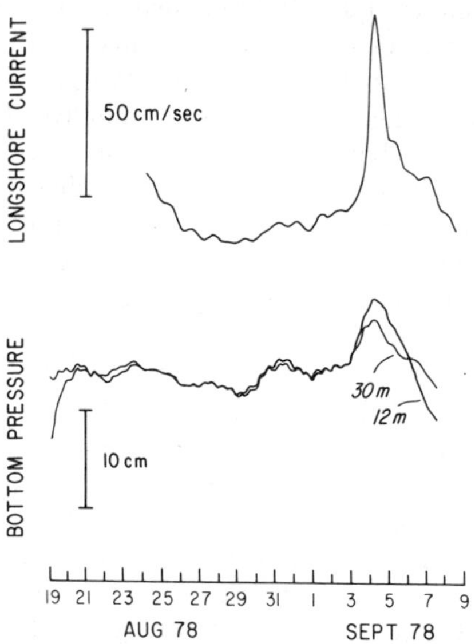

Figure 6 Longshore current as determined from the E.O.F. analysis of a vertical array of 5 current meters, and bottom pressure records from instruments deployed on a common shelf transect with the current meter array, in 30-m and 12-m water depth.

Table 1 Correlation and regression between first E.O.F. mode for longshore current at the mid-shelf location and sea level (A) and alongshore wind stress (B) from Huyer, Smith & Sobey (1978)

A. Correlation with sea level

Season	Correlation Coefficient at zero lag	Correlation Coefficient at lag L (maximum correlation)	L (hrs)	Regression coefficient R (s^{-1})	R_1 (s^{-1})
Winter	0.91	0.91	0	4.5	
Spring	0.81	0.81	0	4.7	3
Summer	0.80	0.83	6	3.5	

B. Correlation with wind stress

Season	Correlation Coefficient at zero lag	Correlation Coefficient at lag L (maximum correlation)	L (hrs)	Regression coefficient R ($cm^3\ s^{-1}\ dyn^{-1}$)	R_2 ($cm^3\ s^{-1}\ dyn^{-1}$)
Winter	0.60	0.69	12	30.5	
Spring	0.53	0.73	12	32.1	20
Summer	0.49	0.59	18	23.7	

longshore velocities and sea level determined by Huyer et al (1978) in Table 1. Clearly the agreement is good.

If the cross shelf momentum balance is geostrophic (on time scales greater than t_s), $U_t = 0$ and $U = \text{constant} = 0$. The extent to which this conclusion is supported by the data is unclear, because of the difficulties involved in estimating U from the data. If alongshore variations are not negligible, neither V_y nor U_x need be zero. Indeed, this is a common feature in many models of coastal circulation. It is, however, a result that is incompatible with the view adopted here, in which longshore gradients in anything but P are neglected. Thus the predominant balance in the vertically integrated momentum equation is taken to be between acceleration, forcing, and bottom friction.

It is useful, in order to estimate time scales involved in such a balance, to use a linear parameterization of the bottom drag,

$$\tau_b^y = \rho r V, \tag{11}$$

as suggested by Csanady (1976, 1978) and Scott & Csanady (1976). The value r is clearly a depth-dependent parameter, but in depths of order 30 m, estimates range around 5×10^{-2} cm s^{-1} (Winant & Beardsley 1979). With this simplification the longshore momentum equation takes the form

$$V_t + r/H\, V = F \tag{12}$$

where F represents the combined forcing due to surface stress and pressure gradients. If F is constant the velocity has an e-folding time $t_f = Hr^{-1}$. Thus very shallow shelves achieve the frictional balance rapidly. For typical shelf configurations, with $H = 100$ m, $t \sim 2$ days. For

Table 2 Correlation and regression coefficients between wind stress (τ_s^y) and two parameterizations of bottom stress from Winant & Beardsley (1979)

Location	Period over which data was collected	Local water depth (m)	Height of current meter above bottom (m)	Comparison of τ_s^y and $\rho/u/u$: Correlation	Comparison of τ_s^y and $\rho/u/u$: Regression C_D	Comparison of τ_s^y and ρu: Correlation	Comparison of τ_s^y and ρu: Regression r (cm s^{-1})
36°45′N, 75°10′W (off Chesapeake Bay)	16 Jan–12 Feb 1974	38.0	6.6	0.82	2.6×10^{-3}	0.85	5.8×10^{-2}
			19.2	0.85	1.2×10^{-3}	0.86	3.9×10^{-2}
			31.2	0.83	1.25×10^{-3}	0.86	4.0×10^{-2}
40°44′N, 72°25′W (11 km S of Long Island coast)	5 Sep–28 Sep 1975	32.5	2.8	0.87	11.5×10^{-3}	0.85	17.9×10^{-2}
			16.0	0.88	2.2×10^{-3}	0.85	7.6×10^{-2}
			26.3	0.93	2.1×10^{-3}	0.87	6.7×10^{-2}
40°47′N, 72°30′W (6 km S of Long Island coast)	5 Feb–21 Mar 1976	27.8	2.8	0.74	14.3×10^{-3}	0.76	20.8×10^{-2}
			12.1	0.76	3.7×10^{-3}	0.76	9.5×10^{-2}
			20.4	0.71	4.5×10^{-3}	0.74	10.4×10^{-2}
			24.3	0.74	3.8×10^{-3}	0.75	9.5×10^{-2}
32°33′N, 117°13′W (2 km off Southern California coast)	3 Aug–19 Sep 1977	30.0	3.0	0.33	5.9×10^{-3}	0.24	10.5×10^{-2}
			10.0	0.39	2.1×10^{-3}	0.29	5.1×10^{-2}
			20.0	0.36	2.0×10^{-3}	0.27	5.0×10^{-2}
			28.0	0.55	1.1×10^{-3}	0.42	3.0×10^{-2}

wind forcing that is steady over long periods ($t_w > t_f$) the longshore momentum balance will become frictional, but for shorter period forcing, the acceleration term should remain substantial. That the longshore momentum balance is frictional is borne out to some degree by Winant & Beardsley (1979), in which wind-driven coastal currents measured in different locations in 30 m water depth are compared based on the premise that the dominant longshore momentum balance is frictional, or that wind stress is balanced by bottom stress. The results of that study are reproduced in Table 2. Two parameterizations of the bottom stress were used, one using a quadratic drag law, $\tau_b^y = \rho C_d V|V|$, and the other using a linear friction law, as in Equation (11). Although the correlation between surface stress and longshore currents varies widely between locations (because of the variability in wind forcing at the different locations), values of either the friction factor r or the drag coefficient C_d deduced from the various measurements vary by a factor of 2, and fall in the expected range for such coefficients. A further test for the existence of frictional balance in the longshore direction can also be found in the results of Huyer et al (1978), in which correlations at various lags between the dominant longshore mode and wind stress are computed. These results, which are reproduced in the lower half of Table 1, indicate that the maximum correlation occurs for a lag of 12 to 18 hours, which is not inconsistent with the t_f estimated previously. If the frictional balance were complete, and a linear bottom-friction parameterization assumed, the ratio of longshore current to wind stress, R_2, is

$$R_2 = V/\tau_s^y = 1/\rho r = 20 \text{ cm}^3 \text{ s}^{-1} \text{ dyn}^{-1}.$$

This value is compared with the regression estimates determined by Huyer et al (1978) in Table 1. The agreement here is surprising in view of the neglect of any longshore pressure gradients in the forcing.

To a certain extent the vertically integrated picture concentrated on so far begs the important issues of upwelling or downwelling, which, as their names imply, involve vertical motion. It is this vertical motion that historically generated interest in upwelling as a mechanism by which nutrients in deeper waters are brought near the surface, increasing productivity. The classical definition of upwelling is a motion resulting from "... a wind in the Northern Hemisphere which blows parallel to the coast, with the coast on the left hand side. In this case the light surface water will be transported away from the coast and must, owing to the continuity of the system, be replaced near the coast by heavier subsurface water" (Sverdrup, Johnson & Fleming 1942), a definition which immediately brings forth the notion of boundary layers. These have so far been avoided through the use of vertically integrated equations. Classical conceptions of coastal upwelling emphasize this boundary-layer approach, ever since Ekman (1905) showed that the wind-induced surface stress acting over the ocean can be balanced by the Coriolis term in the alongwind momentum equation. Integrating the momentum equation across the Ekman layers yields the result

$$\rho f \delta_E V_E = 0 \tag{13}$$

$$\rho f \delta_E U_E = \tau_s^y \tag{14}$$

where U_E and V_E are the vertical averages of cross wind and down wind velocity and δ_E is the thickness of the Ekman layer. This is the classical Ekman layer formulation in which bottom stress plays no role. Ekman (1905) considered the complications introduced by a finite depth ocean in his original work and pointed to the obvious conclusion: the standard Ekman layer is only relevant in waters of depth much greater than δ_E. More recent models of coastal circulation support this conclusion. As an example, Csanady (1978) shows that for time-independent motions, the arrested topographic wave solution satisfies the frictional balance inshore, and becomes a classical Ekman layer far offshore. Conventional estimates of the thickness of the neutral turbulent Ekman layer, δ_E, range around $0.4\,u^*/f$ where u^* is the friction velocity $[u^* = (\tau/\rho)^{\frac{1}{2}}]$ for $u^* \sim 1$ cm s^{-1}, $\delta_E \sim 40$ m. Thus most shelves are only a few Ekman layers deep at the shelf break. Smith & Long (1976) indicate that the entire water depth on the Oregon-Washington shelf is frictionally dominated, consisting of two superposed Ekman layers. Kundu (1977) compares current meter records obtained at different depths near the coast of Oregon and northeast Africa and concludes that the entire column is frictional off Africa while that off Oregon is not. Allen & Kundu (1978) have used the CUE-2 data to test the Ekman balance (Equation 14). Although the correlation between the offshore mass flux in the mixed layer and the surface stress has the proper sign, the correlation is low, and the magnitude of $\delta_E V_E$ does not match that of $\tau_s^y/\rho f$.

In view of these arguments, it seems plausible that the essential vertically integrated dynamics discussed earlier are pertinent, at least to depths of order several δ_E. It is clear, however, that these dynamics must eventually merge with classical Ekman dynamics some distance offshore. Experimental evidence available to date does not allow the clear definition of what this offshore distance might be, although estimates are available from models such as that proposed by Csanady (1978).

It is important to realize that even in cases where surface and bottom stress balance each other, the cross shelf circulation may still develop. That such a circulation exists implies the existence of longshore vorticity.

Potential solutions exist for corner flows similar in geometry to the flow in a cross shelf plane; however, these are irrotational because the tangential gradient of radial velocity and the radial gradient of tangential velocity just balance to yield zero vorticity. Observations demonstrate such is not the case in classical up or downwelling, in which the cross shore gradients of vertical velocity are small enough that the longshore vorticity is well approximated by the vertical gradient of the cross shore velocity. Thus, in order to determine the cross shelf circulation the vorticity equation is required. The longshore component of Equation (7) is

$$\frac{\partial \omega_y}{\partial t} = f\frac{\partial v}{\partial z} + g/\rho\frac{\partial \rho}{\partial x} + \frac{\partial}{\partial z}\left(1/\rho\frac{\partial \tau^x}{\partial z}\right) \tag{15}$$

which, as stated earlier, relates the rate of change of longshore vorticity to the tilting of planetary vorticity by vertical gradients in the longshore velocity, to cross shore density gradients, and to diffusion. Vertical integration of this equation is not as fruitful as in the case of the momentum equation, since not much is known about the vertical gradients of stress at the surface and bottom. Thus it appears more useful to attempt to discuss observations of cross shelf circulation in terms of Equation (15). So little is known about the diffusion term that it seems futile to consider it at all. In the surface boundary layer, the forcing of Equation (15) comes from the tilting of planetary vorticity. As soon as the water column is set in motion by longshore wind stress, vertical gradients in longshore velocity develop which lead to longshore vorticity. The circulation thus set up will distort lines of constant density until the thermal wind balance is reached. This process bears some similarity to the geostrophic balance in the cross shore momentum equation and it is no surprise that the time, t_i, required for the thermal wind balance to become dominant is $t_i \sim L(gH_i\Delta\rho/\rho)^{-\frac{1}{2}}$, or the time required for an internal wave to cross the shelf. With $\Delta\rho/\rho \sim 10^{-3}$, t_i varies from five hours to two days for shelf widths between 10 km and 100 km. Thus t_i is of the same order as the inertial period, as well as periods characteristic of the forcing.

With this value for t_i, Equation (15) suggests that the maximum longshore vorticity will be of order

$$\frac{\partial u}{\partial z} \sim \frac{L}{\lambda}\frac{\partial v}{\partial z} \tag{16}$$

where

$$\lambda = (gH_i\Delta\rho/\rho)^{\frac{1}{2}}/f$$

is the Rossby radius of deformation.

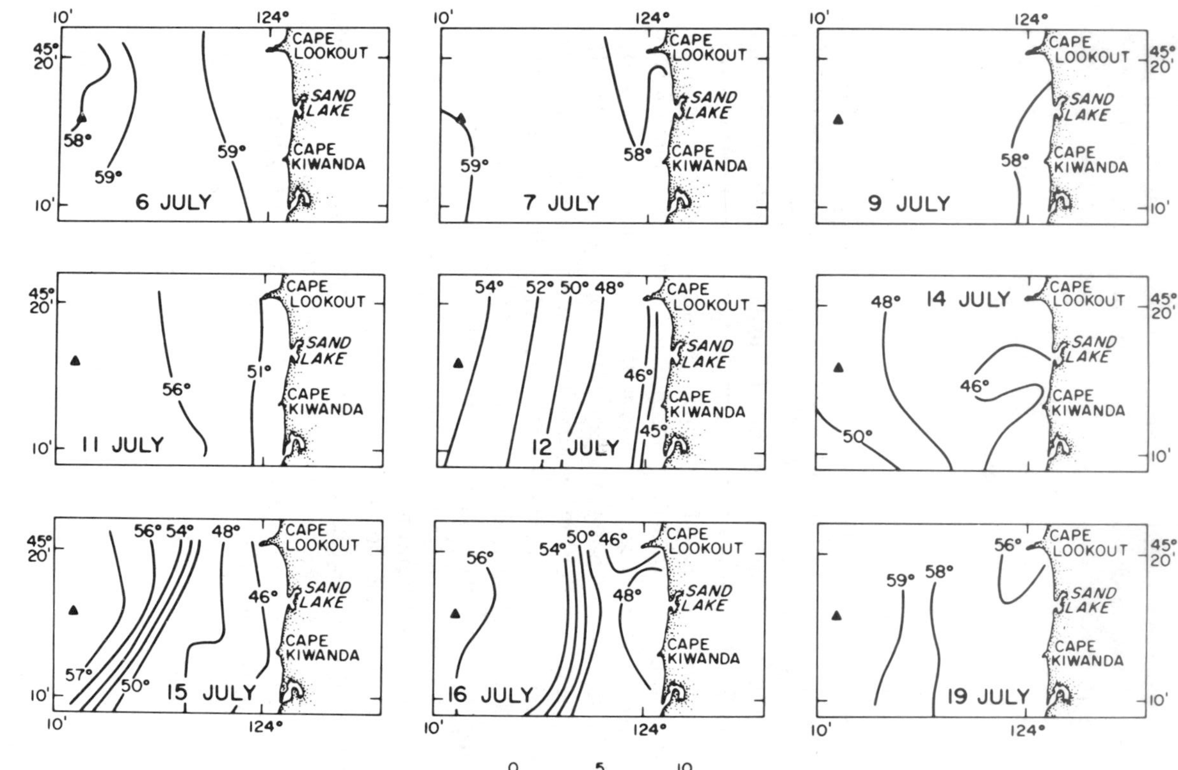

Figure 7 Time sequence of sea surface temperature measured off Oregon, July 6–July 19, 1973. From Halpern (1976).

The observations of Halpern (1976) in the upper boundary layer are useful in examining these questions. Those observations are reproduced in Figures 7 and 8, which show sea surface temperature and vertical profiles of u and v in the surface layer between 6 July and 19 July 1973.

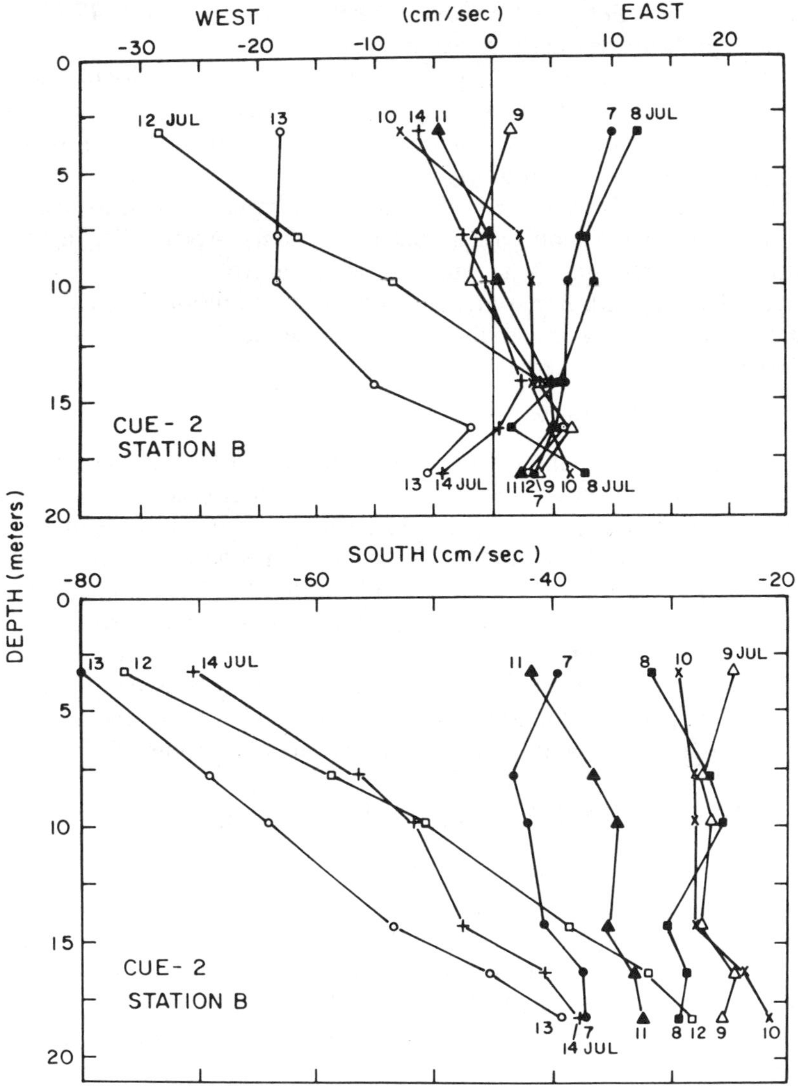

Figure 8 Vertical distribution of low-pass filtered currents at 0000 GMT on successive days between 7 and 14 July 1973. The instruments span the upper 20 m of a 100-m deep water column. From Halpern (1976).

During this time a strong upwelling event occurred off the Oregon coast. τ_s^y were sustained for two days, and the wind then decreased and reversed direction on 16 July. The sea surface temperature observations clearly reflect the upwelling. On 11 July cold water begins to appear at the coast, and on 12 July when the wind peaked, the surface cross shore density gradient, fairly uniform over 10 km, is $1.6 \times 10^{-6}\ \mathrm{s}^{-2}$. The large vertical gradient of longshore velocity ($\Delta v/\Delta z \sim -3 \times 10^{-2}\ \mathrm{s}^{-1}$) is coincident with the large vertical gradient in cross shore velocity ($\Delta u/\Delta z \sim -2 \times 10^{-2}\ \mathrm{s}^{-1}$). These terms are compared in Table 3, and clearly they are not inconsistent with the balance suggested by Equation (15). Indeed the agreement seems fortuitously good.

Huyer et al (1978) use the WISP data (Gilbert et al 1976) to test the thermal wind balance off Oregon during February–April 1975. In these observations continuous records of temperature and salinity were obtained, along with horizontal currents, from the moored array. Temperature and salinity are combined to compute density and the thermal wind balance is estimated as

$$\Delta v/\Delta z \sim -(g/\rho f)\Delta\rho/\Delta x$$

where Δz and Δx are the vertical and horizontal instrument separation. Thirteen pairs of time series of $\Delta v/\Delta z$ and $-\Delta\rho/\rho g f \Delta x$ are obtained from the various instrument pairs. In all but four of the thirteen cases, significant correlation is found between the time series. In the uncorrelated cases, the level of $\Delta v/\Delta z$ is small, so that the lack of correlation may be ascribed to signal-to-noise problems. Where the correlation is significant, the regression coefficient is of order one, ranging between 0.2 and 13.7. No attempts to compute lagged correlations or to include the rate of change of longshore vorticity are reported, but these results at least confirm the validity of the steady form of Equation (15).

3 BOUNDARY LAYERS AND MIXING

The major difficulties associated with surface and bottom boundary-layer observations are caused by the presence of high frequency surface waves. Because these occur at much higher frequencies than those considered here, they may be viewed as a noise, although, through nonlinear interactions, surface waves may have important effects on the lower frequency components of coastal circulation. Drift currents induced by waves are a most obvious example of this interaction. The recent models of bottom friction of J. D. Smith (1977) and Grant & Madsen (1979) illustrate the influence that surface waves can have on the frictional energy loss at lower frequencies. The greatest difficulty, however, is technical; surface waves

Table 3 Estimation of the terms in Equation (15) from the data of Halpern (1976)

Term in Equation (15)	$\frac{\partial \omega_y}{\partial t}$	$f\frac{\partial v}{\partial z}$	$g/\rho\frac{\partial \rho}{\partial x}$
Estimated as	$\frac{1}{t_i}\frac{\Delta u}{\Delta z}$	$f\frac{\Delta v}{\Delta z}$	$-g\beta\frac{\Delta T}{\Delta x}\Big\vert_{\text{surface}}$
Numerical value	$-2 \times 10^{-6}\ \text{s}^{-2}$	$-3 \times 10^{-6}\ \text{s}^{-2}$	$1.6 \times 10^{-6}\ \text{s}^{-2}$

induce large fluctuations in velocity with small means. These high frequency components place severe constraints on the linearity, frequency response, and cosine response required of instruments designed to measure proper averages of currents over many wave periods. It is only in the past few years that proper attention has been given such problems (McCullough 1974, 1977, 1978, Halpern et al 1974, Halpern 1977, Beardsley et al 1977a) in relation to the measurement of horizontal currents. As a result a new generation of current meters has become available, specifically designed to measure currents in the presence of a wave field. Where vertical currents are concerned, and more particularly in dealing with the estimation of the vertical components of Reynolds stress, the large wave-induced vertical velocities are a notable problem, and such estimations represent the current state of the art.

These considerations provide some excuse for the relative lack of information available where boundary layers are concerned, and underline the clever planning and careful execution of experiments performed to date. The current observations of Halpern (1976), reproduced in Figure 8, were obtained with Vector Averaging Current Meters in the upper 20 m of the water column. The large gradients observed attest to the importance of good resolution in the upper layer if proper dynamical inferences are to be made, and questions raised in the previous section will have to await further investigation of both boundary layers to be fully answered.

Since the effects of waves are relatively smaller near the bottom, more attention has been devoted to the experimental study of the bottom boundary layer. The dynamics of both surface and bottom boundary layers are usually taken to be turbulent Ekman dynamics (e.g. Wimbush & Munk 1971, Deardorf 1970, Caldwell, Van Atta & Helland 1972), although these must be somehow modified in shallow waters. Two obvious questions of importance here concern the vertical extent of the layers and the value of the shear stress at the boundary. Observations are not available to satisfactorily answer either question. With regard to vertical extent, most reports to date have emulated the analysis of

Wimbush & Munk (1971), which attempts to determine whether the logarithmic layer extends to some height over the bottom where current measurements are available. Measurements from two depths are used to infer the friction velocity $[u^* = (\tau/\rho)^{\frac{1}{2}}]$ using the usual logarithmic velocity distribution

$$u/u^* = \frac{1}{k} \log (z/z_0) \tag{17}$$

with an estimated z_0. Although z_0 is not known with any degree of certitude, in many cases the bottom will be smooth (but surface waves may make it appear rough) and $z_0 = 0.1\nu/u^*$, where ν is the kinematic viscosity. It is argued that because of the logarithmic dependence in Equation (17), u^* is relatively insensitive to the estimated z_0. Difficulties associated with these estimates are typified in Kundu's (1977) determination of the logarithmic thickness, δ_L, off Oregon. The computation is premised on the lowest observations of current (5 m off the bottom) in the logarithmic layer, but δ_L is then estimated to be only 2 m.

Bottom boundary-layer observations using profiling instruments are somewhat more successful in attempting to define the vertical structure. Weatherly & Van Leer (1977) report on temperature and current profiles obtained using a cyclosonde on the western Florida shelf, and show that stratification and lateral advection play significant roles in the boundary-layer structure. Weatherly & Martin (1978) suggest that the thickness of the bottom layer in the presence of stratification is given by

$$\delta_E = A\frac{u^*}{f}\,[1+(N^2/f^2)]^{-\frac{1}{4}}$$

where N represents the buoyancy frequency and A is a constant. This is based on the observations of Weatherly & Van Leer (1977) and on numerical model studies. The value δ_E converges to the usual form depending on u^*/f in the absence of stratification, and to the form $u^*/(Nf)^{\frac{1}{2}}$ given by Pollard, Rhines & Thomson (1973) for the upper mixed layer in the presence of strong stratification ($N \gg f$).

A detailed study of the bottom boundary layer off Oregon, reported by Caldwell & Chriss (1979), uses a heated thermistor traversed vertically. These results are characterized by extremely good vertical resolution of velocities very close to the bottom. A vertical profile of velocity from that study is reproduced in Figure 9. Most of the shear takes place in the lowest half centimeter, the viscous sublayer. The drag coefficient referred to the velocity 1 m above the bottom is 1.1×10^{-3}, which is close to previous estimates for smooth flow regimes.

Experimental studies of the atmospheric boundary layers, reviewed by Garratt (1977), have attempted to correlate mean flow properties (e.g. wind some distance over the ground) with the bottom stress (as estimated from Reynolds stress measurements). A number of similar investigations in the oceanic boundary layer are in preparation. Until these are concluded, the estimation of bottom stress by the usual quadratic drag law will remain a procedure for which it is difficult to estimate errors, although the experience in the atmospheric boundary layer suggests that such errors are not overly severe.

Of great importance to the biological food chain, mixing is another question that remains to be fully addressed. Few observations of vertical mixing in coastal waters are reported and this contrasts with the deeper ocean, where more effort has been expended, using instrumentation capable of resolving very fine scales, to resolve the broad band of mixing scales. The last decade has seen a number of laboratory investigations of this problem, reviewed by Turner (1973), Sherman, Imberger & Corcos (1978), and Linden (1979).

Physical and biological tracer studies support the contention that

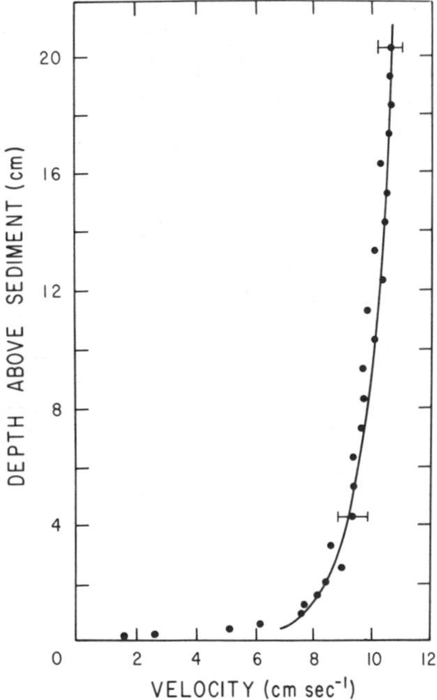

Figure 9 Vertical profile of velocity in the laminar sublayer (bottom four points) and the logarithmic layer, measured off the bottom in 200 m of water off Oregon. Redrawn from Caldwell & Chriss (1979).

upwelling and downwelling events are accompanied by mixing. In these events, the stabilizing effect of stratification cannot be neglected. The observations are further complicated to the extent that it is the combined effect of the vertical distribution of velocity and density that determines the amount of mixing that will occur. The mixed layer observations of Halpern (1976), which include current and temperature, can be used to estimate the vertical stability of the water column. The results are reproduced in Figure 10, in which the Richardson number between adjacent current meters is computed as

$$\mathrm{Ri} = -\frac{g}{\rho}\frac{\Delta P}{\Delta z}\Bigg/\left[\left(\frac{\Delta u}{\Delta z}\right)^2 + \left(\frac{\Delta y}{\Delta z}\right)^2\right].$$

The value Δz is the vertical separation between instruments, $\Delta\rho$ the density difference, and Δu and Δv are the cross shelf and longshore velocity differences. Clearly the actual Richardson number is only approximated by this estimate. Nonetheless Ri decreases by over two orders of magnitude during the upwelling event, reaching values well beneath 1. Although the evidence is not conclusive, it is strongly suggestive of vertical mixing.

As tropical storm Norman swept up along the coast of California in September 1978 (Winant 1980), a strong downwelling situation developed on the shelf, with the mixed layer deepening at a rate of order 0.5 cm s^{-1}. As the interface sweeps down by instruments, the time series they record may be interpreted as vertical profiles of temperature and current, if the

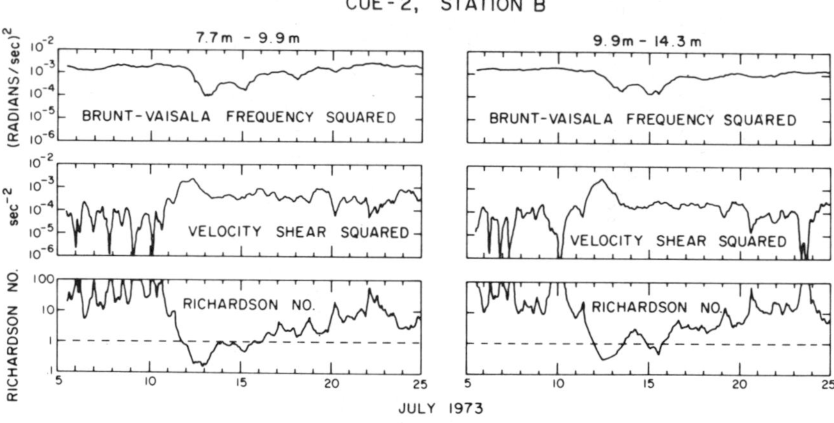

Figure 10 Time series of 25 hour running mean values of the depth averaged Brunt-Väisälä frequency squared, current shear squared, and Richardson number. Value of shear squared less than 10^{-6} was arbitrarily set at 10^{-6}; values of Richardson number greater than 100 were arbitrarily set at 100. From Halpern (1976).

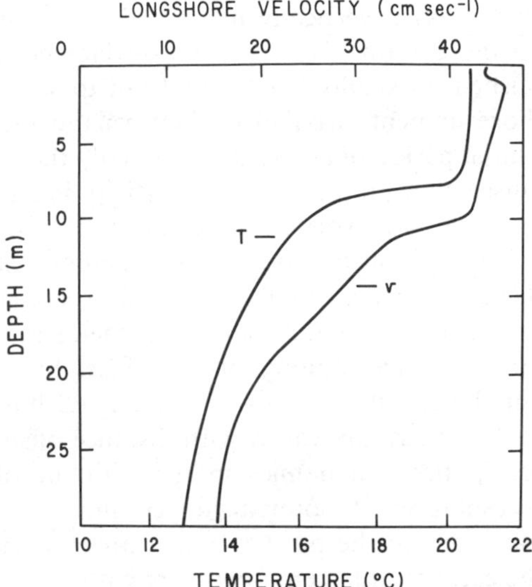

Figure 11 Vertical profiles of velocity and temperature observed during a strong downwelling event off Southern California, 5 September 1978. The velocity and temperature fields are assumed to be advected at 0.5 cm s^{-1} by a stationary current and temperature sensor.

vertical depth is computed as $z = wt$ where $w = 0.5$ cm s^{-1}. Smoothed profiles of temperature and current, obtained in this way, are reproduced in Figure 11. The particular instrument chosen here was near the middle of the water column in 60-m water depth. In this instance the overall temperature difference is 8°C and the overall velocity difference is 25 cm s^{-1}. The thickness of the mixing area is of order 5 m and thus an overall Richardson number is of order 1 or less. It is interesting to note that the maximum shear in velocity occurs in a layer approximately 1 m thick, as does the maximum temperature gradient, but that those layers are offset with respect to each other. The Richardson number has a strong maximum coincident with the maximum temperature gradient, but diminishes and reaches a minimum at the level of maximum velocity shear. The situation is reminiscent of laboratory experiments in which the turbulent mixing and shear are observed to be eating away at the interface, either from beneath or above.

FUTURE WORK

The observations reported to date support certain dynamical interpretations of the response of coastal waters to wind stress. Under the influence of longshore wind stress, a vertically integrated cross shore momentum balance between the Coriolis force and sea-surface elevation gradients is rapidly established. In shallow water and out to some undetermined depth the longshore momentum balance is between the surface and bottom stress, after an initial period of acceleration. Finally there exists evidence that is not inconsistent with a longshore vorticity balance between the rate of change of longshore vorticity, the tilting of planetary vorticity, and the tilting of lines of constant density in the cross shelf plane. To the extent that models of coastal circulation have proceeded well beyond these simple notions, it is clear that new experiments are required. Such experiments should resolve, among other things, the coastal bottom pressure field simultaneously with the current field, how the vertically integrated cross shelf mass flux varies with distance offshore, and finally how the inshore frictional dynamics merge with the offshore Ekman dynamics. The resolution of boundary-layer and internal shear-layer dynamics is clearly one of the most pressing areas where observations are required. The early results mentioned here with regard to mixing are exciting not only because they have significant biological implications, but also because they suggest that a confirmation, on very large scales, of laboratory investigations of the stability of stratified flows may be possible.

ACKNOWLEDGMENTS

The support of the National Science Foundation under Grant No. OCE76-19295 is gratefully acknowledged.

Literature Cited

Allen, J. S. 1973. Upwelling and coastal jets in a continuously stratified ocean. *J. Phys. Oceanogr.* 3(3):245

Allen, J. S., Kundu, P. K. 1978. On the momentum, vorticity and mass balance on the Oregon Shelf. *J. Phys. Oceanogr.* 8(1):13–27

Beardsley, R. C., Butman, B. 1974. Circulation on the New England continental shelf: Response to strong winter storms. *Geophys. Res. Lett.* 1(4):181–84

Beardsley, R. C., Winant, C. D. 1979. On the mean circulation in the mid-Atlantic bight. *J. Phys. Oceanogr.* 9(3):612–19

Beardsley, R. C., Boicourt, W. C., Hansen, D. V. 1976. Physical Oceanography of the Middle Atlantic Bight. *Limnol. & Oceanogr. Spec. Symp. Ser.* 2:20–34. Lawrence, Kansas: Am. Soc. Limnol. & Oceanogr.

Beardsley, R. C., Boicourt, W. C., Huff, L. C., Scott, J. 1977a. CMICE 76: A current meter intercomparison experiment conducted off Long Island in February–March 1976. *Tech. Rep. WHOI 77-62.* Woods Hole, Mass.: Woods Hole Oceanogr. Inst. 123 pp.

Beardsley, R. C., Mofjeld, H., Wimbush, M., Flagg, C. N., Vermersch, J. A. Jr. 1977b. Ocean tides and weather-induced bottom pressure fluctuations in the Middle Atlantic Bight. *J. Geophys. Res.* 82(21):3175–82

Bennett, J. R., Magnell, B. A. 1979. A dynamical analysis of currents near the New Jersey coast. *J. Geophys. Res.* 84(C3):1165–75

Boicourt, W. C., Hacker, P. W. 1976. Circulation on the Atlantic continental shelf of the United States, Cape May to

Cape Hatteras. *Mem. Soc. R. Sci. Liège 6e Ser.* X: 187–200

Brooks, D. A., Mooers, C. N. K. 1977. Wind-forced continental shelf waves in the Florida current. *J. Geophys. Res.* 82(18): 2569–76

Brown, W. S., Beardsley, R. C. 1978. Winter circulation in the Western Gulf of Maine: Part 1: cooling and water mass formation. *J. Phys. Oceanogr.* 8(2): 265–77

Butman, B., Noble, M., Folger, D. W. 1979. Long term observations of bottom current and bottom sediment movement on the mid-Atlantic continental shelf. *J. Geophys. Res.* 84(C3): 1187–1205

Caldwell, D. R., Chriss, T. M. 1979. The viscous sublayer at the sea floor. *Science*. Submitted

Caldwell, D. R., Van Atta, C. W., Helland, K. N. 1972. A laboratory study of the turbulent Ekman layer. *Geophys. Fluid Dyn.* 3(2): 125

Clarke, A. J. 1977. Observational and numerical evidence for wind-forced coastal trapped long waves. *J. Phys. Oceanogr.* 7(2): 231–46

Collins, C. A., Pattullo, J. G. 1970. Ocean currents above the continental shelf off Oregon as measured with a single array of current meters. *J. Mar. Res.* 28(1): 51–68

Csanady, G. T. 1976. Mean circulation in shallow seas. *J. Geophys. Res.* 81(30): 5389–99

Csanady, G. T. 1978. The arrested topographic wave. *J. Phys. Oceanogr.* 8(1): 47–62

Cutchin, D. L., Smith, R. L. 1973. Continental shelf waves: Low-frequency variations in sea level and currents over the Oregon continental shelf. *J. Phys. Oceanogr.* 3(1): 73–82

Daddio, E., Wiseman, W. J., Murray, S. P. 1978. Inertial current over the inner shelf near 30°N. *J. Phys. Oceanogr.* 8(4): 728–33

Deardorff, J. W. 1970. A three dimensional numerical investigation of the idealized planetary boundary layer. *Geophys. Fluid Dyn.* 1: 377

Ekman, V. W. 1905. On the influence of the earth's rotation on ocean currents. *Ark. Mat. Astron. Fys.* 2(11): 1

Forristal, G. Z., Hamilton, R. C., Cardone, V. J. 1977. Continental shelf currents in tropical storm Delia: Observations and theory. *J. Phys. Oceanogr.* 7(4): 532–46

Garratt, J. R. 1977. Review of drag coefficients over ocean and continents. *Mon. Weather Rev.* 105: 915–29

Garrett, C. 1979. Topographic Rossby waves off east Australia: Identification and role in shelf circulation. *J. Phys. Oceanogr.* 9(2): 244–53

Gilbert, W. E., Huyer, A., Barton, E. D., Smith, R. L. 1976. Physical oceanographic observations off the Oregon coast, 1975. *WISP & UP-75, Ref. 76-4*. Oregon State Univ., Corvallis: Sch. Oceanogr. 189 pp.

Gonella, J. 1972. A rotary-component method for analysing meteorological and oceanographic vector time series. *Deep-Sea Res.* 19(12): 833–46

Grant, W. D., Madsen, O. S. 1979. Combined wave and current interaction with a rough bottom. *J. Geophys. Res.* 84(C4): 1797–1808

Halpern, D. 1971. Semidurnal internal tides in Massachusetts Bay. *J. Geophys. Res.* 76(27): 6573–84

Halpern, D. 1976. Structure of a coastal upwelling event observed off Oregon during July 1973. *Deep-Sea Res.* 23(6): 495

Halpern, D. 1977. Review of intercomparisons of moored current measurements. *Oceans 77 Conf. Rec. Comb. Meet. IEEE Conf. Eng. Ocean Environ. 13th Ann. Meet. Mar. Technol. Soc.* 2: 46D1–46D6

Halpern, D., Pillsbury, R. D., Smith, R. L. 1974. An intercomparison of three current meters operated in shallow water. *Deep-Sea Res.* 21(6): 489–97

Hamon, B. V., Godfrey, J. S., Greig, M. A. 1975. Relation between mean sea level, current and wind stress on the east coast of Australia. *Aust. J. Mar. Freshwater Res.* 26: 389–403

Haury, L. R., Birscoe, M. H., Orr, M. H. 1979. Tidally-generated internal wave packets in Massachusetts Bay. *Nature* 278: 312–16

Hayes, S. P. 1979. Variability of current and bottom pressure across the continental shelf in the northeast Gulf of Alaska. *J. Phys. Oceanogr.* 9(1): 88

Hayes, S. P., Schumacher, J. D. 1976. Description of wind, current, and bottom pressure variations on the continental shelf in the northeast Gulf of Alaska from February to May 1975. *J. Geophys. Res.* 81(36): 6411–19

Hendershott, M. C., Rizzoli, P. 1976. The winter circulation of the Adriatic Sea. *Deep-Sea Res.* 23(5): 353–70

Hickey, B. M. 1979. The California current system—Hypotheses and facts. *Prog. Oceanogr.* In press

Huyer, A., Hickey, B. M., Smith, J. D., Smith, R. L., Pillsbury, R. D. 1975. Alongshore coherence at low frequencies in currents observed over the continental shelf off Oregon and Washington. *J. Geophys. Res.* 80(24): 3495–3505

Huyer, A., Gagner, J., Huggett, S. 1976.

Observations from current meters moored over the continental shelf off Vancouver island, 28 November 1974 to 8 April 1975, and related oceanographic and meteorological data. *Tech. Rep. No. 4*. Environment Canada, Fisheries & Marine Service

Huyer, A., Smith, R. L., Sobey, E. J. C. 1978. Seasonal differences in low-frequency current fluctuations over the Oregon continental shelf. *J. Geophys. Res.* 83(C10): 5077–89

Kroll, J., Niiler, P. P. 1976. The transmission and decay of barotropic topographic Rossby waves incident on a continental shelf. *J. Phys. Oceanogr.* 6(4): 432–50

Kundu, P. K. 1977. On the importance of friction in two typical continental waters: Off Oregon and Spanish Sahara. In *Bottom Turbulence, Proc. 8th Liège Colloq. Ocean Hydrodyn.*, ed. J. Nihoul. New York: Elsevier. 309 pp.

Lee, C. Y., Beardsley, R. C. 1974. The generation of long, nonlinear internal waves in a weakly stratified shear flow. *J. Geophys. Res.* 79(3): 453–62

Lee, T. N. 1975. Florida current spin-off eddies. *Deep-Sea Res.* 22(11): 753–66

Lee, T. N., Mayer, D. A. 1977. Low-frequency current variability and spin-off eddies along the shelf off southeast Florida. *J. Mar. Res.* 35(1): 193–220

Linden, P. F. 1979. Mixing in stratified fluids. *Geophys. Astrophys. Fluid Dyn.* Submitted

Maxworthy, T. 1978. A note on the internal solitary waves produced by tidal flow over a three-dimensional ridge. *J. Geophys. Res.* 84(C1): 338–46

McCullough, J. R. 1974. *In search of moored current sensors.* Paper presented at 10th Ann. Buoy Conf., Mar. Technol. Soc., Washington DC

McCullough, J. R. 1977. Problems in measuring currents near the ocean surface. *Oceans 77 Conf. Rec. Comb. Meet. IEEE Conf. Eng. Ocean Environ. 13th Ann. Meet. Mar. Technol Soc.* 2: 46A1–46A7

McCullough, J. R. 1978. Near-surface ocean current sensors: Problems and performance. *Proc. Working Conf. Current Measurements, Tech. Rep. DEL-SG-3-78.* Newark, Del.: Coll. Mar. Studies, Univ. Del. 372 pp.

Munk, W. H. 1950. On the wind-driven ocean circulation. *J. Meteorol.* 7: 79–93

Munk, W. H., Snodgrass, F., Carrier, G. 1956. Edge waves on the continental shelf. *Science* 123: 127–37

Munk, W. H., Snodgrass, F., Gilbert, F. 1964. Long waves on the continental shelf: An experiment to separate trapped and leaky modes. *J. Fluid Mech.* 20(4): 529–54

Munk, W. H., Snodgrass, F., Wimbush, M. 1970. Tides offshore: Transition from California coastal to deep sea waters. *Geophys. Fluid Dyn.* 1(1, 2): 161–236

Murray, S. P. 1975. Trajectories and speeds of wind-driven currents near the coast. *J. Phys. Oceanogr.* 5(3): 347

Niiler, P. P. 1975. A report on the continental shelf circulation and coastal upwelling. *Rev. Geophys. Space Phys.* 13(3): 609–59

Noble, M., Butman, B. 1979. Low frequency wind-induced sea level oscillations along the east coast of North America. *J. Geophys. Res.* 84(C6): 3227–36

Pedlosky, J. 1974. Longshore currents, upwelling and bottom topography. *J. Phys. Oceanogr.* 4(2): 214–26

Petrie, B. 1975. M2 surface and internal tides on the Scotian shelf and slope. *J. Mar. Res.* 33(3): 303–23

Petrie, B., Smith, P. C. 1977. Low frequency motions on the Scotian shelf and slope. *Atmosphere* 15: 117–40

Pillsbury, R. D., Bottero, J. S., Still, R. E., Gilbert, W. E. 1974. A compilation of observations from moored current meters, Vol. 7, Oregon continental shelf, July–August 1973. *Data Rep. 58 Ref. 74-7.* Corvallis: Sch. Oceanogr., Oregon State Univ. 87 pp.

Pollard, R. T., Rhines, P. B., Thomson, R. O. R. Y. 1973. The deepening of the wind mixed layer. *Geophys. Fluid Dyn.* 4(4): 381–404

Rattray, M. 1960. On the coastal generation of internal tides. *Tellus* XII(1): 54–62

Scott, J. T., Csanady, G. T. 1976. Nearshore currents off Long Island. *J. Geophys. Res.* 81(30): 5401–9

Semtner, A. J., Mintz, Y. 1977. Numerical simulation of the Gulf Stream and mid-ocean eddies. *J. Phys. Oceanogr.* 7: 208–30

Sherman, F. S., Imberger, J., Corcos, G. M. 1978. Turbulence and mixing in stably stratified waters. *Ann. Rev. Fluid Mech.* 10: 267–88

Smith, J. D. 1977. Modelling of sediment transport on continental shelves. In *The Sea* 6: 539–77. New York: Wiley

Smith, J. D., Long, C. E. 1976. The effect of turning in the bottom boundary layer on continental shelf sediment transport. *Mem. Soc. R. Sci. Liège 6e Ser.* X: 369–96

Smith, N. P. 1978. Longshore currents on the fringe of hurricane Anita. *J. Geophys. Res.* 83(C12): 6047–51

Smith, P. C. 1978. Low-frequency fluxes of momentum, heat, salt and nutrients at the edge of the Scotian shelf. *J. Geophys. Res.* 83(C8): 4079

Smith, R. L. 1974. A description of current,

wind and sea level variations during coastal upwelling off the Oregon coast, July–August 1972. *J. Geophys. Res.* 79(3): 435–43

Smith, R. L. 1978. Poleward propagating perturbations in currents and sea levels along the Peru coast. *J. Geophys. Res.* 83(C12): 6083–92

Sverdrup, H. U., Johnson, M. W., Fleming, R. H. 1942. *The Oceans*. New York: Prentice-Hall. 1087 pp.

Turner, J. S. 1973. *Buoyancy Effects in Fluids*. Cambridge Univ. Press. 367 pp.

Vermersch, J. A., Beardsley, R. C., Brown, W. S. 1979. Winter circulation in the Western Gulf of Maine: Part 2: current and pressure observations. *J. Phys. Oceanogr.* 9(4): 768–84

Wang, D. P. 1979. Low frequency sea level variability on the middle Atlantic bight. *J. Mar. Res.* In press

Wang, D. P., Mooers, C. N. K. 1977. Long coastal-trapped waves off the west coast of the United States, summer 1973. *J. Phys. Oceanogr.* 7(6): 856–64

Weatherly, G. L., Martin, P. J. 1978. On the structure and dynamics of the oceanic bottom boundary layer. *J. Phys. Oceanogr.* 8(4): 557–70

Weatherly, G. L., Van Leer, J. 1977. On the importance of stable stratification to the structure of the bottom boundary layer on the western Florida shelf. In *Bottom Turbulence, Proc. 8th Liège Colloq. Ocean Hydrodyn.*, ed. J. Nihoul. New York: Elsevier

Welander, P. 1961. Numerical predictions of storm surges. *Adv. Geophys.* 8: 315

Wimbush, M., Munk, W. H. 1971. The benthic boundary layer. In *The Sea*. Part 1, 4: 731–58. New York: Wiley

Winant, C. D. 1979. Coastal current observations. *Rev. Geophys. Space Phys.* 17(1): 89–98

Winant, C. D. 1980. Downwelling events over the Southern California shelf. *J. Phys. Oceanogr.* Submitted

Winant, C. D., Beardsley, R. C. 1979. A comparison of shallow currents induced by wind stress. *J. Phys. Oceanogr.* 9(1): 218–20

Winant, C. D., Olson, J. R. 1976. The vertical structure of coastal currents. *Deep-Sea Res.* 23(10): 925–36

Wunsch, C. 1975. Internal tides in the ocean. *Rev. Geophys. Space Phys.* 13(1): 167–82

Ann. Rev. Fluid Mech. 1980. 12: 303–34

INSTABILITIES OF WAVES ON DEEP WATER

Henry C. Yuen and Bruce M. Lake
Fluid Mechanics Department, TRW Defense and Space Systems Group,
One Space Park, Redondo Beach, California 90278

INTRODUCTION

Waves on deep water provide one of the most vivid examples of wave motion found in nature and men have long sought to understand this familiar yet complex phenomenon. It is not surprising, therefore, that the problem of the generation of deep-water waves by a local disturbance was the first hydrodynamic problem to be investigated systematically through the general equations of fluid motion; it was posed as a prize topic by the French Academy in 1816, and solved in the same year by Poisson and Cauchy. The Cauchy-Poisson solution describing the radiation of waves from a local source served as one of the most outstanding examples of mathematics applied to a widely observed physical process.

A comprehensive treatise on the general subject of waves on deep water was published by Stokes in 1849. The mathematical techniques introduced therein became the cornerstone for modern theories of linear and nonlinear dispersive waves. It is noteworthy that this pioneering theoretical study was motivated by physical observations (Russell 1844). The continued close coupling between experimental and theoretical efforts is probably most responsible for the widespread interest and significant progress in the subject over the past century.

For some, the novelty of the Stokes expansion raised questions of convergence in the mathematical sense (Burnside 1916, Nekrasov 1919, Levi-Civita 1925), while for others (Michell 1893, Wilton 1913, Rayleigh 1917), the existence of finite-amplitude steady periodic waves seemed practically certain, and determining the profiles of very steep waves presented the real challenge. Until recently, the stability of a train of steady periodic deep-water waves, regardless of steepness, remained unquestioned. Thus, the theoretical discovery by Lighthill (1965) using

0066-4189/80/0115-0303$01.00

Whitham's theory, and the theoretical and experimental confirmation and extension by Benjamin & Feir (1967), that weakly nonlinear deep-water wavetrains are unstable to modulational perturbations came as a surprise to workers in the field and aroused great interest in the problem of the unsteady time evolution of deep-water wavetrains and packets. This discovery coincided with a remarkable advance in the theory of nonlinear partial differential equations. Specifically, it was found that the time evolution of weakly nonlinear deep-water wavetrains obeys the nonlinear Schrödinger equation (Zakharov 1968, Hasimoto & Ono 1972, Davey 1972), which was solved exactly by Zakharov & Shabat (1971) using the inverse scattering technique of Gardner et al (1967). The exact solution predicted the existence of deep-water wave envelope solitons and this prediction was verified by experiments (Yuen & Lake 1975). The experimental observations then suggested recurrence in the evolutionary process of an unstable wavetrain, which was confirmed by numerical solutions of the nonlinear Schrödinger equation (Lake et al 1977). Once again, theoretical and experimental efforts have mutually stimulated and benefited one another.

In this article, we provide a brief description of some interesting properties of deep-water waves that have been discovered as investigators have sought an increasingly complete understanding of the instabilities of waves on deep water. The presentation first highlights the developments that led to the present stage of understanding of the weakly nonlinear wavetrain, and then examines phenomena associated with very steep waves, with emphasis on some significant progress made quite recently. In conclusion, we present a brief summary and point to several areas presently undergoing active investigations which appear likely to produce significant new results in the near future.

The Governing Equations

The equations taken for the study of gravity waves on deep water are the Euler equations for an irrotational flow of an incompressible inviscid fluid with a free surface:

$$\Delta\phi = 0 \qquad -\infty < z < \eta(x,y,t) \tag{1}$$

$$\left.\begin{aligned} &\frac{\partial\phi}{\partial t} + \tfrac{1}{2}(\nabla\phi)^2 + gz = p \\ &\frac{\partial\eta}{\partial t} + \nabla\phi\cdot\nabla\eta - \frac{\partial\phi}{\partial z} = 0 \end{aligned}\right\} \qquad z = \eta(x,y,t) \tag{2}$$

$$\frac{\partial\phi}{\partial z} \to 0 \qquad z \to -\infty, \tag{3}$$

where ϕ is the velocity potential, η is the free surface, g is the gravitational acceleration, p is an external surface pressure, the horizontal coordinates are (x,y) and the vertical coordinate is z, pointing upwards; $\Delta = \partial^2/\partial x^2 + \partial^2/\partial y^2 + \partial^2/\partial z^2$ is the Laplacian operator, and $\nabla = (\partial/\partial x, \partial/\partial y)$ is the horizontal gradient operator. The density is normalized to unity with no loss of generality. For a wave to be considered a deep-water wave, the ratio of its wavelength λ to the water depth must be small. For surface tension to be negligible (i.e. for a wave to be considered a gravity wave), the wavelength must be substantially longer than 1.7 cm, at which point surface tension and gravitational effects are equal. Strictly speaking, water is not inviscid. However, viscosity is most effective only for small-scale motions and is negligible for most of the gravity-wave phenomena considered here. Therefore, potential flow provides a good approximation.

The dispersion relation for the linearized problem can be obtained by neglecting the nonlinear terms in (2) and applying the resulting linear boundary conditions at the undisturbed surface $z = 0$ with periodic conditions in the horizontal coordinates. We obtain a spatially periodic solution,

$$\eta(\mathbf{x},t) = a \cos(\mathbf{k}\cdot\mathbf{x} - \omega t), \tag{4}$$

$$\phi = \frac{ga}{\omega} \sin(\mathbf{k}\cdot\mathbf{x} - \omega t) e^{|\mathbf{k}|z}, \tag{5}$$

where $\mathbf{x} = (x,y)$ is the horizontal spatial vector, $\mathbf{k} = (k_x,k_y)$ is the wave vector, and ω the wave frequency in radians. The linear dispersion relation, which relates the frequency to the wave vector, is given as

$$\omega = \sqrt{g|\mathbf{k}|}. \tag{6}$$

The phase velocity, which gives the speed of advance of the individual crests, is

$$\mathbf{C} = \frac{\omega}{\mathbf{k}} = \frac{\omega}{|\mathbf{k}|^2}(k_x,k_y) \tag{7}$$

and the group velocity, which describes the speed and direction of energy propagation, is

$$\mathbf{C}_g = \frac{\partial \omega}{\partial \mathbf{k}} = \left(\frac{\partial \omega}{\partial k_x}, \frac{\partial \omega}{\partial k_y}\right). \tag{8}$$

The fact that $\mathbf{C}$ is a nontrivial function of $\mathbf{k}$ reflects the dispersive nature of the system.

Henceforth, unless stated otherwise, we shall be discussing one-dimensional waves that are independent of y. For simplicity of notation, we shall denote k_x by k in these one-dimensional problems.

THE WEAKLY NONLINEAR WAVETRAIN

Concept of a Wavetrain

The expression for the free surface given by Equation (4) describes a linear wavetrain. Despite its obvious idealization, including perfect periodicity and infinite extent, the linear wavetrain occupies a uniquely important position in the study of properties of linear wave systems. Perturbations are first performed on the single wavetrain, and their effects on an entire wave system are determined by superposition. Unfortunately, the principle of linear superposition no longer holds in nonlinear systems. Even so, the wavetrain still provides the simplest entity that exhibits much of the richness of nonlinear wave systems.

A wavetrain can be characterized by three parameters: its amplitude a, wavenumber k, and frequency ω. Generally, its evolution is described by equations for the conservation of wavenumber, the conservation of wave action, and the dispersion relation. For weakly nonlinear waves, they are respectively

$$\frac{\partial k}{\partial t}+\frac{\partial \omega}{\partial k}=0, \tag{9}$$

$$\frac{\partial a^2}{\partial t}+\frac{\partial}{\partial x}\left(\frac{\partial \omega}{\partial k}\,a^2\right)=0, \tag{10}$$

$$\omega=\sqrt{gk}\,(1+\tfrac{1}{2}k^2a^2). \tag{11}$$

These are the leading-order equations derivable from the averaged variational principle of Whitham (1965, 1967, 1970) and are sometimes known as Whitham's conservation equations. Equivalent systems can be derived by a variety of methods, including the multiple-scales method (Chu & Mei 1970).

Note that the dispersion relation contains a nonlinear correction that was first obtained by Stokes (1849). The absolute magnitude of the correction is small, and does not exceed 11% even for the steepest possible waves. Nevertheless, its existence couples the three equations nonlinearly, and leads to some remarkable phenomena during the evolution of the weakly nonlinear wavetrain.

Modulational Instability

Equations (9)–(11) were used by Lighthill (1965) to study the evolution of weakly nonlinear waves on deep water. Two sets of initial conditions were examined: a wave packet with a Gaussian envelope and a slightly modulated Stokes wavetrain. In the case of the wave packet, the envelope developed a cusp at its peak within a finite time. For the wavetrain, the modulation grew and again the envelope developed into a cusped shape. Smooth solutions ceased to exist beyond this stage of evolution, and the time taken to reach the singularities was found to be inversely proportional to the steepness of the waves.

These results were the first indications of modulational instabilities for weakly nonlinear waves on deep water. A detailed perturbation analysis of the uniform wavetrain performed on the Euler equations (1)–(3) was presented by Benjamin & Feir (1967) that confirmed the instability and extended the results to a wider range of perturbation wavelengths. It was found that a uniform wavetrain with amplitude a_0, wavenumber k_0, and frequency ω_0 is unstable to perturbations with wavenumber K in the range

$$0 < K < 2\sqrt{2}\, k_0^2 a_0. \tag{12}$$

The maximum instability occurs when $K = K_{\max}$, where

$$K_{\max} = 2k_0^2 a_0, \tag{13}$$

and the maximum growth rate $(\operatorname{Im} \Omega)_{\max}$ is

$$(\operatorname{Im} \Omega)_{\max} = \tfrac{1}{2}\omega_0 k_0^2 a_0^2. \tag{14}$$

Benjamin & Feir (1967) also reported experimental data in fairly good agreement with the predictions regarding the wavenumber and growth rate of the instability. An example of such a modulation growing on an initially uniform wavetrain is shown in Figure 1. The waveforms in the figure correspond to the typical experimental situation, where the wavetrain evolves as it propagates along a distance x, whereas the theory is usually formulated for evolution in time. The spatial and temporal cases can be related simply by transforming coordinates using the group velocity of the carrier wave.

The Nonlinear Schrödinger Equation

Identification of the instability led to the question of the evolution of the unstable nonlinear wavetrain. The analysis of Benjamin & Feir was valid only for the initial growth period, and Whitham's equations, as

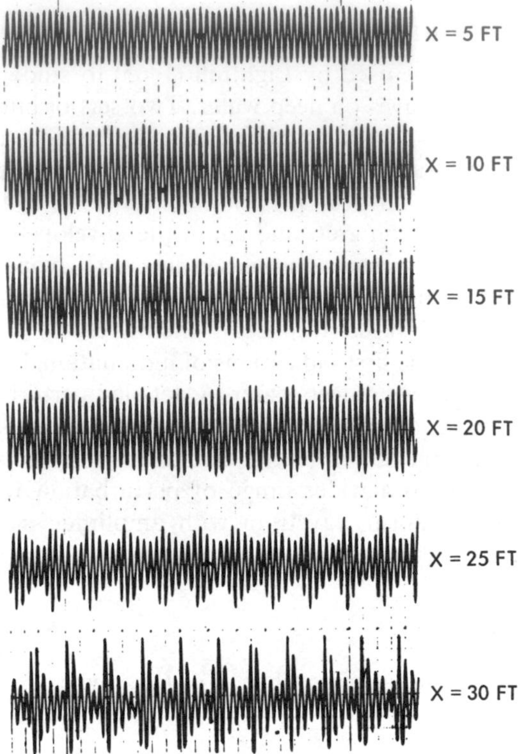

Figure 1 Wave-amplitude measurements showing the onset and growth of the Benjamin-Feir modulational instability on an initially unmodulated nonlinear wavetrain. Each record shows the wave amplitude measured versus time at fixed locations along a wave-tank. The distance x is measured from the wavemaker. Although the output of every gauge was linearly proportional to wave amplitude, each gauge had a slightly different sensitivity. As a result, the oscillograph record of the output from any given gauge is an accurate representation of the waveform of the waves measured by that gauge, but the absolute magnitudes of the waveforms recorded by different gauges cannot be used to compare actual wave amplitudes at different stations unless differences in probe sensitivity are taken into account.

applied by Lighthill, yielded singularities in the solutions within a finite time. Effort was therefore directed toward derivation of equations that would be valid for large times. Chu & Mei (1970, 1971) used the multiple-scales method to show that an additional correction term, effective when the curvature of the envelope becomes large, must be included in the dispersion relation. They showed further that the singularity Lighthill encountered can be eliminated with the inclusion of the "curvature dispersion" term, but their computation met with another numerical difficulty at a later time, the physical significance of which was unclear. Equivalent correction terms in the dispersion relation were also obtained by Benney & Roskes (1969) and Hayes (1970a,b, 1973). Meanwhile, Hasimoto & Ono (1972) used the derivative-expansion method to show that the evolution of a nonlinear deep-water wavetrain obeys the nonlinear Schrödinger equation, and they recovered the results of Benjamin & Feir by a stability analysis of its uniform solution. Yuen & Lake (1975) showed that the nonlinear Schrödinger equation is a consequence of Whitham's equations when second variations in the dispersion relation are included. Soon it was realized that all methods, under the same set of assumptions, lead to the same nonlinear Schrödinger equation for the complex wavetrain envelope $A(x,t) = a(x,t)e^{i\theta(x,t)}$:

$$i\left(\frac{\partial A}{\partial t} + \frac{\omega_0}{2k_0}\frac{\partial A}{\partial x}\right) - \frac{\omega_0}{8k_0^2}\frac{\partial^2 A}{\partial x^2} - \tfrac{1}{2}\omega_0 k_0^2 |A|^2 A = 0 \tag{15}$$

where $\partial\theta/\partial t = \omega_0 - \omega$, $\partial\theta/\partial x = k - k_0$, and the free surface of water is $\eta(x,t) = \mathrm{Re}\{A \exp i(k_0 x - \omega_0 t)\}$.

This equation was first discussed in Benney & Newell (1967) in the general context, but, unknown to many workers in the field at the time, it (and its two-space-dimensional extension) was originally derived for deep-water waves by Zakharov (1968) via a spectral method. Identification of the nonlinear Schrödinger equation as the correct equation for describing the time evolution of weakly nonlinear wavetrains did not immediately resolve all the questions regarding the nature of wavetrain evolution, but it was a major step in that direction. For reasons of theoretical and experimental advantage, the next step actually involved consideration of finite wave packets rather than continuous wavetrains.

Envelope Solitons

One of the most important characteristics of the nonlinear Schrödinger equations is that it can be solved exactly for initial conditions that decay sufficiently rapidly as $|x| \to \infty$. This was done by Zakharov & Shabat (1971) using what was then the newly discovered inverse scattering method (Gardner et al 1967). They showed that any initial wave packet eventually evolves into a number of "envelope solitons" and a dispersive tail. The bulk of the energy is contained in the solitons, which propagate with permanent form once produced. Solitons also survive interactions with other solitons or wave packets. Since the nonlinear Schrödinger equation describes the envelope of deep-water waves with a carrier frequency, the theory predicts the existence of packets of deep-water waves with soliton properties. The existence of these envelope soliton properties would hardly have been expected on the basis of experience with linear deep-water wave systems in which wave components are uncoupled and highly dispersive.

The theoretical predictions of soliton properties were confirmed by experiments (an example is shown in Figure 2), and numerical solutions of the nonlinear Schrödinger equation were found to compare well with experimentally measured wave-envelope profiles (Yuen & Lake 1975). In the case of wave packets and solitons, the description of nonlinear deep-water wave dynamics provided by the nonlinear Schrödinger

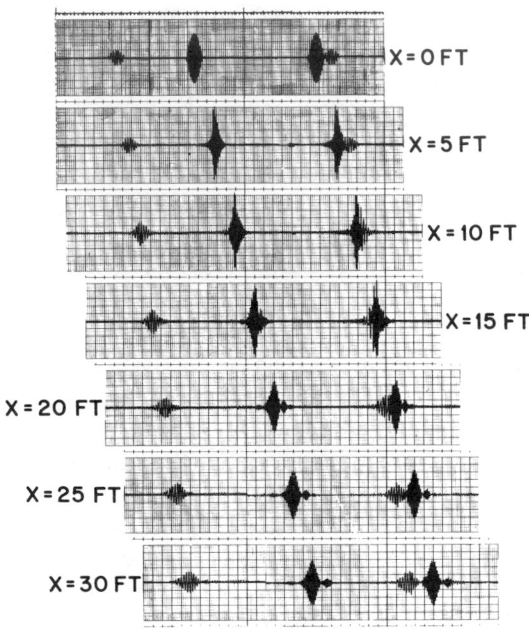

Figure 2 Example of one wave pulse overtaking and passing through another wave pulse. *Left-hand trace*: first pulse alone $\omega_0 = 1.6$ Hz, initial $(ka)_{\max} \simeq 0.1$, 6-cycle pulse. *Center trace*: second pulse alone $\omega_0 = 3$ Hz, initial $(ka)_{\max} \simeq 0.2$, 12-cycle pulse that disintegrates into two solitons. *Right-hand trace*: the two pulses together. At $x = 0$, the 3-Hz pulse is generated ahead of the 1.6-Hz pulse. As they propagate to $x = 30$ ft, the faster 1.6-Hz pulse passes through the 3-Hz pulse and emerges ahead of it while the 3-Hz pulse is disintegrating into two solitons. Note that at 30 ft the pulses emerge from the interaction relatively unchanged from the forms they have at $x = 30$ ft when each evolves in the absence of the other. Strictly speaking, this type of interaction between pulses of different carrier-wave frequencies cannot be described by a single equation such as (15). It has been shown, however, by Oikawa & Yajima (1974) that solitons with different carrier frequencies also survive interactions with no permanent change other than position and phase shifts. The governing equations for these multiwave systems can likewise be obtained by Whitham's method by use of more than one phase function.

equation was found to be both qualitatively and quantitatively correct. Although the practical significance of these envelope solitons is questionable because of their one-dimensional nature, their mathematical and physical existence increased confidence in the nonlinear Schrödinger equation description and provided a dramatic demonstration of the potential significance of nonlinear effects on the dynamics of deep-water waves. The soliton investigations are also another example of the close relationship between experiment and theory that has become almost a tradition in this subject, and that has recently led to new insights into phenomena in more realistic wave systems, particularly, phenomena influenced by the degree of coherence of the waves.

Long-Time Evolution of an Unstable Wavetrain

There have been a variety of propositions regarding the possible end-state of a wavetrain undergoing modulational instability. These range from a complete breakdown of the wavetrain with equipartition of energy among all modes, to a train of stable soliton envelopes (see, for example, Benjamin 1967, Hasselmann 1967, Hasimoto & Ono 1972). The answer, however, turned out to be somewhat a surprise.

Experiments in the laboratory by Lake et al (1977) indicated that the unstable modulations grew to a maximum and then subsided (Figure 3). Furthermore, at some stage of the evolution the wavetrain actually became nearly uniform again. Numerical computations using the nonlinear Schrödinger equation, which proved to be satisfactory for describing the long-time evolution of the wave packets, confirmed this interesting phenomenon. Thus, in the absence of viscous dissipation there are no steady end-states, but a series of modulation and demodulation cycles, known as the Fermi-Pasta-Ulam recurrence phenomenon (Fermi, Pasta & Ulam 1955; see also Scott, Chu & McLaughlin 1973).

The existence of the Fermi-Pasta-Ulam recurrence is especially significant for the usefulness of the wavetrain concept in the study of deep-water waves. If the end-state was one of complete breakdown (or thermalization), it would indicate that the wavetrain is a totally idealized and artificial entity in that it would not occur naturally, and even if it was somehow created, it could not sustain its identity as such. Moreover, the nonlinear Schrödinger equation, which was derived based on the concept of a wavetrain, would have truly limited applicability. Instead, because the nonlinear wavetrain maintains its coherence during its evolution, the wavetrain concept and the nonlinear Schrödinger equation are expected to be applicable to investigations of more realistic wave problems.

Relation Between Initial Conditions and Long-Time Evolution

The existence of the Fermi-Pasta-Ulam recurrence phenomenon also indicates that the solutions of the nonlinear Schrödinger equation have strong "memory" of their initial conditions. A positive link between the character of the long-time evolution and the initial conditions was reported by Yuen & Ferguson (1978a). They showed by numerical experiment that the long-time evolution of an unstable wavetrain is governed by the unstable modes and their harmonics contained in the initial condition. The stable harmonics do receive energy, but they never appear as the dominant mode at any stage of the evolution. The results of this view are summarized in Figure 4.

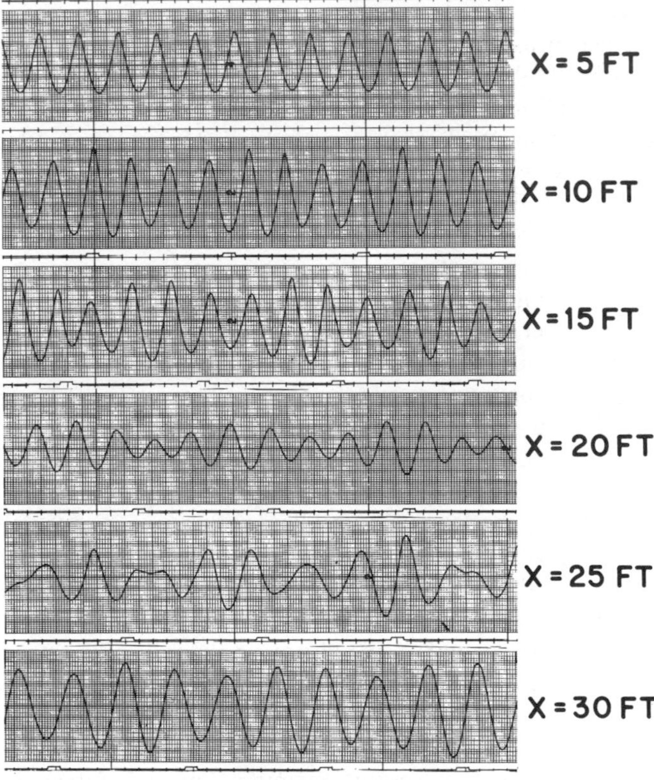

Figure 3 Evolution of an initially uniform weakly nonlinear wavetrain through one cycle of modulation and demodulation. Oscillograph records shown on expanded time scale to display individual wave shapes; wave shapes in each modulation period are not exact repetitions because the modulation period does not contain an integral number of waves.

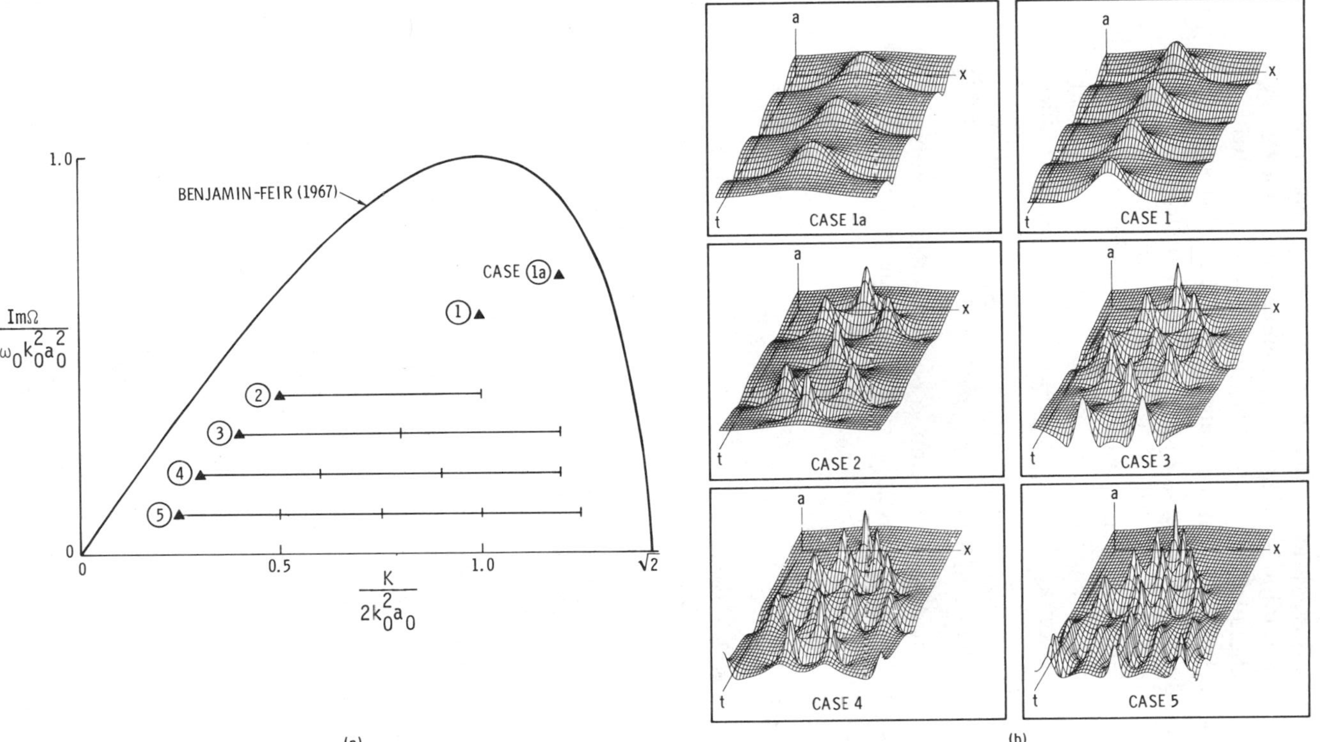

Figure 4 Relationship between initial conditions and long-time evolution of solutions of the nonlinear Schrödinger equation. Figure 4*a* gives the perturbation wavenumber K of the various cases and Figure 4*b* shows their corresponding time evolution. The initial conditions consist of a uniform wavetrain with wavenumber k_0 and amplitude a_0 subject to a 1% perturbation. The numerals in circles in Figure 4*a* identify the cases in Figure 4*b*, and also correspond to the number of harmonics of the perturbation (including the primary) that lie within the unstable regime according to the stability analysis. Note that the number of unstable harmonics corresponds exactly to the number of modes that dominate the evolution. For example, Case 4 shows an evolution in which the 1st, 2nd, 3rd, 4th, 1st and 2nd harmonics, in that order, took turns dominating the evolution as indicated by the number of peaks at various stages of evolution in the amplitude plot.

A corollary to this observation is that if the instability diagram that governs the eventual evolution has a high-wavenumber cutoff, the energy in the solution for subsequent times must be effectively confined within the unstable modes. This rules out thermalization in the classical sense, which corresponds to equipartition of energy among all modes.

The nonthermalization of solutions of the nonlinear Schrödinger equation for smooth initial conditions was actually proved by Thyagaraja (1979) who constructed an algorithm for finding N, given any δ, so that the energy of the solution at all times is confined to the first N modes within an error of δ.

Wavetrain in Two Space Dimensions

Thus far we have confined our discussion to one-dimensional problems. The equation for the complex envelope of a wavetrain propagating in two space dimensions is

$$i\left(\frac{\partial A}{\partial t}+\frac{\omega_0}{2k_0}\frac{\partial A}{\partial x}\right)-\frac{\omega_0}{8k_0^2}\frac{\partial^2 A}{\partial x^2}+\frac{\omega_0}{4k_0^2}\frac{\partial^2 A}{\partial y^2}-\tfrac{1}{2}\omega_0 k_0^2 |A|^2 A=0 \tag{16}$$

where the free surface of water is

$$\eta(x,y,t)=\mathrm{Re}\{A(x,y,t)e^{i(k_0 x-\omega_0 t)}\}. \tag{17}$$

When there is no y-dependence in the wave envelope, the equation reduces to the nonlinear Schrödinger Equation (15). Note that the second-y-derivative term and the second-x-derivative term are of opposite sign, leading to a basically hyperbolic, rather than an elliptic, equation. Also note that the carrier wave still propagates in the x-direction, but now its envelope entertains two-dimensional variations.

In the following we briefly review the concepts of envelope solitons, modulational instability, and recurrence in the two-dimensional context.

ENVELOPE SOLITONS Plane envelope solitons exist for Equation (16) provided that the angle between the directions of propagation of the carrier wave and of the envelope profile is less than 35.26°. An illustration of an oblique plane soliton is given in Figure 5. These two-dimensional plane solitons are infinite in extent. In addition, they are unstable to two-dimensional disturbances (Zakharov & Rubenchik 1973, Saffman & Yuen 1978). Thus far, no fully two-dimensional solutions exhibiting true soliton behavior have been found. Therefore the possibility of applying the concept of solitons as fundamental entities to construct fully two-dimensional wave fields has not yet been realized, though there have been some attempts in this direction (Cohen, Watson & West 1976, Mollo-Christensen & Ramamonjiarisoa 1978).

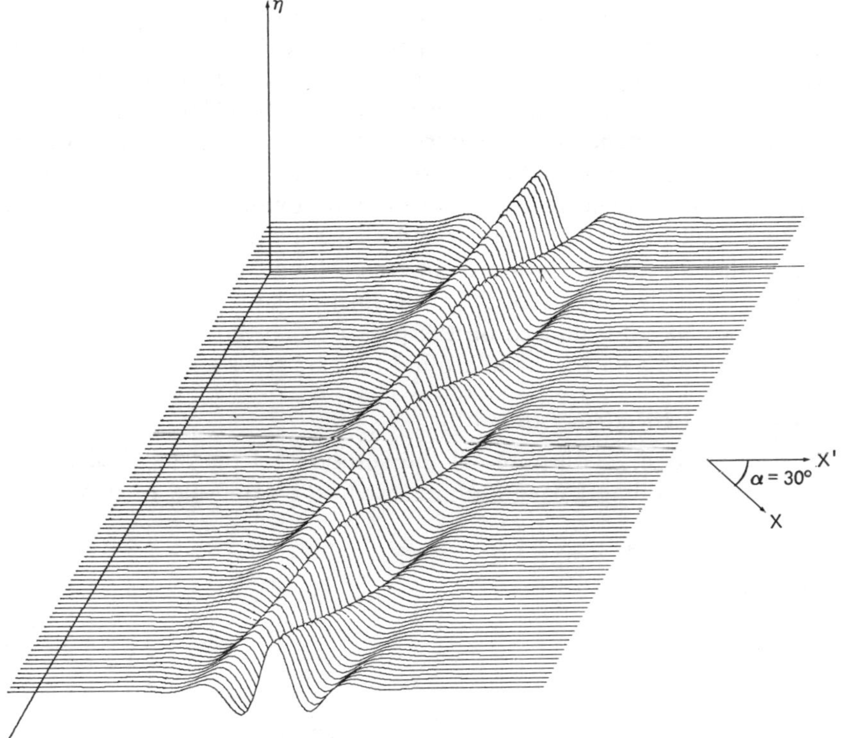

Figure 5 A sketch of the free surface corresponding to an oblique plane envelope soliton in two space dimensions (not to scale). The individual waves in the envelope propagate in the x direction at the phase speed and the envelope propagates in the x' direction at the group velocity.

MODULATIONAL INSTABILITY Generalization of the stability analysis of a uniform wavetrain to two space dimensions yields the result illustrated in Figure 6. The Benjamin-Feir results are recovered by setting K_y to zero. Of particular significance is the fact that the two-dimensional instability does not have a high-wavenumber cutoff. This result has important implications when the long-time evolution of an unstable wavetrain in two dimensions is considered.

LONG-TIME EVOLUTION AND RECURRENCE For a wide variety of initial conditions, the phenomenon of Fermi-Pasta-Ulam recurrence still exists for wavetrains in two space dimensions (Yuen & Ferguson 1978b). When one examines the relationship between initial conditions and long-time evolution for the two-dimensional nonlinear Schrödinger equation, however, one concludes that, unlike the one-dimensional result, there exists for any fully two-dimensional unstable perturbation vector (K_x, K_y) an infinite set of integers (m,n) such that (mK_x, nK_y) lie in the unstable regime and are eligible to participate actively in the evolution. Furthermore, it has been confirmed by Martin & Yuen (1980) that an unstable wavetrain does recur, but that during each new cycle unstable modes with larger values of m and n are excited. Thus, the long-time evolution consists of a periodic return to the initial condition together with a gradual "leak" of energy to higher and higher modes. Although this may not be the classical definition of thermalization, such behavior clearly indicates that the validity of the two-dimensional equation ceases at a finite time, since it is derived on the premise that most of the energy is confined within a narrow band around the carrier wavenumber.

Zakharov's Integral Equation

Most of the desirable features of the solutions of the one-dimensional nonlinear Schrödinger equation are lost in the extension to two space

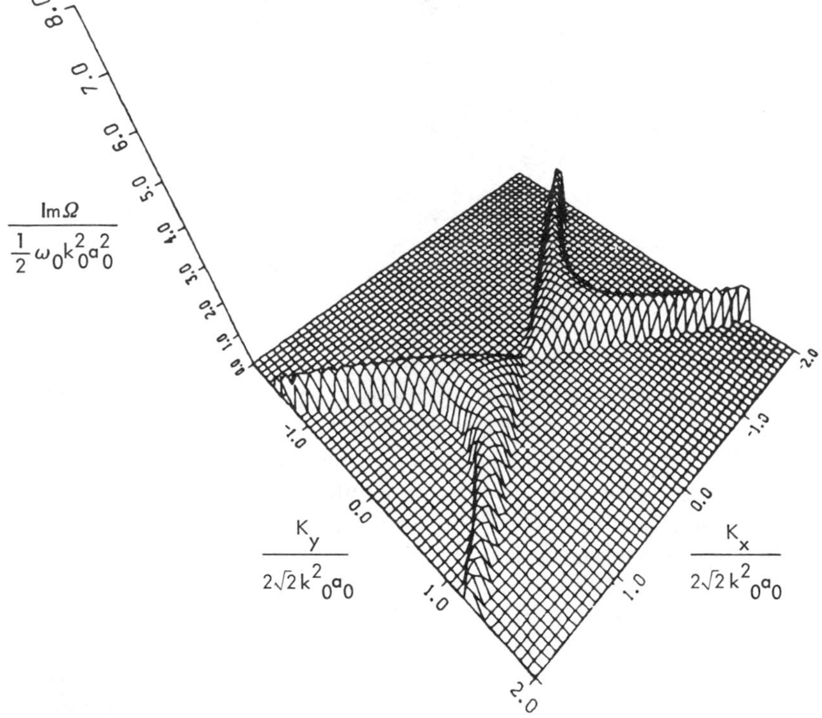

Figure 6 Stability diagram of a uniform wavetrain subject to infinitesimal two-dimensional plane-wave perturbations.

dimensions. The inverse-scattering method can no longer be applied. No localized solitons exhibiting stability and immunity from interaction effects have been found. The concept of wave coherence, which was prominent and important in the one-dimensional case, appears to be of questionable relevance in light of the results in the two-dimensional case. Most perturbing of all, however, is the fact that certain solutions of the two-dimensional nonlinear Schrödinger equation experience a progressive leakage of energy to high modes, thus eventually extending beyond the domain of validity of the equation.

As these problems with the two-dimensional theoretical results were identified, the need for a more accurate description than that provided by the nonlinear Schrödinger equation became apparent. A nonlinear integro-differential equation, first derived by Zakharov (1968), was identified by Crawford et al (1978) as being the desirable replacement. Zakharov's equation describes the time evolution of the complex envelope spectral function $\bar{A}(\mathbf{k},t)$,

$$i\frac{\partial}{\partial t}\bar{A}(\mathbf{k},t) = \iiint_{-\infty}^{\infty} T(\mathbf{k},\mathbf{k}_1,\mathbf{k}_2,\mathbf{k}_3)\delta(\mathbf{k}+\mathbf{k}_1-\mathbf{k}_2-\mathbf{k}_3)$$

$$\times\, e^{i[\omega(\mathbf{k})+\omega(\mathbf{k}_1)-\omega(\mathbf{k}_2)-\omega(\mathbf{k}_3)]t}\bar{A}^*(\mathbf{k}_1)\bar{A}(\mathbf{k}_2)\bar{A}(\mathbf{k}_3)\,d\mathbf{k}_1\,d\mathbf{k}_2\,d\mathbf{k}_3, \qquad (18)$$

where $\bar{A}(\mathbf{k},t)$ is related to the free surface by the expression

$$\eta(\mathbf{x},t) = \frac{1}{2\pi}\int \frac{|\mathbf{k}|^{\frac{1}{4}}}{\sqrt{2}\,g^{\frac{1}{4}}}\{\bar{A}(\mathbf{k},t)e^{i(\mathbf{k}\cdot\mathbf{x}-\omega t)}+\bar{A}^*(\mathbf{k},t)e^{-i(\mathbf{k}\cdot\mathbf{x}-\omega t)}\}\,d\mathbf{k} \qquad (19)$$

and $T(\mathbf{k},\mathbf{k}_1,\mathbf{k}_2,\mathbf{k}_3)$ is a real interaction coefficient first calculated by Zakharov (1968) with minor algebraic corrections made by Crawford, Saffman & Yuen (1980).

The nonlinear Schrödinger equation can be recovered, in one or two dimensions, by expanding the frequencies $\omega(\mathbf{k}_i)$ about a carrier-wave vector to second order, and replacing the interaction coefficient $T(\mathbf{k},\mathbf{k}_1,\mathbf{k}_2,\mathbf{k}_3)$ by its value when all four arguments are evaluated at the carrier-wave vector $\mathbf{k}_0 = (k_0,0)$. The resultant integral equation is then the Fourier-transformed equation for the complex amplitude $A(x,y,t)$ that satisfies the nonlinear Schrödinger equation. This simplification is equivalent to retaining only the leading-order terms in nonlinearity and dispersion.

Results obtained from the Zakharov integral equation were found to be very encouraging. The stability analysis for one-dimensional modulational perturbation yields an instability growth rate lower than that predicted by the nonlinear Schrödinger equation (and by the Benjamin

& Feir analysis). The difference becomes appreciable for moderate wave steepness ($ka = 0.15$ and higher), and the growth rate predicted by the Zakharov equation is in better agreement with experimental data (Crawford et al 1978; see Figure 7). For very large values of wave steepness, Zakharov's equation predicts a restabilization, which is in qualitative agreement with the exact numerical results of Longuet-Higgins (1979) and Peregrine & Thomas (1979).

Because the Zakharov equation does not require that there be a single constant wave vector, it can be used to investigate wave dynamics as a function of the bandwidth of components in the spectrum as well as of wave steepness. As an example of the dependence of dynamics upon those parameters, one can calculate dispersive characteristics for such wave systems. An example of such a result taken from D. R. Crawford and co-workers (in preparation) is shown in Figure 8, where component

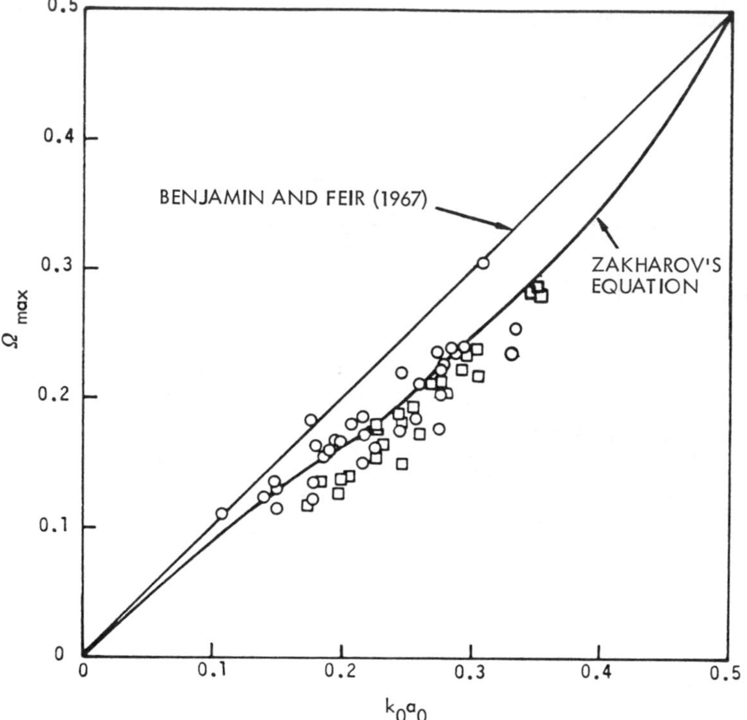

Figure 7 Plot of most unstable modulation frequency Ω_{max} against carrier wave steepness $k_0 a_0$. ○, □ experimental data; (*heavy line*) results from stability analysis based on Zakharov's equation; (*light line*) results from stability analysis based on nonlinear Schrödinger equation (or from Benjamin & Feir 1967).

phase speeds are plotted against component frequencies for a particular wave system. The experimental data are component phase speeds measured using a two-point filter and correlation technique for a wave system of finite bandwidth and steepness. The theoretical results were obtained using the Zakharov equation for a spectrum of waves with the same bandwidth and dominant frequency as those in the experiments. Depending upon bandwidth and steepness, such wave systems may exhibit component dispersion ranging from that given by the linear dispersion relation to that of an effectively nondispersive phase-locked system in which components propagate at essentially a single speed. Most realistic wave systems would be expected to fall somewhere between the two limits. An example is given in Figure 8. The dispersive properties of deep-water wave systems, therefore, vary continuously over

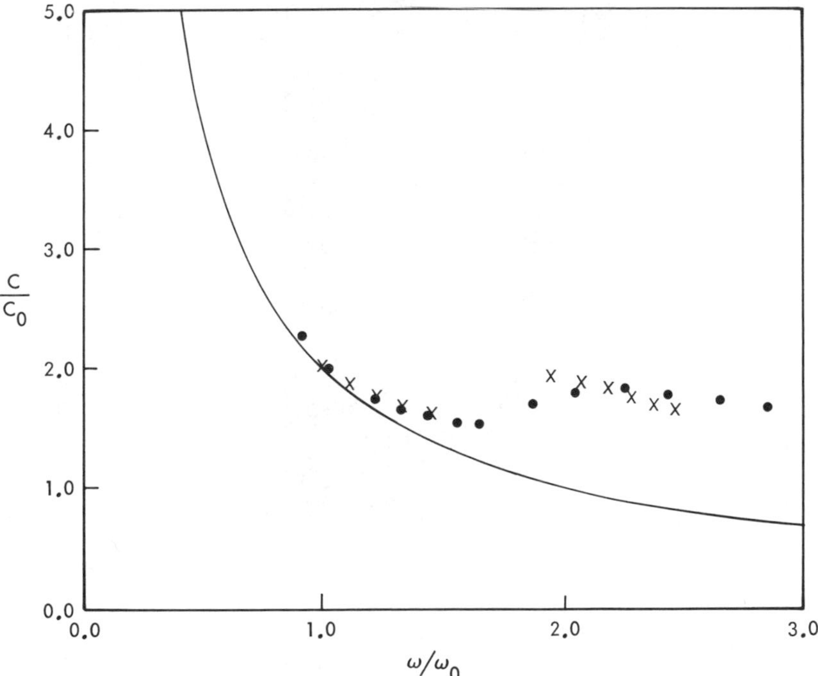

Figure 8 Plot of normalized component phase speed C/C_0 against normalized frequency ω/ω_0 for a spectrum of waves with characteristic bandwidth (half-width normalized by peak frequency) of 0.07, characteristic nonlinearity (root-mean-square wave slope) of 0.098. The phase speed and frequency are normalized by their values at the spectral peak. × numerical computation; ● experimental data from laboratory (no wind); ——— phase-speed result from linear dispersion relation.

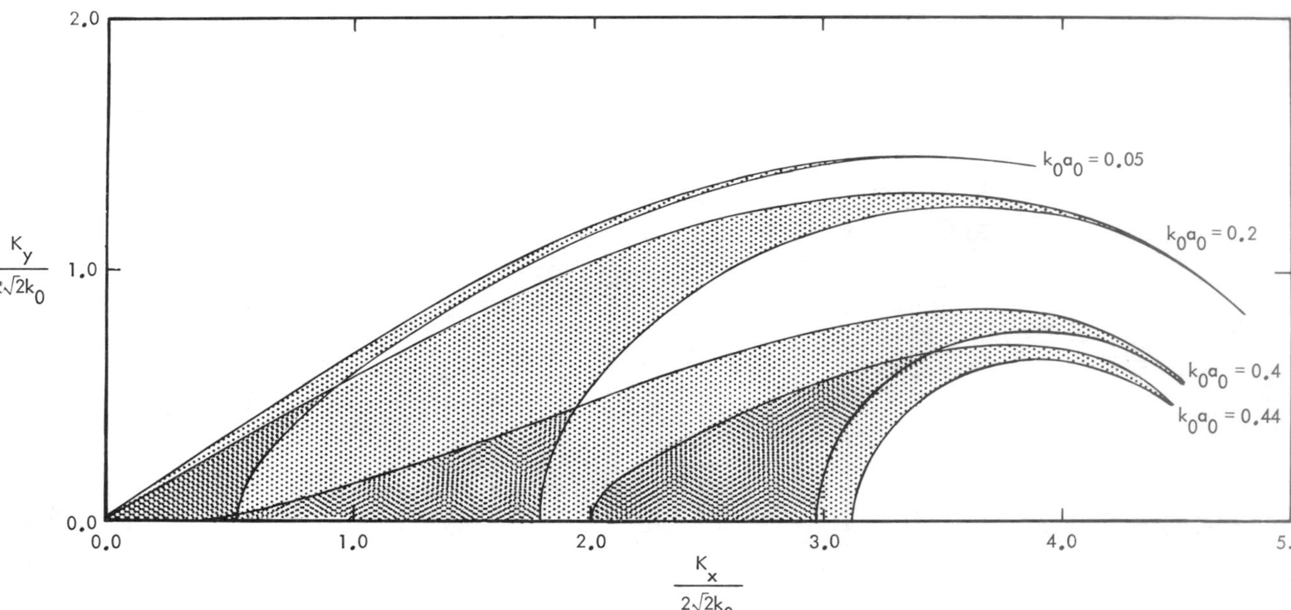

Figure 9 Stability boundary for a uniform wavetrain subject to infinitesimal two-dimensional plane-wave perturbations based on Zakharov's equation. Note that the stability boundary depends strongly on the carrier wave steepness k_0a_0. The shaded regions represent instability. The long waves (low wavenumber) have become stable for large values of k_0a_0 (0.4 and 0.44). For k_0a_0 greater than 0.495, the entire system is stable again. Whereas this value of k_0a_0 is unphysically large, the restabilization agrees with the trend exhibited by the exact calculations of Longuet-Higgins for the one-dimensional case.

this range of possibilities, with deviations from the linear results increasing with increasing steepness and decreasing bandwidth.

The Zakharov equation can also be used to study the interaction of more than one wave component traveling in different directions. In particular, the problem of the generation of a third wavetrain by two resonantly interacting and intersecting wavetrains was examined by D. R. Crawford and H. C. Yuen (in preparation), and the results compare very well with the experimental data of McGoldrick et al (1966) and Longuet-Higgins & Smith (1966).

The most important result, however, is the stability analysis in two space dimensions. D. R. Crawford and H. C. Yuen (in preparation) showed that, contrary to the results from the two-dimensional nonlinear Schrödinger equation, the instability diagram for the uniform wavetrain possesses a high-wavenumber cutoff (see Figure 9). Therefore, the leakage to high modes experienced by certain solutions of the two-dimensional nonlinear Schrödinger equation need not occur in wave systems described by the Zakharov equation. The concept of wave coherence appears to be relevant to wave systems in two space dimensions after all, although positive confirmation has yet to be produced.

Effects of Randomness

The effects of randomness on the stability properties of a nonlinear wavetrain were examined by Alber & Saffman (1978) and extended to two dimensions by Alber (1978) using the nonlinear Schrödinger equation as a starting point. It was shown that the presence of phase and/or amplitude randomness reduces the growth rate of the modulational instability, and acts to suppress the growth of modulations having wavelength substantially shorter than the decorrelation length associated with the randomness. For sufficiently strong randomness, the modulational instability could be eliminated. More generally, Alber & Saffman (1978) showed that the evolution of a random wavetrain obeys a transport equation similar to the Vlasov equation for weak turbulence in a plasma:

$$\frac{\partial F}{\partial t} + \left(\frac{\omega_0}{2k_0} - \frac{\omega_0}{4k_0^2}p\right)\frac{\partial F}{\partial x} - \omega_0 k_0^2 \sin\left(\frac{1}{2}\frac{\partial}{\partial p}\frac{\partial}{\partial x'}\right) \times \left[F(p,x,t)\int_{-\infty}^{\infty} F(p',x',t')\,dp'\right]\Bigg|_{x'=x} = 0, \qquad (20)$$

where $F(p,x,t)$ is the Fourier transform of the two-point, one-time correlation function

$$F(p,x,t) = \frac{1}{2\pi} \int_{-\infty}^{\infty} e^{-ipr} \left\langle A\left(x + \frac{r}{2}, t\right) A^*\left(x - \frac{r}{2}, t\right) \right\rangle dr \tag{21}$$

with $\langle \ \rangle$ denoting the ensemble average.

The significant point is that this equation predicts an evolutionary time scale of $O(F)$ or $O(k^2a^2)$, which is a much faster time scale than the weak nonlinear energy transfer time scale of $O(F^2)$ or $O(k^4a^4)$ found by Hasselmann (1962, 1963) for a random, homogeneous ocean. This apparent conflict was resolved by Crawford, Saffman & Yuen (1980) who derived a more general equation for a random wave field based on the Zakharov equation, of which Equation (20) is a special case when constant peak frequency and narrowbandedness are assumed, and showed that the $O(k^2a^2)$ term represents effects of inhomogeneity and vanishes for a homogeneous ocean. The presence of the term does not lead to net energy transfer. In other words, the evolution of the random wave field consists of a relatively fast process of redistribution of inhomogeneity that is reversible, and a much slower process of net energy transfer among wavenumbers that is irreversible. The former process is intimately related to the Fermi-Pasta-Ulam recurrence phenomenon associated with a single nonlinear wavetrain and it is responsible for the modulation scales and patterns of the wave field. The latter process is related to the much slower evolution of the overall spectral distribution of wave energy among components.

VERY STEEP WAVES

The Limiting Wave

The occurrence of instabilities and nonlinear effects in deep-water waves is associated with finite amplitude or, more accurately, with finite wave steepness, k_0a_0. As amplitude or steepness increases from the infinitesimal amplitude linear wave limit, the nature and variety of instabilities and nonlinear effects increase. In the following we describe some phenomena of wavetrains and individual waves that occur as the wave amplitude or steepness becomes very large. First, however, we recall that Stokes (1880) suggested that there is a maximum possible wave steepness for steady waves on deep water, so that "very large" refers to values of wave steepness approaching that of the limiting wave. The existence of a limiting wave and identification of its physical characteristics have been the subjects of numerous investigations ever since.

One line of investigation pertains to the calculation of profiles and flow fields of waves of finite and large amplitudes. Stokes (1880) used an argument locally valid near the crest to demonstrate that the limiting wave profile can have a 120° corner at its crest, which is a stagnation point. Michell (1893) fitted a Fourier series to the leading-order singular term used by Stokes to arrive at the 120° result and determined that the height-to-wavelength ratio of the wave of greatest height is 0.142 ($ka = 0.446$). His analysis was refined and extended by Wilton (1913) and Havelock (1919), and in a modified form by Yamada (1957), but with only slight changes in the value of the limiting steepness. These calculations were very laborious, and the complexity increased rapidly with the increasing number of terms retained in the expansion. Recently, Grant (1973) and Norman (1974) reported difficulties in obtaining further terms in the expansion. Using the Padé approximants to accelerate convergence, Schwartz (1974) performed calculations that have accuracy equivalent to retaining 117 terms in the Stokes expansion and found that the nature of the singularity near the crest changes from square-root type to cube-root type when the steepness approaches the limiting value. Accurate profiles were also given for the entire range of wave steepness leading to the near-limiting wave. Similar calculations were performed by Longuet-Higgins (1975) and Cokelet (1977) with results in good agreement with those of Schwartz (1974). The nature of the profile and the flow field near the crest of the near-limiting wave was examined analytically by Longuet-Higgins & Fox (1977) using matched asymptotic expansions. They showed that the maximum slope is slightly greater than the 30° implied by the Stokes conjecture, being 30.37°. Subsequently, Longuet-Higgins & Fox (1979) matched this inner solution to third order to an outer solution, providing the first definite link between the local corner flow at the crest as suggested by Stokes and a full-period wave profile.

The calculations of Schwartz (1974) clearly identified an intriguing feature of large-amplitude waves: it was found that the phase speed, kinetic energy, potential energy, and wave impulse each reach an absolute maximum before the wave reaches its maximum steepness. The more detailed investigation by Longuet-Higgins & Fox (1979) indicates that these quantities, after attaining their maxima, oscillate an infinite number of times before they reach the values for the steepest wave. The exact implications of this rather puzzling behavior have not yet been completely explored.

The other line of investigation concentrates on existence in the mathematical sense. The existence of small-amplitude, steady, periodic waves was proved by Nekrasov (1919) and later by Levi-Civita (1925). The existence of finite-amplitude and very steep waves was not established, however, until very recently. Toland (1978), extending the work of Krasovskii (1961) and Keady & Norbury (1978), proved the existence of a "limiting wave" that has a stagnation point at the crest and is smooth everywhere except at the crest. He showed that this wave is indeed the limit of a sequence of smooth, periodic steady waves, and that the maximum slope of some waves in the sequence must be at least 30°. However, the properties of the limiting wave near its crest have not yet been ascertained. Depending on the type of singularity associated with the limiting wave solution at the crest, the crest can either be a corner of 120° (and hence the Stokes corner flow) or an infinite number of ripples of large steepness.

Restabilization of the Uniform Wavetrain at Large Amplitude

The modulation instability for weakly nonlinear wavetrains as calculated by Benjamin & Feir (1967), and from the nonlinear Schrödinger equation, is "similar" in the wave steepness (here taken to be $k_0a_0 = \pi h/\lambda$), in the sense that when plotted in scaled variables, with the modulational wavenumber scaled by k_0a_0 and the growth rate scaled by $k_0^2a_0^2$, the results are no longer dependent on k_0a_0. This is not true for the Zakharov equation, which has a stronger dependence on k_0a_0. In fact, the Zakharov equation indicates that the modulational instability disappears for sufficiently large values of k_0a_0. The Zakharov equation predicts wavetrain restabilization at $k_0a_0 \doteq 0.5$. Since this is greater than the k_0a_0 of the limiting wave, the prediction is only suggestive of the correct behavior. The value of k_0a_0 at which restabilization actually occurs is around 0.34, as shown by the exact numerical calculations of Longuet-Higgins (1979). Longuet-Higgins also showed that a second, more violent instability occurs at a larger value of k_0a_0, but he was unable to obtain accurate results in the required range.

Bifurcation of Large Amplitude Waves

Compared with the question of existence, the question of uniqueness of steady periodic water waves has received much less attention. The main results are contained in a little known paper by Garabedian (1965), who used topological methods to prove the nonexistence of asymmetric waves and the uniqueness of symmetric waves, provided that their crests and troughs are of the same height. In other words, perfectly uniform, symmetric wavetrains are unique. These results apply to waves of arbitrary amplitudes.

The proviso that uniqueness is guaranteed only if the waves have uniform crests and troughs turns out to be critical. In fact, Chen & Saffman (1979a) have demonstrated numerically that for sufficiently large amplitudes, the uniform wavetrain bifurcates into a train of waves with unequal crests or troughs, with the individual waves in the train remaining symmetric about their crests. These bifurcated wavetrains acquire a periodicity with periods longer than those of the unbifurcated solutions; hence, the bifurcation is subharmonic or fractional subharmonic. Consistent with the results of Garabedian (1965), Chen & Saffman (1979a) reported no superharmonic bifurcations.

The critical amplitudes at which bifurcation occurs lie in the range of amplitudes where the wavetrain is stable to modulational perturbations. In fact, they coincide with points of neutral stability in the modulational perturbation calculations. In the one-dimensional case, these points were calculated by Longuet-Higgins (1979) for several selected perturbational wavelengths and found to lie around $k_0 a_0 = 0.4$.

For the two-dimensional case, the neutral stability curve in the limit of infinitely long perturbations can be obtained from the calculations of Peregrine & Thomas (1979) to demonstrate that neutral stability occurs for smaller and smaller amplitudes for increasingly oblique perturbations. For finite perturbation wavelengths, the same trend is found from the Zakharov equation (D. R. Crawford & H. C. Yuen, in preparation). Thus it should be possible for a uniform wavetrain to bifurcate into a steady, fully two-dimensional, periodic wave pattern even at small wave amplitudes. This question has been pursued by D. U. Martin (in preparation) who studied the bifurcation from the uniform solution of the two-dimensional nonlinear Schrödinger equation which, despite its limitations at large amplitudes, should be adequate for small amplitude situations. A two-parameter degenerate family of bifurcated solutions was found—one example of which is illustrated in Figure 10.

Generation of Capillary Waves by Steep Gravity Waves

Thus far we have neglected the effects of surface tension on the premise that they are small compared to the effects of gravity for sufficiently long waves. Though this assumption holds for the bulk properties of the gravity waves, it breaks down locally near the crest of very steep gravity waves where the local curvature is large and the surface-tension pressure strong. In fact, short capillary waves have been observed near the crests of steep gravity waves, with and without the presence of wind (Cox 1958).

Theoretical considerations for the generation of short capillary waves by steep gravity waves have been given by Longuet-Higgins (1963). Based on an argument by Munk (1955), Longuet-Higgins modeled the crest of a steep, steady gravity wave by a localized pressure source. The situation is analogous to the fish-line problem considered by Rayleigh (1883), in which capillary waves are found ahead of, and gravity waves behind, the steadily traveling pressure source. The relative locations of the capillary

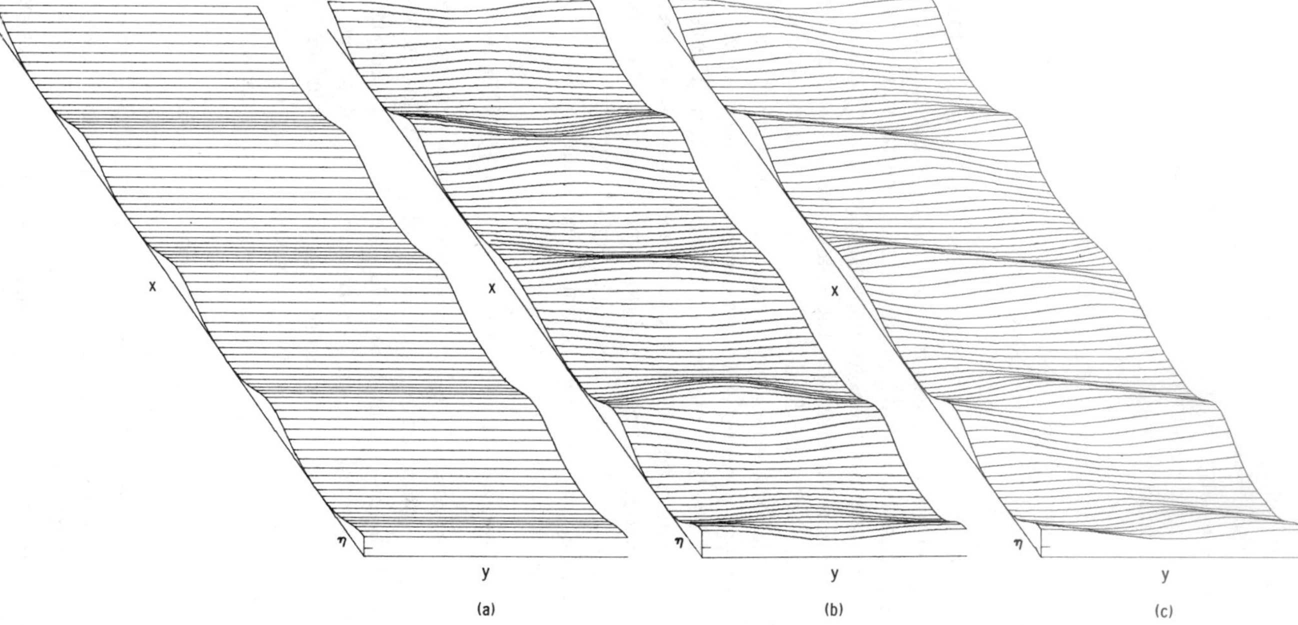

Figure 10 Example of steady two-dimensional bifurcated solutions from the uniform wavetrain solution of the nonlinear Schrödinger equation. Free surface $\eta(x,y)$ as a function of x and y. Figure 10a: unbifurcated uniform solution. Figure 10b: symmetric bifurcated solution. Figure 10c: asymmetric bifurcated solution.

and gravity waves are caused by the fact that the group velocity is greater than the phase velocity in capillary waves and vice versa for gravity waves. The respective wavelengths of the capillary and gravity waves are determined by the requirement that the wave pattern remain steady with respect to the pressure source.

Applying this concept to the capillary wave generation problem, Longuet-Higgins concluded that 1. capillary waves are generated at the crest of the gravity wave and propagated down its front face; 2. the steepness of the capillary wave nearest to the crest is proportional to $\exp(-g/6T'\kappa^2)$, where T' is the surface tension coefficient and κ is the curvature of the underlying gravity wave at its crest; 3. the frequency and wavelength of the capillary waves are determined by the requirement that they remain steady with respect to the crest, with the effect of the gravity-wave orbital velocity taken into account; 4. viscous dissipation acts to destroy the capillary waves more than a fraction of wavelength down the front face of the gravity wave. These results were compared with observations by Cox (1958) and qualitative agreement was found except that there seemed to be an underprediction of the steepness of the nearest capillary wave.

The problem was re-examined by Crapper (1970) who sought steady solutions of Whitham's equation describing capillary waves riding on a current field created by the steady gravity wave, with a source representing the surface tension pressure excess at the crest and with viscous dissipation. The result was substantially the same as in Longuet-Higgins's model, except that the steepness of the first capillary wave was overpredicted, rather than underpredicted. Some capillary waves were also permitted on the back side of the gravity wave.

Detailed measurements of capillary wave generation by steep steady gravity waves were made with a high-resolution laser slope gauge under controlled conditions by Chang, Wagner & Yuen (1978). The results support the model of Longuet-Higgins (1963) and suggest that the cause for underprediction of the lead capillary wave may be due to the fact that the crest curvature required in Longuet-Higgins's model is not given by the measured value, but is the idealized higher value in the absence of surface tension.

Note that the model of Longuet-Higgins represents a dynamic equilibrium reached when dissipation balances the production of capillary waves, and is not a steady situation in the usual sense. (In fact, steady capillary-gravity wave solutions obtained as bifurcated states from pure gravity waves have capillary waves riding in the *troughs* of the gravity waves (Chen & Saffman 1979b; see also the experimental observations of Schooley 1960) and bear little resemblance to the Longuet-Higgins model.) To maintain this dynamic equilibrium of the model, energy must be extracted from the gravity waves and effectively dissipated as capillary waves. Longuet-Higgins (1963) showed that through this mechanism the presence of capillary waves can result in a dissipation rate many times that caused by molecular viscosity acting on the pure gravity wave, and proposed that this may be partly responsible for limiting the steepness that a real gravity wave can reach. In fact, Ferguson, Saffman & Yuen (1978) showed by a model equation that the properties of capillary waves riding on gravity waves are strongly affected by the relative importance of viscosity and the degree of unsteadiness in each case.

Wavebreaking

That deep-water waves "break" under certain circumstances has never been in dispute. The precise meaning of "wavebreaking," however, is still unsettled. For example, one definition of wavebreaking is the loss of a smooth wave profile, such as the attainment of a cornered crest in the Stokes limiting wave. Another definition is the development of a locally vertical slope so that the wave profile becomes double-valued. Still another definition, more functional than mathematical, is the occurrence of a drastic, irreversible change in the important wave properties. This may include capillary-wave generation although it more commonly refers to the appearance of turbulent patches with air entrainment.

In view of the lack of understanding of this complicated process, a precise, accepted definition of wavebreaking may be too much to expect. For the present it may actually be best to define wavebreaking according to the particular requirements of the specific problem being addressed at the time, as long as the definition is reasonable and clearly stated.

The concept of "incipient wavebreaking" was used by Banner & Phillips (1974) to study the effects of wind-drift on wavebreaking. Incipient wavebreaking is defined as occurring for a steady progressive wave when the particle velocity at the crest of the wave becomes equal to its phase velocity. Beyond this stage, they argued, the water would outrun the wave, initiating wavebreaking in the sense that water particles might run down the front face of the wave, or the wave profile might curl over. Note that this condition is also met by the Stokes limiting wave at the crest. Banner & Phillips (1974) noted that in the presence of wind, a surface drift layer is set up with thickness typically small compared to the wavelength of the steady wave under consideration. This layer has little effect on the phase velocity, but strongly affects the local particle velocity at the crest. Thus the presence of the wind-drift layer can significantly increase the probability of incipient wavebreaking. The same argument was applied to the situation of short steady waves in the presence of a long wave by Phillips & Banner (1974). In this case, the orbital velocity of the long wave plays the role of the wind drift.

In reality, however, deep-water gravity waves are seldom, if ever, steady. Even for the simplest situation of a single weakly nonlinear wavetrain, the existence of the Benjamin-Feir instability indicates that the wavetrain is normally in a modulated state. Since the phase velocity of the gravity wave is greater than its group velocity, the individual waves have continuously changing amplitude, slope, and profile as they move and are therefore unsteady. In more complicated situations, where more than one wave system is involved, the concept of a steady wave system simply does not apply. It is under these inherently unsteady conditions that wavebreaking is most likely to occur.

Experimental and theoretical investigations of unsteady wavebreaking of waves on deep water are difficult and very few quantitative results are available. In order to study the details of unsteady wavebreaking theoretically, the full set of Equations (1)–(3) must be solved. This has been an impossible task analytically, and serious numerical attempts did not begin until the advent of large computers.

The first high-resolution calculations of unsteady deep-water waves of large amplitude were performed by Longuet-Higgins & Cokelet (1976). They took advantage of the fact that the velocity potential is completely determined by its values at the boundary, and rewrote the equations solely in terms of the free surface and tangential and normal derivatives of the velocity potential. All tangential derivatives could be evaluated by numerical differentiation, and the normal-derivative term determined by inverting an integral equation. As a consequence, only the boundary values of the velocity potential needed to be stored and updated. The savings in storage and operations permitted the use of a sufficient number of grid points to insure high resolution while staying within the realm of economic reality. However, like other similar calculations (Chan & Street 1970, Kulyaev 1975, 1977), the time stepping led to grid-scale oscillations that had to be suppressed by numerical smoothing.

Longuet-Higgins & Cokelet (1976) presented computations of a steady wave evolving in response to an asymmetric pressure applied for half a wave period. After the wave was allowed to run free, the potential and kinetic energies oscillated strongly and the wave profile eventually achieved a vertical slope near the front of the crest and then curled over. The computation broke down shortly after this point. A subsequent paper by Longuet-Higgins & Cokelet (1978) followed the development of individual waves in a wavetrain undergoing modulational instability to the point where some of them curled over. In all these calculations, the energy loss due to numerical smoothing remained acceptably small.

A generally accepted interpretation of these results is that the waves break because they are forced to contain more energy than would be contained in the steepest steady wave with the same wavelength. Calculations made by Yuen (1977), however, using essentially the same method as that proposed by Longuet-Higgins & Cokelet (1976) have shown that unsteady waves can break even when their steepness is much less than the limiting steepness, and their energy much less than the corresponding limiting wave energy. In particular, it was found that an initially symmetric sinusoidal profile, with a steepness (h/λ) of 0.095 and an energy content approximately 60% of the maximum allowable in a steady wave with the same wavelength, "broke" (curled over) within one wave period. The proposition that waves can break at values of steepness and energy less than the limiting wave levels is also supported by experiments (B. M. Lake and H. Rungaldier, in preparation) in which high-resolution laser slope gauges were used to measure unsteady waves breaking as they propagated through strong wavetrain modulations. These results indicate that the breaking of deep-water waves is strongly dependent upon the extent to which they are unsteady.

We must note that the numerical results described above do not take into account the effects of surface tension, which may be very important just as the wave begins to break (as described in the previous section). Since surface tension introduces a second length scale into the problem (one that is drastically different from the first), the inclusion of surface-tension effects in the unsteady wavebreaking problem would make an already difficult numerical task even more difficult.

DISCUSSION

In the relatively few years since the discovery of the modulational instability of weakly nonlinear, deep-water wavetrains by Lighthill (1965) and Benjamin & Feir (1967), new phenomena associated with nonlinear deep-water waves have been identified at an accelerating rate. The first stage of investigations involved the properties of weakly nonlinear waves in one space dimension, and led to the discovery of the nonlinear Schrödinger equation, the envelope solitons, and the Fermi-Pasta-Ulam recurrence and its relationship to initial conditions. The second stage consisted of the attempts to extend to two space dimensions, resulting in exposition of the limitations of the nonlinear Schrödinger equation and identification of Zakharov's equation. Concurrent with this line of investigation was the study of the properties of very steep waves. Much progress was made in this area, notably the stability calculations of large amplitude waves by Longuet-Higgins (1978) and the calculation of unsteady evolution and wavebreaking by Longuet-Higgins & Cokelet (1976), with the most recent and stimulating topic being the bifurcation of the uniform solution at large amplitude (Chen & Saffman 1979a).

These advances, however, still leave many interesting observations unexplained. A good example is the recent measurement in the ocean and laboratory of component phase speeds of wave spectra which indicate discrepancies between the measured values and the values predicted by the linear, free-wave theory. Although it was shown earlier in this article that nonlinearity might account for the disagreement in one dimension, there have been a variety of alternate explanations ranging from two-dimensional effects to wind-induced drift-current effects. More detailed investigations, both experimental and theoretical, are necessary to determine which of the mechanisms are mainly responsible and under what circumstances. Another example is the process by which the peak frequency in a wind-wave spectrum continuously shifts to lower values. This downshifting was previously attributed entirely to wind action. However, laboratory results of Lake et al (1977) clearly show that downshifting can occur even in the absence of wind. In fact, Figure 3 of this article displays this observation—the wavetrain at 30 ft has a lower frequency than that at 5 ft, as evidenced by the difference in the number of waves in a fixed time interval. No satisfactory explanation has yet been found for this observation of downshifting without wind in the gravity wave range, although Chen & Saffman (1979b) have proposed bifurcation as a possible candidate for the downshifting in the capillary-gravity wave regime.

There are some areas in which theory has outrun experiment. One example is unsteady wavebreaking. Additional high-resolution measurements of breaking waves in controlled experimental environments would be of great value. Another example is the quantitative determination of two-space-dimensional modulation patterns for which experimental measurements are required to test the theoretical results. In general, it is expected that more quantitative experimental investigations of isolated mechanisms will be performed as new phenomena are uncovered either theoretically or by observations.

In this article, we have tried to present a sampling of the many areas in which recent efforts have uncovered new and often unexpected properties of waves on deep water. The topics that we have chosen to highlight are those related to instabilities of deep-water waves and those with which we are most familiar; they are clearly not meant to be exhaustive. We have not been able to mention many interesting and relevant results, and new discoveries are almost certain to appear in the very near future. We hope, however, that this article may serve to stimulate new interest in this old, but vital, subject.

ACKNOWLEDGMENTS

We are grateful to Professor P. G. Saffman for many valuable suggestions. We also wish to thank Miss Janet Nay for the expert preparation of the manuscript. This work is supported by the Applied Physics Laboratory of the Johns Hopkins University under Contract No. 601038.

Literature Cited

Alber, I. E. 1978. The effects of randomness on the stability of two-dimensional surface wavetrains. *Proc. R. Soc. London Ser. A* 363: 525–46

Alber, I. E., Saffman, P. G. 1978. Stability of random nonlinear deep water waves with finite bandwidth spectra. *TRW Rep. No. 31326-6035-RU-00*

Banner, M. L., Phillips, O. M. 1974. On the incipient breaking of small scale waves. *J. Fluid Mech.* 65: 647–57

Benjamin, T. B. 1967. Instability of periodic wavetrains in nonlinear dispersive systems. *Proc. R. Soc. London Ser. A* 299: 59–75

Benjamin, T. B., Feir, J. E. 1967. The disintegration of wavetrains on deep water. Part 1. Theory. *J. Fluid Mech.* 27: 417–30

Benney, D. J., Newell, A. C. 1967. The propagation of nonlinear wave envelopes. *J. Math. Phys.* 46: 133–39

Benney, D. J., Roskes, G. 1969. Wave instabilities. *Stud. Appl. Math.* 48: 377–85

Burnside, W. 1916. On periodic irrotational waves at the surface of deep water. *Proc. London Math. Soc.* (2)15: 26–30

Chan, R. K. C., Street, R. L. 1970. A computer study of finite-amplitude water waves. *J. Comp. Phys.* 6: 68–94

Chang, J. H., Wagner, R. N., Yuen, H. C. 1978. Measurement of high frequency capillary waves on steep gravity waves. *J. Fluid Mech.* 86: 401–13

Chen, B., Saffman, P. G. 1979a. Numerical evidence for the existence of new types of gravity waves of permanent form on deep water. *Stud. Appl. Math.* 61(2): In press

Chen, B., Saffman, P. G. 1979b. Steady gravity-capillary waves on deep water. II. Numerical results for finite amplitude. *Stud. Appl. Math.* 61(2): In press

Chu, V. H., Mei, C. C. 1970. On slowly-varying Stokes waves. *J. Fluid Mech.* 41: 873–87

Chu, V. H., Mei, C. C. 1971. The evolution of Stokes waves in deep water. *J. Fluid Mech.* 47: 337–51

Cohen, B. I., Watson, K. M., West, B. J. 1976. Some properties of deep water solitons. *Phys. Fluids* 19: 345–50

Cokelet, E. D. 1977. Steep gravity waves in water of arbitrary uniform depth. *Philos. Trans. R. Soc. London Ser. A* 286: 183–230

Cox, C. S. 1958. Measurements of slopes of high-frequency wind waves. *J. Mar. Res.* 16: 199–225

Crapper, G. D. 1970. Nonlinear capillary waves generated by steep gravity waves. *J. Fluid Mech.* 40: 149–59

Crawford, D. R., Lake, B. M., Saffman, P. G., Yuen, H. C. 1978. Nonlinear deep water waves. VI. Higher order results from discretized Zakharov integral equation. *TRW Rep. No. 31326-6029-RU-00*

Crawford, D. R., Saffman, P. G., Yuen, H. C. 1980. Evolution of a random inhomogeneous field of nonlinear deep-water gravity waves. *J. Wave Motion* 1: In press

Davey, A. 1972. The propagation of a weak nonlinear wave. *J. Fluid Mech.* 53: 769–81

Ferguson, W. E., Saffman, P. G., Yuen, H. C. 1978. A model equation to study the effects of nonlinearity, surface tension and viscosity in water waves. *Stud. Appl. Math.* 58: 165–85

Fermi, E., Pasta, J., Ulam, S. 1955. Studies of nonlinear problems. *Collected Papers of Enrico Fermi* 2: 978–88. Chicago: Univ. Chicago Press

Garabedian, P. R. 1965. Surface waves of finite depth. *J. Analyse Math.* 14: 161–69

Gardner, C. S., Greene, J. M., Kruskal, M. D., Miura, R. M. 1967. Method for solving the Korteweg-de Vries equation. *Phys. Rev. Lett.* 19: 1095–97

Grant, M. A. 1973. The singularity at the crest of a finite amplitude progressive Stokes wave. *J. Fluid Mech.* 59: 247–62

Hasimoto, H., Ono, H. 1972. Nonlinear modulation of gravity waves. *J. Phys. Soc. Jpn.* 33:805–11

Hasselmann, K. 1962. On the nonlinear energy transfer in a gravity-wave spectrum. 1: General theory. *J. Fluid Mech.* 12:481–500

Hasselmann, K. 1963. On the nonlinear energy transfer in a gravity-wave spectrum. 2: Conservation theorems, wave-particle correspondence, irreversibility. *J. Fluid Mech.* 15:273–81

Hasselmann, K. 1967. Discussion. *Proc. R. Soc. London Ser. A* 299:76

Havelock, T. H. 1919. Periodic irrotational waves of finite height. *Proc. R. Soc. London Ser. A* 95:38–51

Hayes, W. D. 1970a. Conservation of action and modal wave action. *Proc. R. Soc. London Ser. A* 320:187–208

Hayes, W. D. 1970b. Kinematic wave theory. *Proc. R. Soc. London Ser. A* 320:209–26

Hayes, W. D. 1973. Group velocity and nonlinear dispersive wave propagation. *Proc. R. Soc. London Ser. A* 332:199–221

Keady, G., Norbury, J. 1978. On the existence theory for irrotational water waves. *Proc. Cambridge Philos. Soc.* 76:345–58

Krasovskii, Yu. P. 1961. On the theory of steady-state waves of finite amplitude. *Zh. Vychisl. Mat. Mat. Fiz.* 1:836–55. Translated in *USSR Comput. Math. Math. Phys.* 1:996–1018

Kulyaev, R. L. 1975. Internal waves of finite amplitude. *Zh. Prikl. Mekh. Tekh. Fiz.* 1975, no. 1:96–105

Kulyaev, R. L. 1977. Calculation of nonstationary finite amplitude wind waves. *Zh. Prikl. Mekh. Tekh. Fiz.* 1977, no. 6: 82–86

Lake, B. M., Yuen, H. C., Rungaldier, H., Ferguson, W. E. Jr. 1977. Nonlinear deep-water waves: Theory and experiment. Part 2: Evolution of a continuous wave train. *J. Fluid Mech.* 83:49–74

Levi-Civita, T. 1925. Détermination rigoureuse des ondes permanentes d'ampleur finie. *Math. Ann.* 93:264–314

Lighthill, M. J. 1965. Contributions to the theory of waves in nonlinear dispersive systems. *J. Inst. Math. Appl.* 1:269–306

Longuet-Higgins, M. S. 1963. The generation of capillary waves by steep gravity waves. *J. Fluid Mech.* 16:138–59

Longuet-Higgins, M. S. 1975. Integral properties of periodic gravity waves of finite amplitude. *Proc. R. Soc. London Ser. A* 342:157–74

Longuet-Higgins, M. S. 1978. The instabilities of gravity waves of finite amplitude in deep water. II. Subharmonics. *Proc. R. Soc. London Ser. A* 360:489–505

Longuet-Higgins, M. S., Cokelet, E. D. 1976. The deformation of steep surface waves on water. I. A numerical method of computation. *Proc. R. Soc. London Ser. A* 350:1–36

Longuet-Higgins, M. S., Cokelet, E. D. 1978. The deformation of steep surface waves on water. II. Growth of normal-mode instabilities. *Proc. R. Soc. London Ser. A* 364:1–28

Longuet-Higgins, M. S., Fox, M. J. H. 1977. Theory of the almost-highest wave: The inner solution. *J. Fluid Mech.* 80:721–42

Longuet-Higgins, M. S., Fox, M. J. H. 1979. Theory of the almost-highest wave. II. Matching and analytic extension. *J. Fluid Mech.* In press

Longuet-Higgins, M. S., Smith, N. D. 1966. An experiment on third order resonant wave interactions. *J. Fluid Mech.* 25:417–36

Martin, D. U., Yuen, H. C. 1980. Quasi-recurring energy leakage in the two-space-dimensional nonlinear Schrödinger equation. *Phys. Fluids.* In press

McGoldrick, L. F., Phillips, O. M., Huang, N. E., Hodgson, T. H. 1966. Measurements of third-order resonant wave interactions. *J. Fluid Mech.* 25:437–56

Michell, J. H. 1893. The highest waves in water. *Philos. Mag.* (5)36:430–37

Mollo-Christensen, E., Ramamonjiarisoa, A. 1978. Modeling the presence of wave groups in a random wave field. *J. Geophys. Res.* 83:4117–22

Munk, W. 1955. High frequency spectrum of ocean waves. *J. Mar. Res.* 14:302–14

Nekrasov, A. I. 1919. On Stokes' wave. *Izv. Ivanovo-Voznesensk. Politekh. Inst.* 2:81–91 (in Russian)

Norman, A. C. 1974. Expansions for the shape of maximum amplitude Stokes waves. *J. Fluid Mech.* 66:261–65

Oikawa, M., Yajima, N. 1974. A perturbation approach to nonlinear systems. II. Interaction of nonlinear modulated waves. *J. Phys. Soc. Jpn.* 37:486–96

Peregrine, D. H., Thomas, G. P. 1979. Finite-amplitude deep-water waves on currents. *Proc. R. Soc. London Ser. A*. In press

Phillips, O. M., Banner, M. L. 1974. Wave breaking in the presence of wind drift and swell. *J. Fluid Mech.* 66:625–40

Rayleigh, Lord. 1883. The form of standing waves on the surface of running water. *Proc. London Math. Soc.* 15:69–78

Rayleigh, Lord. 1917. On periodic irrotational waves at the surface of deep water. *Philos. Mag.* (6)33:381–89

Russell, J. S. 1844. Report on Waves.

Rep. 14th Mtg. Brit. Assoc. Adv. Sci., pp. 311–90, plates XLVII–LVII

Saffman, P. G., Yuen, H. C. 1978. Stability of a plane soliton to infinitesimal two-dimensional perturbations. *Phys. Fluids* 21:1450–51

Schooley, A. H. 1960. Double, triple, and higher-order dimples in the profiles of wind-generated water waves in the capillary-gravity transition region. *J. Geophys. Res.* 65:4075–79

Schwartz, L. W. 1974. Computer extension and analytic continuation of Stokes expansion for gravity waves. *J. Fluid Mech.* 62:553–78

Scott, A. C., Chu, F. Y. F., McLaughlin, D. W. 1973. The soliton: a new concept in applied science. *Proc. IEEE* 61:1443–91

Stokes, G. G. 1849. On the theory of oscillatory waves. *Trans. Cambridge Philos. Soc.* 8:441–55. *Math. Phys. Pap.* 1:197–229

Stokes, G. G. 1880. Supplement to a paper on the theory of oscillatory waves. *Math. Phys. Pap.* 1:314–26. London: Cambridge Univ. Press

Thyagaraja, A. 1979. Recurrent motions in certain physical systems. *Phys. Fluids.* 22(2): In press

Toland, J. F. 1978. On the existence of a wave of greatest height and Stokes's conjecture. *Proc. R. Soc. London Ser. A* 363:469–85

Whitham, G. B. 1965. A general approach to linear and nonlinear dispersive waves using a Lagrangian. *J. Fluid Mech.* 22:273–83

Whitham, G. B. 1967. Nonlinear dispersion of water waves. *J. Fluid Mech.* 27:399–412

Whitham, G. B. 1970. Two-timing, variational principles and waves. *J. Fluid Mech.* 44:373–95

Wilton, J. R. 1913. On the highest wave on deep water. *Philos. Mag.* (6)26:1053–58

Yamada, H. 1957. Highest waves of permanent type on the surface of deep water. *Rep. Res. Inst. Appl. Mech., Kyushu Univ.* 5, no. 18:37–52

Yuen, H. C. 1977. Nonlinear deep-water waves. V. Unsteady wavebreaking. *TRW Rep. No. 31326-6013-RU-00*

Yuen, H. C., Ferguson, W. E. Jr. 1978a. Relationship between Benjamin-Feir instability and recurrence in the nonlinear Schrödinger equation. *Phys. Fluids* 21:1275–78

Yuen, H. C., Ferguson, W. E. Jr. 1978b. Fermi-Pasta-Ulam recurrence in the two-space dimensional nonlinear Schrödinger equation. *Phys. Fluids* 21:2116–18

Yuen, H. C., Lake, B. M. 1975. Nonlinear deep water waves: Theory and experiment. *Phys. Fluids* 18:956–60

Zakharov, V. E. 1968. Stability of periodic waves of finite amplitude on the surface of a deep fluid. *Zh. Prikl. Mekh. Tekh. Fiz.* 1968, no. 2:86–94. Translated in *J. Appl. Mech. Tech. Phys.* 1968, no. 2:190–94

Zakharov, V. E., Rubenchik, A. M. 1973. Instability of waveguides and solitons in nonlinear media. *Zh. Eksp. Teor. Fiz.* 65:997–1011. Translated in *Sov. Phys. JETP* 38:494–500 (1974)

Zakharov, V. E., Shabat, A. B. 1971. Exact theory of two-dimensional self-focusing and one-dimensional self-modulating waves in nonlinear media. *Zh. Eksp. Teor. Fiz.* 61:118–134. Translated in *Sov. Phys. JETP* 34:62–69 (1972)

Ann. Rev. Fluid Mech. 1980. 12:335–63

STOKESLETS AND EDDIES IN CREEPING FLOW

Hidenori Hasimoto and Osamu Sano
Department of Physics, University of Tokyo, Tokyo, Japan

1 INTRODUCTION

Flow at low Reynolds number $\mathrm{Re} = \rho UL/\mu$ is characterized by the smallness of representative quantities in the flow, i.e. the density ρ, the velocity U, and the length L, as well as by large values of the viscosity μ. If we let $\mathrm{Re} \to 0$ and neglect the molecular structure of the fluid that is studied in rarefied fluid dynamics as well as the Brownian motions that appear for extremely small ρ and L, we are led to a universal feature of a real fluid. We have a perfectly laminar creeping flow governed by the Stokes equations of motion, which are linear and reversible, and contain no parameters to complicate the structure of the flow field.

The most essential factor is the geometry of the flow. As is well known, the Stokes equations are not uniformly valid in an unbounded fluid, leading to various paradoxes, which however are remedied by matching to an outer region where the neglected inertia terms are dominant. Further, in real situations, the flow is usually bounded and is dominated by the so-called Stokes regions if we let $\rho U/\mu$ be sufficiently small.

In this review certain limited aspects of creeping flow will be surveyed, since extensive surveys before 1964 are found in Happel & Brenner's treatise (1965) and in several articles in the *Annual Review of Fluid Mechanics* (Cox & Mason 1971, Brennen & Winet 1977). First, we neglect unsteady effects on the assumption that the typical frequency is not larger than U/L, as well as the effect of free surfaces such as bubbles and suspensions. In Section 2 Stokes flow and various singularities are introduced as factors as free of geometries as possible, leading to general solutions of the Stokes equations proposed by Imai (1973). In Section 3 two-dimensional solutions of the Stokes equations are discussed as the inner solutions near a cylindrical body. The complex velocity at a distance is characterized by the logarithmic term and constants depending on the size and ellipticity of the cylinder. In Section 4 typical stokeslet constants depending on the geometry of a particle, i.e. the

0066-4189/80/0115-0335$01.00

force and torque, are surveyed. In Section 5, as an example of showing the effect of a boundary, the classical Lorentz formulas yielding the image field due to the presence of a plane wall are newly derived in terms of the solutions in Section 2. The effect of a cylindrical surface is considered in Section 6, leading to various changes in the far field of a stokeslet including the appearance of eddies. The application to the nonlinear effect for a translating particle is sketched. In Section 7, recent studies of various eddies formed on the surface of walls and moving bodies are reviewed.

2 STOKESLET

Let us consider the Stokes equations for the creeping motion of an incompressible viscous fluid in the presence of a concentrated body force $\mathbf{K}_\mathrm{P}\delta(\mathbf{x}-\mathbf{X}_\mathrm{P})$ at $\mathbf{x}=\mathbf{X}_\mathrm{P}$, where δ denotes Dirac's delta function. Then the Stokes equations are given by

$$\nabla\cdot\mathbf{v}=0, \tag{2.1}$$

$$\nabla\cdot\mathbf{T}=-\nabla p+\mu\Delta\mathbf{v}=-\mathbf{K}_\mathrm{P}\delta(\mathbf{x}-\mathbf{X}_\mathrm{P}), \tag{2.2}$$

with

$$\begin{aligned} T_{ij}&=-p\delta_{ij}+T'_{ij},\\ T'_{ij}&=\mu(\partial_i v_j+\partial_j v_i), \end{aligned} \tag{2.3}$$

where $\mathbf{v}$ (v_i) is the velocity, $\mathbf{T}$ (T_{ij}) the stress tensor, p the pressure, μ the viscosity, and ∂_i denotes $\partial/\partial x_i$.[1]

Taking the divergence of (2.2) and using (2.1) we have

$$\Delta p=(\mathbf{K}_\mathrm{P}\cdot\nabla)\delta(\mathbf{x}-\mathbf{X}_\mathrm{P}). \tag{2.4}$$

A particular solution p_s of (2.4) is given by

$$p_\mathrm{s}=(\mathbf{K}_\mathrm{P}\cdot\nabla)\phi_\mathrm{P}, \tag{2.5}$$

where

$$\Delta\phi_\mathrm{P}=\delta(\mathbf{x}-\mathbf{X}_\mathrm{P}), \tag{2.6}$$

i.e.

$$\phi_\mathrm{P}=\begin{cases}-1/4\pi R & \text{(three-dimensional case)},\\ (2\pi)^{-1}\log R & \text{(two-dimensional case)},\end{cases} \tag{2.7}$$

with $R=|\mathbf{x}-\mathbf{X}_\mathrm{P}|$.

[1] Hereafter we make use of vector and tensor notation freely and the summation convention is adopted.

Equation (2.5) shows that the induced pressure p is represented by a harmonic dipole whose axis is parallel to $\mathbf{K}_\mathrm{P}$. Substituting (2.5) into (2.2) and noticing (2.6) and the identity

$$\Delta[(\mathbf{x}-\mathbf{X}_\mathrm{P})\cdot\mathbf{K}_\mathrm{P}\phi_\mathrm{p}] = 2(\mathbf{K}_\mathrm{P}\cdot\nabla)\phi_\mathrm{p}, \tag{2.8}$$

we have

$$\begin{aligned}\mu\mathbf{v}_\mathrm{s} &= \tfrac{1}{2}\nabla[(\mathbf{x}-\mathbf{X}_\mathrm{P})\cdot\mathbf{K}_\mathrm{P}\phi_\mathrm{p}]-\mathbf{K}_\mathrm{P}\phi_\mathrm{p},\\ &= \tfrac{1}{2}\nabla(\mathbf{x}\cdot\mathbf{K}_\mathrm{P}\phi_\mathrm{p})-\mathbf{K}_\mathrm{P}\phi_\mathrm{p}-\tfrac{1}{2}(\mathbf{K}_\mathrm{P}\cdot\mathbf{X}_\mathrm{P})\nabla\phi_\mathrm{p}.\end{aligned} \tag{2.9}$$

The fundamental solution represented by (2.5) and (2.9) is called a stokeslet, following Hancock (1953). Its long range effect characterized by ϕ_p should be noted. Taking the curl of (2.9), we have for the vorticity

$$\mu\omega_\mathrm{s} = \mathbf{K}_\mathrm{P}\times\operatorname{grad}\phi_\mathrm{p} = -\operatorname{curl}(\mathbf{K}_\mathrm{P}\phi_\mathrm{p}). \tag{2.10}$$

Integrating (2.3) over an arbitrary domain containing $\mathbf{X}_\mathrm{P}$ surrounded by the closed surface S and using Gauss's theorem, we have

$$\int \mathbf{T}\cdot d\mathbf{S} = -\mathbf{K}_\mathrm{P}, \tag{2.11}$$

which shows that the force $\mathbf{F}$ acting on this domain from the outside is just $-\mathbf{K}_\mathrm{P}$, i.e. the reaction to the stokeslet. As in electrostatics, the general solution of the Stokes equation is regarded as the superposition of stokeslets with various intensity $\mathbf{K}_\mathrm{P}$ and position $\mathbf{X}_\mathrm{P}$ over the obstacles and walls outside of the flow field. Summing with respect to $\mathbf{X}_\mathrm{P}$ we get

$$p = 2\nabla\cdot\boldsymbol{\phi}, \qquad \mu\omega = -2\nabla\times\boldsymbol{\phi}, \tag{2.12}$$

and

$$\mu\mathbf{v} = \operatorname{grad}(\mathbf{x}\cdot\boldsymbol{\phi})-2\boldsymbol{\phi}-\operatorname{grad}\chi, \tag{2.13}$$

which yield the viscous stress

$$T'_{ij} = 2[x_k\partial_i\partial_j\phi_k-\partial_i\partial_j\chi], \tag{2.14}$$

where

$$\begin{aligned}\boldsymbol{\phi} &= \tfrac{1}{2}\int \mathbf{K}_\mathrm{P}\phi_\mathrm{p}\,d\mathbf{X}_\mathrm{P},\\ \chi &= \tfrac{1}{2}\int (\mathbf{K}_\mathrm{P}\cdot\mathbf{X}_\mathrm{P})\phi_\mathrm{p}\,d\mathbf{X}_\mathrm{P},\end{aligned} \tag{2.15}$$

which are harmonic outside of the domain of summation:

$$\Delta\boldsymbol{\phi} = 0, \qquad \Delta\chi = 0. \tag{2.16}$$

Table 1 Fundamental singularities at the origin

	Stokeslet	Di-stokeslet: Stresslet	Di-stokeslet: Rotlet (Couplet)	Di-stokeslet: Di-stokeslet
ϕ_i	$-\frac{1}{2}F_i\phi_0$	$-\frac{1}{2}S_{ij}\partial_j\phi_0$	$-\frac{1}{2}M_{ij}\partial_j\phi_0$	$-\frac{1}{2}D_{ij}\partial_j\phi_0$
p	$-\text{div}\,(\mathbf{F}\phi_0)$	$-S_{ij}\partial_i\partial_j\phi_0$	0	$-D_{ij}\partial_i\partial_j\phi_0$
$\mu\omega_i$	$\text{curl}\,(\mathbf{F}\phi_0)$	$\varepsilon_{ijk}S_{kl}\partial_j\partial_l\phi_0$	$\frac{1}{2}M_j\partial_i\partial_j\phi_0$	$\varepsilon_{ijk}D_{kl}\partial_j\partial_l\phi_0$
T'_{ij}	$-(\mathbf{F}\cdot\mathbf{x})\partial_i\partial_j\phi_0$	$-x_k\partial_i\partial_j S_{kl}\partial_l\phi_0$	Similar to the left.	
$F_i = \int T_{ij}\,dS_j$	F_i	0	0	0
$M_i = \int \varepsilon_{ijk}x_j T_{kl}\,dS_l$	0	0	$M_i = \varepsilon_{ijk}M_{jk}$	M_i

Equations (2.12) and (2.13) are nothing but the general solution of the Stokes equation proposed by Imai (1972).[2]

Let us consider a rigid body and calculate the force **F** and the couple **M** acting on the body by the formulas,

$$F_i = \int T_{ij}\,dS_j, \qquad M_i = \int \varepsilon_{ijk}x_j T_{kl}\,dS_l, \tag{2.17}$$

where $d\mathbf{S}$ is the surface element of an arbitrary closed surface enclosing the body and ε_{ijk} is the alternating unit tensor, defined to be 1 for an even permutation of $(i,j,k) = (1,2,3)$ and -1 if it is an odd permutation. For the evaluation of (2.17), it is convenient, as in electrostatics, to decompose ϕ_i into a regular part Φ_i and a singular part, i.e. a source, doublet, . . ., which are singular at the origin inside of the body,

$$\phi_i = \Phi_i - \tfrac{1}{2}[F_i + D_{ij}\partial_j + \cdots]\phi_0, \qquad \phi_0 = \phi_{\rm p}(X_{\rm P} = 0).$$

Table 1 shows various singularities at the origin obtained in this manner (Batchelor 1970a, Imai 1972, Chwang & Wu 1975), where S_{ij} and M_{ij} are respectively the symmetric and antisymmetric parts of D_{ij}, i.e. $\frac{1}{2}(D_{ij} \pm D_{ji})$.

In this manner a stokeslet and antistokeslet with opposite **F** yield a di-stokeslet at a distance, with $|\mathbf{v}| = O(r^{-2})\,[O(r^{-1})$ for two dimensions$]$. Such situations occur for the aggregate of many bodies having only

[2] Though the function χ may be eliminated by a kind of gauge transformation $\tilde{\phi} = \phi - \text{grad}\,\chi$ (Imai 1973), it is rather convenient to retain this freedom.

relative motions without total net force, as in the locomotion of micro-organisms, or in the motion of a body in the presence of large boundaries exerting image forces on the flow (Sections 5 and 6). The first-order effect of the proximity of a boundary on the Stokes resistance of an arbitrary body is easily obtainable by the use of such image forces (Happel & Brenner 1965, Williams 1966). Sometimes, an exponentially small field at a distance is possible. In such cases the Stokes equations become uniformly valid even at a distance, and the Stokes paradox in two-dimensional flow disappears. In addition, the superposition introduces recirculation of the flow, which sometimes leads to the occurrence of stationary eddy patterns (Section 7). Sometimes, there appears a situation where the superposition of infinitely many stokeslets of the same sign is necessary, as in the case of the flow through an infinite array of bodies (Hasimoto 1959, 1974, Saffman 1973). In such cases, the occurrence of a pressure drop (in a planar array) or a mean pressure gradient (in a space array) rescues the divergence of the summation. Extensions to the flow through semi-infinite periodic space arrays of small spheres are possible (Ishii 1979).

3 TWO-DIMENSIONAL CASE

It is instructive to start from the case of unidirectional flow $\mathbf{v} = v_3(x_1, x_2)\mathbf{e}_3$ as is induced by an infinite cylinder C moving parallel to the x_3-axis, though it is not the proper two-dimensional case. If we put $x_1 = x_2 = 0$ in (2.9) and note $\mathbf{K}_\mathrm{P} = -F_3\mathbf{e}_3$, we have

$$p_\mathrm{s} = 0 \quad \text{and} \quad \mu\mathbf{v}_\mathrm{s} = F_3\phi_\mathrm{p}\mathbf{e}_3,$$

i.e.

$$\mu v_{3\mathrm{s}} = (F_3/2\pi)\log r, \qquad r = |\mathbf{x}|. \tag{3.1}$$

Though this expression is divergent for $r \to \infty$ this solution may be regarded as the asymptotic form of the inner solution for the axial velocity near a straight slender body, where F_3 is a slowly varying function of x_3 except near the tips (Batchelor 1970b). The flow near an arbitrary cylinder making quasi-steady axial motion belongs to this type (Hasimoto 1954, Batchelor 1954).

If we solve the two-dimensional Laplace equation $\Delta v_3 = 0$, on the condition that $v_3 \to v_{3\mathrm{s}}$ at a distance and $v_3 = v_{30}$ on C, we have (Hasimoto 1954, Batchelor 1954)

$$\mu(v_3 - v_{30}) = (F_3/4\pi)\log\zeta\bar{\zeta} \sim (F_3/4\pi)[\log(z\bar{z}) + \lambda_0], \tag{3.2}$$

with $\lambda_0 = -2\log a$, where $z = x_1 + ix_2$, and we have mapped conformally the domain outside of C onto the domain outside of a unit circle $\zeta\bar{\zeta} = 1$ by use of the analytic function

$$z = a[\zeta + h(\zeta)] \sim a[\zeta + 0(\zeta^{-1})]. \tag{3.3}$$

Here a may be regarded as the radius of an equivalent circular cylinder. If we find by some means $\mu\mathbf{v}_\infty$, the inner expansion of the outer solution, F_3 can be determined by the method of matching (Van Dyke 1964). That is, if the outer solution has an inner expansion of the form

$$\mu v_{3\infty} = (F_3/4\pi)[\log(z\bar{z}) - 2\log L + \lambda_\infty], \tag{3.4}$$

F_3 is given by

$$F_3 = -4\pi\mu v_{30}/[2\log(L/a) - \lambda_\infty]. \tag{3.5}$$

For ordinary two-dimensional flow with $\mathbf{v} = (v_1, v_2) = (u, v)$ in the $(x_1, x_2) = (x, y)$ plane, it is convenient to introduce the stream function ψ defined by

$$u = \partial\psi/\partial y, \qquad v = -\partial\psi/\partial x, \tag{3.6}$$

and the complex variables

$$z = x + iy \quad \text{and} \quad \bar{z} = x - iy. \tag{3.7}$$

Since the vorticity $\omega = -\Delta\psi$ is harmonic according to (2.12), we have a biharmonic equation for ψ,

$$\Delta^2\psi = 4\partial^2\psi/\partial z\partial\bar{z} = -\Delta\omega = 0, \tag{3.8}$$

whose integral yields

$$\mu\psi = \text{Im}\,[f(z) + \bar{z}g(z)], \tag{3.9}$$

where f and g are analytic functions of z. Making use of (3.6) and (3.9) we have for the complex velocity $w \equiv u - iv$

$$\mu w = 2i\mu\partial\psi/\partial z = f'(z) + \bar{z}g'(z) - \bar{g}(\bar{z}). \tag{3.10}$$

Noticing further that

$$\mu\omega = -\mu\Delta\psi = -4\text{Im}\,g'(z) \tag{3.11}$$

is the harmonic conjugate to p, we have

$$p - i\mu\omega = 4g'(z). \tag{3.12}$$

We may remark the relation of f and g to ϕ and χ in (2.12) and (2.13):

$$\phi_1 + i\phi_2 = g(z) \quad \text{and} \quad \chi = -\text{Re}\,f(z). \tag{3.13}$$

Then for the stokeslet we may take

$$g_s(z) = -c \log z, \qquad f_s'(z) = \bar{c} \log z, \tag{3.14}$$

with

$$c = (X + iY)/8\pi, \tag{3.15}$$

(X, Y) being the force acting on the stokeslet.

If we consider the flow past a cylinder and satisfy the inner boundary condition $w = w_0$ on the cylinder, the flow induced by the cylinder at a distance is given by

$$\mu(w - w_0) = \mu w_s + \lambda \bar{c} + \kappa c + O(z^{-1}),$$

with (3.16)

$$\mu w_s = \bar{c} \log (z\bar{z}) - c(\bar{z}/z),$$

where λ and κ are constants depending only on the size and shape of the cylinder. The strength of c may be determined by matching with the outer flow[3] (Imai 1954, Tamada 1956, Proudman & Pearson 1957, Hasimoto 1963). Let the outer solution have an inner expansion of the form

$$\mu w = \mu w_s + \lambda_\infty \bar{c} + \kappa_\infty c. \tag{3.17}$$

Then matching (3.16) with this expression and solving with respect to c we have

$$c = \mu[(\lambda_\infty - \lambda)\bar{w}_0 - (\bar{\kappa}_\infty - \bar{\kappa})w_0]/[|\lambda - \lambda_\infty|^2 - |\kappa_\infty - \kappa|^2]. \tag{3.18}$$

In the case of an elliptic cylinder whose half axes are $a(1+\sigma^2)$ and $a(1-\sigma^2)$, λ and κ are known to be (Berry & Swain 1923) $\lambda = \lambda_0 \equiv -2 \log a$, $\kappa = \sigma^2$. For this reason λ may be regarded as a constant of an equivalent circular cylinder and κ as that of an equivalent elliptic cylinder, in a similar manner as Batchelor (1970b), who used a tensor K_{ij} relating ψ and the force. Unfortunately, there are few examples in which λ and κ are explicitly known, in contrast to axial flow problems. In the case when $h(\zeta)$ is a rational function of ζ, they as well as f and g are obtained similarly to (3.2) taking into account (3.10), (3.14), (3.16), and the no-slip condition on C.

Table 2 gives the values of λ and κ obtainable in this manner. It should be noted that $\lambda - \lambda_0$ and κ are independent of a. The $(n+1)$-indented cylinder has $(n+1)$-fold symmetry about the origin, with $a(1 \pm \beta)$ as maximum and minimum radii, and cusps for $n\beta = 1$. The fact that $\kappa = 0$ may be attributed to its almost circular cross section.

[3] The outer field may be inertial (Oseen type), a three-dimensional Stokes flow, or a two-dimensional Stokes field satisfying the boundary condition on the wall.

Table 2 Constants of an equivalent circular cylinder λ and an equivalent elliptic cylinder κ

	$h(\zeta)$	$\lambda-\lambda_0$	κ
Elliptic cylinder	$\frac{\sigma^2}{\zeta}, \sigma^2 \leqq 1$	0	σ^2
$(n+1)-$fold indented cylinder	$\frac{\beta}{\zeta^n}, \; n\beta \leqq 1, \; n \geqq 2$	$-(n-1)\beta^2$	0
Joukowski profile	$\frac{\sigma^2}{\zeta+\alpha}, \; \lvert\sigma \pm \alpha\rvert \leqq 1, \; \lvert\alpha\rvert \leqq 1$	$-\frac{\sigma^4 \alpha\bar{\alpha}}{Q}$[a]	$\frac{\sigma^2[(1-\alpha\bar{\alpha})^2-\sigma^2\alpha^2]}{Q}$[a]
Circular arc	$\frac{\sigma^2}{\zeta+is}, \; \sigma = \cos(\gamma/2), \; s = \sin(\gamma/2)$	$\frac{-\sigma^2 s^2}{1+s^2}$	$\frac{\sigma^2}{1+s^2}$

[a] $Q = (1-\alpha\bar{\alpha})^2 - \sigma^2(\alpha^2+\bar{\alpha}^2)$.

The circular arc is given by $r = a/s$ and $|\theta| \leqq \gamma$ in cylindrical coordinates, and its chord length is $4a\sigma$, being obtained as a special case of the Joukowski profile (Hasimoto 1979). It may be noted in these examples that $\lambda - \lambda_0$ is not zero and the equivalent radius is larger than a except for the case of the elliptic cylinder, where $\lambda = \lambda_0$.

Many studies of slender bodies have been made in connection with the motion of microorganisms, leading to Keller & Rubinow's (1976) study for arbitrary deformation of non-uniform strings with variable circular cross section and Geer's (1976) expansion in $\varepsilon^m (\log \varepsilon)^{-n}$ (ε being the slenderness) for arbitrary flow past an arbitrary body of revolution. The methods of matching as well as the integral equation for the stokeslet distribution (two- or three-dimensional for a near or far field) have been widely adopted. Since extensive reviews by Lighthill (1975, 1976) and Brennen & Winet (1977) are available no repetition will be made here.

4 FORCE AND TORQUE ON A PARTICLE

The force $\mathbf{F}$ and the torque $\mathbf{M}$ on a particle moving with velocity $\mathbf{U}$ and rotating with angular velocity $\mathbf{\Omega}$ are determined by the solution of the Stokes equations under the boundary condition

$$\mathbf{v} = \mathbf{U} + \mathbf{\Omega} \times \mathbf{X}_s \text{ on the surface } \mathbf{x} = \mathbf{X}_s \text{ of the particle,} \tag{4.1}$$

and

$$\mathbf{v} \to 0 \text{ as } |\mathbf{x}| \to \infty. \tag{4.2}$$

F and **M** are given by

$$F_i = -\mu(K_{ij}U_j + C_{ij}\Omega_j),$$
$$M_i = -\mu(C^T_{ij}U_j + L_{ij}\Omega_j), \tag{4.3}$$

where $K_{ij} = K_{ji}$, $L_{ij} = L_{ji}$, and $C^T_{ij} = C_{ji}$ depend on the size and shape of the particle (Happel & Brenner 1965).[4]

Explicit analytical forms of these tensors have been known only for the classical case of the ellipsoid and its special cases. For a dumbbell of two equal spheres of radius a in contact, $K_s(=K_{33})$ and $K_{11}(=K_{22})$ are respectively 0.64514 (Goldman et al 1966) and 0.72426 (O'Neill 1969) multiplied by $12\pi a$. It is remarkable that these tensors have recently been determined for the spherical cap ($r = a, |\theta| \leqq \pm\alpha$), which is a body without fore-aft symmetry. In a coordinate system with the origin at $r = 0$ and the x_3-axis along the axis of symmetry ($\theta = 0$), their non-vanishing elements are given by

$$K_{11} = K_{22} = a[6A - S^2/(6A)] \text{ (Takagi 1974a, Dorrepaal 1976)},$$
$$K_s = K_{33} = a(6A+S) \text{ (Collins 1963)},$$
$$C_{12} = -C_{21} = a^2S[2+S/(3A)] \text{ (Takagi 1974a, Dorrepaal 1976)},$$
$$L_{11} = L_{22} = (2/3)a^3(12\alpha + 9 \sin\alpha + \sin 3\alpha - S^2/A) \text{ (Takagi 1974a)},$$
$$L_s = L_{33} = 4a^3[2\alpha - \sin 2\alpha + (4/3)\sin^3\alpha] \text{ (Collins 1959)},$$

where

$$A = \alpha + \sin\alpha,$$
$$S = 2\sin\alpha + \sin 2\alpha.$$

Takagi solved, in a manner similar to Collins who solved the axisymmetric case, three pairs of dual series equations in associated Legendre functions. Dorrepaal applied the integral transform technique invented by Ranger (1973) who gave explicit results for the hemispherical cap (with some calculational errors). Numerical values of these tensors are given in Takagi's paper. Takagi (1974b) and Dorrepaal (1978a) discuss the motion of the cap in the presence of shear flow with special reference to the case of free motion (**F** = **M** = 0) and the latter author calculates the trajectory in a manner similar to Nir & Acrivos (1973) for an asymmetric spherical dumbbell.

[4] The symmetries of the extended matrices to include the stresslet and the pure strain are discussed by Hinch (1972).

If we appeal to numerical methods, the numerical values of these tensors may be in principle obtainable by solving a system of linear equations of the first kind for a distribution of stokeslets over the particle surface (Youngren & Acrivos 1975). Purely three-dimensional examples are still awaited. For an arbitrary convex body of revolution, we may cite the paper by Gluckman et al (1972). In this connection we should mention that the axisymmetric body of the smallest drag for a given volume was proposed to be that which has fore-aft symmetry and conical ends of angle 120° (Pirronneau 1973), and the ratio C_f of K_s to that on the sphere of equal volume is numerically determined to be 0.95425 (Bourot 1974). Wakiya's (1976) corresponding value for the spindle of angle 120° is 0.960.

For the value of L_s we mention Chwang & Wu's (1974) results for prolate bodies by the use of axial distribution of rotlets, some of which may be derived simply by the theorem of Collins (1955) and Kanwal (1961) for axisymmetric bodies,

$$L_s = 4(M+V),$$

where M is the virtual mass of the same body of volume V translating with unit velocity in a perfect fluid of density 1.

If we confine ourselves to axisymmetric flows, several new values of the drag coefficient K_s have been obtained since Payne & Pell's paper (1960) which is cited in Happel & Brenner (1965) and the value for the spherical cap is corrected by Collins (1959). We may mention Masuda's (1970) value $8\pi^2 b/[\frac{1}{2}+\ln(8b/a)]$ for a thin ring of thickness $2a$ and diameter $2b$ as well as $11.22\pi a$ for a closed torus (Takagi 1973). The values for general b/a are given by Wakiya (1974) and Majumdar & O'Neill (1977). The flux through the opening is also given by Wakiya. Wakiya (1976) also corrected the numerical values for spindles of various thickness given by Stasiw et al (1974). For a cardioid of revolution $r = a(1-\cos\theta)$, Bourot (1975) and Richardson (1977) obtained the correct value $7.68474\pi a$ (a misprint in Bourot is corrected). The case of an asymmetrical spherical dumbbell is treated by Cooley & O'Neill (1969) as well as by Goren (1970). It is seen that the value of K_s for one body which can be completely inside of the surface of other body is less than that for the latter body (compare the values $16:5.61\pi:6\pi$ for circular disk, closed torus, and sphere with equal radius). Such inequalities have been predicted by Hill & Power's extremum principles (1956), starting from the upper and lower bounds of dissipation. They are extended by Keller et al (1967), Skalak (1970), and Nir et al (1975) so as to treat suspensions, drops, and shearing flows.

5 LORENTZ FORMULA

As a simple example treating the wall effect, a simple derivation of the Lorentz formula (1907) for the image system due to the presence of the wall $x_1 = 0$ is presented here.

Let v_i and p be the flow field represented by (2.12) and (2.13). It is required to find $\tilde{v}_i$ and $\tilde{p}$ which make

$$v_i + \tilde{v}_i = 0,$$

at $x_1 = 0$.

For this purpose it is convenient to introduce the reflection of $f(x_1,x_2,x_3)$ by $f^*(x_1,x_2,x_3) = f(-x_1,x_2,x_3)$ and observe the following lemmas for any harmonic function ϕ, which are easily proved.

i) $\Delta\phi^* = 0$,

ii) $\Delta(x_i\partial_j\phi - x_j\partial_i\phi) = 2\partial_i\partial_j\phi - 2\partial_j\partial_i\phi = 0$,

iii) $\phi^* = \phi,\ \partial_1\phi^* = -(\partial_1\phi)^* = -\partial_1\phi$, at $x_1 = 0$.

Then the four harmonic functions $\tilde{\phi}_j$ and $\tilde{\chi}$ yielding the reflected field $\tilde{v}_j$ and $\tilde{p}$ are found as follows.

The tangential velocities v_α at $x_1 = 0$ are given by

$$v_\alpha = x_\beta\partial_\alpha\phi_\beta - \phi_\alpha - \partial_\alpha\chi.$$

In this section, Greek-letter suffixes are used to denote 2 or 3. If we note that ϕ_1 and x_1 derivatives are not contained in this expression, we have only to put

$$\tilde{\phi}_\beta = -\phi_\beta^*,\ \tilde{\chi} = -\chi^*, \tag{5.1}$$

in order to satisfy the condition $v_\alpha + \tilde{v}_\alpha = 0$ at $x_1 = 0$.

Introducing these expressions into the condition $v_1 + \tilde{v}_1 = 0$ at $x_1 = 0$,

$$\tilde{\phi}_1 = -\phi_1 + x_\alpha(\partial_1\phi_\alpha + \partial_1\tilde{\phi}_\alpha) - \partial_1\chi - \partial_1\tilde{\chi},$$

and rewriting by use of lemma iii), we have

$$\tilde{\phi}_1 = -\phi_1^* + 2(x_\alpha\partial_1\phi_\alpha - x_1\partial_\alpha\phi_\alpha - \partial_1\chi)^* \text{ at } x_1 = 0,$$

where the vanishing term multiplied by x_1 is appended for the sake of analytic continuation. Then $\tilde{\phi}_1$ for $x_1 \geqq 0$ is found to be

$$\tilde{\phi}_1 = -[\phi_1 + 2(x_1\partial_\alpha\phi_\alpha - x_\alpha\partial_1\phi_\alpha) + 2\partial_1\chi]^*, \tag{5.2}$$

which is harmonic owing to the lemma i) and ii). The results for the image system of a stokeslet at $(l,0,0)$

$$\phi_j = c_j/r \quad \text{and} \quad \chi = lc_1/r,\ c_j = F_j/(8\pi) \tag{5.3}$$

are obtained as follows:

$$\tilde{\phi}_1 = -[c_1/r + 2lc_j\partial_j(1/r)]^*, \quad \tilde{\phi}_\beta = -c_\beta/r^*,$$
$$\tilde{\chi} = -lc_1/r^*,$$

where

$$r^2 = (x_1 - l)^2 + x_2^2 + x_3^2,$$

and (5.4)

$$r^{*2} = (x_1 + l)^2 + x_2^2 + x_3^2,$$

which are in accordance with those obtained by Blake (1971) by the use of Fourier integrals. The appearance of a di-stokeslet in addition to an anti-stokeslet is clear. The image system of a rotlet

$$\mu v_i = -2\phi_i = \varepsilon_{ijk}N_j\partial_k(1/r), \quad \chi = 0, \quad N_j = M_j/(8\pi) \tag{5.5}$$

satisfying $\partial_i\phi_i = x_i\phi_i = 0$, is

$$\tilde{\phi}_1 = -3\phi_1^*$$

and (5.6)

$$\tilde{\phi}_\beta = -\phi_\beta^*,$$

which yield

$$\mu\tilde{v}_i = -\mu v_i^* - 4x_1\partial_i\phi_1^*,$$

and (5.7)

$$\tilde{p} = -8\partial_1\phi_1^*,$$

yielding a di-stokeslet in addition to an antirotlet, in accordance with Blake & Chwang (1974).

The derivation of the Lorentz formula proceeds as follows:

$$\begin{aligned}\tilde{\phi}_1 &= [\phi_1 + 2(x_\alpha\partial_1\phi_\alpha - \phi_1 - \partial_1\chi) - 2x_1\partial_\alpha\phi_\alpha]^*,\\ &= \phi_1^* + (2\mu v_1 - x_1 p)^*,\end{aligned} \tag{5.8}$$

which as well as (5.1) yields for $\tilde{p}$

$$\begin{aligned}\tilde{p} = 2\partial_j\tilde{\phi}_j &= [-2\partial_j\phi_j - 2\partial_1(2\mu v_1 - x_1 p)]^*,\\ &= (p + 2x_1\partial_1 p - 4\mu\partial_1 v_1)^*.\end{aligned} \tag{5.9}$$

Similarly, we obtain

$$\mu\tilde{v}_1 = (-\mu v_1 + 2\mu x_1\partial_1 v_1 - x_1^2\partial_1 p)^*,$$
$$\mu\tilde{v}_\beta = (-\mu v_\beta - 2\mu x_1\partial_\beta v_1 + x_1^2\partial_\beta p)^*. \tag{5.10}$$

These are nothing but the classical Lorentz formulas.

6 FLOW BOUNDED BY A CYLINDRICAL SURFACE

The analysis in Section 4 may be extended to the flow in a domain bounded by a surface whose generators are parallel to one of the co-ordinate axes (say x_3). This surface S is represented parametrically as $x_\beta = X_\beta(s)$ $(\beta = 1, 2)$ and $x_3 = X_3$, where s is the length along the cross-sectional profile C and X_3 is the length along S. Let us suppose S is rigid and at rest. If we impose the boundary condition (Hasimoto 1976)

$$\phi_3 = 0 \qquad \text{on } S, \tag{6.1}$$

the vanishing of v_3 and the tangential velocity is satisfied by the boundary condition

$$X_\beta\phi_\beta - \chi = 0 \quad \text{and} \quad X'_\beta(s)\phi_\beta = 0 \qquad \text{on } S, \tag{6.2}$$

or

$$W\phi_1 = X'_2\chi, \qquad W\phi_2 \doteq -X'_1\chi, \tag{6.2'}$$

with

$$W = X_1X'_2 - X'_1X_2.$$

Then the vanishing of the normal component yields

$$\partial_n\chi + W^{-1}\chi = X_j\partial_n\phi_j \quad \text{with} \quad \partial_n = X'_2\partial_1 - X'_1\partial_2 \qquad \text{on } S. \tag{6.3}$$

One exceptional case arises when $W \equiv 0$, i.e. a plane surface through the origin: $X_1 = s\cos\alpha$, $X_2 = s\sin\alpha$. In this case (6.2) and (6.3) are replaced by (Sano & Hasimoto 1976, 1977, 1978)

$$\chi = 0 \quad \text{and} \quad X_j\partial_n\phi_j = n_\beta\phi_\beta + \partial_n\chi.$$

We may note as examples the Poiseuille flow in a circular cylinder of radius a. Since $X_1 = a\cos(s/a)$, $X_2 = a\sin(s/a)$, $W = a$, we have $\phi_\beta = -(1/4)Px_3x_\beta$, $\phi_3 = 0$, $\chi = -(1/4)a^2Px_3$, which yield

$$p = 2\,\mathrm{div}\,\boldsymbol{\phi} = -Px_3, \; v_i = (1/4)P(a^2 - x_\beta x_\beta)\delta_{i3}. \tag{6.4}$$

Similarly for the wedge domain $|\theta| \leqq \alpha$ we have

$$\phi_1/x_1 \sin^2 \alpha = -\phi_2/x_2 \cos^2 \alpha = (P/2 \cos 2\alpha)x_3, \tag{6.5}$$

and

$$v_3 = (P/2 \cos 2\alpha)(x_1^2 \sin^2 \alpha - x_2^2 \cos^2 \alpha). \tag{6.6}$$

It is easily seen that the boundary-value problem is simple for ϕ_3 (and χ for the exceptional case). Other conditions are rather complex. For this reason, even in the case of the reflection of stokeslets, integral transforms are adopted in dealing with a canal $|x_1| < l$ (Ho & Leal 1974, Liron & Mochon 1976), wedge-shaped domain $|\theta| < \alpha$ (Sano & Hasimoto 1976, 1977, 1978), and a circular tube (Hasimoto 1976, Liron & Shahar 1978). In contrast to the classical treatments surveyed in Happel & Brenner (1965), the direction and position of the stokeslets are arbitrary. Liron & Mochon (1976) discussed the field far from the stokeslet and showed that if it is perpendicular to the wall the total field is exponentially small in contrast to the parallel case yielding algebraically small Hele-Shaw flow. On the basis of Sano & Hasimoto's results on the force and torque for a small particle, Sano (1978) discussed the drifting effect of a corner on a sedimenting particle. Hasimoto (1976) calculated the correction to the Stokes drag for a small sphere and showed that on the axis $r = 0$ the drag D_r on a small sphere moving perpendicularly to the wall is less than the drag D_3 for the motion parallel to the wall (D_r continually increases to exceed D_3 at $r \sim 0.3a$. and D_3 has a famous minimum at $r \sim 0.4a$), in contrast to the parallel wall case, where $D_3 < D_1$ and in accordance with classical results $D_3 > D_1$ due to Faxen and Westberg for a small circular cylinder midway between plane walls. It is interesting to notice that the far field for the parallel motion decays exponentially as in Moffatt's type vortices (Section 7) as shown explicitly by Liron & Shahar (1978) in the case of a circular tube. In the classical two-dimensional elastostatical case similar results had been obtained by Buchwald (1964).

Ho & Leal (1974) calculated the lateral force on a small sphere moving parallel to canal walls taking into account the nonlinear effect. The presence of two-dimensional Poiseuille flow or shear flow was included. The general theory for the lateral force in a tube had been presented by Cox & Brenner (1968). However, their double integral expressions for the side-force contain the expression for the stokeslet perpendicular to the axial direction. Ho & Leal showed that a freely rotating neutrally buoyant rigid sphere in two-dimensional Poiseuille flow and shear flow has a stable lateral equilibrium respectively at $|x_1| \sim 0.6a$ and $|x_1| = 0$, supporting many experiments since Segré & Silberberg (1961, 1962a,b) (for a survey see Cox & Mason 1971). The inaccuracy of Ho & Leal's numerical values near the wall was pointed out by Vasseur & Cox (1976) in comparison with the asymptotic value predicted by Cox & Hsu (1977) in the study of a single wall case for many characteristic situations, which have been extensively studied by Vasseur & Cox in two-plane cases. As regards equilibrium, similar results were obtained for the same cases. Vasseur & Cox (1977) gave the side force for a small sphere sedimenting parallel to the canal walls by use of the Oseen equations, which are linear and are considered to be valid when the wall is outside of the Stokes region. It is remarkable to notice that the side force, which is always directed to the central lines, seems to tend to the values obtained by Cox & Brenner's integral in the limit of $\mathrm{Re} \to 0$. Experimental verifications are also made including studies for the presence of another sphere. Shinohara & Hasimoto (1979) made a similar analysis in the case of a circular tube and obtained larger drift velocity towards the central axis than that in the two-plane case. Further, a slight maximum of the migration velocity is found near the wall. In both cases, however, the same value $3\rho U^2 a/(32\mu)$ is attained in the vicinity of the wall.

7 CONFINED EDDIES

One of the striking phenomena in creeping flow is the appearance of confined eddies near the boundary of a wall or body. Owing to the reversibility of the Stokes flow, these eddies can appear even in front of the bodies.

Preliminary Considerations

Taking into account that these eddies are confined by dividing streamlines starting from the boundary, we want to study the local behavior of these lines near the boundary. We consider two-dimensional slow viscous flow in or around a wedge. Since the appropriate solution of the Stokes equations in this case is given by substitution of $f'(z) = \mu A z^{\lambda}$ and $g(z) = \mu B z^{\lambda}$ $(\lambda > 0)$ into Equation (3.10), the resulting complex velocity becomes

$$\begin{aligned} w &= Az^{\lambda} + \lambda B \bar{z} z^{\lambda-1} - \bar{B}\bar{z}^{\lambda}, \\ &= r^{\lambda}\{A \exp(i\lambda\theta) + \lambda B \exp[i(\lambda-2)\theta] - \bar{B}\exp(-i\lambda\theta)\}, \end{aligned} \tag{7.1}$$

where we take polar coordinates (r,θ) with their origin at the vertex and the plane $\theta = 0$ on the bisector of the wedge. The no-slip condition on the walls $\theta = \pm\alpha$ $(|\alpha| \leq \pi)$ imposes the relation between A and B,

$$A \exp(\pm i\lambda\alpha) + \lambda B \exp[\pm i(\lambda-2)\alpha] - \bar{B}\exp(\mp i\lambda\alpha) = 0. \tag{7.2}$$

The dividing streamline starting from the origin is expected to occur along the line $\theta = \varphi$ when

$$\operatorname{Im}\left[w(\theta = \varphi)\exp(i\varphi)\right] = 0, \tag{7.3}$$

that is when the velocity is locally in the radial direction.

FLOW NEAR THE VERTEX OF A WEDGE OF ARBITRARY ANGLE We shall rewrite Equation (7.2) in the following form:

$$A \sin 2\alpha + \bar{B} \sin 2(\lambda - 1)\alpha = 0, \tag{7.2a}$$

$$\lambda B \sin 2\alpha - \bar{B} \sin 2\lambda\alpha = 0, \tag{7.2b}$$

and consider the case $\sin 2\alpha \neq 0$ (i.e. $\alpha \neq \pi/2, \pi$) in the first place. Then we have

$$A = -\bar{B} \sin 2(\lambda - 1)\alpha/\sin 2\alpha, \quad B = \bar{B} \sin 2\lambda\alpha/\lambda \sin 2\alpha. \tag{7.4}$$

Since $|B| = |\bar{B}|$, we must have

$$\sin 2\lambda\alpha \pm \lambda \sin 2\alpha = 0. \tag{7.5}$$

This is the equation studied by Dean & Montagnon (1949) and by Moffatt (1964) in obtaining the flow within a wedge. For the upper sign in (7.5), B becomes purely imaginary and the flow is antisymmetric with respect to the bisector of the wedge. For the lower sign, on the other hand, B becomes real and the flow is symmetric. It is found that no real solution of (7.5) exists for 2α less than 146.3° or 159.1° (from Collins & Dennis 1976) according as the flow is antisymmetric or symmetric.

For 2α less than these critical angles, λ becomes complex $(= \xi + i\eta)$, and its complex conjugate $\xi - i\eta$ may be taken at the same time. In this case we must take (Imai 1973)

$$\begin{aligned} f' &= \pm\kappa C z^{\lambda} + \bar{\kappa}\bar{C} z^{\bar{\lambda}}, \quad g = C z^{\lambda} \pm \bar{C} z^{\bar{\lambda}}, \\ \kappa &= -\sin 2(\lambda - 1)\alpha/\sin 2\alpha, \end{aligned} \tag{7.6}$$

for symmetric (upper sign) and antisymmetric (lower sign) flow, respectively, where C is a complex constant. These flows are interpreted by Moffatt (1964) as representing an infinite sequence of vortices towards the corner. The dimensions and intensities of adjacent eddies decrease in geometric progression with common ratios $\exp(\pi/\eta)$ and $\exp(\pi\xi/\eta)$ respectively. Except for the dividing streamline that is present on the bisector in the symmetric flow, separation of streamlines occurs only on the smooth boundary of the wall.

We shall now consider the flow for 2α greater than these critical angles. The separation condition (7.3) requires

$$\cos(\lambda-1)\alpha\cos(\lambda+1)\varphi-\cos(\lambda+1)\alpha\cos(\lambda-1)\varphi=0 \tag{7.7}$$

for the antisymmetric flow. This equation has no solution in the fluid region $|\varphi|<\alpha$, so that separation does not occur. In the case of symmetric flow, the condition (7.3) becomes

$$\sin(\lambda-1)\alpha\sin(\lambda+1)\varphi-\sin(\lambda+1)\alpha\sin(\lambda-1)\varphi=0, \tag{7.8}$$

which has only one solution $\varphi=0$ in the region $|\varphi|<\alpha$. The dividing streamline occurs only along the line of symmetry near the vertex. These conclusions are in agreement with those of Michael & O'Neill (1977).

When α approaches $\pi/2$ or π, the same value of λ satisfies both equations of (7.5), which suggests the possibility of arbitrariness in separation angle. This point will be examined in the following section.

SEPARATION ON A SMOOTH BOUNDARY When $\alpha=\pi/2$, we have $\sin 2\lambda\alpha = 0$ from (7.2a,b), so that any integer values of $\lambda=n(=1, 2, 3, \ldots)$ are allowed. Returning to our original boundary condition (7.2), we have

$$A=nB+(-1)^n\bar{B}.$$

If we assume that the total flow is in general represented by a superposition of the component flows, the one corresponding to the smallest value of n is dominant sufficiently near $r=0$. For the smallest value, $n=1$, we have $w=(B-\bar{B})(z+\bar{z})$. This is a shear flow, which has no separation. If the shear stress vanishes at $r=0$, the strongest remaining component is given by $n=2$. In this case w and ψ are

$$\begin{aligned} w &= 2|B|r^2\cos\theta(2e^{i(\theta+\beta)}+e^{i(\theta-\beta)}-e^{-i(\theta+\beta)}),\\ \psi &= \tfrac{4}{3}|B|(1+8\cos^2\beta)^{\frac{1}{2}}r^3\cos^2\theta\sin(\theta-\varphi), \end{aligned} \tag{7.9}$$

where the angle of separation φ is given by

$$\varphi=-\arctan(\tfrac{1}{3}\tan\beta),$$

with $\beta=\arg B$. Thus separation occurs at any angle on a smooth boundary, in accordance with the conclusion given by Dean (1950), whose derivation, however, is quite different from ours.

SEPARATION AT A CUSPED EDGE This is another special case when $\sin 2\alpha=0$, and (7.2a,b) require half integer values of $\lambda=m/2$ ($m=1, 2, 3, \ldots$). Applying the no-slip condition on the wall we have

$$A=-\frac{m}{2}B+(-1)^m\bar{B}.$$

For $m = 1$, w and ψ are

$$w = -|B|r^{\frac{1}{2}} \cos\frac{\theta}{2}(e^{i\beta} - e^{i(\beta-\theta)} + 2e^{-i\beta}),$$

$$\psi = -\tfrac{4}{3}|B|(1+8\cos^2\beta)^{\frac{1}{2}} r^{\frac{3}{2}} \cos^2\frac{\theta}{2} \sin\left(\frac{\theta-\varphi}{2}\right). \tag{7.10}$$

Separation may occur at any angle $\varphi = 2\arctan(\frac{1}{3}\tan\beta)$ at the edge. Consequently if the flow is symmetric, the dividing streamline occurs only along the line $\varphi = 0$. The antisymmetric flow around an edge does not separate, which agrees with the conclusion obtained by Dean (1944a,b).

Confined Eddies

We now present some illustrative flows with confined eddies. In the following section we look in detail at two-dimensional, axisymmetric, and three-dimensional flows, although this classification is a matter of convenience.

TWO-DIMENSIONAL FLOW Classical work related to confined eddies includes that of Dean and Wannier. Dean (1944a,b) examined the two-dimensional shearing motion of a fluid over a projection with a round or a sharp edge in a plane. He showed that the flow does not separate from a projection that possesses fore-aft symmetry with respect to the flow, but that if the projection has reentrant geometry as well as a cusped edge, the flow separates from the edge and a single eddy forms in a closed region of a flow adjacent to the concave part of the boundary. Wannier (1950) considered the flow between two eccentrically placed circular cylinders rotating about their own axes. Separation on the smooth surface of the outer cylinder (at rest) occurs for sufficiently large eccentricity.

Since the publication of Moffatt's paper (1964), many investigations have been made of the infinite set of corner eddies. Numerical computation of streamlines of this type was given by Pan & Acrivos (1967), who made calculations of the cavity flow, and a more detailed one by Collins & Dennis (1976), who studied the secondary flow in a curved tube of triangular cross section. Experimental verification of Moffatt's asymptotic theory seems to need at least three out of an infinite number of eddies in order to compare relative positions. Such experimental verification has not yet been successfully made.

Moffatt's flow has such a universal character near a sharp corner that it appears in many situations. Schubert (1967) considered the shear flow past a circular cylinder in contact with a plane wall. The asymptotic flow near a cusped corner consists of a sequence of antisymmetric eddies corresponding to $\alpha = 0$. The general case when a circular cylinder intersects a plane at an arbitrary angle was studied by Wakiya (1975), who demonstrated the existence of antisymmetric Moffatt eddies in the vicinity of the line of intersection if the angle is less than 146.3°. Davis & O'Neill (1977a) and O'Neill (1977) calculated the separation streamlines $\psi = 0$ in this case. A slightly different situation where a cylinder is not in contact with the plane so that there is a flux of fluid between the cylinder and the plane was considered by Davis & O'Neill (1977b). They showed that when the gap is less than approximately 0.685 times the cylinder radius, single eddies form alternately on the plane and the cylinder. (See also Figure 1.)

O'Neill (1977) and Wakiya (1975, 1978) also considered the shear flow past a plane with a cylindrical depression. They showed that separation of the flow from the wall occurs if the angle of intersection between the trough and the plane is greater than 245.15°. A single eddy forms in the closed region of flow. The separation points are different from the edge except when the subtended angles become 360° (i.e. when the trough becomes a half-space bounded by two coplanar half-planes). Davis & O'Neill (1977a) who, on the other hand, examined the flow caused by a two-dimensional stagnation point flow on the plane with depression, showed that if the fluid region subtends angles greater than about 224.46° at the intersection of the trough and plane, two symmetric line vortices appear near the lowest point of the trough, and that the separation point on the wall does not coincide with the edge except when the subtended angles become 360°.

For the two-dimensional flow past a single body, the inner solution near the cylinder (Section 3) should be investigated as the universal flow in the limit of $\mathrm{Re} \to 0$. Ranger (1977) formulated the two-dimensional flow around a smooth body containing a concave region by making use of the inversion method, and Dorrepaal (1978b) extended the same calculation. They pointed out the existence of attached line vortices in the concave region. Hasimoto (1979) demonstrated, as a special case of the Joukowski profile considered in Section 3, streamlines for both symmetric and asymmetric flow past a cylinder whose cross section is a circular arc. For the symmetric case the flow always separates at the two edges of the arc and twin eddies are confined to the concave side except when the arc reduces to a strip. The calculated streamlines shown in Figure 2*a* are in agreement with Taneda's (1979a,b) experiment for a semicircle (Figure 2*b*), among extensive visualization of the confined eddies predicted by theorists. When the cylinder is turned by 90°, the

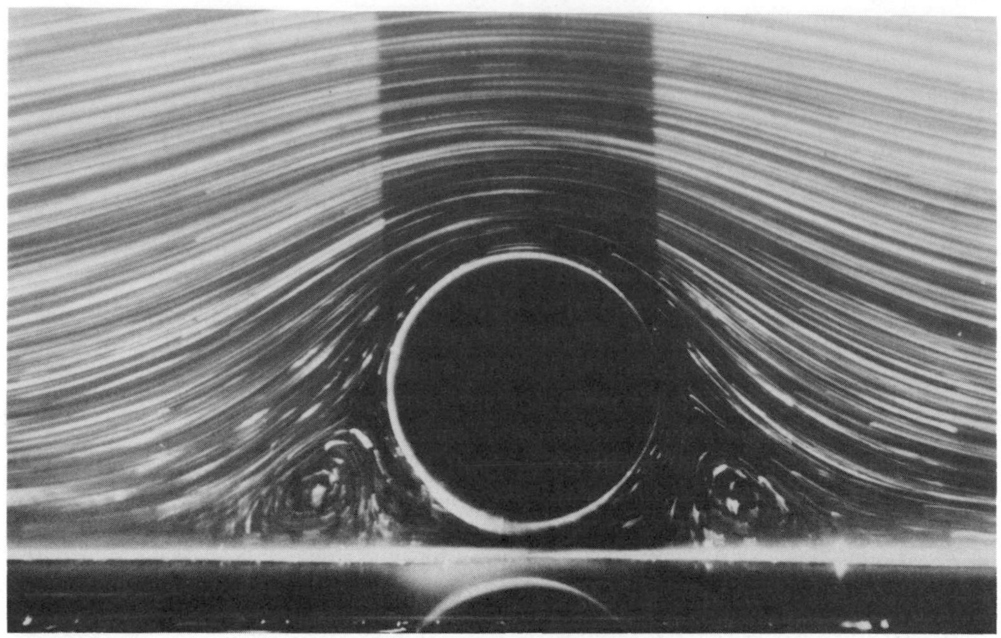

a

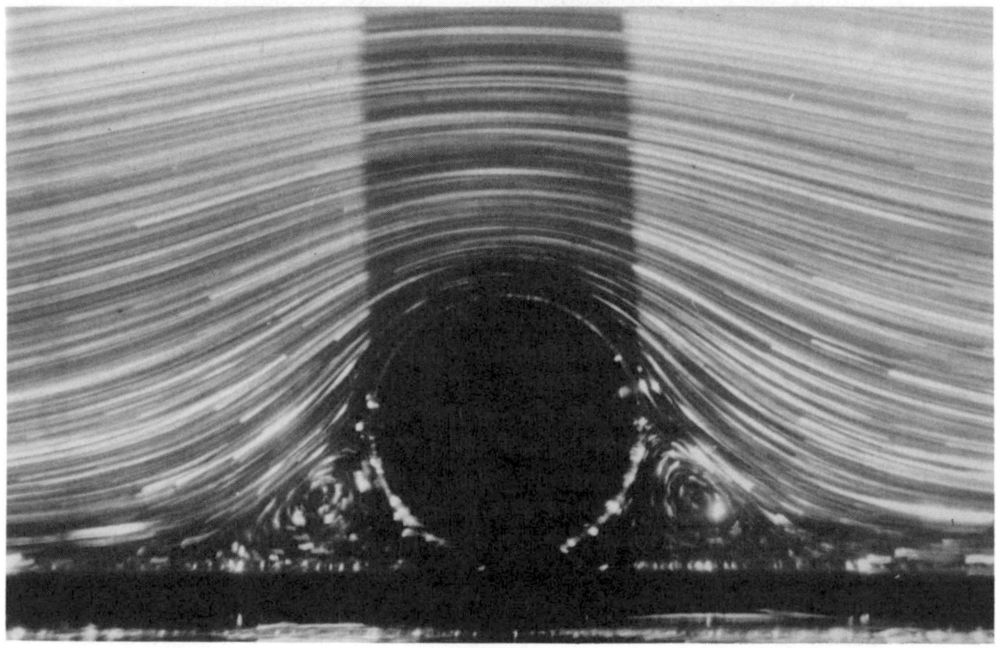

b

flow pattern is rather complicated. If we denote the semi-angle of the circular arc by γ, the flow patterns fall into the following categories: (*a*) for $\gamma < 65.88°$ no eddies are present, (*b*) at $\gamma = 65.88°$ a single line vortex appears, (*c*) for $65.88° < \gamma < 80.12°$ the eddy grows and the separation points lie on the smooth surface of the arc, (*d*) for $\gamma \geqq 80.12°$ the separation points coincide with the edges, and the eddies are confined to the concave side.

Viscous flow in a rectangular cavity driven by uniform translation of the top wall gives separated flows even in the low-Reynolds-number range. The geometry of the boundary has both depression and sharp corners, so that the two types of separated flow mentioned above are expected. Kawaguti (1961) gave numerical calculation of the flow in two-dimensional rectangular cavities of several aspect ratios (width/depth) with Reynolds numbers from 0 to 64. He showed the existence of

←*Figure 1* Streamline patterns around a circular cylinder placed near a flat plate ($Re = 1.1 \times 10^{-2}$). (*a*) $\varepsilon/a = 0.2$, (*b*) $\varepsilon/a = 0$, where ε is the gap between the plate and the cylinder with radius a. (We are indebted to Dr. Taneda for these photographs.)

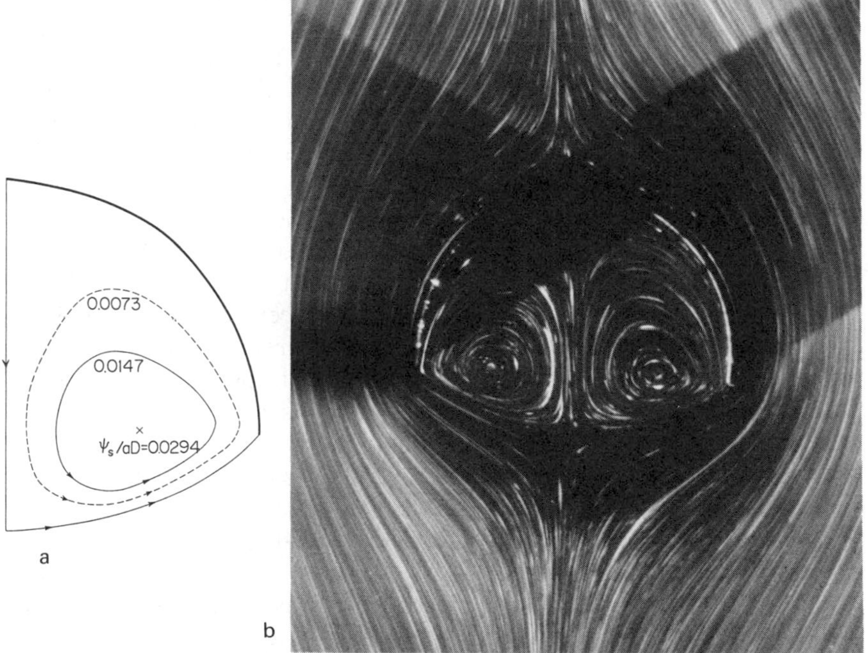

Figure 2 Streamlines around a semicircular arc. (*a*) Calculation (Hasimoto 1979), and (*b*) experiment at $Re = 3.1 \times 10^{-2}$ (Taneda 1979b).

a circulating flow inside the cavity. Takematsu (1965, 1966) treated analytically the two-dimensional cavity of infinite depth over which a shear flow prevails. He showed that in creeping flow the outer flow penetrates to a considerable depth into the cavity, and that separation (or reattachment) of the flow occurs on the cavity wall. The detailed structure of the flow was examined by Burggraf (1966), who calculated the flow in a square cavity for Reynolds numbers up to 400, and by Pan & Acrivos (1967), who analysed numerically the flow in cavities having aspect ratios from 1/4 to 5 on the basis of the Stokes equations. Pan & Acrivos demonstrated a finite number of eddies as well as Moffatt-type eddies near the bottom corners or within that portion of the flow outside the primary vortex in the case of a cavity with high aspect ratio. They also tested experimentally in a Reynolds number range from 20 to 4000.

Viscous flow due to a train of equally spaced strips of the same widths moving broadside between two parallel flat plates was analysed by Fukuyu (1975), which reveals separated regions with pairs of line vortices between the strips as long as the spacing is not large.

AXISYMMETRIC FLOW Since axisymmetric flows are qualitatively analogous to two-dimensional ones, one may expect similar eddies in axisymmetric boundaries. In fact, Bourot (1975) and Dorrepaal et al (1976a) found an infinite sequence of ring vortices of diminishing sizes and intensities toward the conical cusp of a cardioid of revolution and a closed torus, respectively, when the obstacles translate along the axis of revolution. Axisymmetric flow due to the translation of two spheres of equal radii in contact (Davis et al 1976) provides another example of Moffatt-type flow. They also investigated the flow due to the axisymmetric translation of two spheres of equal radii not in contact. They showed that (*a*) when the distance of the sphere centers is less than 3.57 times the sphere radius, wakes form on both spheres and the fluid within the wakes moves in closed eddy motion, (*b*) when the distance between the centers is less than 3.22 times the sphere radius, ring vortices attached to both spheres coalesce, and (*c*) the number of vortices increases as the distance between the spheres decreases. Photographs of streamlines corresponding to this case as well as the two-dimensional one (Figure 3) were provided by Taneda (1979a,b). The general axisymmetric Stokes flow near the vertex of a cone as well as the examples of a circular cylinder and spindles including a closed torus, was examined by Wakiya (1976), who showed the existence of an infinite set of ring vortices near the apex provided that the semi-angle of the cone is less than about 80.9°. The same critical angle was obtained independently by Kim (1976,

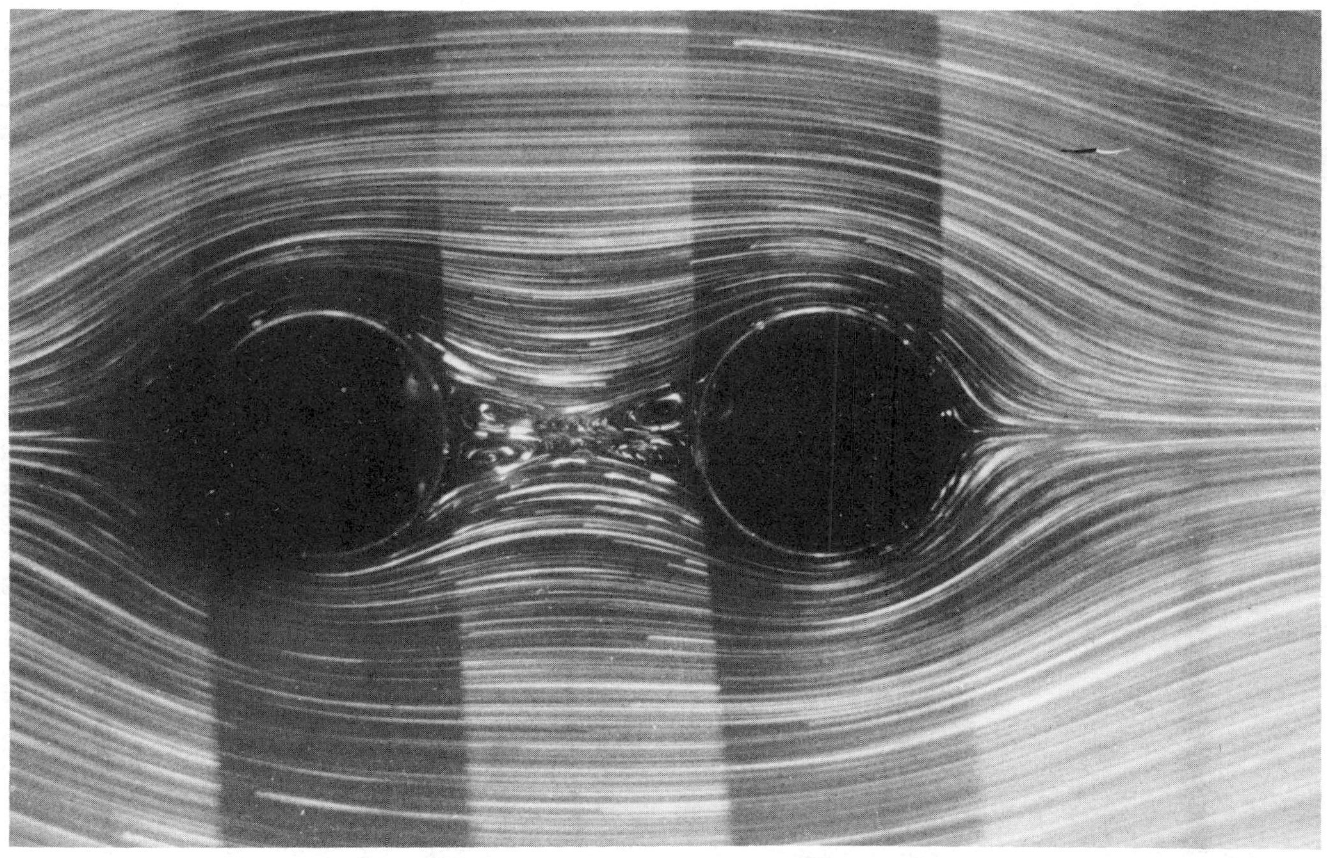

Figure 3 Streamline pattern around two equal circular cylinders placed streamwise in a uniform flow with $\varepsilon/a = 2$, $\mathrm{Re} = 1.0 \times 10^{-2}$, where ε is the gap between the two cylinders and a is the radius of the cylinder. (We are indebted to Dr. Taneda for this photograph.)

1979) who considered the axisymmetric flow due to a small sphere translating along the axis of a right circular cone. Liron & Shahar (1978) showed the first one of the eddy rings induced in a circular pipe when a point force is applied along the axis of the cylinder.

Next we consider the axisymmetric flow appearing in the concave region. The axisymmetric flow past a spherical cap, described by $r =$ constant, $0 \leqq \theta \leqq \gamma^*$, and $0 \leqq \varphi < 2\pi$ in spherical polar coordinates, was calculated by Dorrepaal et al (1976b) by use of the closed-form expression for the stream function. A single vortex ring attached to the concave surface of the cap is shown. The experiment by Collins (1979) for axisymmetric flow past a spherical cap shows good agreement with theory in that the flow separates from the rim. Michael & O'Neill (1977) analysed the flow around a spherical lens and confirmed their already mentioned general discussion about the local solution near the corner. For a thin asymmetric lens placed axisymmetrically in a uniform stream, separation occurs on its flatter or concave side with a single toroidal vortex adjacent to the axis of symmetry and a point of separation in the meridian plane lies on the flatter or concave side of the lens. As the lens shrinks to a spherical cap the separation point moves towards the rim. For a symmetric biconvex lens no separation occurs.

"Bolus flow" was originally found in blood flow through narrow capillaries (microcirculation), where the plasma makes a circulating motion relative to a train of red blood cells. This flow was studied experimentally by Prothero & Burton (1961), numerically by Bugliarello & Hsiao (1970), and analytically by Wang & Skalak (1969) and Fitz-Gerald (1972). Circulating toroidal motion in the gaps between successive red blood cells is obtained in a reference frame moving with them. These results, however, vary among themselves depending on the choice of models and the methods of solution.

THREE-DIMENSIONAL FLOW General flow patterns in the three-dimensional case are rather complicated, in spite of Takami & Kuwahara's (1974) numerical calculation of the flow with $\mathrm{Re} = 0 \sim 400$ in a cubic cavity, where the effect of the side wall is restricted to a relatively thin layer. Tokuda's (1972, 1975) boundary-layer analysis incorporated corner eddies with secondary flow. Modification of Moffatt-type eddies is anticipated by Sano (1977) for the stokeslet perpendicular to the bisector of a wedge whose angle 2α is less than 146.3°. An infinite sequence of deformed vortex rings forming a family of eccentric circles on the bisector is shown. For $\alpha = 0$, the case of parallel planes by Liron & Mochon (1976) arises. As mentioned in Section 6, for other orientations of a stokeslet, the flow at a distance is rather dominated by a two-dimensional Hele-Shaw doublet. These facts may be illustrated by a new type of eigenfunction in the corner flow

$$\text{(A1)}\ \mu v_3 = \phi = r^\nu \cos \nu\theta, \quad p = 0, \quad \nu\alpha = \pm\pi/2,$$

$$\text{(A2)}\ \mu v_3 = \Phi = \tfrac{1}{4}r^2(1 - \cos 2\theta/\cos 2\alpha), \quad p = x_3,$$

$$\text{(B1)}\ \mu v_i = [x_1\phi, x_2\phi, -(\nu+2)x_3\phi], \quad p = 2\phi,$$

$$\text{(B2)}\ \mu v_i = [x_1\Phi, x_2\Phi, -4x_3\Phi], \quad p = -2x_3^2 + \tfrac{1}{2}(x_1^2 + x_2^2) + 2\Phi.$$

(A2) corresponds to (6.6). A dividing surface $x_3 = 0$ appears for (B1) and (B2), yielding streamlines $r^{\nu+2}x_3 =$ constant and $r^4 x_3 =$ constant on the plane $\theta =$ constant. At $\alpha = \pi/4$, a new eigenfunction characterized by $\phi = \mathrm{Re}(z^2 \log z)$ arises. If the value of α exceeds $\pi/4$ the relative magnitude of (A1) and (B1) to (A2) and (B2) becomes opposite. Such a situation may be found in the classical source-sink flow in a wedge and cone (Harrison 1920, Bond 1925, Sternberg & Koiter 1958, Ackerberg 1965, Kerchman 1972); in the latter case we find flow reversal for $\alpha \geqq \pi/2$ and logarithmic flow for $\alpha = 2\pi/3$.

Dorrepaal (1976) has succeeded in expressing the transverse Stokes flow past a spherical cap in a long closed form involving elliptic integrals, starting from Ranger's series solution in spherical harmonics. The explicit flow field is discussed only in the azimuthal plane of symmetry where the flow is two-dimensional. There appear two critical angles $\gamma_1^* = 60.90°$ and $\gamma_2^* = 80.85°$, yielding a pattern similar to the case of a circular arc. The drag and the flow due to the motion of a sphere in a direction perpendicular to the axis of the cone when it is on the axis are also analysed by Kim (1979). He showed that the flow in a meridian plane reverses its direction infinitely many times near the vertex if the semi-angle of the cone is less than 74.5°.

Literature Cited

Ackerberg, R. C. 1965. The viscous incompressible flow inside a cone. *J. Fluid Mech.* 21:47–81

Batchelor, G. K. 1954. The skin friction on infinite cylinders moving parallel to their length. *Quart. J. Mech. Appl. Math.* 7:179–92

Batchelor, G. K. 1970a. The stress system in a suspension of force-free particles. *J. Fluid Mech.* 41:545–70

Batchelor, G. K. 1970b. Slender-body theory for particles of arbitrary cross-section in Stokes flow. *J. Fluid Mech.* 44:419–40

Berry, A., Swain, L. M. 1923. On the steady motion of a cylinder through infinite viscous fluid. *Proc. R. Soc. London Ser. A* 102:766–78

Blake, J. R. 1971. A note on the image system for a stokeslet in a no-slip boundary. *Proc. Cambridge Philos. Soc.* 70:303–10

Blake, J. R., Chwang, A. T. 1974. Fundamental singularities of viscous flow. Part 1. The image systems in the vicinity of a stationary no-slip boundary. *J. Eng. Math.* 8:23–29

Bourot, J. M. 1974. On the numerical computation of the optimum profile in Stokes flow. *J. Fluid Mech.* 65:513–15

Bourot, J. M. 1975. Sur le calcul de l'écoulement irrotationnel et de l'écoulement de Stokes autour d'un obstacle de révolution de méridienne cardioide; sur la structure du champs au voisinage du point de rebroussement. *C. R. Acad. Sci. Ser. A* 281:179–82

Bond, W. N. 1925. Viscous flow through wide-angled cones. *Philos. Mag. Ser. 6* 50:1058–66

Brennen, C., Winet, H. 1977. Fluid mechanics of propulsion by cilia and flagella. *Ann. Rev. Fluid Mech.* 9:339–98

Buchwald, V. T. 1964. Eigenfunctions of plane elastostatics. I. The strip. *Proc. R. Soc. London Ser. A* 277:385–400

Bugliarello, G., Hsiao, G. C. 1970. A mathematical model of the flow in the axial plasmatic gaps of the smaller vessels. *Biorheology* 7:5–36

Burggraf, O. R. 1966. Analytical and numerical studies of the structure of steady separated flows. *J. Fluid Mech.* 24:113–51

Chwang, A. T., Wu, T. Y. 1974. Hydromechanics of low-Reynolds-number flow. Part 1. Rotation of axisymmetric prolate bodies. *J. Fluid Mech.* 63:607–22

Chwang, A. T., Wu, T. Y. 1975. Hydromechanics of low-Reynolds-number flow. Part 2. Singularity method for Stokes flows. *J. Fluid Mech.* 67:787–815

Collins, R. 1979. Separation from spherical caps in Stokes flow. *J. Fluid Mech.* 91: 493–95

Collins, W. D. 1955. On the steady rotation of a sphere in a viscous fluid. *Mathematika* 2:42–47

Collins, W. D. 1959. On the solution of some axisymmetric boundary value problems by means of integral equations. I. Some electrostatic and hydrodynamic problems for a spherical cap. *Quart. J. Mech. Appl. Math.* 12:232–41

Collins, W. D. 1963. A note on the axisymmetric Stokes flow of viscous fluid past a spherical cap. *Mathematika* 10:72–79

Collins, W. M., Dennis, S. C. R. 1976. Viscous eddies near a 90° and a 45° corner in flow through a curved tube of triangular cross-section. *J. Fluid Mech.* 76:417–32

Cooley, M. D. A., O'Neill, M. E. 1969. On the slow motion of two spheres in contact along their line of centres through a viscous fluid. *Proc. Cambridge Philos. Soc.* 66:407–15

Cox, R. G., Brenner, H. 1968. The lateral migration of solid particles in Poiseuille flow. I. Theory. *Chem. Eng. Sci.* 23:147–73

Cox, R. G., Hsu, S. K. 1977. The lateral migration of solid particles in a laminar flow near a plane. *Int. J. Multiphase Flow* 3:201–22

Cox, R. G., Mason, S. G. 1971. Suspended particles in fluid flow through tubes. *Ann. Rev. Fluid Mech.* 3:291–316

Davis, A. M. J., O'Neill, M. E. 1977a. Separation in a Stokes flow past a plane with a cylindrical ridge or trough. *Quart. J. Mech. Appl. Math.* 30:355–68

Davis, A. M. J., O'Neill, M. E. 1977b. Separation in a slow linear shear flow past a cylinder and a plane. *J. Fluid Mech.* 81: 551–64

Davis, A. M. J., O'Neill, M. E., Dorrepaal, J. M., Ranger, K. B. 1976. Separation from the surface of two equal spheres in Stokes flow. *J. Fluid Mech.* 77:625–44

Dean, W. R. 1944a. On the shearing motion of fluid past a projection. *Proc. Cambridge Philos. Soc.* 40:19–36

Dean, W. R. 1944b. Note on the shearing motion of fluid past a projection. *Proc. Cambridge Philos. Soc.* 40:214–22

Dean, W. R. 1950. Note on the motion of liquid near a position of separation. *Proc. Cambridge Philos. Soc.* 46:293–306

Dean, W. R., Montagnon, P. E. 1949. On the steady motion of viscous liquid in a corner. *Proc. Cambridge Philos. Soc.* 45: 389–94

Dorrepaal, J. M. 1976. Asymmetric Stokes flow past a spherical cap. *Z. Angew Math. Phys.* 27:739–48

Dorrepaal, J. M. 1978a. The Stokes resistance of a spherical cap to translational and rotational motions in a linear shear flow. *J. Fluid Mech.* 84:265–79

Dorrepaal, J. M. 1978b. Stokes flow past a smooth cylinder. *J. Eng. Math.* 12:177–85

Dorrepaal, J. M., Majumdar, S. R., O'Neill, M. E., Ranger, K. B. 1976a. A closed torus in Stokes flow. *Quart. J. Mech. Appl. Math.* 29:381–97

Dorrepaal, J. M., O'Neill, M. E., Ranger, K. B. 1976b. Axisymmetric Stokes flow past a spherical cap. *J. Fluid Mech.* 75:273–86

Fitz-Gerald, J. M. 1972. Plasma motions in narrow capillary flow. *J. Fluid Mech.* 51:463–76

Fukuyu, A. 1975. Un modèle d'écoulement dans un vaisseau capillaire. *C. R. Acad. Sci. Ser. A* 280:829–32

Geer, J. 1976. Stokes flow past a slender body of revolution. *J. Fluid Mech.* 78: 577–600

Gluckman, M. J., Weinbaum, S., Pfeffer, R. 1972. Axisymmetric slow viscous flow past an arbitrary convex body of revolution. *J. Fluid Mech.* 55:677–709

Goldman, A. J., Cox, R. G., Brenner, H. 1966. The slow motion of two identical arbitrarily oriented spheres through a viscous fluid. *Chem. Eng. Sci.* 21:1151–70

Goren, S. L. 1970. The normal force exerted by creeping flow on a small sphere touching a plane. *J. Fluid Mech.* 41:619–25

Hancock, G. J. 1953. The self-propulsion of microscopic organisms through liquids. *Proc. R. Soc. London Ser. A* 217:96–121

Happel, J., Brenner, H. 1965. *Low Reynolds Number Hydrodynamics*. Englewood Cliffs, N.J.: Prentice-Hall. 553 pp.

Harrison, W. J. 1920. The pressure in a viscous liquid moving through a channel with diverging boundaries. *Proc. Cambridge Philos. Soc.* 19:307–12

Hasimoto, H. 1954. Rayleigh's problem for a cylinder of arbitrary shape. *J. Phys. Soc. Jpn.* 9:611–19

Hasimoto, H. 1959. On the periodic fundamental solutions of the Stokes equations and their application to viscous flow past a cubic array of spheres. *J. Fluid Mech.* 5:317–28

Hasimoto, H. 1963. Incompressible flows in magneto-fluid dynamics. *Prog. Theor. Phys. (Kyoto) Suppl.* 24:35–85

Hasimoto, H. 1974. Fundamental solutions for the flow past bodies with special reference to the transient flow past a body and the Stokes force on an obstacle in a periodic array. *Theor. Appl. Mech.* 22:287–96

Hasimoto, H. 1976. Slow motion of a small sphere in a cylindrical domain. *J. Phys. Soc. Jpn.* 41:2143–44, 42:1047

Hasimoto, H. 1979. Creeping viscous flow past a circular arc with special reference to separation. *J. Phys. Soc. Jpn.* 47:347–48

Hill, R., Power, G. 1956. Extremum principles for slow viscous flow and the approximate calculation of drag. *Quart. J. Mech. Appl. Math.* 9:313–19

Hinch, E. J. 1972. Note on the symmetries of certain material tensors for a particle in Stokes flow. *J. Fluid Mech.* 54:423–25

Ho, B. P., Leal, L. G. 1974. Inertial migration of rigid spheres in two-dimensional unidirectional flows. *J. Fluid Mech.* 65:365–400

Imai, I. 1954. A new method of solving Oseen's equations and its application to the flow past an inclined elliptic cylinder. *Proc. R. Soc. London Ser. A* 224:141–60

Imai, I. 1972. Some applications of function theory to fluid dynamics. *2nd Int. JSME Symp., Fluid Machinery and Fluidics*, (Tokyo):15–23

Imai, I. 1973. *Ryutai Rikigaku (Fluid Dynamics)*. Tokyo: Syokabo. 428 pp.

Ishii, K. 1979. Viscous flow past multiple planar arrays of small spheres. *J. Phys. Soc. Jpn.* 46:675–80

Kanwal, R. P. 1961. Slow steady rotation of axially symmetric bodies in a viscous fluid. *J. Fluid Mech.* 10:17–24

Kawaguti, M. 1961. Numerical solution of the Navier-Stokes equations for the flow in a two-dimensional cavity. *J. Phys. Soc. Jpn.* 16:2307–15

Keller, J. B., Rubenfeld, L. A., Molyneux, J. E. 1967. Extremum principles for slow viscous flows with applications to suspensions. *J. Fluid Mech.* 30:97–125

Keller, J. B., Rubinow, S. I. 1976. Slender-body theory for slow viscous flow. *J. Fluid Mech.* 75:705–14

Kerchman, V. I. 1972. Slow flow of viscous fluid in a conical diffuser. *Izv. Akad. Nauk SSSR Mekh. Zhidk. Gaza* No. 2: 41–47 (in Russian)

Kim, M. 1976. *Axisymmetric slow viscous flow due to a sphere moving along the axis of a circular cone*. PhD thesis. Univ. Tokyo, Tokyo. 18 pp.

Kim, M. 1979. Slow viscous flow due to the motion of a sphere on the axis of a circular cone. *J. Phys. Soc. Jpn.* 46:1929–34

Lighthill, M. J. 1975. *Mathematical Biofluiddynamics*. Philadelphia: SIAM. 281 pp.

Lighthill, M. J. 1976. Flagellar hydrodynamics. *SIAM Rev.* 18:161–230

Liron, N., Mochon, S. 1976. Stokes flow for a stokeslet between two parallel flat plates. *J. Eng. Math.* 10:287–303

Liron, N., Shahar, R. 1978. Stokes flow due to a stokeslet in a pipe. *J. Fluid Mech.* 86:727–44

Lorentz, H. A. 1907. Ein allgemeiner Satz, die Bewegung einer reibenden Flüssigkeit betreffend, nebst einigen Anwendungen desselben. *Abhand. theor. Phys., Leipzig* 1:23–42

Majumdar, S. R., O'Neill, M. E. 1977. On axisymmetric Stokes flow past a torus. *Z. Angew. Math. Phys.* 28:541–50

Masuda, H. 1970. Stokes' approximation for a ring vertical to the uniform flow. *Bull. Fac. Gen. Ed. Utsunomiya Univ.* (Japan) 3:11–32

Michael, D. H., O'Neill, M. E. 1977. The separation of Stokes flows. *J. Fluid Mech.* 80:785–94

Moffatt, H. K. 1964. Viscous and resistive eddies near a sharp corner. *J. Fluid Mech.* 18:1–18

Nir, A., Acrivos, A. 1973. On the creeping motion of two arbitrary-sized touching spheres in a linear shear field. *J. Fluid Mech.* 59:209–23

Nir, A., Weinberger, H. F., Acrivos, A. 1975. Variational inequalities for a body in a viscous shearing flow. *J. Fluid Mech.* 68:739–55

O'Neill, M. E. 1969. On asymmetrical slow viscous flows caused by the motion of two equal spheres almost in contact. *Proc. Cambridge Philos. Soc.* 65:543–56

O'Neill, M. E. 1977. On the separation of a

slow linear shear flow from a cylindrical ridge or trough in a plane. *Z. Angew. Math. Phys.* 28:439–48

Pan, F., Acrivos, A. 1967. Steady flows in rectangular cavities. *J. Fluid Mech.* 28: 643–55

Payne, L. E., Pell, W. H. 1960. The Stokes flow problem for a class of axially symmetric bodies. *J. Fluid Mech.* 7:529–49

Pirronneau, O. 1973. On optimum profiles in Stokes flow. *J. Fluid Mech.* 59:117–28

Prothero, J., Burton, A. C. 1961. The physics of blood flow in capillaries. I. The nature of the motion. *Biophys. J.* 1:565–79

Proudman, I., Pearson, J. R. A. 1957. Expansions at small Reynolds numbers for the flow past a sphere and a circular cylinder. *J. Fluid Mech.* 2:237–62

Ranger, K. B. 1973. The Stokes drag for asymmetric flow past a spherical cap. *Z. Angew. Math. Phys.* 24:801–9

Ranger, K. B. 1977. The Stokes flow round a smooth body with an attached vortex. *J. Eng. Math.* 11:81–88

Richardson, S. 1977. Axisymmetric slow viscous flow about a body of revolution whose section is a cardioid. *Quart. J. Mech. Appl. Math.* 30:369–74

Saffman, P. G. 1973. On the settling speed of free and fixed suspensions. *Stud. Appl. Math.* 52:115–27

Sano, O. 1977. *Slow motion of a small sphere in a viscous fluid bounded by two plane walls.* PhD thesis. Univ. Tokyo, Tokyo. 98 pp.

Sano, O. 1978. Sedimentation of a small sphere in a viscous fluid bounded by two perpendicular walls. *J. Phys. Soc. Jpn.* 45:1034–37

Sano, O., Hasimoto, H. 1976. Slow motion of a spherical particle in a viscous fluid bounded by two perpendicular walls. *J. Phys. Soc. Jpn.* 40:884–90

Sano, O., Hasimoto, H. 1977. Slow motion of a small sphere in a viscous fluid in a corner. I. Motion on and across the bisector of a wedge. *J. Phys. Soc. Jpn.* 42:306–12

Sano, O., Hasimoto, H. 1978. The effect of two plane walls on the motion of a small sphere in a viscous fluid. *J. Fluid Mech.* 87:673–94

Schubert, G. 1967. Viscous flow near a cusped corner. *J. Fluid Mech.* 27:647–56

Segré, G., Silberberg, A. 1961. Radial particle displacements in Poiseuille flow of suspensions. *Nature* 189:209–10

Segré, G., Silberberg, A. 1962a. Behaviour of macroscopic rigid spheres in Poiseuille flow. Part 1. Determination of local concentration by statistical analysis of particle passages through crossed light beams. *J. Fluid Mech.* 14:115–35

Segré, G., Silberberg, A. 1962b. Behaviour of macroscopic rigid spheres in Poiseuille flow. Part 2. Experimental results and interpretation. *J. Fluid Mech.* 14:136–57

Shinohara, M., Hasimoto, H. 1979. The lateral force on a small sphere sedimenting in a viscous fluid bounded by a cylindrical wall. *J. Phys. Soc. Jpn.* 46:320–27

Skalak, R. 1970. Extensions of extremum principles for slow viscous flows. *J. Fluid Mech.* 42:527–48

Stasiw, D. M., Cook, F. B., Detraglia, M. C., Cerny, L. C. 1974. The drag and sphericity index of a spindle. *Quart. Appl. Math.* 32: 351–54

Sternberg, E., Koiter, W. T. 1958. The wedge under a concentrated couple: A paradox in the two-dimensional theory of elasticity. *J. Appl. Mech.* 25:575–81

Takagi, H. 1973. Slow viscous flow due to the motion of a closed torus. *J. Phys. Soc. Jpn.* 35:1225–27

Takagi, H. 1974a. Slow motion of a spherical cap in a viscous fluid. I. Uniform translation and rotation. *J. Phys. Soc. Jpn.* 37:229–36

Takagi, H. 1974b. Slow motion of a spherical cap in a viscous fluid. II. Free suspension in a shear field. *J. Phys. Soc. Jpn.* 37: 237–39

Takami, H., Kuwahara, K. 1974. Numerical study of three-dimensional flow within a cubic cavity. *J. Phys. Soc. Jpn.* 37:1695–98

Takematsu, M. 1965. Viscous flow in a two-dimensional cavity. *J. Phys. Soc. Jpn.* 20:283

Takematsu, M. 1966. Slow viscous flow past a cavity. *J. Phys. Soc. Jpn.* 21:1816–21

Tamada, K. 1956. *Proc. 9th Int. Congr. Appl. Mech.* 3:343

Taneda, S. 1979a. Visualization of the Stokes flow. *Nagare* 10(4):1–4 (in Japanese)

Taneda, S. 1979b. Visualization of separating Stokes flows. *J. Phys. Soc. Jpn.* 46:1935–42

Tokuda, N. 1972. Viscous flow near a corner in three dimensions. *J. Fluid Mech.* 53: 129–48

Tokuda, N. 1975. Stokes solutions for flow near corners in three-dimensions. *J. Phys. Soc. Jpn.* 38:1187–94

Van Dyke, M. 1964. *Perturbation Methods in Fluid Mechanics.* New York: Academic. 229 pp.

Vasseur, P., Cox, R. G. 1976. The lateral migration of a spherical particle in two-dimensional shear flow. *J. Fluid Mech.* 78:385–413

Vasseur, P., Cox, R. G. 1977. The lateral migration of spherical particles sedimenting in a stagnant bounded fluid. *J. Fluid Mech.* 80:561–91

Wakiya, S. 1974. On the exact solution of the Stokes equations for a torus. *J. Phys. Soc. Jpn.* 37:780–83

Wakiya, S. 1975. Application of bipolar coordinates to the two-dimensional creeping motion of a liquid. I. Flow over a projection or a depression on a wall. *J. Phys. Soc. Jpn.* 39:1113–20

Wakiya, S. 1976. Axisymmetric flow of a viscous fluid near the vertex of a body. *J. Fluid Mech.* 78:737–47

Wakiya, S. 1978. Application of bipolar coordinates to the two-dimensional creeping motion of a liquid. III. Separation in Stokes flows. *J. Phys. Soc. Jpn.* 45:1756–63

Wang, H., Skalak, R. 1969. Viscous flow in a cylindrical tube containing a line of spherical particles. *J. Fluid Mech.* 38:75–96

Wannier, G. H. 1950. A contribution to the hydrodynamics of lubrication. *Quart. Appl. Math.* 8:1–32

Williams, W. E. 1966. Boundary effects in Stokes flow. *J. Fluid Mech.* 24:285–91

Youngren, G. K., Acrivos, A. 1975. Stokes flow past a particle of arbitrary shape: a numerical method of solution. *J. Fluid Mech.* 69:377–403

Ann. Rev. Fluid Mech. 1980. 12: 365–87

CONTINUOUS DRAWING OF LIQUIDS TO FORM FIBERS

Morton M. Denn
Department of Chemical Engineering, University of Delaware, Newark, Delaware 19711

INTRODUCTION

The continuous stretching of viscous liquids to form fibers is a primary manufacturing process in the textile industry. The mechanics of certain sheet formation and sheet and wire coating operations are quite similar, as are the mechanics of the formation of glass fibers. Most published experimental and theoretical work has been motivated by applications using polymeric liquids, where the interactions between processing conditions and the complex fluid rheology are of particular interest.

The melt spinning process for the manufacture of textile fibers is shown schematically in Figure 1. Molten polymer is extruded through a small hole into cross-flowing ambient air at a temperature below the solidification temperature of the polymer. The solidified polymer is taken up at a higher speed than the mean extrusion velocity, resulting in drawing of the filament. The steady-state ratio of extrusion to takeup area is known as the *draw ratio* (D_R); the draw ratio is also equal to the steady-state ratio of takeup to extrusion velocity (to within the approximation that polymer density is independent of temperature). Typical processing variables for the manufacture of poly(ethylene terephthalate) (PET) fiber are shown in Table 1. Commercial PET, a polyester sold under such trade names as Dacron, Terylene, and Fortrel, is a nearly amorphous material as spun, although PET with forty percent crystalline polymer can be produced. Other commercially important melt-spun polymers include Nylon-6, Nylon-66, and polypropylene. The solidified filament is typically subjected to further downstream processing for property development. A single spinneret, or spinning head, contains many holes, and the individual solidified filaments from each head are taken up together to form a yarn.

0066-4189/80/0115-0365$01.00

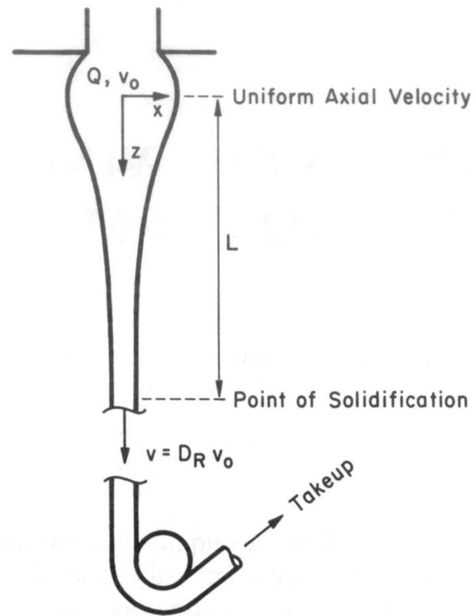

Figure 1 Schematic of continuous filament drawing.

Textile fibers must be produced with uniform diameters and specified physical properties, such as the solid modulus, extension-to-break, and dye acceptance. Many physical properties correlate with the stress at the point of solidification; for noncrystalline polymers like PET the stress is proportional to the optical birefringence, a measure of molecular structure development. The primary goals of a fluid-mechanical analysis are therefore to be able to compute the stress and to define regions of processing sensitivity or instability where large diameter and property fluctuations might occur.

Glass and some carbon and metal fibers can also be formed continuously from the melt, but glass may sometimes be drawn from a rodlike preform instead of a spinneret. Diameter uniformity is particularly important for glass optical fibers. Glass and some polymers, particularly polypropylene, may be spray spun from the melt to form spun-bonded fabrics; here the filament is drawn by contact with high-speed air jets and deposited on a moving substrate. Steel and aluminum fibers are drawn by the combined effect of gravity and a cocurrent gas stream.

There is no mass transfer between the filament and the ambient in the melt drawing process. Other fiber formation processes (*wet spinning*, *dry spinning*) require mass transfer to remove a solvent or to enable a

Table 1 Typical processing variables for manufacture of PET fibers

Processing variable	Typical value
Extrusion temperature	290°C
Solidification temperature	70°C
Ambient air temperature	35°C
Spinneret hole diameter	10–20 mils (0.25–0.50 mm)
Takeup velocity	1000–3500 m/min
Cross-flow air velocity	0.3 m/sec
Draw ratio	200

chemical reaction to occur. Polymer fiber drawing processes are discussed quantitatively in a monograph by Ziabicki (1976), and similar material is contained in Walczak (1977). Some fluid-mechanical aspects of glass fiber formation are described by Geyling (1976). Melt spinning of pitch to form carbon fibers is described by Barr et al (1976), and there is a discussion of spray spinning in Wagner & Roberts (1972). The triple jet (Oliver & Ashton 1976) is a fluid mechanics experiment that is similar in concept to spray spinning. There are many papers on spinning of metal fibers in Mottern & Privott (1978). A compact historical survey of fiber formation is given by White (1976). Moncrieff (1975) contains detailed chemical, manufacturing, and property information about most fibers available at publication. Structure development and structure-stress-property relations are discussed by Ziabicki (1976) and White (1976), and more recent results are given by Danford et al (1978), Nadella et al (1977), and Oda et al (1978).

EXPERIMENTAL OBSERVATIONS

Certain general qualitative experimental observations about continuous drawing of filaments can be made. The most significant is that the stress and rate of diameter attenuation depend strongly on the rheological properties of the liquid being drawn. Figure 2 shows results of laboratory experiments using a Newtonian corn syrup ($\eta = 25$ Pa·s) and a dilute solution of polyacrylamide in corn syrup (Chang & Denn 1979). The polymer solution has a nearly constant viscosity equal to that of the corn syrup, but it has measurable viscoelastic properties. The two liquids were drawn isothermally at the same extrusion velocity and spinline lengths, without solidification, and wiped from the takeup wheel. The force required to draw the viscoelastic liquid is nearly two orders of

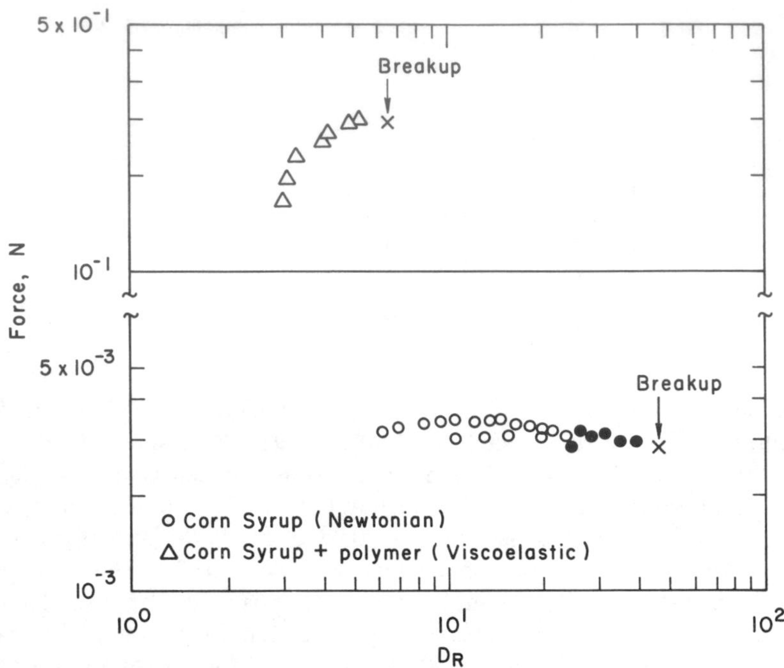

Figure 2 Stress as a function of draw ratio for continuous drawing of a Newtonian corn syrup and a viscoelastic solution of polyacrylamide in corn syrup. Open symbols correspond to steady state and closed symbols to sustained oscillations.

magnitude greater than the force to draw the Newtonian liquid of equal viscosity at the same draw ratio.[1] Similar large takeup forces have been recorded for laboratory experiments on a variety of polymer melts of commercial interest (e.g. Spearot & Metzner 1972; an extensive compilation is given by Petrie 1979a), but direct comparison with a comparable Newtonian liquid is not possible because of the more complex melt rheology. It is important to note that most laboratory spinning experiments have been carried out at takeup speeds that are well below commercial conditions.

[1] A filament extruded at low Reynolds number swells upon leaving the spinneret. The Newtonian corn syrup increased in area by 8.5–9.5% when drawn, independent of takeup speed, while the polymer solution increased in area by less than 2% when drawn. Extrudate swell of 20–50% has been observed in laboratory drawing experiments with polymer melts (White & Roman 1976, Blyler & Gieniewski 1979), and swell of more than 200% has been observed in the absence of a drawing force; Middleman (1977) compiled the data of many investigators on extrudate swell. The draw ratio, D_R, is defined in terms of spinneret area by some authors, and in terms of the maximum area by others. We use the latter definition throughout this chapter.

Sustained oscillations in drawn filament diameter and takeup force are sometimes observed when the drawing zone is maintained close to the extrusion temperature, followed by rapid cooling if solidification is to take place. Diameter fluctuations observed by Ishihara & Kase (1976) in a PET filament are shown in Figure 3. This phenomenon, known as draw resonance, is not associated with viscoelasticity; sustained oscillations were observed by Chang & Denn (1979) in the continuous drawing of a Newtonian corn syrup, corresponding to the filled data points in Figure 2. The onset of draw resonance does depend on the fluid rheology, however; in isothermal experiments there is a critical draw ratio that is of order twenty for fluids with a constant viscosity and of order five for highly shear-thinning melts such as polystyrene and high-density polyethylene. Experimental data through 1976 are summarized by Petrie & Denn (1976).

Filament breakup is the most important processing limitation. In a series of experiments with the viscoelastic polyacrylamide–corn syrup solution at various flow rates and spinning lengths, Chang & Denn (1979) found that filament breakup occurred at an approximately constant stress. Cogswell (1975) has reported a critical cohesive stress for all major commercial polymeric liquids of order 10^6 Pa. Ziabicki (1976) had previously suggested that breakup occurs at a critical stress (which he called *cohesive failure*) for very viscous liquids and by a capillary mechanism for low-viscosity liquids; his correlating variable is the product of viscosity and extrusion speed. The experiments of White & Ide (1978) on low- and high-density polyethylenes, polypropylene, and polystyrene suggest a more complex behavior, including a failure by necking in some cases, but the data reported are not adequate to determine if breakup ever occurs at constant stress.

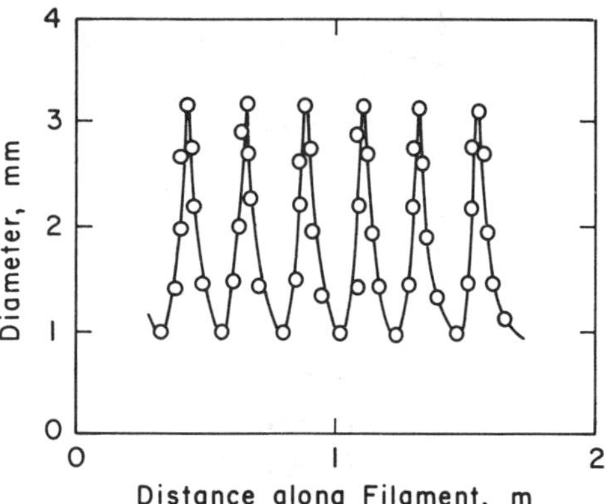

Figure 3 Diameter variations in a drawn filament of PET; data of Ishihara & Kase (1976).

EXTENSIONAL RHEOLOGY

Some insight into stress development in continuous drawing can be obtained by consideration of the steady uniform uniaxial extension of a cylindrical rod of fluid, shown schematically in Figure 4. The stress in a Newtonian fluid is constant in time, equal to $3\eta\Gamma$, where η is the shear viscosity and Γ the extension rate. Figure 5 shows steady stretching data of Laun & Münstedt (1976) on a low-density polyethylene melt. The initial transient, apparent approach to a steady stress, and subsequent upturn (*strain hardening*), is commonly observed in this type of experiment; the approach to a constant stress at long times (i.e. large strains) has not been observed by others, apparently because of the experimental difficulty of obtaining such large extension ratios (400 in this case). [Raible et al (1979) have recently reported a reduction in stress at still larger strains.] The existence of such an asymptote or maximum is of practical importance in analyzing the spinning process. The most complete compilation of experimental studies on extensional flow is given by Petrie (1979a), and other recent surveys include those of Cogswell (1978),

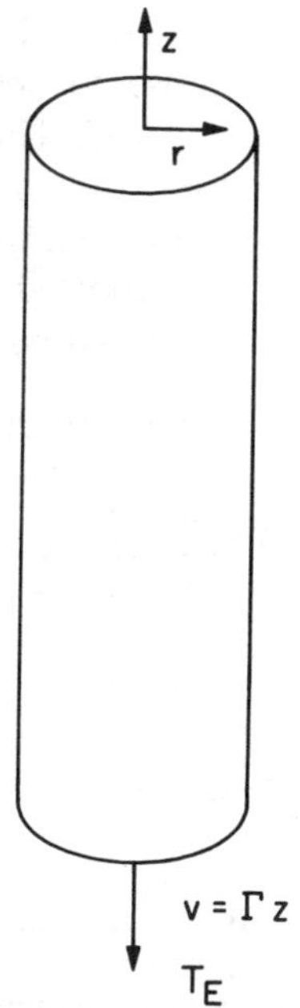

Figure 4 Schematic of uniform uniaxial extension.

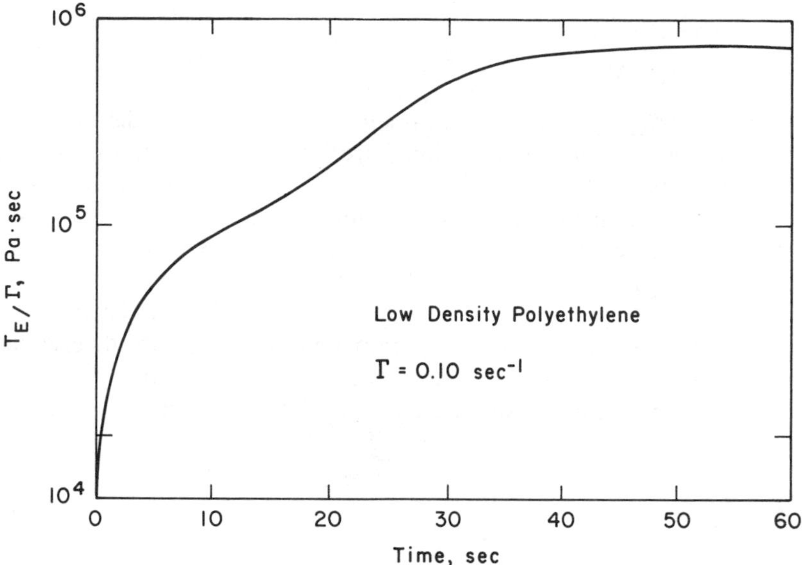

Figure 5 Stress versus time for uniform uniaxial extension of a low-density polyethylene; data of Laun & Münstedt (1976).

Dealy (1978), and Denn (1977). Significantly, such experiments have not been carried out on the polymers most used for fiber manufacture.[2]

Most rheological analyses have been carried out for constitutive relations in which the extra stress is related to the deformation rate through an equation of the form

$$\boldsymbol{\tau} = \sum_{k=1}^{N} \boldsymbol{\tau}_{(k)} \tag{1}$$

$$\boldsymbol{\tau}_{(k)} + \frac{\eta_k}{G_k}\left[\frac{\mathscr{D}\boldsymbol{\tau}_{(k)}}{\mathscr{D}t} - a(\boldsymbol{\tau}_{(k)}\cdot\mathbf{e}+\mathbf{e}\cdot\boldsymbol{\tau}_{(k)})\right] = 2\eta_k\mathbf{e} \tag{2}$$

where

$$\frac{\mathscr{D}\boldsymbol{\tau}_{(k)}}{\mathscr{D}t} = \frac{\partial\boldsymbol{\tau}_{(k)}}{\partial t} + \mathbf{v}\cdot\nabla\boldsymbol{\tau}_{(k)} + \boldsymbol{\omega}\cdot\boldsymbol{\tau}_{(k)} - \boldsymbol{\tau}_{(k)}\cdot\boldsymbol{\omega} \tag{3}$$

[2] Some authors refer to the ratio of extensional stress to stretch rate as an *extensional viscosity*. This terminology should be used only in uniform uniaxial extension, and only when both stress and stretch rate are constant. Some authors have attempted to use the concept of an extensional viscosity in other stretching flows, including spinning, where the ratio stress/stretch rate is not a material property, but is rather a complex interaction between material and experiment.

$$\mathbf{e} = \tfrac{1}{2}[\nabla\mathbf{v} + (\nabla\mathbf{v})^{+}] \tag{4}$$

$$\omega = -\tfrac{1}{2}[\nabla\mathbf{v} - (\nabla\mathbf{v})^{+}]. \tag{5}$$

Here $\{\eta_k\}$, $\{G_k\}$, and a are parameters that may be deformation-dependent, and some modifications are required for nonisothermal flows. The ratio η_k/G_k is often denoted λ_k. N greater than unity reflects the fact that polymeric materials have multiple relaxation mechanisms and are generally represented by a spectrum of relaxation times. Constitutive equations are reviewed by Bird (1976); see also Petrie (1979a,b) for more recent reviews and Acierno et al (1976) and Phan-Thien (1978) for important recent results. The parameters $\{\eta_k\}$, $\{G_k\}$ are determined from linear viscoelastic measurements; the shear viscosity η, shear modulus G, and mean relaxation time $\bar{\lambda}$ are determined in shear and are equal to

$$\eta = \sum_{k=1}^{N} \eta_k, \tag{6a}$$

$$\bar{\lambda} = \sum_{k=1}^{N} (\eta_k^2/G_k)/\eta, \tag{6b}$$

$$G = \eta/\bar{\lambda}. \tag{6c}$$

The most commonly applied equation is the *Maxwell model*, with $N = 1$, $a = 1$. Calculations for uniform uniaxial extension of a Maxwell fluid are shown in Figure 6 (Denn & Marrucci 1971). There is stress

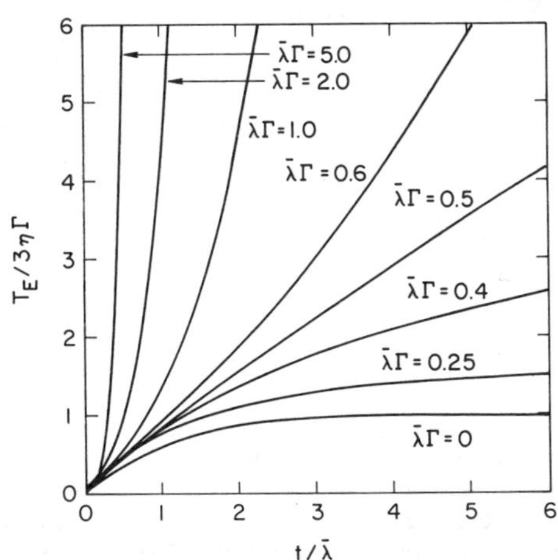

Figure 6 Dimensionless stress as a function of dimensionless time for uniform uniaxial extension of a Maxwell liquid, after Denn & Marrucci (1971).

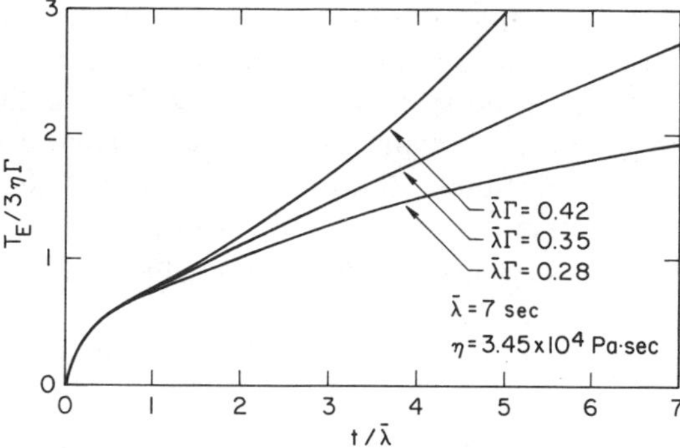

Figure 7 Dimensionless stress as a function of dimensionless time for uniform uniaxial extension of a viscoelastic liquid with two relaxation times, using parameters characteristic of low-density polyethylene, after Denn (1977).

growth with a monotonic approach to an asymptotic value at low stretch rates ($2\bar{\lambda}\Gamma < 1$), and unbounded stress growth at high stretch rates. There is thus some qualitative resemblance to the behavior of real polymers for short times, but the strain hardening and approach to an ultimate asymptote is not described.

Strain hardening is a consequence of multiple relaxation phenomena. Figure 7 shows a calculation for uniform uniaxial extension with $N = 1$, $a = 1$, using parameters characteristic of low-density polyethylene (Denn 1977). The strain hardening is followed by unbounded stress growth when the product of Γ and the maximum relaxation time exceeds $\frac{1}{2}$, and more complex modifications, like those described by Phan-Thien (1978), are required to describe the approach to the ultimate asymptote.

ASYMPTOTIC EQUATIONS

Fluid-mechanical analyses of filament drawing have all been based on asymptotic equations derived by averaging the mass, momentum, and energy equations over a circular filament cross section at each axial position. The most careful derivation is given by Matovich & Pearson (1969). The resulting steady-state momentum equation is

$$\frac{d}{dz}[\pi R^2(\bar{\tau}^{zz} - \bar{\tau}^{xx})] = \rho Q\frac{d\bar{v}}{dz} + \pi R\rho_{\mathrm{a}}\,\bar{v}^2 C_{\mathrm{f}} - \pi R^2\rho g - \pi\sigma\,\frac{dR}{dz}, \tag{7}$$

where z is the drawing direction and x is any direction in the orthogonal plane. The overbar denotes an area average. There is an explicit assumption both here and in the calculation of $\bar{\tau}$ from the constitutive equation that variations in the axial velocity over each cross section can be neglected, and dR/dz has been assumed to be small; the "smallness" of dR/dz has not been made explicit. The best relation for the drag coefficient C_f appears to be that of Matsui (1976). The surface-tension term is never important in polymer melt drawing. [Slattery & Acharya (1979) have recently formulated both spinning and uniform extension taking surface viscosity as well as surface tension into account, and they suggest that surface viscosity could become important at sufficiently high stretch rates.]

The steady-state energy equation, neglecting axial conduction and viscous dissipation, is

$$\rho c_p \bar{v} \frac{d\bar{T}}{dz} = -\frac{2h}{R}(\bar{T} - T_a). \tag{8}$$

The heat-transfer coefficient h depends on the local filament velocity and radius and cross-flow air velocity, and it can be adjusted to account for radiation. The best published correlation for h seems to be that of Kase & Matsuo (1967); other correlations are recorded by Ziabicki (1976). The functional dependence of h is such that the equation for $\bar{T}$ is only weakly dependent on kinematics, and an exponential temperature profile is expected as a first approximation (Fisher & Denn 1977); this result enables an a priori estimate of the location of the point of solidification.

Equations (7) and (8), together with the steady-state equation of conservation of mass, $\rho Q =$ constant, define a boundary-value problem. In most situations the flow rate and temperature will be specified at the beginning of the spinline and the takeup velocity at the end; the takeup force cannot be specified independently unless the takeup velocity or filament length is left unspecified (in which case the system becomes an initial-value problem).

The proper initial conditions for Equations (7) and (8) have not been definitively established, since the asymptotic equations are not applicable at the spinneret. Most authors have taken the area and velocity as given at the point of maximum extrudate swell, which is assumed to be close to the spinneret, and they have assumed that the asymptotic equations apply from that point. George (1979) has used an empirical correlation for extrudate swell as an initial condition. Krishnan & Glicksman (1971) and Glicksman (1974) have done one- and two-dimensional analyses of the upper jet region for conditions relevant in glass fiber spinning, where the initial shape is sensitive to surface-tension forces, but they require an experimental value of the initial diameter. Similarly, Clarke (1968, 1969) and Kaye & Vale (1969) have analyzed the upper jet when the drawing is entirely by gravity, but the solutions do not extend to the jet swell region. The only detailed study of the approach to the asymptotic equations starting from the spinneret is by R. J. Fisher, M. M. Denn, and R. I. Tanner (unpublished) for isothermal drawing of a Newtonian liquid under conditions in which inertia, air drag, gravity, and surface tension can be neglected. The solution to the asymptotic equations in that case is

$$v/v_0 = \exp(z \ln D_R/L) = \exp(zF/3\eta Q). \tag{9}$$

The mean velocity (flow rate/area) obtained by a finite-element solution of the Navier-Stokes equations is shown in Figure 8 for $d_0F/3\eta Q$ ranging from 0.22 to 1.0. The point of maximum extrudate swell is always located

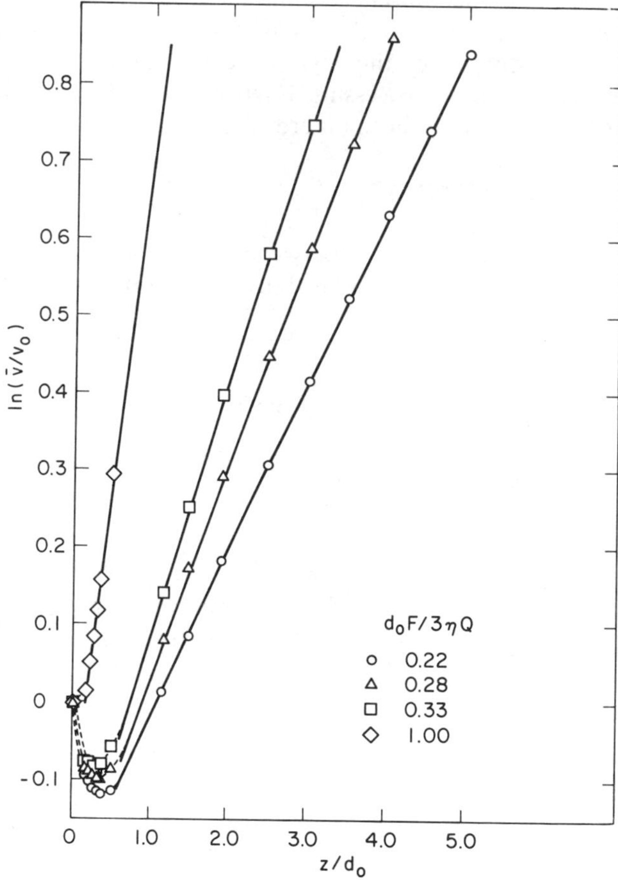

Figure 8 Mean velocity profile development for low-speed isothermal spinning of a Newtonian liquid. The points are a finite-element solution and the straight lines are solutions to the asymptotic equations. Unpublished calculations of R. J. Fisher, M. M. Denn, and R. I. Tanner.

less than one radius from the spinneret, and the asymptotic solution is reached within slightly more than one radius. Extrudate swell ranges from zero to six percent for these calculations, compared with a free jet swell in the absence of drawing of thirteen percent, and the amount of swell decreases with applied force. The only systematic experimental study of extrudate swell with an imposed drawing force is by White & Roman (1976), and they were unable to obtain quantitative agreement between their experiments on polymer melts and their approximate theoretical analysis. No comparable studies have been carried out on the approach to the asymptotic temperature profile, although Matsuo & Kase (1976) have examined the effect of nonuniform radial cooling. In the absence of empirical information, it is likely that use of spinneret conditions as initial conditions for the asymptotic equations will cause little error under conditions of processing interest, where the length of the melt zone is long relative to the spinneret diameter.

NEWTONIAN LIQUIDS

The most detailed comparisons between the asymptotic theory for Newtonian liquids and experimental data are by Glicksman (1968) on

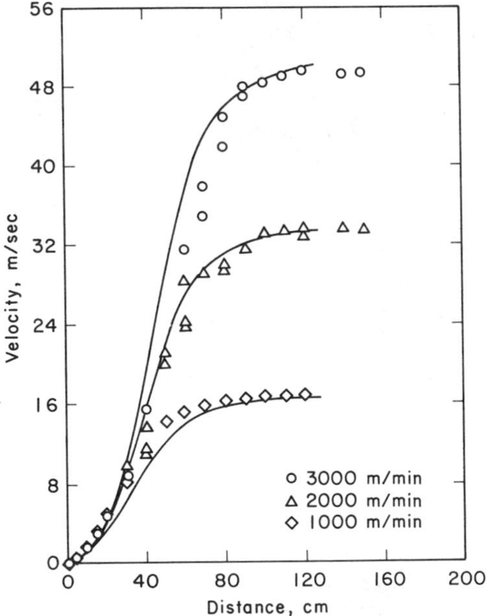

Figure 9 Experimental and computed velocity profiles for spinning of PET at 1000, 2000, and 3000 m/min, after George (1979).

glass fibers and by George (1979) on PET; PET is viscoelastic, but may be approximately Newtonian over at least a portion of the range studied. George used an empirical relation for extrudate swell, while Glicksman used the experimental spinline data at the point where $dR/dz = -0.1$. The polymer was assumed to solidify to an amorphous, inelastic solid at the glass transition temperature.

Agreement between theory and experiment was similar in both cases. Figure 9 shows a comparison by George between computed and experimental mean velocity profiles for PET in the 1000–3000 m/min range, which corresponds to present commercial spinning conditions; air drag and inertia are important here. He has shown that the computed stress at solidification is directly proportional to the optical birefrigence of the spun fiber, indicating that structure development was computed accurately by the model; he reports successful application of his simulation "to a number of commercial problems leading to faster, cheaper, and in some cases, unobvious solutions." Similar studies were carried out by Kase & Matsuo (1967), Hamana et al (1969), Kaye & Vale (1969), and Krishnan & Glicksman (1971); less detailed comparisons for Newtonian liquids under isothermal conditions have been reported by Chang & Denn (1979) and Mewis & Metzner (1974).

VISCOELASTIC LIQUIDS

The calculation of the average stresses $\bar{\tau}^{zz}$ and $\bar{\tau}^{xx}$ in Equation (7) in terms of the gradient of mean velocity $d\bar{v}/dz$ requires the averaging of non-linear terms for any non-Newtonian fluid. The assumption has been made by all investigators that averages of products can be replaced by products of averages; the order of the errors so introduced has not been analyzed. The Maxwell model, Equation (2) with $N = 1$, $a = 1$, then becomes

$$\bar{\tau}^{zz} + \frac{\eta}{G}\left[\bar{v}\,\frac{d\bar{\tau}^{zz}}{dz} - 2\bar{\tau}^{zz}\,\frac{d\bar{v}}{dz}\right] = 2\eta\,\frac{d\bar{v}}{dz} \tag{10a}$$

$$\bar{\tau}^{xx} + \frac{\eta}{G}\left[\bar{v}\,\frac{d\bar{\tau}^{xx}}{dz} + \bar{\tau}^{xx}\,\frac{d\bar{v}}{dz}\right] = -\eta\,\frac{d\bar{v}}{dz}. \tag{10b}$$

This constitutive equation introduces a viscoelastic dimensionless group,

$$\alpha = \bar{\lambda}v_0/L. \tag{11}$$

(For cases in which η or G are functions of deformation rate, α is defined at a deformation rate equal to $\sqrt{3}\,v_0/L$. If the temperature is variable, parameters are evaluated at the initial temperature.)

Computed velocity profiles for the Maxwell fluid are shown in Figure 10 for isothermal drawing to $D_R = 20$ in the absence of inertia, air drag, gravitation, or surface tension (Denn et al 1975). The value $\alpha = 0$ corresponds to the Newtonian liquid, Equation (9). Increasing viscoelasticity (increasing α) results in a more linear velocity profile, and hence more rapid initial diameter attenuation. This is in qualitative agreement with experimental observations by Chang & Denn (1979) and Spearot & Metzner (1972) on viscoelastic liquids. Increasing α also results in rapidly increasing forces, with an asymptotic approach to an infinite force and a linear velocity profile as $\alpha \rightarrow (D_R - 1)^{-1}$. This unbounded force is analogous to the unbounded stress growth for the Maxwell material in uniform uniaxial extension, Figure 6. The structure of this set of equations and of some generalizations is examined in detail by Petrie (1979a).

Comparisons of experimental velocity profiles and force measurements with the solution to Equations (8), (9), and (10), allowing for deformation rate–dependent physical properties, consistently require the use of values of $\bar{\lambda}$ that are two to four times those measured in shear (Denn et al 1975, Fisher & Denn 1977, Chang & Denn 1979). This is a consequence of the multiple relaxation times and is analogous to the strain hardening in Figures 5 and 7; in an extensional flow the longer relaxation times are more important than they are in shear flow, and they lead to more rapid diameter attenuation and higher forces than would be predicted for a fluid with the same mean properties but a single relaxation time (Denn

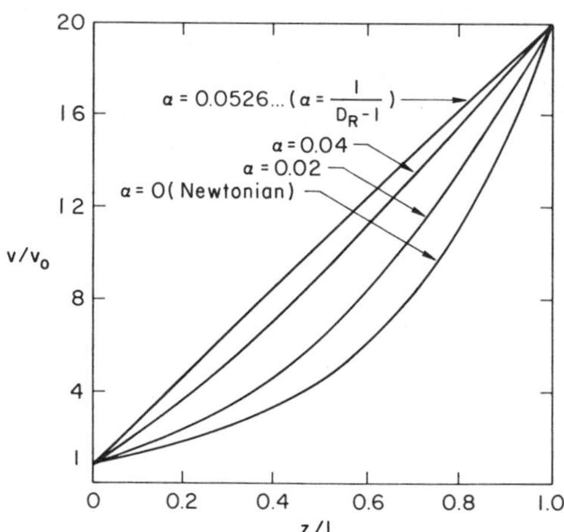

Figure 10 Dimensionless velocity profiles for isothermal low-speed spinning of a Maxwell liquid, $D_R = 20$, after Denn et al (1975).

& Marrucci 1977). [Bankar et al (1977) report that they require a mean relaxation time approximately forty times greater than that measured in shear to obtain agreement with their experiments on Nylon-6 at draw ratios of about five, but they have apparently used the experimental force as an input and allowed the filament length to vary. If the filament length is set at the experimental value, a spot check of a few data points indicates that the agreement between theory and experiment is similar to that found in the other studies cited and is consistent with the small relaxation time of Nylon-6.] Figure 11 shows a calculation by Phan-Thien (1978) to simulate an experiment of Zeichner (1972) on isothermal spinning of polystyrene. The calculation includes multiple relaxation times and uses a constitutive equation that correctly describes the stress-versus-time curve in Figure 5; together with a simulation in the same paper of the polyethylene data of Spearot & Metzner (1972), it represents the most successful attempt to date to use rheological measurements made in shear to predict the behavior in a continuous drawing experiment.

Numerical solution of the asymptotic equations for nonisothermal spinning of a highly viscoelastic liquid has been reported by Fisher & Denn (1977). The only comparison between theory and experiment for nonisothermal spinning of a highly viscoelastic liquid is by Matsui & Bogue (1976). Measured velocity profiles were used as an input, and only stresses were calculated, so interpretation is difficult.

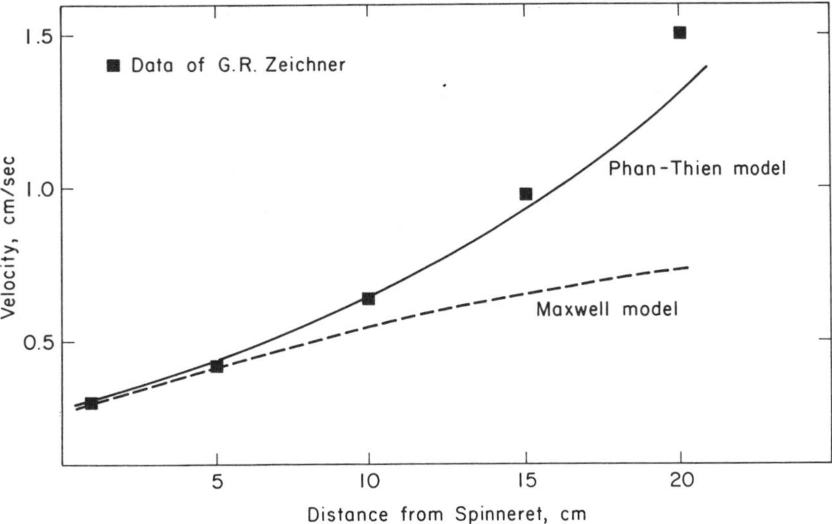

Figure 11 Computed and experimental velocity profile for isothermal spinning of polystyrene, after Phan-Thien (1978).

Gagon & Denn (1978) have solved the asymptotic spinning equations for a Maxwell fluid under conditions typical of commercial production of PET, including takeup speeds in the range 1000–8000 m/min; the latter is more than twice the current technology, but pilot equipment with such takeup speeds currently exists. At low speeds viscoelastic effects become increasingly important with increasing drawdown, and the takeup force becomes unbounded for sufficiently fine filaments, analogous to the isothermal behavior illustrated in Figure 10; this infinite growth is eliminated when the more realistic Phan-Thien (1978) rheological model is used (D. C. Gagon and M. M. Denn, unpublished). Air drag at high speed retards the rate of draw, and this has the surprising effect of slowing the viscoelastic stress growth rate sufficiently to reduce the overall force to a level comparable to (and in fact slightly less than) that computed for a Newtonian fluid. Typical calculations at 500 and 4000 m/min are shown in Figure 12. Calculations of this type represent the present state of the art in steady-state modeling of continuous drawing.

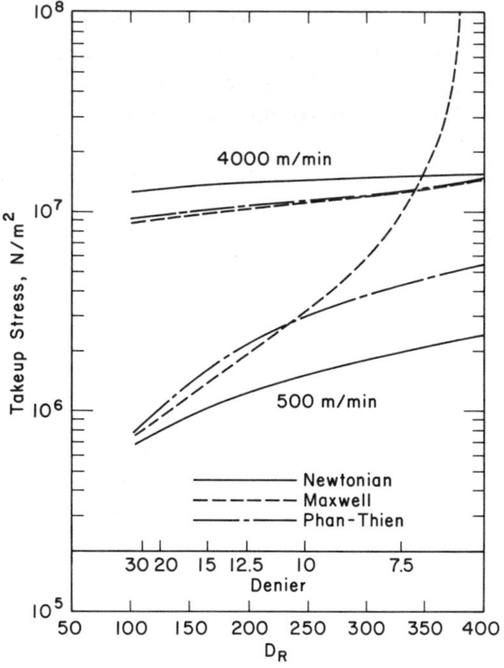

Figure 12 Computed takeup force as a function of draw ratio for spinning of PET at 500 and 4000 m/min. Unpublished calculations of D. C. Gagon and M. M. Denn.

HIGH-SPEED SPINNING

Polyester fibers drawn at takeup speeds of less than 4000 m/min solidify as a noncrystalline material with a density of 1340 kg/m^3. Filaments drawn at 6000 m/min have a density in excess of 1380 kg/m^3, with 20–45% crystalline regions (Heuvel & Huisman 1978, Shimizu et al 1977). This transition takes place over a narrow range of spinning speeds. The semi-crystalline polymer has physical properties that are quite distinct from the usual amorphous as-spun material (Frankfort & Knox 1979).

Theories of stress-induced crystallization have been developed; these are summarized in Ziabicki (1976). Such theories have been coupled with hydrodynamic analyses of the spinning process only in qualitative terms, however, and a complete mechanical theory that includes crystallization has not been developed. D. C. Gagon and M. M. Denn (unpublished) have hypothesized that the major effect of stress-induced crystallization is to introduce a load-bearing crystal phase at a temperature above the glass transition, and hence to raise the effective solidification temperature. A limited number of calculations suggests that the stress at the solidification point is a weak function of solidification temperature. This simple approach will not suffice for a crystalline polymer; Nadkarni & Schultz (1977) show considerable diameter attenuation during crystallization in a spinning study on high-density polyethylene, and Nadella et al (1977) show crystallization occurring over a finite spinline length for isotactic polypropylene.

STABILITY AND SENSITIVITY

The draw-resonance instability illustrated in Figure 3 does not occur under commercial spinning conditions, but it can occur commercially in the similar process of extrusion coating. There appears to be a requirement that the length L of the melt zone be fixed, perhaps by spinning into a cold water bath; that the takeup speed be constant; and that the ambient temperature be maintained close to the extrusion temperature. The oscillations can be explained in terms of an unstable feedback control structure for the maintenance of a constant takeup speed (Kase & Denn 1978). Experimental data through 1976 are reviewed by Petrie & Denn (1976) and Pearson (1976), and more recent experiments are described by Blyler & Gieniewski (1979), Chang & Denn (1979), Matsumoto & Bogue (1978), and White & Ide (1978).

Draw resonance was first analyzed using a linear stability analysis by

Pearson & Matovich (1969) and Kase et al (1966) for inelastic liquids. For a fixed spinning length, isothermal conditions, and no inertia, air drag, gravity, or surface-tension effects, steady spinning is predicted to become unstable beyond a critical draw ratio of 20 for Newtonian liquids, and at lower draw ratios for shear-thinning liquids. Inertia, gravity, and cooling increase the critical draw ratio and are stabilizing, while surface tension is destabilizing. The most complete analysis for inelastic liquids is by Shah & Pearson (1972), but their computed effect of inertia is too large (J. C. Chang and M. M. Denn, unpublished). Allowing the location of the freeze point to adjust dynamically is stabilizing (Pearson et al 1976). The stabilizing effects of cooling and a variable freeze point are apparently the reasons that draw resonance does not occur under commercial spinning conditions. In extrusion coating the location of the freeze point is fixed. Nonlinear analyses by Ishihara & Kase (1975, 1976) and Fisher & Denn (1975) show that sustained oscillations like those observed in draw resonance do occur at draw ratios

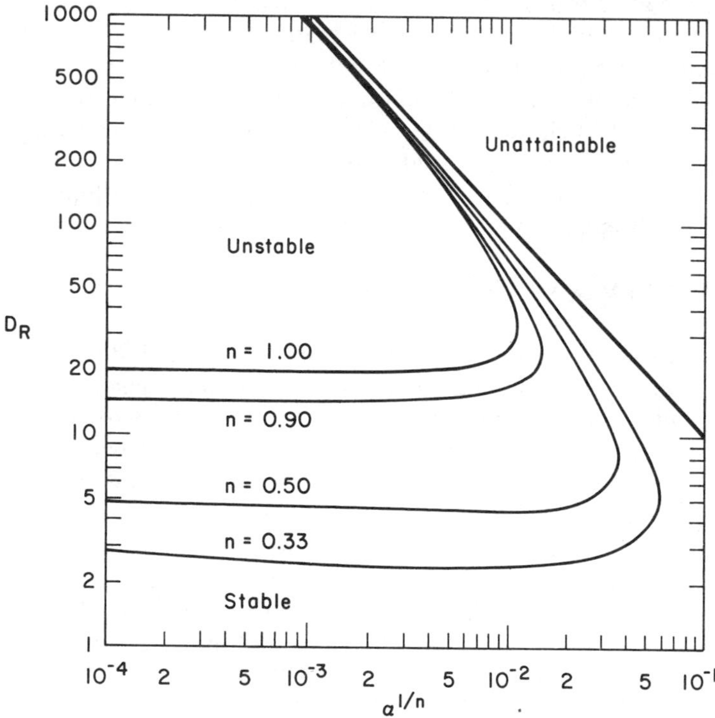

Figure 13 Stability regions for low-speed isothermal spinning of a viscoelastic liquid with a single relaxation time, after Fisher & Denn (1977).

above the critical value computed from linear theory. A nonlinear analysis by Hyun (1978) is incorrect (Denn 1979a).

The effect of viscoelasticity on the onset of draw resonance has been studied by Fisher & Denn (1976, 1977) and Chang & Denn (1979) for a fluid with a single relaxation time. The critical draw ratio as a function of α, Equation (11), is shown in Figure 13 for isothermal spinning. Here n is the power-law index, a measure of viscous shear-thinning, and the region labeled "unattainable" corresponds to infinite takeup forces. The overall effect of viscoelasticity is to increase the region of stable spinning, including a region of stable isothermal spinning at very high draw ratio; a similar stabilizing effect is predicted with cooling, where the contributions of viscoelasticity and heat transfer are synergistic (Fisher & Denn 1977).

Most experimental studies on draw resonance are consistent with the results of the stability analyses, but there are some important exceptions. Chang & Denn (1979) found a throughput effect on the onset of draw resonance at constant α; this anomaly might be a consequence of an inadequate rheological characterization of the fluid, or of failure to satisfy experimentally all boundary conditions, but it is presently unexplained. Matsumoto & Bogue (1978) found that cooling was sometimes destabilizing when an amorphous polymer was cooled below its glass transition temperature prior to the water bath; this observation cannot be reconciled with available theory.

From a processing point of view, spinline *sensitivity* is more important than stability; by sensitivity we mean the amplification of small input disturbances, resulting in a nonuniform product. Ziabicki (1976) shows data on product variations resulting from variations in cross-flow air. Sensitivity was first considered by Pearson & Matovich (1969). Denn (1979b) and Kase & Denn (1978) have reported results of unpublished sensitivity studies by H. H. George, and S. Kase & M. Araki, respectively, for spinning of a Newtonian liquid; some typical results are shown in Figure 14 in the form of a Bode diagram. $|G|$ is the ratio of the amplitude of drawn filament area fluctuations to fluctuations in the initial area, shown over a range of disturbance frequencies, and $\measuredangle G$ is the phase angle; note that the $|G|$ scale is logarithmic. The conditions in Figure 14 correspond to stable spinning, but very large amplifications of input disturbances are still possible. The amplitude ratio is moderated by cooling. Sensitivity studies to date do not include the dynamics of the location of the solidification point in a consistent fashion.

Finally, a complete theoretical understanding of filament breakage under typical processing conditions does not exist, other than the empirical notion of a critical stress that follows from stored energy

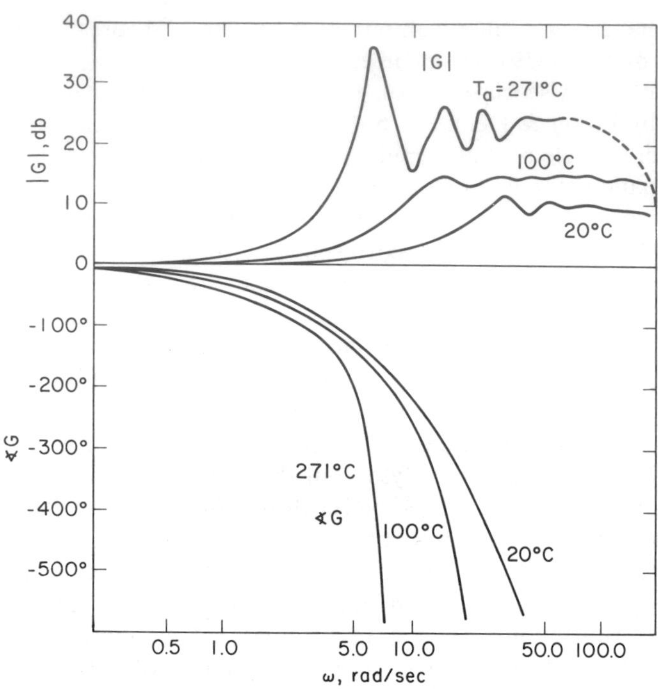

Figure 14 Bode diagram relating drawn-filament area fluctuations to initial area fluctuations. Calculations of S. Kase and M. Araki, from Kase & Denn (1978).

considerations of Reiner & Freudenthal (1938). White & Ide (1978) argue that failure by necking in uniform uniaxial extension tests is analogous to draw resonance, but the relation is not evident. Breakup may be regarded in general as a phenomenon distinct from hydrodynamic instabilities such as draw resonance; more complete discussions of breakup are given by Petrie & Denn (1976) and Ziabicki (1976). Glicksman (1974) discusses an instability in glass fiber spinning that is apparently associated with the heat transfer over the initial portion of the filament.

CONCLUDING REMARKS

The art of steady-state modeling of continuous filament drawing is well developed, and is being applied to the solution of processing problems. The major practical difficulties in obtaining quantitative agreement between theory and experiment are uncertainties in heat transfer and drag correlations, in the temperature dependence of physical properties for glasses, and in the detailed rheological description for polymeric liquids.

The major theoretical uncertainty is in the velocity and temperature profiles near the spinneret.

Dynamical behavior is less well established, although a theoretical basis for most qualitative observations exists; there are, however, recent experiments that differ in important qualitative ways from available theory.

Acknowledgments

The contents of this review have been influenced by my own research program in melt spinning, which has profited from collaboration and discussion with my colleagues D. V. Boger, J. M. Dealy, R. J. Fisher, H. H. George, S. Kase, A. B. Metzner, and C. J. S. Petrie. This assistance over the years is greatfully acknowledged, as is the research support of the National Science Foundation. I am particularly appreciative of detailed critiques of the content and balance of a first draft of this chapter by C. J. S. Petrie and J. L. White.

Literature Cited

Acierno, D., La Mantia, F. P., Marrucci, G., Rizzo, G., Titomanlio, G. 1976. A non-linear viscoelastic model with structure-dependent relaxation times. I. Basic formulation. II. Comparison with LDPE transient stress results. *J. Non-Newtonian Fluid Mech.* 15:125–57

Bankar, V. G., Spruiell, J. E., White, J. L. 1977. Melt spinning dynamics and rheological properties of Nylon 6. *J. Appl. Polym. Sci.* 21:2135–55

Barr, J. B., Chwastick, S., Didchenko, R., Lewis, I. C., Lewis, R. T., Singer, L. S. 1976. High modulus carbon fibers from pitch precursor. *Appl. Polym. Symp.* 29: 161–73

Bird, R. B. 1976. Useful non-Newtonian models. *Ann. Rev. Fluid Mech.* 8:13–34

Blyler, L. L. Jr., Gieniewski, C. 1979. Melt spinning and draw resonance studies on a poly(α-methylstyrene/silicone) block copolymer. *Polym. Eng. Sci.* In press

Chang, J. C., Denn, M. M. 1979. An experimental study of isothermal spinning of a Newtonian and a viscoelastic liquid. *J. Non-Newtonian Fluid Mech.* 5:369–85

Clarke, N. S. 1968. Two-dimensional flow under gravity in a jet of viscous liquid. *J. Fluid Mech.* 31:481–500

Clarke, N. S. 1969. The asymptotic effects of surface tension and viscosity on an axially-symmetric free jet of liquid under gravity. *Quart. J. Mech. Appl. Math.* 22:247–56

Cogswell, F. N. 1975. Polymer melt rheology during elongational flow. *Appl. Polym. Symp.* 27:1–18

Cogswell, F. N. 1978. Converging flow and stretching flow: a compilation. *J. Non-Newtonian Fluid Mech.* 4:23–38

Danford, M. D., Spruiell, J. E., White, J. L. 1978. Structure development in the melt spinning of Nylon 66 fibers and comparison to Nylon 6. *J. Appl. Polym. Sci.* 22:3351–61

Dealy, J. M. 1978. Extensional rheometers for molten polymers: a review. *J. Non-Newtonian Fluid Mech.* 4:9–22

Denn, M. M. 1977. Extensional flows: experiment and theory. In *The Mechanics of Viscoelastic Fluids, AMD-Vol. 22*, ed. R. S. Rivlin. New York: ASME. 126 pp.

Denn, M. M. 1979a. On an analysis of draw resonance by Hyun. *AIChE J.* In press

Denn. M. M. 1979b. Modeling for process control. In *Control and Dynamic Systems*, ed. C. T. Leondes. 15:146–93. New York: Academic

Denn, M. M., Marrucci, G. 1971. Stretching of viscoelastic liquids. *AIChE J.* 17:101–3

Denn, M. M., Marrucci, G. 1977. Effect of a relaxation time spectrum on the mechanics of polymer melt spinning. *J. Non-Newtonian Fluid Mech.* 2:159–68

Denn, M. M., Petrie, C. J. S., Avenas, P. 1975. Mechanics of steady spinning of a viscoelastic liquid. *AIChE J.* 21:791–95

Fisher, R. J., Denn, M. M. 1975. Finite-amplitude stability and draw resonance in isothermal melt spinning. *Chem. Eng. Sci.* 30:1129–34

Fisher, R. J., Denn, M. M. 1976. A theory of isothermal melt spinning and draw resonance. *AIChE J.* 22:236–46

Fisher, R. J., Denn, M. M. 1977. Mechanics of non-isothermal polymer melt spinning. *AIChE J.* 23:23–28

Frankfort, H. R. E., Knox, B. H. 1979. Polyethylene terephthalate filaments. *US Patent No. 4,134,882*

Gagon, D. C., Denn, M. M. 1978. *Simulation of high speed melt spinning.* Presented at 71st Annual AIChE meeting, Miami, Nov.

George, H. H. 1979. *Model of steady state spinning at intermediate take up speeds.* Presented at 2nd Joint US-Japan Societies of Rheology Meeting, Kona, Hawaii, Apr.

Geyling, F. T. 1976. Basic fluid-dynamic considerations in the drawing of optical fibers. *Bell Sys. Tech. J.* 55:1011–56

Glicksman, L. R. 1968. The dynamics of a heated free jet of variable viscosity liquid at low Reynolds number. *J. Basic Eng.* (*Trans. ASME, Ser. D*) 90:343–54

Glicksman, L. R. 1974. A prediction of the upper temperature limit for glass fibre spinning. *Glass Technol.* 15:16–20

Hamana, I., Matsui, M., Kato, S. 1969. The progress of fiber formation processes from melt spinning. *Melliand Textilber* 50:382–88, 499–503

Heuvel, H. M., Huisman, R. 1978. Effect of windup speed on the physical structure of as-spun poly(ethylene terephthalate) fibers, including orientation induced crystallization. *J. Appl. Polym. Sci.* 22:2229–43

Hyun, J. C. 1978. Theory of draw resonance. Part I. Newtonian fluids. Part II. Power-law and Maxwell fluids. *AIChE J.* 24:418–26

Ishihara, H., Kase, S. 1975. Studies on melt spinning. V. Draw resonance as a limit cycle. *J. Appl. Polym. Sci.* 19:557–65

Ishihara, H., Kase, S. 1976. Studies on melt spinning. VI. Simulation of draw resonance using Newtonian and power law viscosities. *J. Appl. Polym. Sci.* 20:169–91

Kase, S., Denn, M. M. 1978. Dynamics of the melt spinning process. *Proc. 1978 Joint Automatic Control Conf.*, pp. II71–84

Kase, S., Matsuo, T. 1967. Studies on melt spinning. II. Steady-state and transient solutions of fundamental equations compared with experimental results. *J. Appl. Polym. Sci.* 11:251–87

Kase, S., Matsuo, T., Yoshimoto, Y. 1966. Theoretical analysis of melt spinning. Part 2: surging phenomena in extrusion casting of plastic films. *Seni Kikai Gakkaishi* 19:T63–72

Kaye, A., Vale, D. G. 1969. The shape of a vertically falling stream of a Newtonian liquid. *Rheol. Acta* 8:1–5

Krishnan, S., Glicksman, L. R. 1971. A two-dimensional analysis of a heated free jet at low Reynolds number. *J. Basic Eng.* (*Trans. ASME, Ser. D*) 93:355–64

Laun, H. M., Münstedt, H. 1976. Comparison of the elongational behavior of a polyethylene melt at constant stress and constant strain rate. *Rheol. Acta* 15:517–24

Matovich, M. A., Pearson, J. R. A. 1969. Spinning a molten threadline–steady-state isothermal viscous flows. *Ind. Eng. Chem. Fundam.* 8:512–20

Matsui, M. 1976. Air drag on a continuous filament in melt spinning. *Trans. Soc. Rheol.* 20:465–73

Matsui, M., Bogue, D. C. 1976. Non-isothermal rheological response in melt spinning and idealized elongational flow. *Polym. Eng. Sci.* 16:735–41

Matsumoto, T., Bogue, D. C. 1978. Draw resonance involving rheological transition. *Polym. Eng. Sci.* 18:564–71

Matsuo, T., Kase, S. 1976. Studies on melt spinning. VII. Temperature profile within the filament. *J. Appl. Polym. Sci.* 20:367–76

Mewis, J., Metzner, A. B. 1974. The rheological properties of suspensions of fibers in Newtonian fluids subjected to extensional deformation. *J. Fluid Mech.* 62:593–600

Middleman, S. 1977. *Fundamentals of Polymer Processing.* New York: McGraw-Hill. 525 pp.

Moncrieff, R. W. 1975. *Man-Made Fibers.* New York: Wiley. 1094 pp. 6th ed.

Mottern, J. W., Privott, W. J. 1978. *Spinning Wire from Molten Metal. AIChE Symp. Ser., Vol. 74, No. 180.* New York: Am. Inst. Chem. Eng. 116 pp.

Nadella, H. P., Henson, H. M., Spruiell, J. E., White, J. L. 1977. Melt spinning of polypropylene: structure development and relationship to mechanical properties. *J. Appl. Polym. Sci.* 21:3003–22

Nadkarni, V. M., Schultz, J. M. 1977. Extensional flow-induced crystallization in polyethylene melt spinning. *J. Polym. Sci., Polym. Phys. Ed.* 15:2151–83

Oda, K., White, J. L., Clark, E. S. 1978. Influence of melt deformation history on orientation in vitrified polymers. *Polym. Eng. Sci.* 18:53–59

Oliver, D. R., Ashton, R. C. 1976. The triple jet: influence of shear history on the stretching of polymer solutions. *J. Non-Newtonian Fluid Mech.* 1:93–104

Pearson, J. R. A. 1976. Instability in non-Newtonian flow. *Ann. Rev. Fluid Mech.* 8:163–81

Pearson, J. R. A., Matovich, M. A. 1969. On spinning a molten threadline–stability. *Ind. Eng. Chem. Fundam.* 8:605–9

Pearson, J. R. A., Shah, Y. T., Mhaskar, R. D. 1976. On the stability of fiber spinning of freezing liquids. *Ind. Eng. Chem. Fundam.* 15:31–37

Petrie, C. J. S. 1979a. *Elongational Flows.* London: Pitman. 254 pp.

Petrie, C. J. S. 1979b. Measures of deformation and convected derivatives. *J. Non-Newtonian Fluid Mech.* 5:147–76

Petrie, C. J. S., Denn, M. M. 1976. Instabilities in polymer processing. *AIChE J.* 22:209–36

Phan-Thien, N. 1978. A nonlinear network viscoelastic model. *J. Rheol.* (formerly *Trans. Soc. Rheol.*) 22:259–83

Raible, T., Demarmels, A., Meissner, J. 1979. Stress and recovery maxima in LDPE melt elongation. *Polym. Bull.* 1:397–402

Reiner, M., Freudenthal, A. 1938. Failure of a material showing creep. (A dynamical theory of strength) *Proc. 5th Int. Cong. Appl. Mech.*, pp. 228–33

Shah, Y. T., Pearson, J. R. A. 1972. On the stability of non-isothermal fibre spinning —general case. *Ind. Eng. Chem. Fund.* 11:150–53

Shimizu, J., Toriumi, K., Tamai, K. 1977. High speed melt spinning of polyester filaments—effect of spinning velocity on the properties and molecular orientation. *Seni Gakkai Shi* 33:T208–14

Slattery, J. C., Acharya, A. 1979. Surface effects in an elongational flow and in spinning. *AIChE J.* Submitted

Spearot, J. A., Metzner, A. B. 1972. Isothermal spinning of molten polyethylenes. *Trans. Soc. Rheol.* 16:495–518

Wagner, W. S., Roberts, J. D. 1972. Method for producing fibrous structure. *US Patent No. 3,634,573*

Walczak, Z. K. 1977. *Formation of Synthetic Fibers.* New York: Gordon & Breach. 333 pp.

White, J. L. 1976. Dynamics and structure development during melt spinning of fibers. *J. Soc. Rheol. Jpn.* 4:137–48

White, J. L., Ide, Y. 1978. Instabilities and failure in elongational flow and melt spinning of fibers. *J. Appl. Polym. Sci.* 22:3057–74

White J. L., Roman, J. F. 1976. Extrudate swell during the melt spinning of fibers—influence of rheological properties and takeup force. *J. Appl. Polym. Sci.* 20: 1005–23

Zeichner, G. R. 1972. *Spinnability of viscoelastic liquids.* MChE thesis. Univ. Del. Newark

Ziabicki, A. 1976. *Fundamentals of Fibre Formation.* New York: Wiley. 488 pp.

Ann. Rev. Fluid Mech. 1980. 12: 389–433

MODELS OF WIND-DRIVEN CURRENTS ON THE CONTINENTAL SHELF

J. S. Allen
School of Oceanography, Oregon State University, Corvallis, Oregon 97331

1 INTRODUCTION

Adjacent to the ocean shoreline, the continental shelves form regions of relatively shallow water for offshore distances that vary with geographical location, but are typically the order of 50–150 km. The sea floor slopes gently across the continental shelf from the coast to water depths of about 200 m, where an abrupt increase in slope generally occurs at the so-called shelf break. Seaward of the break, the continental slope extends downward to the deep ocean floor over an additional horizontal distance of the order of 100 km.

The motion of the oceanic waters over the continental shelf and slope is influenced by the earth's rotation, the density stratification, the offshore current regime, the sloping bottom topography, and the presence of a coastline. In particular, the topographic constraints of the coastline and shallow sloping bottom give the shelf flow field certain characteristics that differ from those typical of the deep ocean. In many instances, flow over the shelf may be considered independent of motions farther offshore.

The physical oceanographic conditions on the shelf influence several important oceanic processes. For example, most of the biological primary productivity of the world's oceans takes place in the relatively fertile surface waters over the continental shelves. The biological phenomena involved there are highly dependent on the fluid mechanical processes on the shelf. Sediment transport and pollutant dispersion are other processes occurring over the shelf that are strongly affected by the properties of the fluid motion.

Observational information on the nature of shelf flow fields has increased tremendously since the early 1970s with the development and

0066-4189/80/0115-0389$01.00

deployment of unattended moored current meters capable of making and recording horizontal velocity measurements for periods of several months. Current measurements have been made by now on many continental shelves and have shown that the nature of the flow in different shelf regions, such as off the northeast and off the northwest coasts of the United States, can vary considerably. The major variations in the flow, to the extent that they are understood at the present, appear to be due in part to differences in the shelf width, to the nature of the local and nearby coastal winds, and to the strength and character of the offshore currents. The observations have also shown, however, that a feature common to most shelf flow fields is a response of currents to coastal winds. It appears, in fact, that a primary driving mechanism for the velocity field on many continental shelves is the alongshore component of wind stress over the shelf. The wind stress typically produces energetic fluctuations in the shelf velocity field on the two- to ten-day time scales that characterize the variability of atmospheric storms and synoptic scale wind events. In addition, on longer time scales, the seasonal variability of the wind field may induce a corresponding seasonal variability in the shelf currents.

Accompanying the increase in current measurements and other observations in coastal regions there has been an increase in effort to model the shelf flow field theoretically. We review here models of wind-driven currents on the continental shelf. In particular, we are concerned with temporal variability on the wind-induced time scales of several days to several weeks. The spatial scales of interest are dictated by the geometry of the continental shelf, by the wind stress forcing function, and by scales intrinsic to the fluid motion. In general, horizontal scales will range from a few kilometers to the order of a hundred kilometers offshore and up to thousands of kilometers alongshore. Vertical scales will range from a few meters to the water depth, which over the shelf and upper slope may be several hundred meters.

2 OBSERVATIONS

Before a discussion of the theoretical models, we refer briefly to observational results to point out some characteristics of wind-driven shelf flow fields. For that purpose, we utilize measurements made off Oregon where the fluid motion on the continental shelf and upper slope is strongly wind-driven. The Oregon observations illustrate several features typical of shelf flow regimes forced by the wind. It should be kept in mind, however, that there will be various differences in behavior on shelves in other geographical locations.

For describing the shelf flow field, a right-handed Cartesian coordinate system (x,y,z) is used with the y-axis aligned alongshore, the x-axis aligned onshore-offshore, and the z-axis vertical. The origin is at the coast and at the mean water level; x is positive onshore. The corresponding velocity components are (u,v,w). The alongshore component of the wind stress at the coast $(x = 0)$ is denoted by $\tau_W^{(y)}(y,t)$ where t is time. Subscripts (x,y,z,t) will be used to denote partial differentiation.

The velocity field on the Oregon shelf has a strong seasonal cycle (Huyer, Pillsbury & Smith 1975). Current measurements from 40-m depth in 100 m of water at a distance of about 15 km from the coast for a time period of over 16 months (Huyer & Smith 1978) are plotted in Figure 1 in the form of a vector stick diagram. The vectors (sticks) represent the magnitude and direction of essentially a daily averaged horizontal velocity. The coast in this location is aligned approximately in a north-south direction, and the vertical axis in the stick diagram corresponds to north-south, with north upward. The water temperature, measured at the same point as the currents, coastal sea level, with the mean value removed, and a vector stick diagram of coastal winds are also plotted.

The seasonal cycle is most apparent in the sea level record, where negative values generally correspond to southward velocities and positive values to northward velocities. The annual cycle in the currents corresponds closely to that in sea level and also to the wind field which evidently is the primary driving mechanism for the flow on the shelf. The presence of energetic fluctuations in the wind on time scales of two to ten days is obvious from the records. Fluctuations on similar time scales occur in the currents and these are presumably driven in some manner by the fluctuations in the winds.

Note that the water temperature is lower in the summer, April–September, than it is in the winter. In the summer, the southward alongshore component of the wind stress drives an offshore transport in a surface Ekman layer of about 20-m depth. In order for mass to be conserved, this offshore surface flow leads in general to a compensating onshore flow below the surface layer and a positive vertical motion near the coast. Off Oregon, the vertical motion evidently takes place over the continental shelf within a distance of 10–15 km from the shore. This process is called coastal upwelling and it leads to the appearance of relatively cold water at the surface near the coast. It also leads to the transport of dissolved nutrients from greater depths to the surface euphotic zone and hence to the large productivity found in the upwelling regions of the world's oceans.

Measurements of density on a section normal to the coast taken during the summer of 1973 are plotted in Figure 2. This figure also clearly

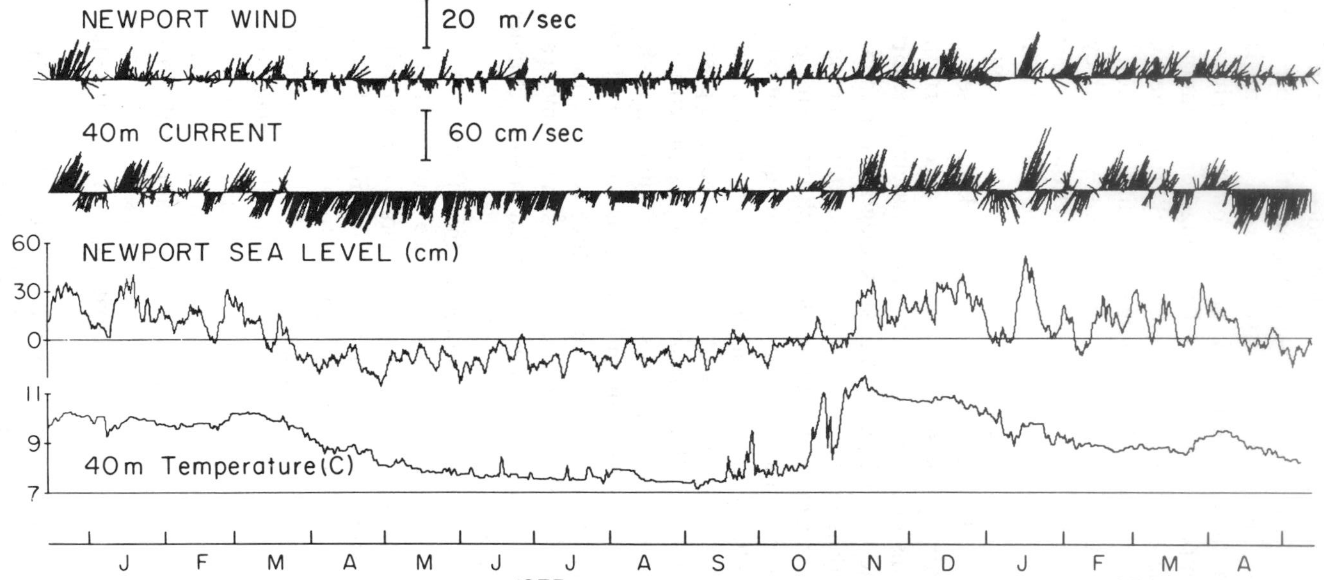

Figure 1 Coastal wind at Newport, Oregon (44°40′N), current at 40-m depth at mooring *P* (44°45′N, water depth 100 m), coastal sea level at Newport, and the water temperature at 40-m depth at mooring *P*, from December 1972 to May 1974 (Huyer & Smith 1978). The observations here, as well as those in all subsequent plots of winds, currents, and sea level, have been low-pass filtered (half-power point 0.6 cpd) to remove inertial, tidal, and other higher frequency oscillations. The horizontal components of the wind and current are plotted as vectors, one per day. The vertical axes of the vector diagrams correspond to the north-south direction, with north upwards. A bathymetric contour map of the Oregon shelf near 45°N is shown in Figure 4. No measurements from mooring *P* are plotted there, but the location of *P* is on the 100-m isobath, west of Cape Foulweather. Note that, although the coastline runs almost north-south, the current vectors tend to be aligned at an angle from North that corresponds approximately to the orientation of the local isobath at *P*.

shows the bottom topography of the continental shelf and slope in this location. The shelf break occurs at a depth of about 200 m at a distance of approximately 30 km from the coast. The density section, expressed in σ_t units [essentially the specific gravity anomaly at atmospheric pressure $\times 10^3$ (Sverdrup, Johnson & Fleming 1942)], shows a pycnocline at about 100-m depth offshore. As the coast is approached, the isopycnals slope upwards so that relatively dense water is present close to the surface near the shore. This is typical during coastal upwelling conditions in this region. As a result, there is considerable density stratification in the water over the shelf (e.g. $\Delta\sigma_t = 2$ over 100-m depth).

We next consider some additional features of the flow during the summer under upwelling-favorable wind conditions. It will be convenient to refer to mean values, over a one or two month time period, with an overbar and to use a prime to denote the fluctuations about the mean, e.g. $v = \bar{v} + v'$, $\bar{v}' = 0$. For the summer, the mean alongshore component of the wind stress is southward, i.e. $\bar{\tau}_W^{(y)} < 0$.

A plot of the alongshore (northward) component of the current v, measured at various depths in 100 m of water during July and August 1973, is given in Figure 3. Also plotted is the alongshore component of the wind and the coastal sea level. Several features of the vertical structure

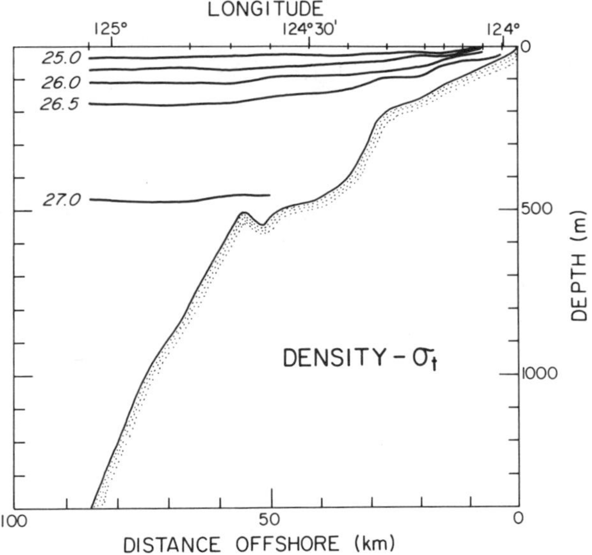

Figure 2 Vertical density section off Oregon at 45°16.5′N, 28–29 June 1973 in σ_t units. The stations at which measurements were made are marked with tics on the top horizontal axis (Allen 1975).

and time dependence of the velocity field are observable from this figure. There is a mean vertical shear in v, i.e. $\bar{v}_z \neq 0$. This shear appears to be balanced by horizontal density gradients, evident in the density section in Figure 2, through the thermal wind equation (Smith 1974; see Section

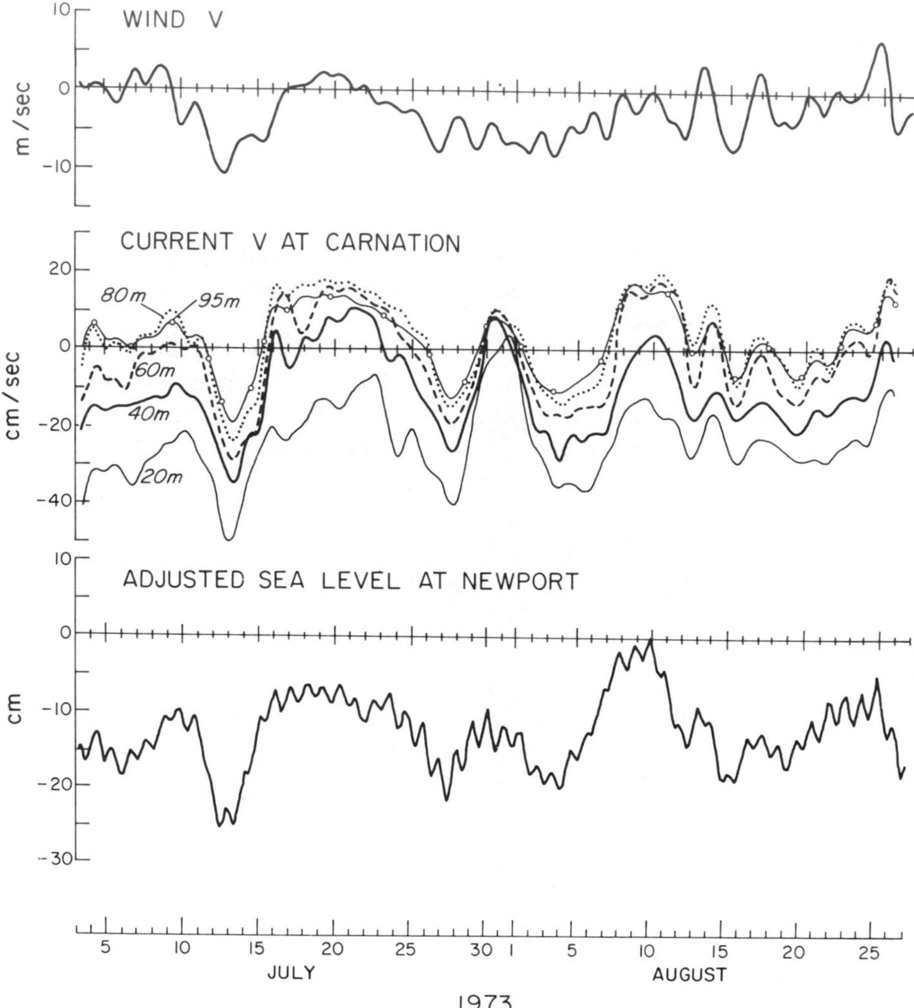

Figure 3 Low-pass-filtered north-south (alongshore) component of wind and currents, and adjusted sea level off Oregon during July and August 1973 (Kundu, Allen & Smith 1975). The currents were measured at mooring C (45°16′N, water depth 100 m) at the depths indicated. The wind and sea level were measured at Newport, Oregon (44°40′N). Measurements from mooring C are plotted on the bathymetric contour map in Figure 4, where the location of C is on the 100-m isobath just north of Cape Kiwanda.

2). The mean values of v above 60 m are southward. In contrast to the mean flow, the fluctuations in v are primarily depth-independent, i.e. $v'_z \simeq 0$. The fluctuations in v are very highly correlated with the fluctuations in sea level, reflecting the apparent geostrophic balance of the alongshore velocity. The fluctuations in v are also clearly correlated with the alongshore component of the local wind stress $\tau_W^{(y)}$ (which has a time variation qualitatively similar to the wind velocity), with a typical correlation coefficient of about 0.6. At times, e.g. 10–20 July, v appears to respond directly to the variation in $\tau_W^{(y)}$, but there are occasions, e.g. 28 July–31 August, when the variation of v seems to be unrelated to that of $\tau_W^{(y)}$.

The horizontal structure of the fluctuating velocity field is illustrated in Figure 4 where velocity scatter diagrams are superimposed on a bathymetric map (Kundu & Allen 1976). In the scatter diagrams, the fluctuating horizontal velocity vector (u',v') is plotted as a point in a hodograph plane at regular time intervals. The origin of the scatter diagram is placed on the map at the horizontal location of the measurements. The data shown were taken at about 40-m depth and include measurements from two summers, 1972 and 1973, at different points. The distribution of the dots in the x and y (northward) directions reflects the magnitudes of the standard deviations u'_{SD} and v'_{SD}, respectively. Note that v'_{SD} is quite a bit larger than u'_{SD} over the shelf and that v'_{SD} increases and u'_{SD} decreases as the coast is approached. Also evident from this plot is the tendency for the horizontal velocity fluctuations to be aligned along the local isobaths.

Other important features of the flow that stand out from the summer observations are the following. The fluctuations in v are well correlated in all directions, i.e. over depth (at least over the range of depths of the measurements 20–200 m), onshore-offshore to at least 50 km from the coast, and over alongshore distances of at least 200 km. The fluctuations in v propagate northward (Huyer et al 1975). During the summer of 1972, this propagation appeared to be nondispersive over a frequency range $0 \leqq \omega \leqq 0.3$ cpd, with a mean speed of about 500 km day^{-1} (Kundu & Allen 1976).

The behavior of the u component of velocity is more complicated than that of v. It has much shorter correlation scales in all directions than v and, in general, a lower correlation with $\tau_W^{(y)}$ than v does (Kundu & Allen 1976). This is presumably due in part to the fact that v' is larger in magnitude than u' and that, consequently, the organized, large scale signal in v' should have a better chance to stand out above background, shorter scale isotropic noise. In addition, the fluctuations in u have a considerable vertical shear, i.e. $u'_z \neq 0$, with the fluctuations near the surface having the larger magnitude.

Accompanying the fluctuations in wind and currents on the several-day time scale are substantial variations in the density field. These are illustrated in Figure 5 where, in particular, the upward slope of the isopycnals toward the coast in response to the strong wind event of

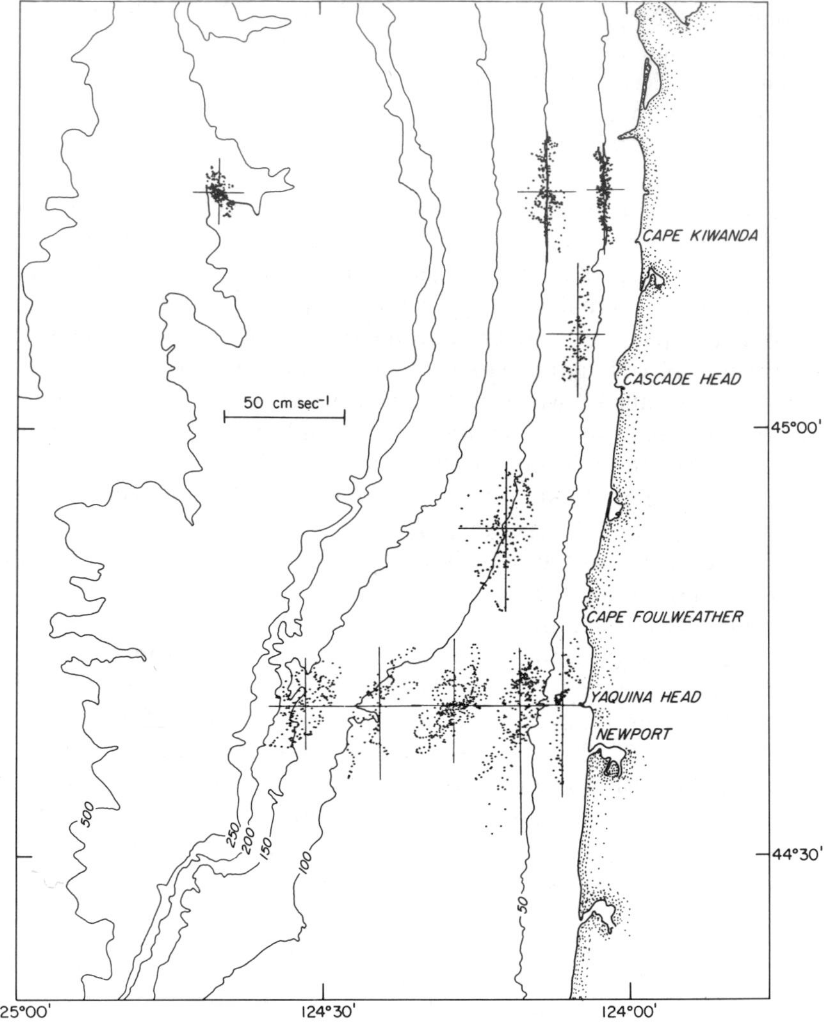

Figure 4 Scatter diagrams of horizontal velocity fluctuations (mean values subtracted) superimposed on a bathymetric contour map, with the origin of the diagram at the horizontal location of the measurements (Kundu & Allen 1976). The currents were measured at about 40-m depth during the summer of 1972 (the six southern diagrams) and during the summer of 1973 (the four northern diagrams). The low-pass-filtered current fluctuations are plotted as a point every six hours. Depths are in meters.

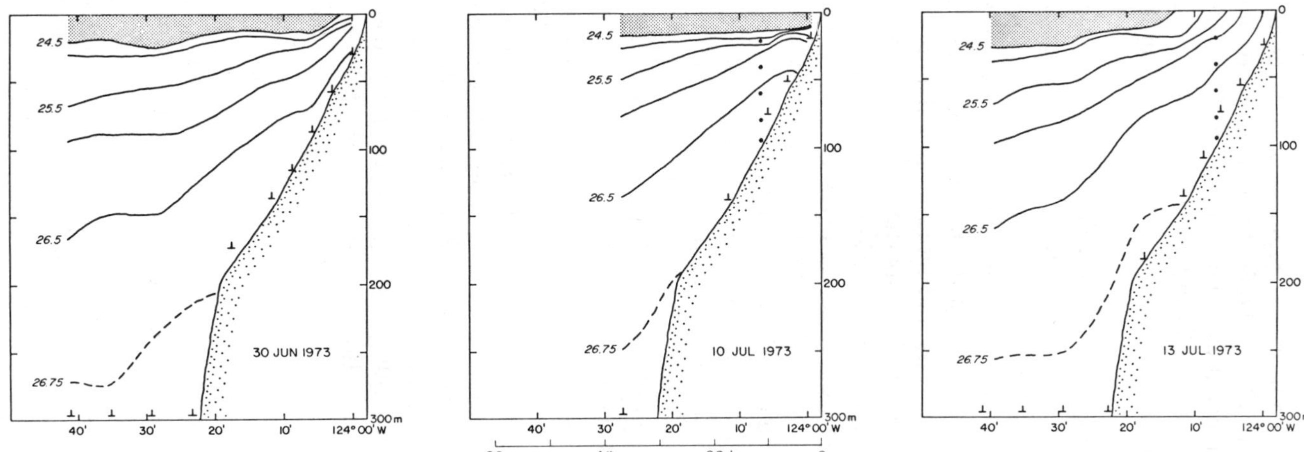

Figure 5 Vertical density sections in σ_t units along 45°16′N during June–July 1973. The positions of the current meters at mooring *C* are marked with solid dots. Contour intervals (0.5 σ_t for $\sigma_t > 24.5$) are not shown for the shaded area ($\sigma_t < 24.5$). Station positions are marked by inverted tees along the bottom. Note the vertical and horizontal scales are different than in Figure 2 (Kundu, Allen & Smith 1975).

10–15 July (Figure 3) may be seen. The density variations seem to occur primarily above 20-m depth or in a region close to shore, i.e. within 10 km of the coast (Smith 1974). The fluctuations in v have an increased shear in this region, compared with those at mid-shelf, but in spite of the relatively large motion of the density surfaces, the major part of the fluctuating v field (measured in 50 m of water, 5 km from the coast) is still depth-independent.

An important question about the flow field over the continental shelf concerns the extent to which the motion may be regarded as two-dimensional in an onshore-offshore vertical plane, i.e. as having negligible gradients in velocity in the alongshore direction. This question is of interest because it bears directly on the formulation of qualitative and quantitative models of the mass and momentum balances on the shelf. The cross-shelf mass balance is of particular concern for studies of coastal upwelling and the simplest situation results, of course, if the flow field is approximately two-dimensional. Investigations of the mass and momentum balances for the fluctuating velocity components during the summer (Allen & Kundu 1978, Smith 1979) indicate, however, that alongshore velocity and pressure gradients probably play a significant role and that the flow is three-dimensional. This result is consistent with the observations of northward propagation of velocity fluctuations.

In the winter, October–March, the characteristics of the flow field are somewhat different (Huyer, Smith & Sobey 1978). In general, the mean wind stress and mean alongshore velocities are toward the north, i.e. $\bar{\tau}_W^{(y)} > 0$, $\bar{v} > 0$. This corresponds to a downwelling situation and the resulting stratification over most of the shelf is substantially weaker than in the summer. Accompanying the weaker stratification on the shelf are smaller values of the vertical shear in $\bar{v}$. The fluctuations in v over the shelf, however, have larger z gradients than in the summer. This shear may be balanced inviscidly by horizontal density gradients through the thermal wind equation (Huyer, Smith & Sobey 1978) or it may be related to bottom frictional effects (Smith & Long 1976) or both.

The wind stress is generally stronger in the winter (see Figure 1), corresponding to the winter storms. In general, the fluctuations in $\tau_W^{(y)}$ and in the alongshore currents v are larger, and the correlation of $\tau_W^{(y)}$ and v is higher, than in the summer. In fact, the strong response of the alongshore currents to the wind is the dominant qualitative feature of the shelf flow field during the winter.

In summary, the most obvious single characteristic of the flow is the relation of the alongshore currents, on both the several-day and the seasonal time scales, to the alongshore component of the wind stress. Variations in the density field over the shelf are also forced by the wind on these time scales. Other features of the flow give indications of the probable dynamic balances. The approximate depth-independence of the fluctuating alongshore velocity component, the tendency for the horizontal velocity fluctuations to align along isobaths, and the general northward propagation of disturbances are all characteristics of an inviscid response. Investigations of the time-dependent mass and momentum balances indicate that alongshore gradients play a substantial role and therefore that the motion is basically three-dimensional. Finally, the behavior of the cross-shelf velocity component u is much more complicated than that of the alongshore component v. This is significant because the cross-shelf velocity field plays a critical role in processes such as coastal upwelling.

3 FORMULATION

The governing equations are those that express the conservation of mass, momentum, energy, and salt, plus an equation of state. In the approximate form frequently used to model wind-driven motion on the shelf, they are

$$\nabla \cdot \mathbf{u} = 0, \tag{3.1a}$$

$$\rho_0(u_t + \mathbf{u} \cdot \nabla u - fv) = -p_x + \tau_z^{(x)} + F^{(x)}, \tag{3.1b}$$

$$\rho_0(v_t + \mathbf{u} \cdot \nabla v + fu) = -p_y + \tau_z^{(y)} + F^{(y)}, \tag{3.1c}$$

$$0 = -p_z - g\rho, \tag{3.1d}$$

$$\rho_t + \mathbf{u} \cdot \nabla \rho = D, \tag{3.1e}$$

where $\mathbf{u}$ is the velocity vector, ρ_0 is a constant reference density such that the total density $\rho_T = \rho_0 + \rho$, p is the modified pressure, f is the Coriolis parameter, g is the acceleration of gravity, and $\tau^{(x,y)}$ are turbulent shear stresses in the (x,y) directions. The Boussinesq and the hydrostatic approximations have been made and the energy and salt conservation equations have been combined into one equation (3.1e) for the density with the assumption of a linear equation of state (Fofonoff 1962). The terms $F^{(x,y)}$ and D represent, respectively, horizontal turbulent stress and density diffusion terms. The stress and diffusion terms are often modeled with eddy coefficients, in which case

$$\tau^{(x)} = \rho_0 A_V\, u_z, \quad \tau^{(y)} = \rho_0 A_V\, v_z, \tag{3.2a,b}$$

$$F^{(x)} = \rho_0[(A_H u_x)_x + (A_H u_y)_y], \tag{3.3a}$$

$$F^{(y)} = \rho_0[(A_H v_x)_x + (A_H v_y)_y], \tag{3.3b}$$

$$D = (K_V \rho_z)_z + (K_H \rho_x)_x + (K_H \rho_y)_y, \tag{3.4}$$

where A_V and A_H are, respectively, the vertical and horizontal eddy viscosities and where K_V and K_H are the vertical and horizontal eddy diffusivities.

The shelf bottom topography is given by $z = -H(x,y)$. The upper water surface may be considered free, but it is common to assume a rigid lid at $z = 0$ and to neglect the effect of surface divergence based on the estimate $f^2L_S^2/g\bar{H} \ll 1$, where L_S is a typical horizontal scale, e.g. the width of the continental shelf, and $\bar{H}$ is the average water depth over the shelf (Gill & Schumann 1974). Unless noted otherwise, the models under discussion may be assumed to have a rigid lid.

Boundary conditions at the upper surface are

$$\tau^{(x,y)} = \tau_W^{(x,y)}, \quad w = 0, \quad \text{at} \quad z = 0, \tag{3.5}$$

where $\tau_W^{(x,y)}$ is the applied wind stress. At the bottom, the velocity components satisfy the usual boundary conditions of no normal flow or no slip for inviscid or frictional models, respectively. Boundary conditions on the density and on the other variables in the offshore direction will be mentioned when needed.

In place of the continuous variation of density in (3.1), the effects of stratification are often represented by layers of homogenous fluids of different densities. As in other areas of oceanography and meteorology, the two-layer model (e.g. Peffley & O'Brien 1976) has proved useful and has been frequently utilized. The layered models lead to simpler formulations in many instances and are especially useful for representing the effects of stratification in inviscid time-dependent problems.

Typical cross-shelf bottom topography is shown in Figure 2. In particular, as the coast is approached the depth H goes to zero to form a wedge-like geometry near $x = 0$ in the (x,z) plane. The complex processes that occur in the shallow-water nearshore zone are not considered. In addition, because of the difficulties in both analytical and numerical models of properly treating a wedge-like corner region, the geometry is frequently simplified. In these cases, it is assumed that the effects of the sloping bottom topography and fluid mechanical processes in the nearshore zone are relatively unimportant in determining the characteristics of the flow field over the rest of the shelf. Consequently, the nearshore region is neglected and the coastal boundary is represented by a vertical wall placed at reasonably shallow depth, e.g. 30–50 m. The extent to which this assumption is valid for modeling specific processes, e.g. coastal upwelling, in different shelf flow regimes is not well understood and needs further study.

Basic Model

The basic dynamical features of time-dependent wind-driven coastal currents may be described with reference to a simple model.

The important forcing function is the alongshore component of the wind stress, which drives a cross-shelf transport in the surface Ekman layer. An expression for the Ekman transport in the x direction may be derived from (3.1c) with the assumption that the Coriolis force term fu balances $\tau_z^{(y)}$ and with the integration of these two terms with respect to z over the depth δ of a surface frictional layer, where $\tau^{(y)}$ $(z = -\delta) \simeq 0$. The result, with (3.5), is

$$\int_{-\delta}^{0} u dz = \tau_W^{(y)}/(\rho_0 f). \tag{3.6}$$

At the coast, the cross-shelf velocity u is zero. If we assume for simplicity that alongshore gradients are small, mass conservation requires that an offshore (onshore) Ekman transport in the surface layer over the shelf be balanced by an onshore (offshore) flow in the interior below the surface layer. If the interior motion is time-dependent and inviscid, the Coriolis force results in an acceleration of fluid in the y direction by the cross-shelf velocities through the $v_t + fu$ balance in (3.1c). Hence, the interior u velocity field forces alongshore currents.

The wind stress is relatively effective in directly driving currents on the continental shelf, compared with the deep ocean, because of the divergence of the Ekman transport at the coast and the shallow shelf topography. The divergence of the Ekman transport leads to relatively large cross-shelf velocities which are further intensified by the shallow topography. The comparatively large magnitudes of u lead, in turn, to large alongshore currents. This process is dependent primarily on the presence of a coastline, the bottom topography, and the wind stress near the coast and is the basic reason that the wind-driven motion on the shelf can often be considered independent of that in the ocean interior.

An appreciation of how an offshore Ekman transport leads to an onshore flow below the surface layer in real shelf flow fields may be obtained from Figure 6. Huyer (1976) has plotted measurements of onshore velocities u at various depths throughout the water column on a time-depth plane. Corresponding values of the wind stress are also shown. The measurements are from mid-shelf locations off Oregon and off Northwest Africa. In both cases, large negative values of $\tau_W^{(y)}$ are accompanied by relatively large values of offshore flow in the surface layer. Note that the depth of the surface layer in which the offshore flow occurs is generally the order of 20 m off Oregon, whereas it is larger off N.W. Africa, extending occasionally to 40 m. This is presumably due to the weaker stratification, generally larger magnitude of $\tau_W^{(y)}$, and smaller value of f off N.W. Africa (Kundu 1977). The fluctuations in the offshore surface layer flow are well correlated with the fluctuations in $\tau_W^{(y)}$. Although there is an obvious tendency, especially off N.W. Africa, for these fluctuations to be balanced

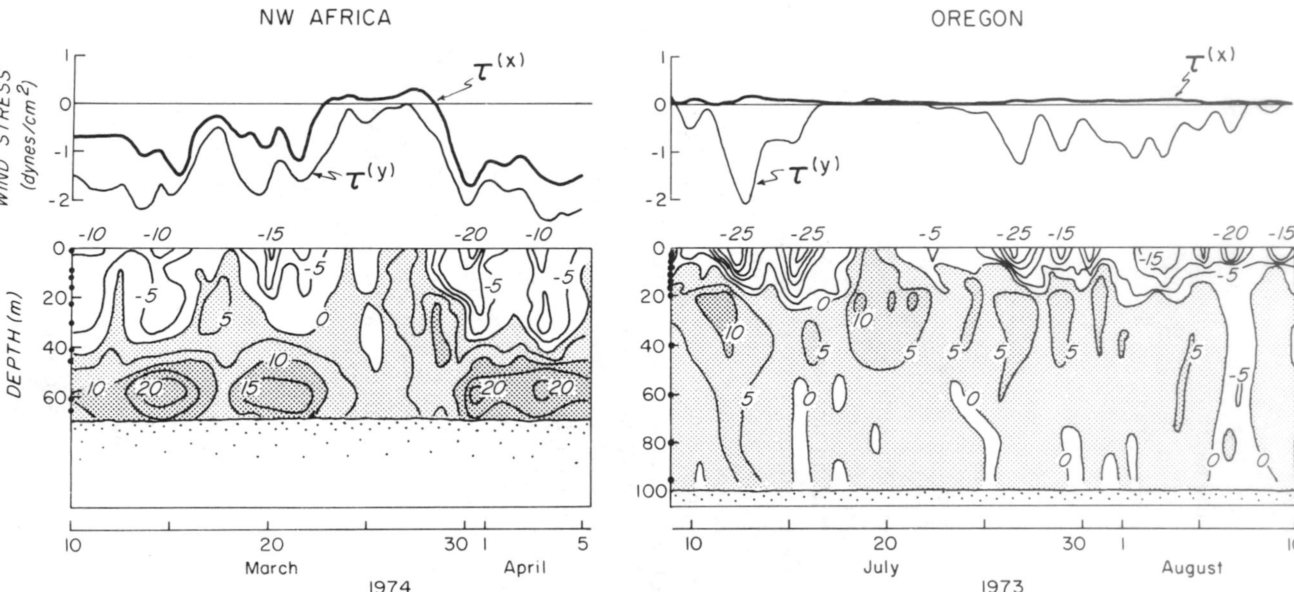

Figure 6 Low-pass filtered components of wind stress (the subscript W is omitted here) and contours of onshore flow from current meters at midshelf on a time-depth plane for Oregon during July–August 1973 and for Northwest Africa during March–April 1974 (Huyer 1976). Wind observations are from Newport, Oregon and from buoys off Northwest Africa. Current observations are from mooring C and buoy B off Oregon and from moorings W, U, and L off Northwest Africa. The contour interval is 5 cm s^{-1} in both cases. The depths at which the currents are measured are marked by black dots on the depth axes.

by fluctuations in the onshore flow in accordance with the above physical argument, it is also clear that the mass balance is not perfectly two-dimensional. Indeed, calculations of the total depth integral of u from these data by Smith (1979) show that alongshore gradients must play a role in the mass balance and that the motion is essentially three-dimensional at these mid-shelf locations.

To complete the basic model, approximate equations for the motion below the surface layer are derived based on several assumptions. First, 1. it is assumed that the time scales of interest δ_t are greater than an inertial period $\delta_t \gg f^{-1}$. This assumption will be utilized in all of the models discussed unless otherwise noted. The inertial period is the order of a day at mid-latitudes, whereas the energetic wind-induced fluctuations generally vary on the several day time scale. 2. The motion is assumed to be linear. This is based on the fact that the Rossby number ε, which reflects the magnitude of the ratio of nonlinear advection to Coriolis force terms, obtained from typical estimates is reasonably small, e.g. $\varepsilon = V(f\delta_x)^{-1} = 0.1$ where $V = 30$ cm s^{-1} is a characteristic alongshore velocity, $\delta_x = 30$ km, and $f = 10^{-4}$ s^{-1}. 3. The interior flow outside of a thin frictional bottom layer is assumed to be inviscid. This simplification is consistent with summer observations off Oregon in water of 100-m depth, but is not necessarily valid for shallow depths or for all shelves. For example, it appears that off N.W. Africa at the mid-shelf location of the measurements in Figure 4 turbulent frictional effects probably extend throughout the water column (Kundu 1977).

With the above assumptions, Equations (3.1b,c) for the interior reduce to

$$\rho_0(u_t - fv) = -p_x, \tag{3.7a}$$

$$\rho_0(v_t + fu) = -p_y. \tag{3.7b}$$

Two common additional assumptions are 4. that the alongshore scales, determined, for instance, by the wind field (≈ 1000 km), are larger than the cross-shelf scale δ_x, determined by the shelf width (≈ 30 km), i.e. that $\delta_y \gg \delta_x$, and 5. that $(\delta_y/\delta_x) \approx f\delta_t$. Assumption 5 is made to retain the maximum amount of physics consistent with 1 and 4. With Assumptions 4 and 5, Equation (3.7a) becomes

$$\rho_0 fv = p_x. \tag{3.7c}$$

The assumption of geostrophic balance (3.7c) for the alongshore component of velocity is supported off Oregon by the very high correlations found for v and coastal sea level. Equation (3.7c) and the hydrostatic balance (3.1d) imply that

$$\rho_0 f v_z = -g\rho_x, \tag{3.8}$$

which is referred to as the thermal wind equation. The validity of this relation is also supported by observations off Oregon (Smith 1974). The terms v_t, fu and u_t, fv, calculated from low-pass filtered measurements at 80 m in 100-m water depth off Oregon, are plotted in Figure 7. It may be seen that u_t is negligible with respect to fv and v_t is the same magnitude as fu, as assumed in (3.7b,c). Furthermore, it may also be seen that there is a clear negative correlation between v_t and fu. Thus, there is a tendency for these two terms to balance each other, as assumed in the last step of the description of the wind-stress driving mechanism.

With regard to Assumption 2, some support has been found from

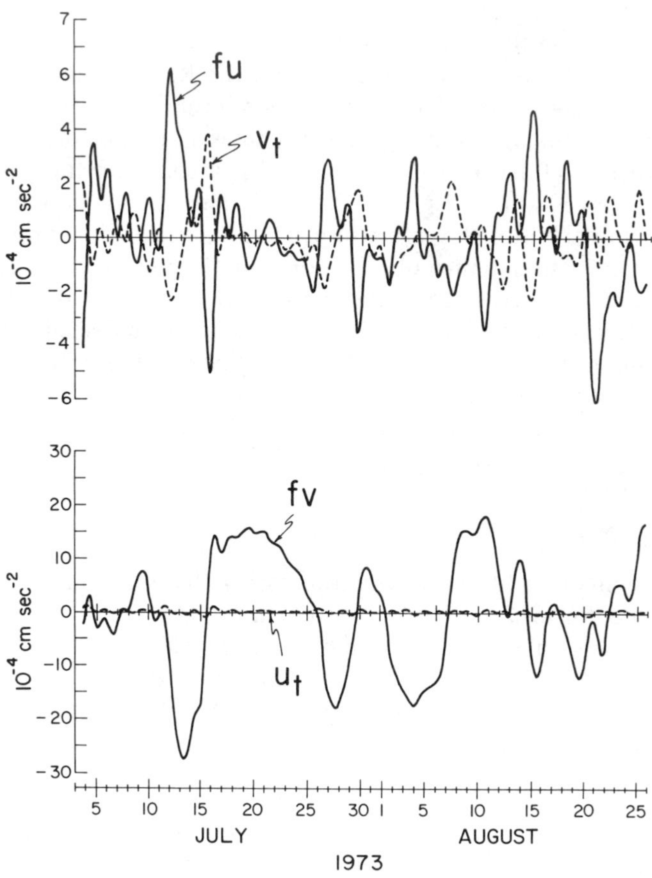

Figure 7 Low-pass filtered time series of v_t, fu, u_t, and fv off Oregon at 80-m depth at mooring C (45°16′N, water depth 100 m) during July and August 1973 (Allen & Kundu 1978).

measurements off Oregon in water of depth 100–200 m for the neglect of the nonlinear advection terms in (3.1b,c) (Allen & Kundu 1978). The observations also indicate, however, that in shallower water (50–100 m) nonlinear effects in the y-momentum equation may not be negligible.

4 ANALYTICAL MODELS

We separate the analytical and numerical studies, and discuss them separately. This is an artificial division in many respects, but it is convenient here because the analytical models may be more easily ordered with respect to the assumptions made and the physical processes involved. The numerical studies may then be discussed in light of the analytical results.

Linear Analytical Models

A large group of models has been developed under the assumption that the nonlinear advection terms in the momentum and density Equations (3.1b,c,e) may be neglected. These models are differentiated by further assumptions that concern the fluid stratification, the extent to which turbulent dissipative processes affect the motion, the nature of the turbulent processes if they are important, and the form of the bottom topography. We start here, therefore, with Assumptions 1 and 2 of Section 3.

A set of simplified models is obtained with the additional assumption that the fluid is homogeneous. This isolates and allows study of the processes that are independent of the density stratification. The horizontal velocities in this case are depth-independent in regions where the flow is inviscid. Results from these models are applied to shelves under relatively unstratified conditions and also to inviscid stratified flow fields with the further assumption that they represent one component of the solution, i.e. the depth-independent barotropic mode. The sloping shelf bottom topography can invalidate the latter assumption by coupling barotropic and baroclinic (stratification) effects (Rhines 1970, Allen 1975). However, the coupling may be weak and/or the baroclinic effects may be confined to regions near the coast, as will be discussed, so that in some cases results from homogenous models may be applicable, at least qualitatively, to substantial regions of the shelf-slope flow field.

In problems for homogeneous fluids, where the motion below the surface layer is assumed to be inviscid or where only the effects of vertical turbulent diffusion are retained, it is frequently convenient to write the equations in depth-averaged form:

$$(UH)_x + (VH)_y = 0, \qquad (4.1a)$$

$$\rho_0(U_t - fV) = -p_x + H^{-1}(\tau_W^{(x)} - \tau_B^{(x)}), \tag{4.1b}$$

$$\rho_0(V_t + fU) = -p_y + H^{-1}(\tau_W^{(y)} - \tau_B^{(y)}), \tag{4.1c}$$

where

$$p_z = 0, \quad (U,V) = \frac{1}{H}\int_{-H}^{0} (u,v)\, dz, \tag{4.2a,b}$$

and where $\tau_B^{(x,y)}$ are bottom stresses in the (x,y) directions.

The form of (4.1a) allows the definition of a transport streamfunction, such that $UH = \psi_y$, $VH = -\psi_x$, and (4.1a,b,c) may be combined into one transport vorticity equation for ψ:

$$\begin{aligned}[\psi_{xx} + \psi_{yy} - (H_x/H)\psi_x - (H_y/H)\psi_y]_t + f(H_x/H)\psi_y - f(H_y/H)\psi_x \\ = (H_x/H)(\tau_W^{(y)} - \tau_B^{(y)}) - (H_y/H)(\tau_W^{(x)} - \tau_B^{(x)}) - (\tau_{Wx}^{(y)} - \tau_{Wy}^{(x)}) \\ + (\tau_{Bx}^{(y)} - \tau_{By}^{(x)}),\end{aligned} \tag{4.3}$$

where it has been assumed that the Coriolis parameter f is a constant since effects from the latitudinal variation of f are typically small relative to those from the variations in topography (Buchwald & Adams 1968). Alternatively, a vorticity equation is often derived in terms of the pressure p.

In the simplest model for wind-forced motion, it is assumed that the flow below the surface layer is inviscid ($\tau_B^{(x,y)} = 0$), that the depth is dependent only on the cross-shelf coordinate $H = H(x)$, and that the scale of the wind-stress forcing is larger than the shelf-slope width. The latter corresponds to Assumption 4 in Section 3. In that case, (4.3) reduces to

$$[\psi_{xx} - (H_x/H)\psi_x]_t + f(H_x/H)\psi_y = (H_x/H)\tau_W^{(y)}. \tag{4.4}$$

In (4.4), the alongshore component of the wind stress $\tau_W^{(y)}$ forces a time rate of change of the vertical component of relative vorticity $\zeta = v_x - u_y \simeq v_x = -H^{-1}[\psi_{xx} - (H_x/H)\psi_x]$ through the stretching of vortex lines by the up and down slope motion of the interior inviscid cross-shelf velocity $H^{-1}[\psi_y - (\tau_W^{(y)}/f)]$. Note that the wind-stress curl term $(\tau_{Wx}^{(y)} - \tau_{Wy}^{(x)})$, which would be the major forcing function if (4.3) were applied to the deep ocean, is neglected relative to the coastal Ekman divergence term $(H_x/H)\tau_W^{(y)}$. This follows formally from the assumptions that the width of the continental margin L_M is smaller than the scale of the wind stress and that $L_M(H_x/H) = 0(1)$.

For stratified flow problems, (3.1e) is linearized about a basic density distribution, i.e. $\rho = \bar{\rho}(z) + \rho'$. In the nondiffusive case, (3.1e) reduces to

$$\rho'_t + w\bar{\rho}_z = 0, \tag{4.5}$$

and ρ' replaces ρ in (3.1d) with a redefinition of p. If the flow is also inviscid, a single governing equation expressing the conservation of potential vorticity may be derived for the pressure from (3.1a,d), (3.7a,b) and (4.5):

$$\{p_{xx}+p_{yy}+[f^2 p_z/N^2(z)]_z\}_t = 0, \tag{4.6}$$

where $N^2(z) = -g\bar{\rho}_z/\rho_0$ is the square of the Brunt-Väisälä frequency. Appropriate boundary conditions for (4.6) are obtained by expressing the velocity conditions in terms of p.

COASTAL TRAPPED WAVES Continental shelf regions, such as that off Oregon, possess two separate properties that can support subinertial frequency ($\omega < f$) coastal trapped wave motion. The first of these is the sloping bottom topography of the continental margin which, in the absence of stratification, can support barotropic continental shelf waves (e.g. Buchwald & Adams 1968). These vorticity waves are trapped along the coast because they depend on the depth variations of the shelf and slope for their restoring mechanism. The other property is the density stratification which, with a vertical boundary and a flat bottom, can support baroclinic internal Kelvin waves (e.g. Gill & Clarke 1974). The internal Kelvin waves are characterized by exponential decay in the offshore direction on the scale of the internal Rossby radius of deformation, $\delta_R \simeq H_0 N_0/f$, where H_0 is a characteristic depth and N_0 is a characteristic Brunt-Väisälä frequency. With the presence of both density stratification and typical continental margin bottom topography, the coastal trapped wave modes are more complicated, with their cross-shelf and vertical structure dependent on the nature of the stratification, the bottom topography, and on f (Allen 1975, Wang & Mooers 1976, Huthnance 1978). Coastal trapped waves naturally play a large role in wind-forced inviscid shelf problems. The properties of these waves are reviewed in the present volume by Mysak (1980). Consequently, we will refer the reader to that article for some of the background information and limit our discussion to the features most relevant to the forced-shelf problem.

Solutions for free barotropic continental shelf waves may be obtained from (4.3) with $\tau_W^{(x,y)} = 0$ and $\tau_B^{(x,y)} = 0$. It is also assumed here, and in the subsequent analyses unless otherwise noted, that $H = H(x)$ and that in the offshore direction, past the continental shelf and slope, $H_x \to 0$. For a given alongshore wave number l, there are an infinite set of cross-shelf modes. The waves are dispersive with frequencies less than f. The dispersion relation for the first three cross-shelf modes, calculated with bottom topography from off Oregon (Cutchin & Smith 1973), is plotted in Figure 8. The phase velocity in the Northern Hemisphere is toward positive y (x positive onshore) for all alongshore wavenumbers, i.e. toward the north off Oregon, and is toward negative y in the Southern Hemisphere. The group velocity is in the same direction as the phase velocity for long waves and in the opposite direction for short waves so that at some l, which depends on the cross-shelf mode number, there is zero group velocity. The points of zero group velocity correspond to the points of zero slope in the plot of the dispersion relation (Figure 8). For long waves, i.e. for those with wavelengths much larger than the width of the continental shelf and slope, the dispersion relation is approximately linear and the propagation is nearly nondispersive. That is the situation implied by Assumptions 4 and 5. The nondispersive long wave speeds for Oregon topography are shown in Figure 8. For realistic bottom topography, the y momentum balance of the lowest few modes is ageostrophic as in (3.7b), with the magnitude of v_t comparable to that of fu

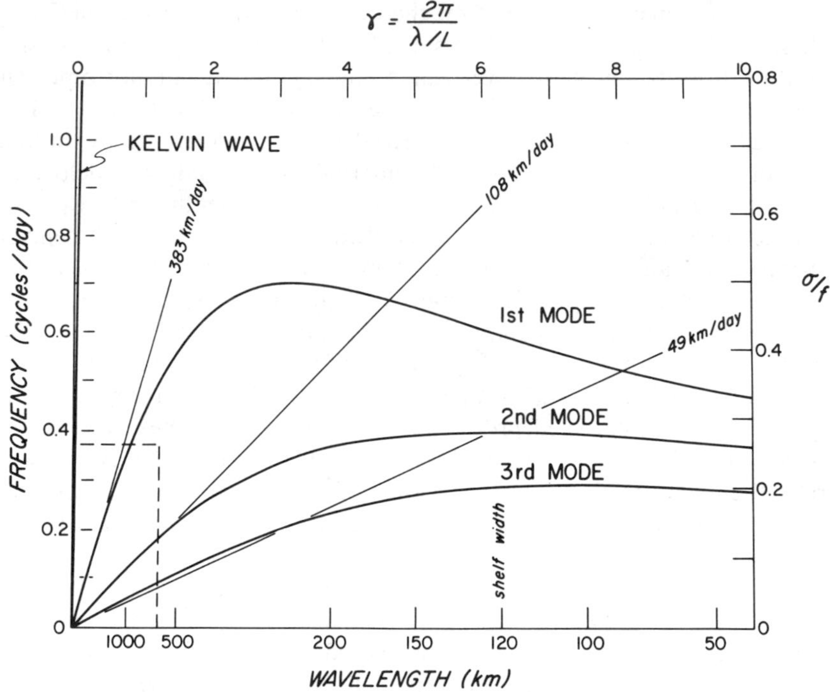

Figure 8 Dispersion curves for the first three barotropic shelf-wave modes off Oregon (Cutchin & Smith 1973). The model includes a free surface and the dispersion curve for the external, barotropic Kelvin wave is shown also. The alongshore wavenumber, nondimensionalized by the shelf width L, is designed by γ here.

(Gill & Schumann 1974). The velocity components have an offshore decay scale over the slope that is related to the scale length of the bottom topography $\delta_B = H/H_x (\approx 30\text{–}40$ km). There is considerable evidence, primarily from the analysis of sea level records, for the alongshore propagation of fluctuations with properties consistent with those expected for free barotropic continental shelf waves (for references see Mysak 1979).

The baroclinic internal Kelvin waves for a flat bottom and vertical coast are obtained from the solution of (4.6) and have an infinite set of vertical modes with zero depth-integrated horizontal mass transport. The way in which the offshore decay scale δ_R arises, for a given vertical scale, may be seen from (4.6). As the vertical mode number n increases, the offshore decay scale decreases. The propagation is nondispersive and toward positive y in the Northern Hemisphere and negative y in the Southern Hemisphere, which is the same direction as for long shelf waves. The propagation velocity c depends on the stratification according to $c \propto H_0 N_0$ and decreases as the mode number increases. Off Oregon, estimates of δ_R and c for the first mode are $\delta_R \simeq 15$ km and $c \simeq 50$ km/day (Kundu, Allen & Smith 1975). The y momentum balance is ageostrophic (3.7b), with $u \equiv 0$ for free internal Kelvin waves over a flat bottom.

For stratified two-layer models with sloping bottom topography, a small vertical coastal wall, and heavy fluid over the entire shelf, the coupling of the baroclinic internal Kelvin wave type mode with barotropic, depth-independent motion depends on the parameter $\lambda = \delta_R/\delta_B$ (Allen 1975). For δ_R smaller than the width of the shelf and $\lambda \ll 1$, the depth variations within δ_R of the coast are small and the coupling is weak. In that case, the internal Kelvin wave and barotropic shelf wave modes exist simultaneously in a weakly perturbed form [except at points in parameter space, which do not concern us here, where the uncoupled phase velocities would be equal (Allen 1975)]. The baroclinic effects are confined to a distance δ_R from the coast, whereas barotropic, depth-independent motions extend offshore to greater distances of order δ_B. Although off Oregon λ is generally $O(1)$ over the shelf (and the coast is not vertical), it may be useful in interpreting observations there to form a conceptual model of the shelf flow field based on the assumption $\lambda \ll 1$. The implications of this model are in qualitative agreement with the properties of the observed flow field. In particular, there is agreement with the facts that the fluctuating alongshore velocity component over the mid- and outer-shelf is nearly depth-independent and that the largest variations in the density field are found within 10 km of the coast.

In the continuously stratified case with bottom topography, general cross-shelf coastal trapped wave modes $\phi_n(x,z)$, which are solutions to

(4.6) with appropriate boundary conditions, have been calculated numerically by Wang & Mooers (1976) and Huthnance (1978). The properties of the modes in Wang & Mooers (1976), calculated with a vertical coast, are similar in many respects to those expected based on the $\lambda \ll 1$ two-layer model. Interesting differences are present, however, in the structure of the modes for some parameter values. In particular, modes that correspond in the unstratified case to shelf waves may have a velocity field that is depth-independent over the shelf, but is bottom-intensified over the slope. The above studies have raised doubts about whether internal Kelvin wave type modes can exist in more realistic geometries with wedge-like regions near the coast at mid-latitude where δ_R is the order of or smaller than the shelf width. In the wedge geometry, the topography does not provide a characteristic vertical scale to determine δ_R. The modes in both of these studies were for the most part calculated without a strong pycnoline over the shelf, however, and it seems possible that in real situations the depth of the pycnocline might provide the appropriate vertical scale to support boundary-trapped baroclinic motions. The question of the possible existence of internal Kelvin waves under realistic oceanic conditions where δ_R, based on the pycnocline depth, is smaller than the shelf width has not been satisfactorily resolved theoretically or observationally. The relatively large motion of density surfaces within 10 km of the coast off Oregon could be caused, for example, by the geometrical constraint of decreasing depth, rather than being related to δ_R. We note, however, that there is good observational evidence for baroclinic coastal trapped wave motion, with properties similar to those of internal Kelvin waves, from Lake Ontario (Csanady & Scott 1974, Csanady 1976a), where the relation of δ_R and topographic scales is similar to that of mid-latitude continental shelves.

At low latitudes where $\lambda \gg 1$ and δ_R is larger than the width of shelf, so that the continental margin appears relatively more like a vertical wall, it certainly seems possible that internal Kelvin wave type modes exist. In fact, measurements of currents and sea level between 10°S and 15°S along the Peru coast during 1976 and 1977 have shown that fluctuations with periods of from 4 to 20 days propagate poleward nondispersively with a velocity of about 200 km/day (Smith 1978). Sea level measurements from 12° and 15°S are shown in Figure 9. The consistent two-day lag of sea level at the southern station and the high correlation of these records at this lag (≈ 0.67) may be seen from the plots. The alongshore velocity fluctuations are baroclinic over the slope, i.e. have large vertical shears that appear to be balanced by horizontal density gradients in agreement with the thermal wind equation (3.8). There is also a high correlation of time rate of change of alongshore velocity (v_t) with alongshore

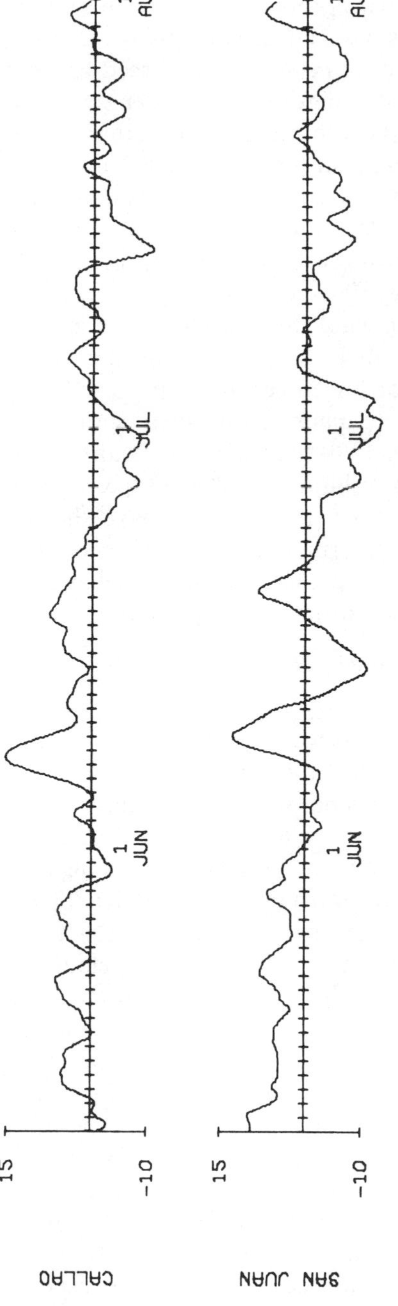

Figure 9 Low-pass filtered adjusted sea level records in cm from Callao (12°04′S) and San Juan (15°20′), Peru, during May–July 1976 (Smith 1978).

sea level slope (p_y) (Brink, Allen & Smith 1978). Both of these characteristics are consistent with propagation in an internal Kelvin wave type mode. The fluctuations associated with this wave motion, although the time scales are typical of atmospherically forced variability, are not well correlated with the local alongshore component of the wind stress. These are presumably free waves forced at some other location equatorward along the coast or perhaps in the equatorial Pacific Ocean. The latter is an interesting possibility since theory indicates that baroclinic waves that propagate eastward in the equatorial waveguide may be transmitted north and south along the eastern boundary as coastal-trapped internal Kelvin waves (Moore & Philander 1977).

When the problem of wind stress driving of an inviscid shelf flow field is considered, it turns out to be identical, on the time scales of interest here, to the problem for forced coastal trapped waves. These have been treated in an illuminating manner, with the assumption that the scale of the wind stress is larger than the shelf-slope width, i.e. with Assumption 4 of Section 3, by Gill & Schumann (1974), Gill & Clarke (1974), Wang & Mooers (1976), and Clarke (1977a). They have shown that if the flow variables are expanded in terms of the free wave modes, e.g. $p = \sum_n Y_n(y,t)$ $\phi_n(x,z)$, the alongshore and time-dependent behavior of each mode is governed by a forced first order wave equation

$$c_n^{-1} Y_{nt} + Y_{ny} = b_n \tau_W^{(y)} (y,t), \tag{4.7}$$

where b_n is a constant.

Equation (4.7) is easily solved by the method of characteristics. The solutions clearly show that the forcing for barotropic and baroclinic disturbances on the shelf is not necessarily local, but that the waveguide nature of continental margins allows disturbances in velocity and density fields, generated at one location, to propagate alongshore in the positive y direction. It may also be noted that, as is usual in forced-wave problems, there is a possibility of a resonant response in (4.7) if $\tau_W^{(y)}$ has components that propagate in the positive y direction with velocity c_n.

In the application of (4.7), it is usually assumed that a single mode will dominate the response. Evidence for the presence of forced, long, coastal trapped waves along the west coast of the United States during the summer of 1973 has been found from sea level, atmospheric pressure, and wind records by Wang & Mooers (1977). These waves presumably are similar to the first mode barotropic shelf wave. Off Oregon, the poleward propagation of nearly depth-independent fluctuations in v and the probable geostrophic balance of v (Sections 2 & 3) are consistent with the existence of coastal trapped waves that are primarily barotropic. In addition, there is some evidence for a resonant response of currents off

Oregon to $\tau_{W}^{(y)}$ at a period of around 7 days during the summer of 1972 (Huyer et al 1975, Kundu & Allen 1976). Some additional support for the validity of (4.7) off Oregon during the summer of 1972 has been found by Clarke (1977a). A review of observational evidence from other locations for forced coastal trapped waves that may satisfy (4.7) is also given by Clarke (1977a).

In the forced problem for the full set of dispersive barotropic shelf waves, i.e. without Assumption 4, the zero group velocity points in the dispersion relation can play a significant role. If it is assumed that the wind stress forcing is represented by a wave number, frequency spectrum and if a small amount of dissipation is added, it is found, by an analysis similar to that of Wunsch & Gill (1976) for equatorially trapped waves, that the response is largest relative to the forcing at the points of zero group velocity. That feature emerged in a study by Brooks (1978) of barotropic shelf waves forced by a stochastic model of atmospheric cold front wind fields. Peaks were found in the response spectra at frequencies corresponding to points of zero absolute group velocity. An accompanying analysis of sea level, atmospheric pressure, and wind stress for the North Carolina coast showed that peaks in the coherence between sea level and the atmospheric variables, and between sea level records from different alongshore locations, occurred at frequencies that coincide with those of the zero absolute group velocity points for the first three modes in the model. Brooks (1978) concludes that the shelf water motion in that location is being selectively forced by the atmosphere at the zero group velocity frequencies and wave numbers. In the forcing function for the model, Brooks (1978) retains the wind stress curl $\tau_{Wx}^{(y)} - \tau_{Wy}^{(x)}$ as well as the wind stress term in (4.4), whose effect we have been emphasizing. With the assumed model for cold fronts, the wind stress curl term $\tau_{Wx}^{(y)}$ contributed significantly to the response spectrum for periods $\leqq 5$ days. The possible contribution of the wind stress curl in some coastal regions is perhaps a feature that deserves further attention.

TWO-DIMENSIONAL MODELS The simplest conceptual picture of the shelf flow field, especially of the cross-shelf circulation, results if it is assumed that the alongshore velocity gradients are zero. The motion is then two-dimensional in the (x, z) plane, but an alongshore pressure gradient p_y, which is a function of time only, may be present. That type of pressure gradient would be set up, for example, in a lake after the abrupt increase in magnitude of a uniform wind stress. The set-up is accomplished by the propagation of long surface gravity waves and is instantaneous in the rigid-lid approximation.

With $v_y \equiv 0$, a streamfunction may be defined such that $u = \hat{\psi}_z$,

$w = -\hat{\psi}_x$. If Assumption 4 is made so that the alongshore velocity is in geostrophic balance (3.7c), the nondissipative stratified case governed by (4.6) may be treated alternatively by deriving a single equation for $\hat{\psi}$ from (3.7b,c), (3.1d), and (4.5):

$$\hat{\psi}_{zz} + [N^2(z)/f^2]\hat{\psi}_{xx} = 0. \tag{4.8}$$

We examine the solution for $\hat{\psi}$ and the other variables in an initial-value problem where a constant wind stress is applied impulsively at $t = 0$ to a fluid at rest and held steady thereafter, i.e. $\tau_W^{(y)} = -\tau_0 H(t)$ where τ_0 is a constant and $H(t)$ is the Heaviside unit function. A model of constant depth H_0 and constant $N^2 = N_0^2$ is utilized. Suction into the surface Ekman layer, which has an asymptotically small thickness in this approximation, is assumed to occur at the coast in a concentrated source or sink (Allen 1973, Pedlosky 1974c). The appropriate boundary conditions for $\hat{\psi}$ are $\hat{\psi}(z=0) = -\tau_0/(\rho_0 f)$, $\hat{\psi}(z = -H_0) = 0$, $\hat{\psi}(x=0) = 0$, and $\hat{\psi}(x \to -\infty) \to -\tau_0[1+(z/H_0)]/(\rho_0 f)$. The baroclinic component of this solution is the same, of course, as the two-dimensional response of forced internal Kelvin waves.

The solution for $\hat{\psi}$ is plotted in Figure 10 in dimensionless variables for the upwelling-favorable case $\tau_0 > 0$, where the cross-shelf coordinate is scaled by the Rossby radius, i.e. $\xi = x/\delta_R$. A (u,w) velocity field of this form is set up on the $O(f^{-1})$ time scale in which the surface Ekman layer is established. In this model, with assumption 1. $\delta_t > f^{-1}$, that set-up is

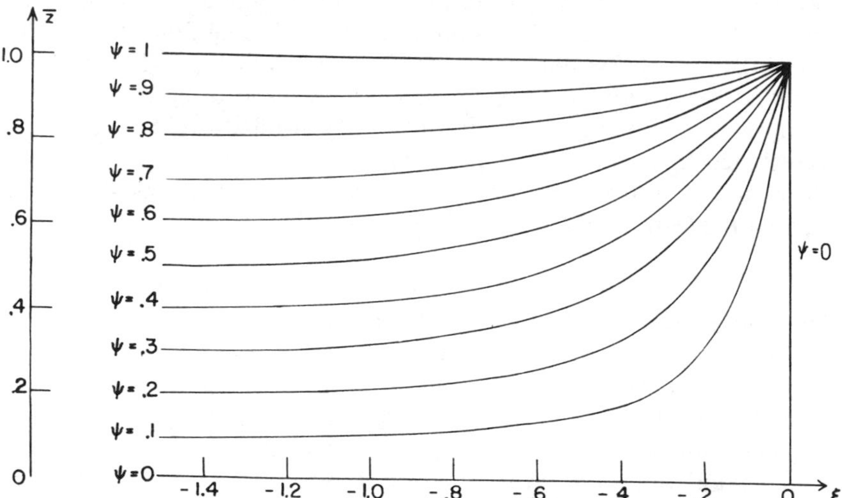

Figure 10 Dimensionless streamfunction $\hat{\psi}$ (shown here as ψ) for the linear, stratified upwelling response problem (Pedlosky 1978b). The variables $(\xi, \bar{z})$ are nondimensionalized as $\bar{z} = 1+(z/H_0)$ and $\xi = x/\delta_R$, where δ_R is the Rossby radius of deformation $\delta_R = H_0 N_0/f$.

instantaneous. With τ_0 a constant and the motion inviscid, $\hat{\psi}$ subsequently does not change with time. The upwelling of fluid within δ_R of coast into the nearshore corner region may be seen from Figure 10.

With $\hat{\psi}$ determined, the alongshore velocity v and the density ρ' are found from (3.7b) and (4.5), respectively, and they vary linearly with time. The constant pressure gradient p_y is specified according to the problem under consideration. In the pure two-dimensional case, $p_y \equiv 0$ and $v_t = -fu$. The increase in magnitude of u near the upwelling corner leads to larger negative values of v in the same region and hence to a baroclinic coastal jet structure in v within δ_R of the coast. With $p_y \equiv 0$, v is in the direction of $\tau_W^{(y)}$ everywhere. In cross-section models for closed basins (Bennett 1974), p_y is chosen so that the net alongshore volume flux is zero. In the present example, that would imply that the constant barotropic component of the onshore velocity u is geostrophically balanced by p_y. In that case, v is composed entirely of baroclinic vertical modes with zero depth-integrated horizontal flux and the structure of the coastal jet is such that near the surface v is in the direction of $\tau_W^{(y)}$, whereas at depth v is in a direction opposite to $\tau_W^{(y)}$ (Pedlosky 1974c).

The baroclinic coastal jet was found originally in a two-layer model by Charney (1955). Implications of the coastal jet results have been applied in the interpretation of observations from the Great Lakes (e.g. Csanady 1977). Off Oregon, however, existing observations of the nearshore baroclinic response are not in particularly good agreement with predictions of the inviscid forced coastal jet, or internal Kelvin wave, theory (Kundu, Allen & Smith 1975, Kundu & Allen 1976).

When two-dimensional models are applied to the continental shelf the proper method for the specification of p_y is not clear. For realistic (except for the vertical coast) bottom topography, where δ_R is less than the shelf width, p_y might be chosen to balance the barotropic onshore flow in the deep water ($H = H_I$) off the continental margin. In this case, the effect of p_y is relatively small in the shallow water near the coast and a barotropic coastal jet will develop on the slope and shelf, outside of δ_R, from the balance $v_t = \tau_W^{(y)}(H^{-1} - H_I^{-1})/\rho_0$. A similar result occurs in cross-section models with bottom topography (Bennett 1974). On the other hand, in shelf problems where $\lambda = \delta_R/\delta_B \ll 1$, $\delta_x \ll \delta_y$ (Assumption 4), and where low mode barotropic shelf wave speeds are much larger than internal Kelvin wave speeds, certain three-dimensional problems might be approximated by the two-dimensional model within δ_R of the coast. This would be possible, for example, if the alongshore scale and forcing frequency were such that the low mode barotropic shelf wave balance in (4.7) was approximately steady $Y_{ny} \approx b_n \tau_W^{(y)}$ while the internal Kelvin wave balance was $c_n^{-1} Y_{nt} \approx b_n \tau_W^{(y)}$.

ALONGSHORE VARIATIONS IN BOTTOM TOPOGRAPHY Probably one of the most common assumptions in models of shelf circulation is that the bottom topography does not vary in the alongshore direction. Actual shelf and slope topography, of course, has significant alongshore variations and these are known to influence the shelf flow field as illustrated in Figure 4. At the present, however, there are only a few analytical models for wind-stress–forced motion over shelf topography with alongshore variations.

For the unforced, inviscid, steady motion of a homogeneous fluid, (4.3) reduces to $H_x\psi_y - H_y\psi_x = 0$ which has the general solution $\psi = \psi(H)$ corresponding to flow along contours of constant depth. This result is a well-known consequence of the constraint of rotation expressed by the Taylor-Proudman Theorem (Greenspan 1968). The influence of this constraint on the shelf flow field off Oregon, even though the motion there is unsteady and the fluid is stratified, is shown in Figure 4 by the tendency of the velocity fluctuations to align along the isobaths.

In the time-dependent problem with $H = H(x,y)$, (4.3) is in general not separable and is consequently difficult to solve analytically. Perturbation methods may be used, however, if it is assumed that the amplitude of the alongshore variations are small enough that $|\psi_x H_y| \ll |\psi_y H_x|$. We outline a perturbation solution to (4.3) (Allen 1976b) which illustrates how the forced flow of an inviscid homogeneous fluid adjusts to alongshore topographic variations. Consider an initial-value problem where a y-independent wind stress $\tau_W^{(y)}$ is applied at $t = 0$. It is assumed that $\psi(t = 0) = 0$, $\tau_W^{(x)} = 0$, $\tau_B^{(x,y)} = 0$. The topography is such that $H = H_0(x) + \hat{\varepsilon}H_1(x,y) + \ldots$ where $\hat{\varepsilon} \ll 1$. The coastline is straight and the contours of constant depth are continuous and uninterrupted. With $\psi = \psi_0 + \hat{\varepsilon}\psi_1 + \ldots$, the lowest order solution is simply

$$\psi_{0x} = -\int_0^t \tau_W^{(y)}\, dt'$$

and the equation for ψ_1 is

$$[\psi_{1xx} + \psi_{1yy} - (H_{0x}/H_0)\psi_{1x}]_t + f(H_{0x}/H_0)\psi_{1y} = f(H_{1y}/H_0)\psi_{0x}. \quad (4.9)$$

If Assumption 4. is made, i.e. $\delta_y \gg \delta_x$, where here δ_y applies to topographic scales, the term ψ_{1yy} may be neglected. Since there are substantial alongshore variations in topography on scales $\delta_y \lesssim \delta_x$, this is not necessarily a good assumption, but it allows a solution to be readily obtained that clearly demonstrates some of the physics involved. As in the wind-stress–forced problem leading to (4.7), (4.9) may then be solved by expanding ψ_1 and the forcing function in terms of the free wave cross-shelf modes $\phi_n(x)$. With $\psi_1 = \sum_n Y_n(y,t)\phi_n$, the resulting equation for Y_n is of the form

$$c_n^{-1} Y_{nt} + Y_{ny} = g_{ny}(y) \int_0^t \tau_W^{(y)}\, dt'. \tag{4.10}$$

With, for example, $\tau_W^{(y)} = \tau_0 H(t)$ the lowest-order alongshore velocity increases linearly with time $\psi_{0x} = -\tau_0 t$. Assume that the alongshore variations are confined to a region of limited extent and let $Y_n = g_n \tau_0 t + Y_n'$. The component Y_n' reaches a steady state in the vicinity of the topographic perturbation after the propagation toward positive y of an unsteady shelf wave front. Consequently, in the region of the topographic variations $Y_n \sim g_n \tau_0 t$ as t increases. The emergence of this component as the dominant part of the solution represents the adjustment of the linearly increasing basic velocity in mode n to flow along contours of constant depth. The adjustment takes place, therefore, on a mode-by-mode basis through the nondispersive propagation of shelf wave disturbances. The characteristic adjustment time is $t_{An} = L_T/c_n$, where L_T is the alongshore scale of the topographic feature. For topographic scales $\delta_y \lesssim \delta_x$, the adjustment involves the generation of dispersive shelf waves with short alongshore scales and consequently with $u \approx v$ (Martell & Allen 1979). This may be a process that contributes to the relatively complicated behavior of u off Oregon (Section 2).

The problem of the effect of alongshore topographic variations on the forced shelf flow field with stratification is more difficult. Some aspects have been examined by Killworth (1978) in a model for linear, inviscid, stratified motion governed by (4.6). The flat bottom, vertical coast solution for the upwelling response problem $\tau_W^{(y)} = -\tau_0 H(t)$ is perturbed by topographic variations that are functions of y only, i.e. where $H = H_0 + \hat{\varepsilon} H_1(y)$, $H_0 =$ constant. The basic effects on the upwelling flow field follow from the fact that the initial vertical motion of density surfaces is modified, in the sense expected, by the vertical velocity induced at the lower boundary through the $w = -vH_y$ boundary condition. The subsequent motion involves the alongshore propagation in the positive y direction of the density and velocity perturbations by internal Kelvin waves. These results may give an indication of the sense in which alongshore topographic variations influence the initial upwelling motion. The topography, however, is very unrealistic in that the contours of constant depth are normal to the coast. It is still not understood exactly how stratification on the shelf affects the adjustment of the velocity and density fields to realistic alongshore variations in bottom topography.

With regard to coastline variations, these are usually accompanied by variations in topography over the shelf. The effect of small-amplitude perturbations in coastline location, for a homogenous fluid with continuous depth contours, is similar to that of small-amplitude alongshore depth variations (Allen 1976b). For stratified, flat bottom, vertical coast models, Killworth (1978) has shown that the effects of small-amplitude or small-curvature coastline variations are much weaker than effects of small topographic perturbations.

BOTTOM FRICTION It is likely that the most important dissipative mechanism for wind-driven time-dependent motion on the shelf is provided by the effects of turbulent bottom friction. We do not go into specific models for bottom boundary layers on the continental shelves here, but rather review the effect of a bottom stress on the behavior of the shelf flow field. To simplify the discussion in this section, Assumption 4, $\delta_y \gg \delta_x$, is utilized. As a result, $v \gg u$ and the dominant effect of the bottom stress, e.g. in (4.1b,c), is from $\tau_B^{(y)}$.

The relation of the bottom stress to velocities in the logarithmic layer above the bottom is commonly expressed by a quadratic relation and drag coefficient $\tau_B^{(y)} = \rho_0 C_D v(v^2+u^2)^{\frac{1}{2}}$ (Lumley & Panofsky 1964). A linear relation of stress and near-bottom velocity has been used frequently in oceanography for analytical convenience, where $\tau_B^{(y)} = \rho_0 rv$ and r is a resistance coefficient with dimensions of velocity. It has been suggested, in fact, that the linear approximation may be justified in representing $\tau_B^{(x,y)}$ for low-frequency flow in the presence of higher-frequency motion (Rooth 1972, Csanady 1976b), e.g. for subtidal frequency motion on the shelf in the presence of energetic tidal motion. A linear stress-velocity relation also follows from the assumption that the bottom boundary layer is governed by linear Ekman dynamics with a constant eddy coefficient A_V: $\tau_B^{(y)} = \frac{1}{2}\rho_0 \delta_E f v$, where $\delta_E = (2A_V/f)^{\frac{1}{2}}$ is the depth scale of the Ekman layer. In any case, linear bottom stress–velocity relations or linear Ekman dynamics are utilized in the analytical models to be discussed here.

Gill & Schumann (1974) suggested that the effect on forced, long coastal trapped waves of dissipative processes, such as bottom friction, might be represented by the addition of a linear drag term in (4.7):

$$c_n^{-1} Y_{nt} + Y_{ny} + (T_{Fn} c_n)^{-1} Y_n = b_n \tau_W^{(y)}(y,t), \tag{4.11}$$

where T_{Fn} is a frictional decay time. Brink & Allen (1978) showed that (4.11) is systematically obtained in an analysis for a homogeneous fluid where the effect of bottom stress is treated as a small perturbation. They calculated a decay time for the first mode of $T_{F1} \approx 4.5$ days with parameter estimates appropriate for Oregon. The problem for the effect of bottom stress on coastal trapped waves with density stratification is more difficult and has yet to be worked out. Note that with $\tau_W^{(y)} = \tau_0 H(t)$, Y_n in (4.11) reaches a steady state in a spin-down time T_{Fn}, which is a familiar result for homogeneous fluids. In two-dimensional stratified response problems with $\tau_W^{(y)} = \tau_0 H(t)$, however, the action of bottom friction alone does not lead to a steady state (Allen 1973, Pedlosky 1974c). It seems likely, therefore, that (4.11) will not hold for a stratified model with bottom stress.

One of the basic effects of bottom friction may be easily seen from (4.11). For the sake of argument assume that the y scales are large, so Y_{ny} may be neglected, and that the wind stress is forcing at frequency ω. In the purely inviscid ($T_{Fn}^{-1} = 0$) or high-frequency case ($\omega \gg T_{Fn}^{-1}$) case, Y_n is $\frac{1}{2}\pi$ out of phase with $\tau_W^{(y)}$, whereas in the frictionally dominated low-frequency case ($\omega \ll T_F^{-1}$), Y_n is in phase with $\tau_W^{(y)}$. The alongshore velocity has the same phase as Y_n. Off Oregon in the summer of 1973, the depth-independent component of v exhibited a phasc lag with $\tau_W^{(y)}$ that was small at frequencies less than 0.1 cpd and increased approximately linearly to about $\frac{1}{2}\pi$ at 0.3 cpd (Kundu, Allen & Smith 1975). It may be seen in Figure 3 that v and the wind appear to be nearly in phase. A similar result has been found during the winter and spring of 1975 off Oregon and Washington (Hickey, Huyer & Smith 1979). This behavior is in qualitative agreement with the effect of friction in (4.11). It should be noted, however, that Y_n and $\tau_W^{(y)}$ will also be approximately in phase for a near resonant response and that this process may possibly contribute to the small phase lags at low frequencies off Oregon.

Brink & Allen (1978) also showed that the effect of bottom friction and increasing depth in the offshore direction leads to a cross-shelf phase lag in v, with the fluctuations nearshore leading those offshore in time. This type of phase lag has been observed off both Oregon (Sobey 1977) and Peru (Brink, Allen & Smith 1978).

The presence of bottom stress can lead to steady, or, as above, quasi-steady forced solutions. For a homogeneous fluid, the steady limit of wind-driven shelf waves with bottom friction was treated in an illuminating manner by Csanady (1978). If a single equation for the pressure is derived from (4.1) with Assumption 4, the steady solution satisfies

$$p_y = \kappa(x) p_{xx}, \tag{4.12}$$

where $\kappa = r(sf)^{-1}$ and $s = -H_x$ $(s > 0)$. If H_x is a constant, (4.12) is the heat diffusion equation and it holds also for the alongshore velocity v. In this case, a vertical coast is not needed and the solution may be obtained in the wedge region with $H = -sx$. A boundary condition on v at the shore ($x = 0$) follows from $Hu \to 0$ as $H \to 0$ and (4.1c):

$$\tau_B^{(y)} = \rho_0 rv = rf^{-1} p_x = \tau_W^{(y)}. \tag{4.13}$$

For coastally trapped motion, it is assumed that offshore, $v \to 0$ as $x \to -\infty$.

Equation (4.12) also arose in the earlier studies of Birchfield (1972, 1973) and Pedlosky (1974b). Birchfield (1973) considered a variable depth model for a closed basin, with $H = 0$ at the coast. A boundary condition similar to (4.13) was derived. In the models of Birchfield (1972) and Pedlosky (1974b), on the other hand, a vertical coast was used and the coastal boundary condition for (4.12) involved p_y rather than p_x. This is one instance of a possible qualitative difference in solutions caused by the vertical coast geometry.

Illustrative solutions for constant H_x may be easily obtained by assuming a sinusoidal dependence of $\tau_W^{(y)}$ on y or by assuming, for instance, $\tau_W^{(y)} = \tau_0$ for $y \geqq 0$, $\tau_W^{(y)} = 0$, for $y < 0$, where τ_0 is a constant. In the first instance, the solutions are confined to a boundary layer of thickness $\delta_x = (\kappa\delta_y)^{\frac{1}{2}}$. The latter case has a solution for v similar to the classical Rayleigh problem: the flow is confined to a boundary layer in x originating at $y = 0$ and growing in width in the positive y direction as $\delta_x \approx (\kappa y)^{\frac{1}{2}}$. The time-like direction, positive y, corresponds of course to the direction of propagation of long shelf waves. With upwelling-favorable stress $\tau_0 < 0$, for example, the fluid beneath the surface layer, which compensates the offshore Ekman transport, flows along the shelf from the positive y direction. The time-dependent development of flow from that direction in the inviscid case is illustrated in Allen (1976a). An estimate of the magnitude of this frictional boundary layer, with $r = 0.1$ cm s^{-1}, $s = 5 \times 10^{-3}$, $f = 10^{-4}$ s^{-1}, $\delta_y = 1000$ km, gives $\delta_x = 45$ km which is larger than, but comparable to, the shelf width off Oregon.

In two-dimensional, flat-bottom, stratified models with inviscid interior motion and an Ekman boundary layer on the bottom, the solution to the response problem $\tau_W^{(y)} = -\tau_0 H(t)$ is modified by frictional effects on the spin-up time scale $T_F = 2(\delta_E f)^{-1} H_0$. The barotropic component outside of a boundary layer of width δ_R in general reaches a steady state on this time scale, but the baroclinic component within δ_R does not, with v and ρ having components that increase linearly with t. An estimate of T_F off Oregon with $H_0 = 100$ m, $\delta_E = 10$ m is $T_F \simeq 2.3$ days. [The larger value of T_{F1} determined for (4.11) in Brink & Allen (1978) is mainly due to the variable depth there.]

The nature of the resulting flow depends on the specified pressure gradient p_y through its effect on the barotropic component. This may be seen from the depth-averaged y-momentum equation (4.1c) applied to the barotropic motion for $|x| > \delta_R$. With $\tau_B^{(y)} = \frac{1}{2}\rho_0 \delta_E f V$, $U \equiv 0$, and $p_y \equiv 0$, (4.1c) reduces to $V_t + (\frac{1}{2}\delta_E f/H_0)V = (H_0\rho_0)^{-1}\tau_W^{(y)}$. The initial balance for small time is $V_t = (H_0\rho_0)^{-1}\tau_W^{(y)}$, which forces a growth of V

and hence of $\tau_B^{(y)}$. A steady state is reached on the T_F time scale with $V_t \to 0$ and $\tau_W^{(y)} = \tau_B^{(y)}$, where the bottom stress is associated with a bottom Ekman layer. The cross-shelf flow changes during the spin-up time from the initial state shown in Figure 10 to one in which the onshore flow outside the upwelling region of width δ_R is confined to the bottom Ekman layer (Allen 1973). The alongshore velocity v forms a modified baroclinic coastal jet structure and is in the direction of $\tau_W^{(y)}$ everywhere. The concentration of onshore flow in a bottom Ekman layer appears to be unrealistic when compared with most observations. It definitely does not seem to be a characteristic of the onshore velocity off Oregon (see Figure 6). For N.W. Africa, it is more difficult to draw a conclusion in this regard from Figure 6, because it appears that an interior region of inviscid flow is not present (Kundu 1977).

If p_y is chosen to geostrophically balance the barotropic component of u, (4.1c) reduces to $p_y = H_0^{-1}\tau_W^{(y)}$ for all t; V_t is not forced and a barotropic alongshore current does not develop. Consequently, outside of δ_R a bottom Ekman layer is not formed and the onshore flow remains depth-independent. The time-dependent component of v is confined to the δ_R boundary layer where it forms a modified coastal jet. The alongshore velocity v is in the direction of $\tau_W^{(y)}$ near the surface and at all depths offshore and is in a direction opposite to $\tau_W^{(y)}$ at depth near the coast. The cross-shelf circulation changes, as shown in Pedlosky (1974c), so that there is motion into and out of the bottom Ekman layer within δ_R. Pedlosky (1974c) actually considered a more general three-dimensional problem, including the effect of a weak bottom slope. Attention was focused on situations where the alongshore scale of the wind stress L_W is large, so that $L_W \gg c_n T_F$, where c_n is an internal Kelvin wave speed. In that case, the baroclinic response within δ_R behaves approximately as a two-dimensional flow over a flat bottom. The barotropic component outside δ_R develops in a three-dimensional manner and, on the T_F time scale, reaches a steady state in which (4.12) is satisfied.

HORIZONTAL FRICTION AND DIFFUSION The role of horizontal eddy stresses and density fluxes in shelf flow fields is not well understood at the present. The processes that may contribute to horizontal eddy diffusion have not been clearly identified or thoroughly studied in field experiments. Constant eddy coefficients are usually used in models, although the validity of that type of representation has not been established from measurements. Even if horizontal eddy coefficients give a reasonable approximate representation of the physics, the values of the eddy coefficients and their possible dependence on other variables and parameters are not well known.

A rough estimate of the relative importance of horizontal and vertical diffusive processes on the shelf may be obtained following Garvine (1973). In the y-momentum equation (3.1c), for example, we approximate $v_z \approx \Delta v/\Delta z$ and $v_x \approx \Delta v/\Delta x$ where Δz and Δx are vertical and horizontal distances, respectively, from an interior point to the shelf bottom, so that $\Delta v = v$ and $\Delta z/\Delta x \approx -H_x$. The ratio R of the horizontal to vertical diffusion terms is then

$$R = A_{\mathrm{H}} v_{xx}/(A_{\mathrm{V}} v_{zz}) \approx (A_{\mathrm{H}}/A_{\mathrm{V}}) H_x^2. \tag{4.14}$$

From observations on continental shelves, equivalent values of A_{V} in the surface layer (Halpern 1976, Halpern 1977) and in the bottom boundary layer (Mercado & Van Leer 1976) are approximately in the range 50–200 $\mathrm{cm}^2\ \mathrm{s}^{-1}$. Although representative values for A_{V} outside of these layers are probably considerably smaller, we utilize a constant $A_{\mathrm{V}} \approx 100\ \mathrm{cm}^2\ \mathrm{s}^{-1}$ for the purpose of making estimates. Values of A_{H} obtained from diffusion experiments increase as the horizontal scale δ_x increases; for $\delta_x \approx 10$ km, an approximate magnitude is $A_{\mathrm{H}} \approx 10^5\ \mathrm{cm}^2\ \mathrm{s}^{-1}$ (Okubo 1971, Murthy 1976). With these values for A_{V} and A_{H} and with $H_x = -5 \times 10^{-3}$, we find $R \approx 2.5 \times 10^{-2}$, which indicates that effects of horizontal momentum diffusion should be small compared with those of vertical diffusion. Based on this type of estimate, horizontal friction is often omitted in model formulations. Different conclusions in this regard may be reached, however, with different magnitudes for the eddy coefficients (e.g. Garvine 1973).

The complete neglect of horizontal diffusion effects may not be totally justified, however, especially with regard to density diffusion. In many upwelling regions, e.g. off Oregon, relatively cold and dense water is brought to the surface near the coast in situations where, in response to the wind stress, there should be a more or less continuous offshore Ekman transport in the surface layer. In the absence of sufficient heat flux from the atmosphere and/or of horizontal density diffusion, the continuous offshore transport of relatively dense water at the surface should lead to convective overturning, which would vertically mix an increasing fraction of the water column and eventually destroy the stratification on the shelf. Except for very near shore, that type of behavior of the density field is not observed off Oregon. In fact, Bryden, Halpern & Pillsbury (1979) have calculated a mean heat budget for the Oregon upwelling region for the summer of 1973 that indicates that onshore eddy heat flux plays a large role in balancing the net offshore heat transport by the mean flow. The eddy heat flux is mainly confined to the surface layer, with roughly two-thirds of it due to fluctuations with periods greater than two days and the other one-third due to higher frequency motion. The estimated value of the eddy diffusity is $K_{\mathrm{H}} \approx 10^6$ $\mathrm{cm}^2\ \mathrm{s}^{-1}$.

A set of stratified shelf models, which include the effects of horizontal momentum and density diffusion through the use of constant eddy coefficients, have been investigated by Allen (1973) and Pedlosky (1974a,b). Although these models are probably unrealistic in many respects, because of the possible importance of horizontal density diffusion we outline their central features.

A flat or weakly sloping bottom with a vertical coast of depth H_0 is utilized. The stratification is assumed to be strong enough that $N^2 f^{-\frac{3}{2}} H_0 > A_{\mathrm{H}}/(A_{\mathrm{V}}^{\frac{1}{2}} \sigma_{\mathrm{H}})$ where $\sigma_{\mathrm{H}} = A_{\mathrm{H}}/K_{\mathrm{H}}$, which appears to be the relevant parameter range for the Oregon shelf. In two- and three-dimensional steady-state upwelling problems, there are two coastal boundary-layer regions. The first, the hydrostatic layer, involves the horizontal diffusion of alongshore momentum and of density. The governing equations for the correction variables are (3.1a,d), (3.7c), and

$$fu = -p_y/\rho_0 + A_{\mathrm{H}} v_{xx}, \quad w\bar{\rho}_z = K_{\mathrm{H}} \rho_{xx}. \tag{4.15a,b}$$

The boundary-layer scale, with constant N^2, is $\delta_{x1} = \sigma_{\mathrm{H}}^{\frac{1}{2}} H_0 N_0/f$. If the horizontal Prandtl number σ_{H} is order one, δ_{x1} is similar in magnitude to the Rossby radius δ_{R}. The second boundary layer is the diffusion layer, which involves the horizontal and vertical diffusion of ρ and v. The governing equation for the correction variables is simply the diffusion equation for the density,

$$O = K_{\mathrm{V}} \rho_{zz} + K_{\mathrm{H}} \rho_{xx}, \tag{4.16}$$

with boundary-layer scale $\delta_{x2} = (K_{\mathrm{H}}/K_{\mathrm{V}})^{\frac{1}{2}} H_0$. As in the inviscid-response problem involving the Rossby radius δ_{R}, both δ_{x1} and δ_{x2} depend on the vertical scale H_0 imposed by the vertical section of coast. The extent to which the physics of these boundary layers would be operative with wedge-like coastal geometries and vertically variable density stratification has not been clarified.

The combined solutions from these two boundary layers yield a steady, baroclinic coastal-jet structure in the alongshore velocity v with the offshore decay scale given by the larger of δ_{x1} and δ_{x2}. In three-dimensional problems, the relation between the wind stress, the alongshore currents, and the upwelling circulation is strongly dependent on the alongshore scale of the wind stress (Pedlosky 1974a). Alongshore currents are most effectively forced by the larger scales of the wind stress. It is worth noting that off Oregon the mean alongshore velocity field during the summer is in the form of baroclinic coastal jet (Kundu & Allen 1976), although the relationship of that observed feature with horizontal diffusive processes has not been established.

Nonlinear Analytical Models

COASTAL TRAPPED WAVES Nonlinear wind-forced internal Kelvin waves in a two-layer model with the upper layer depth small compared to the total depth and with a vertical coast and flat bottom have been studied by Bennett (1973) and Clarke (1977b). Clarke (1977b) investigated, in particular, the effect of a variable coastline. In these nondispersive waves, the nonlinearities lead to a dependence of propagation speed on amplitude, with speed decreasing as the upper layer depth decreases, and to a steepening in the alongshore direction of interface displacements. The interesting question of what the ultimate consequence of the steepening is, i.e. what type of discontinuity might be formed, has evidently not been answered.

TWO-DIMENSIONAL MODELS Pedlosky (1978b) has formulated a two-dimensional, nondissipative, nonlinear model for the onset of coastal upwelling in a continuously stratified fluid. A constant depth H_0 geometry is utilized. The fluid is initially at rest with a constant, stable density gradient ($N^2 = N_0^2$). At $t = 0$ a constant wind stress $\tau_W^{(y)} = -\tau_0 H(t)$ is imposed. As in the linear problem, the motion of fluid into the surface layer is assumed to occur in an upwelling sink at the coast.

It is assumed that the Rossby number ε is small, where $\varepsilon = U/(f\delta_R)$ and $U = \tau_0/(\rho_0 H_0 f)$, so that the alongshore velocity v remains in geostrophic balance (3.7c). The y-momentum and density equations are

$$\frac{Dv}{Dt} + fu = -p_y/\rho_0, \tag{4.17a}$$

$$\frac{D\rho}{Dt} = 0, \tag{4.17b}$$

where $D\rho/Dt = \rho_t + u\rho_x + w\rho_z$, and where $p_y = -\rho_0 f U$ is a constant. The remaining equations are (3.1a), with $v_y = 0$, and (3.1d). A streamfunction is defined such that $u = \hat{\psi}_z$, $w = -\hat{\psi}_x$ and an equation for $\hat{\psi}$ that involves v and ρ in the coefficients is derived. A consideration of the (complex) characteristic curves of that equation leads to the selection of new characteristic coordinates,

$$\xi = x + (vf^{-1} - tU), \quad Z = z, \quad T = t, \tag{4.18}$$

where $D\xi/Dt = 0$. A transformation similar to (4.18) was utilized by Hoskins & Bretherton (1972) in a study of atmospheric frontogenesis.

In terms of the coordinates (ξ, Z, T), the transformed equations for this problem are linear. In particular, $\hat{\psi}$ satisfies Laplace's equation with boundary conditions identical to those given earlier for the linear solution shown in Figure 10. The effects of nonlinearity come in through the transformation back to the original coordinates.

The solution for $\hat{\psi}$ at time $t = 0.1\ (\varepsilon f)^{-1}$ (which, with $H_0 = 10^4$ cm, $\tau_0 = 1$ dyne cm^{-2}, $N_0 = 10^{-2}$ s^{-1}, $f = 10^{-4}$ s^{-1}, corresponds to 1.15 days) is shown in Figure 11. Discontinuities in $\hat{\psi}$ have developed along the coastal boundary $x = 0$ above $\bar{z} = 1 + (z/H_0) = 0.74$. In the region of the discontinuities there is intense upwelling, i.e. infinite vertical velocities in a region of zero width. Corresponding frontal discontinuities exist at the same location in the density and alongshore velocity fields. As time increases, the depth over which the discontinuities occur increases.

The physical interpretation of this process (Pedlosky 1978b) is as

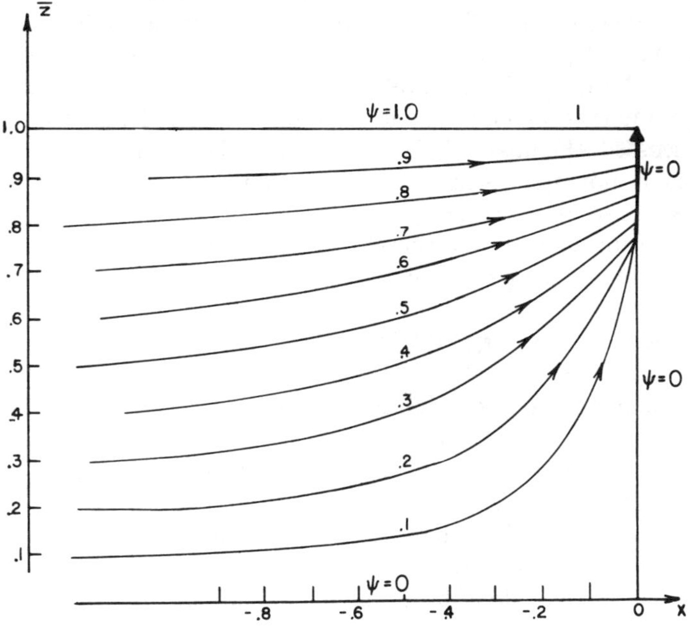

Figure 11 Dimensionless streamfunction $\hat{\psi}$ (shown here as ψ) at time $t = 0.1/(\varepsilon f)$ in the nonlinear, stratified upwelling response problem (Pedlosky 1978b). The variable x here is nondimensionalized by the initial Rossby radius of deformation and $\bar{z} = 1 + (z/H_0)$.

follows. The initial vertical gradients in w, which are largest near the surface at the coast, lead to a stretching apart of the density surfaces. This reduces the local value of $N = (-g\rho_z/\rho_0)^{\frac{1}{2}}$ and hence also reduces the local value of the Rossby radius of deformation $\delta_R = H_0N/f$. Since δ_R is the horizontal scale over which upwelling takes place, a reduction in δ_R increases the magnitude of the vertical motion, which accelerates the separation of density surfaces. As a consequence, a frontal discontinuity forms and increases in strength.

These results show the tendency of the cross-shelf circulation to produce large gradients on scales smaller than the initial Rossby radius. This is consistent with the fact that density fronts are frequently observed in upwelling regions (Mooers, Collins & Smith 1976, Curtin 1979).

Pedlosky (1978a) has also developed a nonlinear inertial model for steady-state coastal upwelling. The motion below the surface layer is nondissipative; density and potential vorticity are conserved following fluid particles. As in the previous study, the effect of the offshore Ekman transport in the surface layer is represented by a concentrated upwelling sink at the coast. Solutions are obtained with a vertical coast and with flat or sloping bottoms. The horizontal scale of the upwelling region in the constant depth case is δ_R (based on the density distribution as $x \to -\infty$). Although the physical situation in the flat-bottom, constant-depth geometry is similar to that in the nonlinear time-dependent problem just discussed (Pedlosky 1978b), the particular steady-state solution obtained is not approached by the unsteady solution as $t \to \infty$. The steady-state solution varies smoothly, except at the upwelling sink, whereas the unsteady solution develops frontal discontinuities. An elucidation of the relationship of these two cases would be of interest.

5 NUMERICAL MODELS

As in other areas of fluid mechanics, numerical models have the advantage of being able to give solutions to complicated, nonlinear problems that are intractable with standard analytical methods. It is convenient to categorize the numerical models here in a slightly different fashion than in Section 4. One of the major problems in numerical modeling of shelf circulation is a proper representation of offshore, and in the case of three-dimensional models, alongshore boundary conditions. This problem may be more or less severe, depending on the particular shelf region under study. We do not review the particular procedures used by different investigators to set these boundary conditions, but refer the reader to the original papers for that aspect of the problem.

Two-Dimensional Models

HOMOGENEOUS-FLUID MODELS Bennett & Magnell (1979) have used a two-dimensional (x,z), nonlinear, homogeneous-fluid model to analyze currents measured in shallow water of 13-m depth off New Jersey. Nonlinear inertial terms are retained and effects of vertical friction are represented with a nonlinear eddy viscosity. The height of the water surface is allowed to vary, to allow the simulation of tidal currents, but it is required that the surface remain level. A constant alongshore pressure gradient p_y, based on an estimate of Scott & Csanady (1976) for mean flow off Long Island, is utilized. The model is driven for a 60-day period with observed values of sea level and wind stress. A comparison of predicted and measured currents shows some agreement, with discrepancies attributed to three-dimensional effects not included in the model.

TWO-LAYER STRATIFIED MODELS Hurlburt & Thompson (1973) considered the near-inviscid response to an abruptly applied wind stress $\tau_W^{(y)} = -\tau_0 H(t)$ with p_y chosen to geostrophically balance the barotropic component of the onshore flow. It was found that with a flat bottom the alongshore velocity of the baroclinic coastal jet in the upper layer was in the direction of the wind stress, whereas in the lower layer it was in the opposite direction, forming a countercurrent. When shelf and slope topography were added, the countercurrent was markedly reduced in strength as discussed in Section 4.

Thompson (1974) has formulated a generalized two-layer model that attempts to include the effect of vertical turbulent mixing in the surface layer and the entrainment of mass, momentum, and energy across the interface between layers. In the solutions to the response problem $\tau_W^{(y)} = -\tau_0 H(t)$, the vertical upwelling velocities initially raise the interface near the coast. As the upper layer depth decreases, vertical turbulent mixing becomes more important and counters the vertical advection, deepening the interface. The upper layer temperature drops rapidly near the coast when vertical mixing becomes important.

CONTINUOUSLY STRATIFIED MODELS A nonlinear, continuously stratified model with a small section of vertical coast and with both flat and sloping bottom topography has been developed by Hamilton & Rattray (1978). Constant, or Richardson-number-dependent, vertical eddy coefficients are utilized when the density stratification is stable, and instantaneous convective overturning with mass conservation is used when the vertical density distribution becomes unstable. The horizontal eddy coefficients are assumed constant. Solutions to the upwelling response problem $\tau_W^{(y)} = -\tau_0 H(t)$ with zero density flux at the surface (i.e. with no heat transfer from the atmosphere) are studied as a function of the value and form of the eddy coefficients. In particular, the effect of the magnitude of the horizontal eddy diffusivity K_H in influencing the density distribution is nicely illustrated. As K_H is reduced, the combined action of upwelling, offshore Ekman transport, and convective overturning results in a relatively deeper mixed layer and in a greater erosion of the initial stratification. This type of behavior was mentioned in Section 4 in connection with horizontal density diffusion.

Three-Dimensional Models

When numerical modeling of shelf flow fields is expanded from two to three dimensions, the demands on computer resources increase greatly, as is familiar from other fluid mechanics problems. Two-layer models have proved especially useful in three-dimensional problems by minimizing calculational needs while still retaining the effect of density stratification.

TWO-LAYER STRATIFIED MODELS Suginohara (1974) has studied the effect of alongshore variations in the wind stress in the context of the upwelling response problem. Both flat-bottom and idealized continental shelf topography are utilized. The results are in accordance with expectations based on the theory for forced coastal trapped waves.

The effects of alongshore variations in coastline and bottom topography in the coastal upwelling response problem have been studied by Hurlburt (1974) and by Kishi & Suginohara (1975) using idealized, isolated, coastline variations and topographic disturbances. For bottom topographic variations, Hurlburt (1974) finds that the initial effects on upwelling are consistent with the strengthening or weakening of the vertical motion of the density interface according to the vH_y contribution to the inviscid bottom boundary condition $w = -vH_y - uH_x$. Coastal trapped wave motion subsequently propagates interface deformations toward positive y.

Peffley & O'Brien (1976) have utilized a smoothed representation of real shelf and slope bottom topography off Oregon to simulate the three-dimensional onset and decay of coastal upwelling in response to a y-independent wind stress. The bottom topography exerts a strong influence on the alongshore distribution of interface displacement, with upwelling relatively intensified over a canyon and, in another location, south of a major cape. After a few days, the alongshore current in the lower layer v_2 is generally in the direction of $\tau_W^{(y)}$ except in nearshore regions north of the cape. At those locations, v_2 is in a direction opposite to $\tau_W^{(y)}$ as in two-dimensional, stratified, flat-bottom models with p_y chosen to balance the barotropic component of u. This may be caused by the setup on the shelf, through the propagation northward of near barotropic shelf waves, of a barotropic alongshore pressure gradient p_y which is supported by the cape.

CONTINUOUSLY STRATIFIED MODELS The upwelling response to an alongshore varying wind stress in a flat-bottom model has been examined by Suginohara (1977). Although the problem is three-dimensional, the main emphasis is on a presentation of the cross-shelf circulation, which develops as the initially level pycnocline upwells near the coast and intersects the sea surface to form a front. Downwelling is found just inshore of the front, with upwelling on the offshore side and at the coast. Some observational evidence suggests that a cross-shelf frontal circulation of a similar type may be present at times off Oregon (Mooers, Collins & Smith 1976).

LAKE CIRCULATION MODELS Some of the results from numerical models of circulation in large lakes are applicable to processes on the continental shelf. Indeed, several of the analytical models mentioned in Section 4 were developed for application to the Great Lakes. Numerical studies of lake currents are not included here as they are discussed in a recent review article by Lick (1976).

DIAGNOSTIC MODELS In diagnostic models, measured values of some variables are specified, either over the field or at appropriate boundary points, and the remaining variables are calculated from the governing equations. The objective is to compute the total flow field over an extended region from a limited number of observations. For example, in a steady-state, linear, stratified model with bottom friction (Hsueh & Peng 1978), measured values of the mean density field over the shelf and mean near-bottom currents along one cross-shelf line are utilized as input data. A steady barotropic vorticity equation for the bottom pressure is solved numerically in a region that extends in the positive y direction, as indicated, for example, by the parabolic nature of (4.12), from the cross-shelf line of the velocity data. The currents are then calculated from the bottom pressure and from the density field through the integration of the thermal wind equations [e.g. (3.8)]. A related procedure has been used to compute time-dependent barotropic currents from the input of velocity measurements along one cross-shelf line and the numerical integration of the unsteady barotropic vorticity equation

(Hsueh & Lee 1978). Both of these models were tested with data from the Oregon shelf and reasonable agreement of calculated and observed currents was claimed.

6 FUTURE WORK

There are many areas where models of wind-driven currents on the continental shelf need improvement or further development. Most of these were mentioned in the main text. A few of the most obvious are the following. A careful resolution of the effect of a vertical coastal boundary on the modeling of various shelf processes is required. Further studies are needed on the effects of bottom friction and alongshore variations in bottom topography on stratified shelf flow fields. The development of an understanding of the mechanisms involved in determining the cross-shelf velocity field remains an important, major task. Better models for the surface frictional layer, in particular, and for frictional and diffusive processes, in general, should be incorporated in shelf circulation models. Most of all, perhaps, there is a need for the close coordination of theoretical and numerical modeling efforts with the conduct of field experiments designed to answer specific dynamical questions.

Acknowledgments

This work was supported by National Science Foundation grants OCE 78-03380 [Coastal Upwelling Ecosystems Analysis (CUEA) program] and OCE 78-26820. The author is grateful to his colleagues who gave permission for the reproduction of figures from their papers and is especially grateful to Dr. Kenneth H. Brink who suggested several improvements to the manuscript.

Literature Cited

Allen, J. S. 1973. Upwelling and coastal jets in a continuously stratified fluid. *J. Phys. Oceanogr.* 3:245–57

Allen, J. S. 1975. Coastal trapped waves in a stratified ocean. *J. Phys. Oceanogr.* 5: 300–25

Allen, J. S. 1976a. Some aspects of the forced wave response of stratified coastal regions. *J. Phys. Oceanogr.* 6:113–19

Allen, J. S. 1976b. Continental shelf waves and alongshore variations in bottom topography and coastline. *J. Phys. Oceanogr.* 6:864–78

Allen, J. S., Kundu, P. K. 1978. On the momentum, vorticity and mass balance on the Oregon shelf. *J. Phys. Oceanogr.* 8:13–27

Bennett, J. R. 1973. A theory of large-amplitude Kelvin waves. *J. Phys. Oceanogr.* 3:57–60

Bennett, J. R. 1974. On the dynamics of wind-driven lake currents. *J. Phys. Oceanogr.* 4:400–14

Bennett, J. R., Magnell, B. A. 1979. A dynamical analysis of currents near the New Jersey coast. *J. Geophys. Res.* 84: 1165–75

Birchfield, G. E. 1972. Theoretical aspects of wind-driven currents in a sea or lake of variable depth with no horizontal mixing.

J. Phys. Oceanogr. 2:355–62

Birchfield, G. E. 1973. An Ekman model of coastal currents in a lake or shallow sea. *J. Phys. Oceanogr.* 3:419–28

Brink, K. H., Allen, J. S. 1978. On the effect of bottom friction on barotropic motion over the continental shelf. *J. Phys. Oceanogr.* 8:919–22

Brink, K. H., Allen, J. S., Smith, R. L. 1978. A study of low-frequency fluctuations near the Peru coast. *J. Phys. Oceanogr.* 8:1025–41

Brooks, D. A. 1978. Subtidal sea level fluctuations and their relation to atmospheric forcing along the North Carolina coast. *J. Phys. Oceanogr.* 8:481–93

Bryden, H., Halpern, D., Pillsbury, R. D. 1979. Importance of eddy heat flux in a heat budget for Oregon coastal waters. In preparation

Buchwald, V. T., Adams, J. D. 1968. The propagation of continental shelf waves. *Proc. R. Soc. London Ser. A* 305:235–50

Charney, J. G. 1955. The generation of oceanic currents by wind. *J. Mar. Res.* 14:477–98

Clarke, A. J. 1977a. Observational and numerical evidence for wind-forced coastal trapped long waves. *J. Phys. Oceanogr.* 7:231–47

Clarke, A. J. 1977b. Wind-forced linear and nonlinear Kelvin waves along an irregular coastline. *J. Fluid Mech.* 83:337–48

Csanady, G. T. 1976a. Topographic waves in Lake Ontario. *J. Phys. Oceanogr.* 6:93–103

Csanady, G. T. 1976b. Mean circulation in shallow seas. *J. Geophys. Res.* 81:5389–99

Csanady, G. T. 1977. The coastal jet conceptual model in the dynamics of shallow seas. In *The Sea*, ed. E. D. Goldberg, I. N. Cave, J. J. O'Brien, J. H. Steele, 6:117–44. New York: Wiley. 1048 pp.

Csanady, G. T. 1978. The arrested topographic wave. *J. Phys. Oceanogr.* 8:47–62

Csanady, G. T., Scott, J. T. 1974. Baroclinic coastal jets in Lake Ontario during IFYGL. *J. Phys. Oceanogr.* 4:524–41

Curtin, T. B. 1979. *Physical dynamics of the coastal upwelling frontal zone off Oregon*. PhD thesis. Univ. Miami, Florida. 338 pp.

Cutchin, D. L., Smith, R. L. 1973. Continental shelf waves: low frequency variations in sea level and currents over the Oregon continental shelf. *J. Phys. Oceanogr.* 3:73–82

Fofonoff, N. P. 1962. Dynamics of Ocean Currents. In *The Sea*, ed. M. N. Hill, 1:323–95. New York: Wiley. 864 pp.

Garvine, R. W. 1973. The effect of bathymetry on the coastal upwelling of homogeneous water. *J. Phys. Oceanogr.* 3:47–56

Gill, A. E., Clarke, A. J. 1974. Wind-induced upwelling, coastal currents and sea-level changes. *Deep-Sea Res.* 21:325–45

Gill, A. E., Schumann, E. H. 1974. The generation of long shelf waves by the wind. *J. Phys. Oceanogr.* 4:83–90

Greenspan, H. P. 1968. *The Theory of Rotating Fluids*, p. 9. Cambridge Univ. Press. 327 pp.

Halpern, D. 1976. Structure of a coastal upwelling event observed off Oregon during July, 1973. *Deep-Sea Res.* 23:495–508

Halpern, D. 1977. Description of wind and of upper ocean current and temperature variations on the continental shelf off Northwest Africa during March and April 1974. *J. Phys. Oceanogr.* 7:422–30

Hamilton, P., Rattray, M. Jr. 1978. A numerical model of the depth dependent, wind-driven upwelling circulation on a continental shelf. *J. Phys. Oceanogr.* 8:437–57

Hickey, B. M., Huyer, A., Smith, R. L. 1979. The alongshore coherence and generation of fluctuations in currents and sea level on the Pacific northwest continental shelf, winter and spring, 1975. *J. Phys. Oceanogr.* Submitted for publication

Hoskins, B. J., Bretherton, F. P. 1972. Atmospheric frontogenesis models: mathematical formulation and solution. *J. Atmos. Sci.* 29:11–37

Hsueh, Y., Lee, C. 1978. A hindcast of barotropic response over the Oregon-Washington continental shelf during the summer of 1972. *J. Phys. Oceanogr.* 8:799–810

Hsueh, Y., Peng, P. Y. 1978. A diagnostic model of continental shelf circulation. *J. Geophys. Res.* 83:3033–41

Hurlburt, H. E. 1974. *The influence of coastline geometry and bottom topography on the eastern ocean circulation.* PhD thesis. Florida State Univ. 103 pp.

Hurlburt, H. E., Thompson, J. D. 1973. Coastal upwelling on a β-plane. *J. Phys. Oceanogr.* 3:16–32

Huthnance, J. M. 1978. On coastal trapped waves: analysis and numerical calculation by inverse iteration. *J. Phys. Oceanogr.* 8:74–92

Huyer, A. 1976. A comparison of upwelling events in two locations: Oregon and Northwest Africa. *J. Mar. Res.* 34:531–46

Huyer, A., Hickey, B. M., Smith, J. D., Smith, R. L., Pillsbury, R. D. 1975. Alongshore coherence at low frequencies in currents observed over the continental shelf off Oregon and Washington. *J. Geophys. Res.* 80:3495–505

Huyer, A., Pillsbury, R. D., Smith, R. L.

1975. Seasonal variation of the alongshore velocity field over the continental shelf off Oregon. *Limnol. Oceanogr.* 20: 90–95

Huyer, A., Smith, R. L. 1978. Physical characteristics of Pacific northwestern coastal waters. In *The Marine Plant Biomass of the Pacific Northwest Coast*, ed. R. Kraus, pp. 37–55. Corvallis: Oregon State Univ. Press

Huyer, A., Smith, R. L., Sobey, E. J. C. 1978. Seasonal differences in low-frequency current fluctuations over the Oregon continental shelf. *J. Geophys. Res.* 83: 5077–89

Killworth, P. D. 1978. Coastal upwelling and Kelvin waves with small longshore topography. *J. Phys. Oceanogr.* 8: 188–205

Kishi, M. J., Suginohara, N. 1975. Effects of longshore variation of coastline geometry and bottom topography on coastal upwelling in a two-layer model. *J. Oceanogr. Soc. Jpn.* 31: 48–50

Kundu, P. K. 1977. On the importance of friction in two typical continental waters: off Oregon and Spanish Sahara. In *Bottom Turbulence*, ed. J. C. J. Nihoul, pp. 187–207. *8th Liège Colloq. Ocean Hydrodyn.*, 1976. Amsterdam: Elsevier

Kundu, P. K., Allen, J. S. 1976. Some three-dimensional characteristics of low frequency current fluctuations near the Oregon coast. *J. Phys. Oceanogr.* 6: 181–99

Kundu, P. K., Allen, J. S., Smith, R. L. 1975. Modal decomposition of the velocity field near the Oregon coast. *J. Phys. Oceanogr.* 5: 683–704

Lick, W. 1976. Numerical modeling of lake currents. *Ann. Rev. Earth Planet. Sci.* 4: 49–74

Lumley, J., Panofsky, H. 1964. *The Structure of Atmospheric Turbulence.* p. 103. New York: Wiley. 239 pp.

Martell, C. M., Allen, J. S. 1979. The generation of continental shelf waves by alongshore variations in bottom topography. *J. Phys. Oceanogr.* 9: 696–711

Mercado, A., Van Leer, J. 1976. Near bottom velocity and temperature profiles observed by Cyclesonde. *Geophys. Res. Lett.* 3: 633–36 (Erratum. 1977. 4: 96)

Mooers, C. N. K., Collins, C. A., Smith, R. L. 1976. The dynamic structure of the frontal zone in the coastal upwelling region off Oregon. *J. Phys. Oceanogr.* 6: 3–21

Moore, D. W., Philander, S. G. H. 1977. Modeling of the tropical ocean circulation. In *The Sea*, ed. E. D. Goldberg, I. N. Cave, J. J. O'Brien, J. H. Steele, 6: 319–61. New York: Wiley. 1048 pp.

Murthy, C. R. 1976. Horizontal diffusion characteristics in Lake Ontario. *J. Phys. Oceanogr.* 6: 76–84

Mysak, L. A. 1979. Recent advances in shelf wave dynamics. *Rev. Geophys. Space Phys.* 17: In press

Mysak, L. A. 1980. Topographically trapped waves. *Ann. Rev. Fluid Mech.* 12: 45–76

Okubo, A. 1971. Oceanic diffusion diagrams. *Deep-Sea Res.* 18: 789–802

Pedlosky, J. 1974a. On coastal jets and upwelling in bounded basins. *J. Phys. Oceanogr.* 4: 3–18

Pedlosky, J. 1974b. Longshore currents, upwelling, and bottom topography. *J. Phys. Oceanogr.* 4: 214–26

Pedlosky, J. 1974c. Longshore currents and the onset of upwelling over bottom slope. *J. Phys. Oceanogr.* 4: 310–20

Pedlosky, J. 1978a. An inertial model of steady coastal upwelling. *J. Phys. Oceanogr.* 8: 171–77

Pedlosky, J. 1978b. A nonlinear model of the onset of upwelling. *J. Phys. Oceanogr.* 8: 178–87

Peffley, M. B., O'Brien, J. J. 1976. A three-dimensional simulation of coastal upwelling off Oregon. *J. Phys. Oceanogr.* 6: 164–80

Rhines, P. B. 1970. Edge-, bottom-, and Rossby waves in a rotating stratified fluid. *Geophys. Fluid Dyn.* 1: 273–302

Rooth, C. 1972. A linearized bottom friction law for large-scale oceanic motions. *J. Phys. Oceanogr.* 4: 509–10

Scott, J. T., Csanady, G. T. 1976. Nearshore currents off Long Island. *J. Geophys. Res.* 81: 5401–9

Smith, J. D., Long, C. E. 1976. The effect of turning in the bottom boundary layer on continental shelf sediment transport. *Mém. Soc. R. Sci. Liège.* 6e ser., tome X. pp. 369–96

Smith, R. L. 1974. A description of current, wind and sea-level variations during coastal upwelling off the Oregon coast, July-August 1972. *J. Geophys. Res.* 79: 435–43

Smith, R. L. 1978. Poleward propagating perturbations in currents and sea levels along the Peru coast. *J. Geophys. Res.* 83: 6083–92

Smith, R. L. 1979. A comparison of the structure and variability of the flow field in three coastal upwelling regions: Oregon, Northwest Africa and Peru. In preparation

Sobey, E. J. C. 1977. *The response of Oregon shelf waters to wind fluctuations: differences and transition between winter and summer.* PhD thesis. Oregon State Univ., Corvallis. 153 pp.

Suginohara, N. 1974. Onset of coastal upwelling in a two-layer ocean by wind stress with longshore variation. *J. Oceanogr. Soc. Jpn.* 30:23–33

Suginohara, N. 1977. Upwelling front and two-cell circulation. *J. Oceanogr. Soc. Jpn.* 33:115–30

Sverdrup, H. U., Johnson, M. W., Fleming, R. H. 1942. *The Oceans. Their Physics, Chemistry, and General Biology*, p. 57. Englewood Cliffs: Prentice Hall. 1087 pp.

Thompson, J. D. 1974. *The coastal upwelling cycle on a beta-plane: hydrodynamics and thermodynamics*. PhD thesis. Florida State Univ., Tallahassee. 141 pp.

Wang, D.-P., Mooers, C. N. K. 1976. Coastal trapped waves in a continuously stratified ocean. *J. Phys. Oceanogr.* 6:853–63

Wang, D.-P., Mooers, C. N. K. 1977. Long coastal trapped waves off the west coast of the United States, Summer 1973. *J. Phys. Oceanogr.* 7:856–64

Wunsch, C., Gill, A. E. 1976. Observations of equatorially trapped waves in Pacific sea level variations. *Deep-Sea Res.* 23: 371–90

Ann. Rev. Fluid Mech. 1980. 12:435–76

PARTICLE MOTIONS IN A VISCOUS FLUID ✕8165

L. G. Leal
Department of Chemical Engineering, California Institute of Technology, Pasadena, California 91125

1 INTRODUCTION

The motion of small particles, drops, and bubbles in a viscous fluid at low Reynolds number is one of the oldest classes of problems in theoretical fluid mechanics, dating at least to Stokes's (1851) analysis of the translation of a rigid sphere through an unbounded quiescent fluid at zero Reynolds number. A reasonably comprehensive review of research published prior to 1965 is contained in the book of Happel & Brenner (1965), and extensive reviews were later written by Goldsmith & Mason (1967), Brenner (1966, 1970), Leal (1979), and Caswell (1977). Additional topics either not mentioned or only briefly mentioned in previous reviews include 1. the use of slender-body approximations to analyze the motions of rod-like particles (cf Cox 1970a, Batchelor 1970, Tillett 1970, Burgers 1938) and elongated drops (Acrivos & Lo 1978, Rallison & Acrivos 1978, Buckmaster 1972, 1973, Buckmaster & Flaherty 1973); 2. the use of singularity methods for Stokes equation to construct solutions for single, finite particles (Chwang & Wu 1974, 1975, 1976, Chwang 1975) and in a truncated form to construct approximate solutions for multiple particles (Gluckman, Weinbaum & Pfeffer 1972); 3. studies of the existence and properties of "wakes" and closed streamline regions in creeping flows (Davis et al 1976, Michael & O'Neill 1977); 4. boundary effects on particle motions near a liquid-liquid interface (Lee, Chadwick & Leal 1979, Lee & Leal 1979); 5. the effects of electrostatic repulsion, van der Waals' attraction, double layers, and other colloidal phenomena, such as Brownian motion, on particle motions (Batchelor 1976, Saville 1977, Russell 1976, 1977a,b, Spielman 1977, Leal & Hinch 1971); 6. the motions of nonspherical, nonaxisymmetric particles in shear flows (Hinch & Leal 1979, Gierszewski & Chaffey 1978); 7. and, finally, the motions of non-

0066-4189/80/0115-0435$01.00

spherical particles near a solid wall (Russell et al 1977, de Mestre & Russell 1975, Caswell 1970, Blake 1971).

It can be seen from this list, as well as from the more general books and reviews cited at the outset, that the vast majority of theoretical research on particle motions in viscous fluids is predicated on three approximations. The most critical of these, in the sense that the Reynolds number can never be exactly zero, is the *complete* neglect of inertial effects. Additionally, it is generally assumed that the particles are fixed in shape, and that the suspending fluid is Newtonian. Implicit in these approximations is the assumption that the actual behavior will deviate only slightly from that predicted theoretically provided that the Reynolds number is small, and that any shape changes or non-Newtonian effects are suitably weak. This assumption is, of course, correct insofar as the *instantaneous* velocity or pressure fields are concerned. Frequently, however, it is the position or orientation of the particles relative either to the boundaries or to the undisturbed flow that is important, and in this case small instantaneous contributions of inertia, non-Newtonian rheology, or flow-induced changes in particle shape may have a significant cumulative effect even when the instantaneous contributions to the particles' motion are small. It is known, for example, that a small degree of inertia will cause a small spherical particle that is suspended in a Poiseuille flow to move across the flow to a specific radial position that is *independent* of its initial position (cf Segre & Silberberg 1962a,b). Such changes in particle configuration (i.e. either its position or orientation) relative to the boundaries, or to the undisturbed velocity field, are not only of fundamental interest, but, when reflected in a statistical sense by the many particles of a suspension, can lead to technologically significant variations in its bulk properties (cf Ho & Leal 1974, Segre & Silberberg 1963). The focus of the present review is thus the motions of bubbles, drops, and small particles in a viscous fluid when any (or all) of the restrictions of zero Reynolds number, Newtonian rheology, and fixed shape are relaxed. We shall refer to these as the "standard conditions" for particle motions in a viscous fluid.

If the departures from the "standard conditions" are large (or, at least, not small), it is obvious that even the instantaneous particle motion may also exhibit a large change and this is certainly a domain of considerable interest. Indeed, experimental studies frequently encompass such conditions, which may also be the *most* important from a technological point of view. Unfortunately, however, theoretical investigation is difficult, and, even if possible, would generally require a new solution for each new combination of conditions. Furthermore, in the case of non-Newtonian fluids, there is no constitutive equation that is *known* to provide an adequate model of the mechanics for this class of flows, when the fluid behavior departs *strongly* from Newtonian (cf Leal 1979).

The particle-motion problem *can* be treated theoretically if the deviations from the "standard conditions" are *small*. Although this limit may seem to be of limited interest, since the instantaneous contributions of inertia, non-Newtonian rheology, or shape deformation must also be small, we have already seen that these small changes can have a strong (or even dominant) cumulative effect on the particle's position or orientation. This occurs for the class of so-called "indeterminate" particle motions, in which *no* position or orientation is intrinsically favored under "standard conditions." Numerous examples of this class exist, though by no means does it encompass all particle motions in Stokes flow.

The distinction between "determinate" and "indeterminate" is whether or not the particle adopts a configuration, under "standard conditions" in a *finite* period of time, that is independent of its "initial" configuration (i.e. position, orientation, or both). An example of "deterministic" behavior is the rotation of a prolate spheroid in a uniaxial extensional flow, $\mathbf{u} = \mathbf{E} \cdot \mathbf{x}$ where the rate-of-strain tensor $\mathbf{E} = E[^{2} \;{}^{-1}\; {}_{-1}]$ and $\mathbf{x}$ is a position vector with origin at the particle center. In this case (cf Bretherton 1962, or Leal & Hinch 1972), the axis of revolution rotates into alignment with the principle axis of $\mathbf{E}$, for any initial orientation, and the presence of *weak* inertia or non-Newtonian fluid properties can only contribute small, generally unimportant changes in the steady-state orientation. We shall be concerned in this review with small departures from "standard conditions" for particle motions that are "indeterminate" in creeping flow. Some well-known examples of this class are the lateral position (relative to walls) of a neutrally buoyant rigid sphere that is translating in a unidirectional "shear" flow, the orientation of a transversely isotropic rigid particle in sedimentation through an unbounded quiescent fluid, the orientation of an axisymmetric rigid particle in simple shearing flow, and the relative positions of two identical spherical particles in sedimentation. For this second class of problems, weak, instantaneous contributions to the particle velocity due to *departures* from "standard conditions" *can* yield a slow drift to some final configuration that differs significantly from, and is independent of, the *initial* configuration.

It will, no doubt, be noted that the discussion of "determinate" and "indeterminate" Stokes-flow problems has been completely qualitative until now. In view of the fundamental distinction between these two classes of particle motions, it is logical to inquire whether a more definitive, analytical scheme could not be achieved for identifying them on the basis of particle or boundary shape, and the nature of the undisturbed flow field. Such a scheme has, in fact, been developed by Bretherton (1962) for the restricted problem of rotation of axisymmetric rigid particles in the general unbounded *linear* flow, $\mathbf{u} = \mathbf{\Gamma} \cdot \mathbf{x}$ with $\mathbf{\Gamma} = \mathbf{E} + \mathbf{\Omega}$ where $\mathbf{E}$ is the rate-of-strain tensor and $\mathbf{\Omega}$ is the vorticity tensor. According to Bretherton's analysis, such a particle will either rotate in a "deterministic" fashion to a fixed orientation, or else rotate in an "indeterminate" configuration, depending upon the existence (or absence) of a positive real part for one of the eigenvalues of the tensor, $\mathbf{\Omega} + B\mathbf{E}$, where B is a constant that depends upon the particle geometry. A similar approach can also be used to classify *translational* creeping motions for particles of fixed shape in either a linear undisturbed velocity field or a quiescent fluid, using the intrinsic resistance tensors that relate the force and torque on a particle to its translational and angular velocities (cf Happel & Brenner 1965). Unfortunately, however, analytical classification schemes of this type are not possible for more general undisturbed fluid motions, especially when the boundaries figure prominently in the particle's motion. Moreover, even the schemes for unbounded linear flows are less useful than might be anticipated since drastic restrictions on particle geometry are required to make them reasonable for application. It is no great loss, therefore, to simply observe that the presence or absence of a configurational indeterminacy for a particle in Stokes flow depends both on the nature of the undisturbed flow, and on the geometry and deformability of the particles and boundaries.

In the sections that follow, we will concentrate on the effects of weak inertia, viscoelasticity, and/or small flow-induced particle deformation for the class of "indeterminate" Stokes-flow problems. It is, of course, obvious that no *real* particle will exhibit "indeterminate" behavior if an arbitrarily small contribution from inertia, viscoelasticity, or deformation is sufficient to yield a preferred orientation and/or position. Inertia is always present, if only to a limited degree (and so too, quite often, non-Newtonian fluid properties and particle deformation), and a *real* particle in a real fluid would thus eventually approach an intrinsically "preferred" position and/or orientation. However, as the magnitude of the Reynolds number is decreased (or alternatively, the non-Newtonian or shape deformation parameters are decreased) the characteristic *time scale* for the approach to equilibrium will increase. When this time scale exceeds that of a given experiment, the particle motion will *appear* to be "indeterminate," in spite of the fact that the particle motion is *always* deterministic. In comparing the small-parameter theories for weak nonlinear contributions to the particle's motion with available experiments, which generally are *not* similarly restricted, one important question is the range (if any) over which the theory can yield qualitatively (or quantitatively) correct results.

2 METHODS OF ANALYSIS

We have indicated, in the previous section, that small departures from the assumptions of zero Reynolds number, Newtonian rheology, and fixed shape can lead to a significant, cumulative change in the orientation or position of a particle with time. Three specific problems involving this type of effect have received the majority of attention to date. These are

(*a*) The orientation of transversely isotropic rigid particles in sedimentation through an unbounded quiescent fluid (the orientation is predicted to remain fixed indefinitely at its initial value for motion in a Newtonian fluid at zero Reynolds number; Cox 1965).
(*b*) The lateral position of spherically isotropic particles relative to bounding walls, either in sedimentation through a quiescent fluid or in translation in a unidirectional shear flow (the lateral position is predicted to remain fixed for all time at its initial value in a Newtonian fluid at zero Reynolds number; Bretherton 1962).
(*c*) The orbit for a transversely isotropic rigid particle that is freely rotating in a simple shear flow (it is predicted that the orbit is determined for all time in a Newtonian fluid at zero Reynolds number by the orientation at some initial instant; Jeffrey 1922).

The indicated predictions in each case are in marked contrast to experimental observations. Specifically, in the presence of *a small degree of inertia in a Newtonian fluid*, the available experiments, which we shall consider in more detail in Section 3, show: 1. rotation of axisymmetric particles in sedimentation through an "unbounded" fluid to an orientation in which the symmetry axis is either vertical (oblate particles) or horizontal (prolate particles); 2. "lateral" migration of rigid spheres to a fixed position midway between the bounding walls in sedimentation and between the walls and the centerline in a unidirectional shear flow; and 3. cross-orbital drift in simple shear flow to a final "orbit" of rotation which, for slender axisymmetric particles, lies entirely in the plane of the flow. *Weakly non-Newtonian fluid properties* also lead to "preferred" orientations or positions, though these are generally different from those induced by inertia in a Newtonian fluid. Specifically, observations show: 1. rotation of axisymmetric particles in sedimentation through a viscoelastic fluid to an orientation that is horizontal for oblate particles and vertical for prolate ones; 2. "lateral" migration, which is generally to the centerline in viscoelastic fluids, but is toward the boundaries in a purely viscous (i.e. nonelastic) fluid; and 3. orbital drift in simple shear flow to a final orbit for axisymmetric particles that depends on the particle axis ratio, but is the orbit of "minimum dissipation" for very large or small axis ratios. Finally, a *deformable* drop in a Newtonian fluid at zero Reynolds number is observed to migrate laterally in the direction of the centerline in a unidirectional shear flow.

Although the equilibrium configuration will, in some circumstances, depend on the magnitude of the deviation from the "standard" theoretical conditions, a common situation is that it is the *rate of approach* to this equilibrium that is changed, rather than the final state itself. This, plus the indeterminate nature of the Stokes-flow problem, gives some hope that useful predictions of at least the steady-state configuration might be achieved by a theory for asymptotically *small* departures from the standard conditions. Detailed comparison of predictions and experimental data shows that this is in fact the case. Indeed, virtually all of the "successful" theories for this class of problems have been based on an asymptotic expansion around the limit of creeping motion, Newtonian fluid, and fixed shape. One advantage of such a theory is that it allows small but nonzero inertia effects ($\mathrm{Re} \ll 1$), weakly non-Newtonian fluid properties, and small deformations of the particle shape to be included within a common theoretical framework. Our purpose in the remainder of this section is to describe this general framework, both as a prelude to discussion of detailed results in Section 3, and as a guideline for analysis of new problems by future researchers.

We will begin by considering the case of a single rigid particle suspended in a fluid whose properties, under the conditions of motion, are slightly non-Newtonian. Although we will later focus on a specific constitutive approximation for "weak, slow" flows of a viscoelastic fluid, the general theory can be developed without need for such specialization and we follow this approach. After considering rigid particles in slightly non-Newtonian fluids, we will show that the same basic development can be adapted in most cases to the motion of rigid particles with weak inertia effects in a Newtonian fluid. Finally, we describe the generalization necessary to treat drops (or bubbles), which may either be spherical or slightly deformed.

A *Rigid Particles in a Slightly Non-Newtonian Fluid*

We begin by considering the case of a rigid particle, of arbitrary shape, translating and rotating relative to its "center of rotation," in a fluid that is slightly non-Newtonian. The fluid motion in this case is then described by the differential equations and boundary conditions

$$\nabla \cdot \mathbf{S} = 0, \quad \nabla \cdot \mathbf{u} = 0 \tag{1}$$

$$\mathbf{S} = -p\mathbf{I} + (\nabla\mathbf{u} + \nabla\mathbf{u}^{\mathrm{T}}) + \lambda\boldsymbol{\Sigma}(\mathbf{u}) \tag{2}$$

with

$$\mathbf{u} \to \mathbf{V} \quad \text{at } \infty \text{ and on the walls}$$

and (3)

$$\mathbf{u} = \boldsymbol{\Omega}_{\mathrm{p}} \wedge \mathbf{r} + \mathbf{U}_{\mathrm{p}} \quad \text{at the particle surface.}$$

Here $\mathbf{V}$ represents the undisturbed fluid motion, which is assumed to satisfy the *same* Equations (1) and (2) as the velocity field in the presence of the particle, plus appropriate boundary conditions at any walls. We shall denote the stress field corresponding to this undisturbed motion by $\mathbf{T}$. The constitutive equation for the fluid is assumed to be explicit in the stress, as indicated in (2), and the non-Newtonian contribution is denoted by $\lambda\boldsymbol{\Sigma}(\mathbf{u})$ where $\boldsymbol{\Sigma}(\mathbf{u})$ is a nonlinear function (or functional) of $\mathbf{u}$, and λ is a *small* parameter that measures the magnitude of this contribution relative to the Newtonian stress. The translational and rotational velocities of the particle are denoted by $\mathbf{U}_{\mathrm{p}}$ and $\boldsymbol{\Omega}_{\mathrm{p}}$, respectively, and are specified relative to a *fixed* reference frame whose origin is instantaneously coincident with the center of rotation of the particle. Generally speaking, the objective of the theories we will describe is to determine the non-Newtonian contributions, at $O(\lambda)$ for $\lambda \ll 1$, to $\mathbf{U}_{\mathrm{p}}$ and $\boldsymbol{\Omega}_{\mathrm{p}}$, with note being taken of the fact that the "indeterminate" nature of the Stokes problem for $\lambda \equiv 0$ is a reflection of certain "components" of $\mathbf{U}_{\mathrm{p}}$ or $\boldsymbol{\Omega}_{\mathrm{p}}$ being zero at leading order. The determination of $\mathbf{U}_{\mathrm{p}}$ and $\boldsymbol{\Omega}_{\mathrm{p}}$ requires the solution of (1)–(3), together with the equations of motion for the particle.

The obvious method of obtaining these solutions for $\lambda \ll 1$ is via an asymptotic expansion of the form

$$\begin{aligned} \mathbf{u} &= \mathbf{u}_0 + \lambda\mathbf{u}_1 + \dots \\ p &= p_0 + \lambda p_1 + \dots \\ \boldsymbol{\Omega}_{\mathrm{p}} &= \boldsymbol{\Omega}_{\mathrm{p}}^{(0)} + \lambda\boldsymbol{\Omega}_{\mathrm{p}}^{(1)} + \dots \\ \mathbf{U}_{\mathrm{p}} &= \mathbf{U}_{\mathrm{p}}^{(0)} + \lambda\mathbf{U}_{\mathrm{p}}^{(1)} + \dots \end{aligned} \tag{4}$$

Implicit in this form is the assumption that a *regular* expansion exists, and this is true for slightly non-Newtonian fluids. The first terms in (4) represent the solutions for particle motion in a *Newtonian* fluid. In order to proceed further, it is, of course, necessary to assume that this $O(1)$ problem *can* be solved. Thus, $\boldsymbol{\Omega}_{\mathrm{p}}^{(0)}$ and $\mathbf{U}_{\mathrm{p}}^{(0)}$, as well as the detailed velocity and pressure fields, $\mathbf{u}_0$ and p_0, are assumed to be known. The main objective of the theory is then to determine $\mathbf{U}_{\mathrm{p}}^{(1)}$ and $\boldsymbol{\Omega}_{\mathrm{p}}^{(1)}$. Two approaches are possible. The straightforward one is simply to determine the velocity and pressure fields at $O(\lambda)$, and from them and the equations of motion for the particle to determine its motion. If the expansion in λ were singular, this would, in fact, be the most effective way to proceed.[1] However, the perturbation to slightly non-Newtonian fluids is *regular* and a second, less cumbersome method can be used, which was originally derived for rigid particles in the presence of inertia by Cox (1965) and Cox & Brenner (1968), and later "rediscovered" by Ho & Leal (1974). This method uses the "reciprocal theorem" of Lorentz (1907) to derive general formulae from which $\mathbf{U}_p^{(1)}$ and $\mathbf{\Omega}_p^{(1)}$ can be calculated knowing $\mathbf{u}_0$ and p_0, plus the solution of a second Stokes-flow problem sometimes called the "complementary problem," but *without* the need to determine $\mathbf{u}_1$ and p_1. When it can be used, the procedure is advantageous since the necessary calculations are generally easier than solution of the full problem at $O(\lambda)$. In addition, the complementary problem can be solved once and for all for a given particle and wall geometry, and given boundary conditions at the particle surface, independent of the type of nonlinearity (i.e. small non-Newtonian effect, inertia, or even particle deformation) we wish to consider. Application of the reciprocal-theorem approach to the motion of a particle in a non-Newtonian fluid was first reported, in a somewhat less general context than outlined here, by Ho & Leal (1976), Leal (1975), and Brunn (1976a,b).

The general formulae for $\mathbf{U}_p^{(1)}$, $\mathbf{\Omega}_p^{(1)}$ are most conveniently stated in terms of the "disturbance flow" velocity, pressure, and stress fields

$$\hat{\mathbf{u}} = \mathbf{u} - \mathbf{V}$$
$$\hat{p} = p - Q \tag{5}$$
$$\hat{\mathbf{S}} = \mathbf{S} - \mathbf{T}.$$

The governing equations for $(\hat{\mathbf{u}}, \hat{p})$ can be obtained directly from (1)–(3), together with the equations for $\mathbf{V}$, which in this case are

$$\nabla^2 \mathbf{V} - \nabla Q = -\lambda \nabla \cdot \mathbf{\Sigma}(\mathbf{V}) \tag{6}$$

$$\nabla \cdot \mathbf{V} = 0. \tag{7}$$

In order to derive formulae for $\mathbf{U}_p^{(1)}$ and $\mathbf{\Omega}_p^{(1)}$ that allow them to be evaluated without solving the complete disturbance-flow problem at $O(\lambda)$, we introduce the "complementary" Stokes-flow problem.

The "complementary" problem, in this case, consists simply of the translation and rotation of the same particle with unit velocity in exactly

[1] We shall return to this point shortly in connection with the calculation of inertia effects.

the same geometric configuration, but with the fluid *stationary* apart from the motion induced by the particle. Thus,

$$\nabla \cdot \boldsymbol{\tau} = 0, \quad \nabla \cdot \mathbf{v} = 0 \tag{8}$$

$$\boldsymbol{\tau} = -q\mathbf{I} + (\nabla \mathbf{v} + \nabla \mathbf{v}^{\mathrm{T}}) \tag{9}$$

with boundary conditions

$$\mathbf{v} = \mathbf{e} + (\hat{\mathbf{e}} \wedge \mathbf{r}) \quad \text{on the particle surface}$$

and (10)

$$\mathbf{v} \to 0 \quad \text{at } \infty \text{ and on the walls.}$$

It is convenient to consider the translation and rotation problems separately, i.e.

$$\mathbf{v} = \mathbf{v}_{\mathrm{T}} + \mathbf{v}_{\mathrm{R}}, \quad q = q_{\mathrm{T}} + q_{\mathrm{R}}, \quad \boldsymbol{\tau} = \boldsymbol{\tau}_{\mathrm{T}} + \boldsymbol{\tau}_{\mathrm{R}},$$

and to express the solutions in terms of generalized stress, velocity, and pressure fields, $(\mathbf{T}_{\mathrm{T}}, \mathbf{U}_{\mathrm{T}}, \mathbf{Q}_{\mathrm{T}})$ and $(\mathbf{T}_{\mathrm{R}}, \mathbf{U}_{\mathrm{R}}, \mathbf{Q}_{\mathrm{R}})$, which are third-, second-, and first-order tensors, respectively.[2] These are defined by

$$\mathbf{v}_{\mathrm{T}} = \mathbf{U}_{\mathrm{T}} \cdot \mathbf{e} \qquad \mathbf{v}_{\mathrm{R}} = \mathbf{U}_{\mathrm{R}} \cdot \hat{\mathbf{e}} \tag{11}$$

$$q_{\mathrm{T}} = \mathbf{Q}_{\mathrm{T}} \cdot \mathbf{e} \qquad q_{\mathrm{R}} = \mathbf{Q}_{\mathrm{R}} \cdot \hat{\mathbf{e}} \tag{12}$$

$$\boldsymbol{\tau}_{\mathrm{T}} = \mathbf{T}_{\mathrm{T}} \cdot \mathbf{e} \qquad \boldsymbol{\tau}_{\mathrm{R}} = \mathbf{T}_{\mathrm{R}} \cdot \hat{\mathbf{e}}$$

and have the advantage that they are independent of the orientations of the translation and rotation axes, $\mathbf{e}$ and $\hat{\mathbf{e}}$. The governing equations for these higher-order fields can be derived directly from (8)–(10).

We can now briefly outline the derivation of general formulae for the translational and rotational velocity of the particle at $O(\lambda)$. One such expression can be obtained by combining the governing equation for the disturbance flow at $O(\lambda)$ and the translational complementary problem in the form

$$\int_{V_{\mathrm{f}}} [(\nabla \cdot \hat{\mathbf{S}}_1) \cdot \mathbf{U}_{\mathrm{T}} - (\nabla \cdot \mathbf{T}_{\mathrm{T}}) \cdot \hat{\mathbf{u}}_1] \, dV = 0 \tag{13}$$

where V_{f} is the entire fluid volume bounded by the particles and walls. This can be written alternatively as

$$\int_{A_{\mathrm{p}}} [\hat{\mathbf{S}}_1 \cdot \mathbf{U}_{\mathrm{T}} - \mathbf{T}_{\mathrm{T}} \cdot \hat{\mathbf{u}}_1] \cdot \mathbf{n} \, dA + \int_{V_{\mathrm{f}}} [\hat{\mathbf{S}}_1 : \nabla \mathbf{U}_{\mathrm{T}} - \mathbf{T}_{\mathrm{T}} : \nabla \hat{\mathbf{u}}_1] \, dV = 0 \tag{14}$$

[2] A first-order tensor is a vector.

where A_p is the particle surface. Hence, using the equation of continuity, the reciprocal theorem in the form

$$\int_{V_f} [(-\hat{p}_1\mathbf{I}+(\nabla\hat{\mathbf{u}}_1+\nabla\hat{\mathbf{u}}_1^T)):\nabla\mathbf{U}_T-(-\mathbf{Q}_T\mathbf{I}+(\nabla\mathbf{U}_T+\nabla\mathbf{U}_T^T)):\nabla\hat{\mathbf{u}}_1]\,dV=0,$$

and the divergence theorem, we obtain[3]

$$\int_{A_p} [\hat{\mathbf{S}}_1\cdot\mathbf{U}_T-\mathbf{T}_T\cdot\hat{\mathbf{u}}_1]\cdot\mathbf{n}dA=\int_{V_f}\{\boldsymbol{\Sigma}(\hat{\mathbf{u}}_0+\mathbf{V})-\boldsymbol{\Sigma}(\mathbf{V})\}:\nabla\mathbf{U}_T\,dV. \tag{15}$$

Finally, applying the boundary conditions on $\mathbf{U}_T$ and $\hat{\mathbf{u}}_1$ at the particle surface gives

$$-\int_{A_p}\hat{\mathbf{S}}_1\cdot\mathbf{n}dA+\mathbf{K}_T\cdot\mathbf{U}_p^{(1)}+\mathbf{K}_c^T\cdot\boldsymbol{\Omega}_p^{(1)}=\int_{V_f}\{\boldsymbol{\Sigma}(\hat{\mathbf{u}}_0+\mathbf{V})-\boldsymbol{\Sigma}(\mathbf{V})\}:\nabla\mathbf{U}_T\,dV. \tag{16}$$

Note that the first integral is just the contribution to the force on the particle at $O(\lambda)$. Normally, for a body force of $O(1)$, this term is *zero*. Furthermore, the coefficients $\mathbf{K}_T$ and $\mathbf{K}_c^T$, defined by

$$\mathbf{K}_T\equiv\int_{A_p}(\mathbf{T}_T\cdot\mathbf{n})\,dA$$

and (17)

$$\mathbf{K}_c^T=\int_{A_p}\hat{\mathbf{r}}\wedge(\mathbf{T}_T\cdot\mathbf{n})\,dA,$$

are the standard "translational" and "coupling" resistance tensors of low-Reynolds-number flow for this specific particle and wall geometry. Provided the "complementary" translation problem can be solved, as we have implicitly assumed, these coefficients can be considered as knowns in Equation (16). It may be noted that the term $\mathbf{K}_T\cdot\mathbf{U}_p^{(1)}$ is, in fact, just the hydrodynamic force that the particle would experience if it were translating with a velocity $\mathbf{U}_p^{(1)}$ through a Newtonian fluid at zero Reynolds number in an identical geometric configuration. The term $\mathbf{K}^T\cdot\boldsymbol{\Omega}_p^{(1)}$ is the force under similar conditions except that the particle is assumed to be rotating with angular velocity $\boldsymbol{\Omega}_p^{(1)}$ rather than translating. For particles of simple shape—say spherical or axisymmetric—with no

[3] The formula (15) depends critically on the fact that the expansion (4) is *regular* so that the far-field condition, $\hat{\mathbf{u}}\rightarrow 0$ at ∞ and on any container walls, is satisfied in the expansion at order λ.

walls (or with walls normal to the axis of rotation), this "coupling" term is zero. Finally, we note that the integral on the right-hand side of (16) can be evaluated from the solution $\hat{\mathbf{u}}_0$ of the original disturbance-flow Stokes problem at $O(1)$ plus the complementary velocity field. Thus, (16) is a vector equation relating $\mathbf{U}_p^{(1)}$ and $\mathbf{\Omega}_p^{(1)}$, in which all coefficients and other terms are known from the complementary and $O(1)$ Stokes solutions. An additional relationship of similar form to (16) can be derived by replacing $\mathbf{U}_T$ and $\mathbf{T}_T$ in (13) with $\mathbf{U}_R$ and $\mathbf{T}_R$.

The procedure of the preceding paragraph thus yields two equations relating $\mathbf{U}_p^{(1)}$ and $\mathbf{\Omega}_p^{(1)}$

$$\begin{aligned} -\mathbf{F}_1+\mathbf{K}_T\cdot\mathbf{U}_p^{(1)}+\mathbf{K}_c^T\cdot\mathbf{\Omega}_p^{(1)} &= \int_{V_f}[\mathbf{\Sigma}(\hat{\mathbf{u}}_0+\mathbf{V})-\mathbf{\Sigma}(\mathbf{V})]:\nabla\mathbf{U}_T\,dV \\ -\mathbf{G}_1+\mathbf{K}_c\cdot\mathbf{U}_p^{(1)}+\mathbf{K}_R\cdot\mathbf{\Omega}_p^{(1)} &= \int_{V_f}[\mathbf{\Sigma}(\hat{\mathbf{u}}_0+\mathbf{V})-\mathbf{\Sigma}(\mathbf{V})]:\nabla\mathbf{U}_R\,dV. \end{aligned} \tag{18}$$

Normally, $\mathbf{F}_1=\mathbf{G}_1=0$. The terms $\mathbf{K}_c\cdot\mathbf{U}_p^{(1)}$ and $\mathbf{K}_R\cdot\mathbf{\Omega}_p^{(1)}$ are the hydrodynamic torque that would occur with translation or rotation, $\mathbf{U}_p^{(1)}$ or $\mathbf{\Omega}_p^{(1)}$, through a quiescent Newtonian fluid in an identical geometrical configuration. The coefficient matrices, $\mathbf{K}_c$ and $\mathbf{K}_R$, are dependent only on the geometry of the particle plus walls,

$$\mathbf{K}_c\equiv\int_{A_p}\mathbf{T}_R\cdot\mathbf{n}\,dA \qquad \mathbf{K}_R\equiv\int_{A_p}\mathbf{r}\wedge(\mathbf{T}_R\cdot\mathbf{n})\,dA.$$

Thus, the translational and angular velocity of an arbitrary rigid particle at $O(\lambda)$, i.e. $\mathbf{U}_p^{(1)}$ and $\mathbf{\Omega}_p^{(1)}$, can in principle be calculated from (18) using the solution of the complementary Stokes-flow problem defined by (8)–(12) (in order to determine $\mathbf{K}_r$, $\mathbf{K}_c$, $\mathbf{K}_R$, and the "velocity" gradients $\nabla\mathbf{U}_T$ and $\nabla\mathbf{U}_R$), plus the solution of the original disturbance-flow Stokes problem at $O(1)$ [in order to evaluate $\mathbf{\Sigma}(\hat{\mathbf{u}}_0+\mathbf{V})$ on the right-hand side of (18)]. Although the relationships (18a) and (18b) are completely general as regards particle and container geometry, a severe constraint on the geometry *will* be required if the disturbance and complementary Stokes-flow problems are to be solved analytically for arbitrary particle positions and orientations (as is required) rather than at a few discrete points where symmetry could help. The problem is particularly difficult in *bounded* domains where boundary conditions must be satisfied on both the particle and container surfaces. For this reason, all published theories have assumed that $\varepsilon\equiv a/d\ll 1$, where a is a particle dimension and d a cross-flow dimension of the flow channel; in addition, it is assumed that the particle is not too close to the walls, i.e. $a/l\ll 1$ where l is the distance from the particle center to the nearest wall. Two different methods have been used to solve the Stokes-flow problems when these conditions are satisfied. The more common approach has been to use the "method of reflections," cf Happel & Brenner (1965). This approach takes explicit account of the particle geometry at each level in an asymptotic expansion in a/d, but breaks down if the particle is too close to the walls. The second approach was described recently by Vasseur & Cox (1976) for the analysis of *inertial* effects (next subsection) and involves the use of integral transforms with the particle represented as a point-force singularity. This technique appears to be advantageous in terms of simplicity, but can only be used if the volume integrals in (18) (or their counterparts in the case of inertia or deformation effects) are dominated by the region near the walls so that the detailed flow *near* the particle need not be calculated. For *non-Newtonian effects*, it is found (cf Ho & Leal 1976) that the integrals in (18) are actually dominated by the region *near* a particle so that it is essential to use the method of reflections.

The results of calculations using (18), or the direct but arduous solution of the complete disturbance-flow problems at $O(\lambda)$, will be presented in Section 3.

B *Rigid Particles with Weak Inertial Effects*

A second class of problems that can be analyzed using almost the same theoretical framework is the motion of a rigid particle in a Newtonian fluid with weak fluid inertia. In this case, the governing equations are identical to Equations (1)–(7) of the previous section with the exception that the non-Newtonian stress $\lambda\mathbf{\Sigma}$ is identically zero, and is replaced by the inertia term

$$\mathrm{Re}\,\mathbf{f}(\mathbf{u}) = \mathrm{Re}\left[\frac{\partial \mathbf{u}}{\partial t} + \mathbf{u}\cdot\nabla\mathbf{u}\right] \tag{19}$$

which appears on the right-hand sides of (1a) and (6). The Reynolds number that appears in (19) is for the disturbance flow produced by the particle, and thus contains the length scale a of the particle and a characteristic velocity that is either the relative translational velocity of the particle, or the shear velocity Ga, whichever is larger. Here, G is a shear rate typical of the undisturbed flow, normally U/d where U is the characteristic velocity of the undisturbed flow, and d is a cross-flow dimension of the flow channel.

In one important respect, however, the inertial problem does differ from that for weakly non-Newtonian fluids, and that is that the asymptotic expansion in Reynolds number, corresponding to (4), is *singular* and thus generally valid only in an "inner" region near the particle. Higher-order terms, beyond $\hat{\mathbf{u}}_0$, are obtained by matching with a second "outer" expansion, which is relevant to the "Oseen" domain in which inertia and viscous terms are comparable in magnitude as $\mathrm{Re} \to 0$. Thus, in general, the far-field condition, $\hat{\mathbf{u}} \to 0$ at ∞, is *not* valid and the procedure outlined in the preceeding subsection for obtaining $\mathbf{U}_\mathrm{p}^{(1)}$ and $\mathbf{\Omega}_\mathrm{p}^{(1)}$ from the reciprocal theorem breaks down and is normally replaced by a solution of the full singular perturbation problem for small Re. One important class of problems exists, however, where the simpler reciprocal theorem approach retains its validity, namely those in a *bounded* domain which satisfy the condition

$$\mathrm{Re} \ll \left(\frac{a}{d}\right)^m \tag{20}$$

where the constant m depends on the flow type ($m = 1$, translation; $m = 2$, shear flow, etc). With an appropriate choice of m, the condition (20) insures that any walls lie *inside* the "inner" region, and the asymptotic expansion in Re is therefore regular. In this case, the reciprocal theorem, Equations (18a) and (18b), can be carried over almost without change. Obviously, particle motions in an *unbounded* shear flow cannot satisfy the condition (20) and are therefore usually analyzed via a full matched asymptotic solution of the governing equations and boundary conditions.

When (20) is satisfied, the relevant complementary problem is again (8)–(12), and the disturbance-flow problem at $O(\mathrm{Re})$ is the same as that for weak non-Newtonian effects, except that $\nabla \cdot \mathbf{\Sigma}$ is replaced by $\mathbf{f}(\mathbf{u})$ as already noted. Thus, it follows that

$$\begin{aligned} -\mathbf{F}_1 + \mathbf{K}_\mathrm{T} \cdot \mathbf{U}_\mathrm{p}^{(1)} + \mathbf{K}_\mathrm{c}^\mathrm{T} \cdot \mathbf{\Omega}_\mathrm{p}^{(1)} &= \int_{V_\mathrm{f}} \{\mathbf{f}(\hat{\mathbf{u}}_0 + \mathbf{V}) - \mathbf{f}(\mathbf{V})\} \cdot \mathbf{U}_\mathrm{T}\, dV \\ -\mathbf{G}_1 + \mathbf{K}_\mathrm{c} \cdot \mathbf{U}_\mathrm{p}^{(1)} + \mathbf{K}_\mathrm{R} \cdot \mathbf{\Omega}_\mathrm{p}^{(1)} &= \int_{V_\mathrm{f}} \{\mathbf{f}(\hat{\mathbf{u}}_0 + \mathbf{V}) - \mathbf{f}(\mathbf{V})\} \cdot \mathbf{U}_\mathrm{R}\, dV. \end{aligned} \tag{21}$$

Here $\mathbf{U}_\mathrm{p}^{(1)}$ and $\mathbf{\Omega}_\mathrm{p}^{(1)}$ represent the *inertial* contributions to the particle velocity at $O(\mathrm{Re})$, while $\mathbf{f}$ is the nonlinear inertial term defined in (19), evaluated using the velocity fields $(\hat{\mathbf{u}}_0 + \mathbf{V})$ and $\mathbf{V}$, respectively. The force and torque at $O(\mathrm{Re})$, $\mathbf{F}_1$ and $\mathbf{G}_1$, are normally zero since body forces or torques are $O(1)$ and accelerations at least $O(\mathrm{Re}^2)$. As in the non-Newtonian problem, the first inertial correction to the particle's motion can thus be obtained in this case without need to determine the detailed velocity and pressure fields at the same order of approximation.

C *Deformable Drops*

A third class of problems that has received considerable attention in the literature is the motion of deformable drops, with or without non-Newtonian or inertial effects present. In this case, the main feature of interest is the possibility of *cross-stream migration*, either in shear flow or in a quiescent fluid, to a "preferred" lateral position.

An analytical approach similar to that outlined in the preceeding two sections provides the most efficient method of determining the lateral migration velocity. Indeed, by limiting the analysis to the component of $\mathbf{U}_p$ normal to the local undisturbed velocity, $\mathbf{V}$, the complementary Stokes-flow problem can be reduced to the translation of the drop in the lateral direction, here called $\mathbf{e}_3$, through an otherwise quiescent Newtonian fluid. Thus, for the fluid outside the drop,

$$\begin{aligned} &\nabla\cdot\boldsymbol{\tau}=0 \qquad \nabla\cdot\mathbf{v}=0 \\ &\boldsymbol{\tau}=-q\mathbf{I}+(\nabla\mathbf{v}+\nabla\mathbf{v}^{\mathrm{T}}) \end{aligned} \tag{22}$$

and inside

$$\begin{aligned} &\nabla\cdot\tilde{\boldsymbol{\tau}}=0 \qquad \nabla\cdot\tilde{\mathbf{v}}=0 \\ &\tilde{\boldsymbol{\tau}}=-\tilde{q}\mathbf{I}+(\nabla\tilde{\mathbf{v}}+\nabla\tilde{\mathbf{v}}^{\mathrm{T}}) \end{aligned} \tag{23}$$

with boundary conditions

$$\mathbf{v}\to 0 \qquad \text{at } \infty \text{ and on the walls}$$

$$\left.\begin{aligned} &\mathbf{v}=\tilde{\mathbf{v}} \\ &\mathbf{n}\cdot\mathbf{v}=\mathbf{n}\cdot\tilde{\mathbf{v}}=\mathbf{n}\cdot\mathbf{e}_3 \\ &\text{and} \\ &\mathbf{t}\cdot\mathbf{n}\cdot\boldsymbol{\tau}=\kappa\mathbf{t}\cdot\mathbf{n}\cdot\tilde{\boldsymbol{\tau}} \end{aligned}\right\} \quad \text{at the drop interface.} \tag{24}$$

Here, κ is the viscosity ratio, and $\mathbf{n}$ and $\mathbf{t}$ are unit vectors normal and tangential to the drop surface. In general, the reciprocal-theorem approach requires that the complementary Stokes-flow problem be solved for a drop of the same shape as the "real" one under consideration—which must itself be calculated from the normal stress balance at the interface in the full disturbance-flow Stokes problem of drop motion in the undisturbed velocity field, $\mathbf{V}$. With this shape "given" insofar as the complementary problem is concerned, the solution of (22) and (23) is completely determined from the conditions of continuity of velocity and of the *tangential* component of stress at the drop interface as indicated in (24).

The analysis required to obtain a counterpart of (18) or (21) for this case has been outlined in considerable detail in the recent paper of Chan & Leal (1979), and will thus be displayed here only in outline. The basic idea is to consider each of the possible effects, i.e. non-Newtonian rheology in one or both fluids, inertia, or deformations in shape, as *small* corrections to the basic problem of the motion of a spherical drop in a Newtonian fluid at zero Reynolds number. Under these circumstances the effects on drop motion are simply additive at first order and can be considered independently. Thus, we can consider motion of a spherical drop at zero Reynolds number in a slightly non-Newtonian fluid, motion of a spherical drop in a Newtonian fluid with slight inertia, and finally, motion of a slightly deformed drop at zero Reynolds number in a Newtonian fluid.

The first two problems can be analyzed in a fashion analogous to that outlined in the preceeding sections based on an asymptotic expansion in λ or Re, except that we must take account of the equations of motion inside the drop as well as outside, with matching of the velocities and the tangential stress components at the interface.

Let us consider, for the moment, the motion of the spherical drop in a slightly *non-Newtonian* fluid. In this case, we can easily derive an equation analogous to (15) for the fluids inside *and* outside the drop, i.e.

$$-\int_{A_p} [\hat{\mathbf{S}}_1 \cdot \mathbf{v} - \boldsymbol{\tau} \cdot \hat{\mathbf{u}}_1] \cdot \mathbf{n}\, dA = \int_{V_f} \{\boldsymbol{\Sigma}(\hat{\mathbf{u}}_0 + \mathbf{V}) - \boldsymbol{\Sigma}(\mathbf{V})\} : \nabla \mathbf{v}\, dV \tag{25}$$

and

$$\int_{A_p} [\tilde{\mathbf{S}}_1 \cdot \tilde{\mathbf{v}} - \tilde{\boldsymbol{\tau}} \cdot \tilde{\mathbf{u}}_1] \cdot \mathbf{n}\, dA = \int_{V_p} \tilde{\boldsymbol{\Sigma}}(\tilde{\mathbf{u}}_0) : \nabla \tilde{\mathbf{v}}\, dV \tag{26}$$

respectively, where $(\tilde{\mathbf{u}}, \tilde{p}, \tilde{\mathbf{S}})$ are the full velocity, pressure, and stress fields inside the drop and $(\hat{\mathbf{u}}, \hat{p}, \hat{\mathbf{S}})$ are the same quantities for the disturbance flow outside. In addition, V_p represents the fluid volume *inside* the drop. When (26) is multiplied by $\kappa(\tilde{\lambda}/\lambda)$ and added to (25) and the interfacial boundary conditions are applied at the sphere surface at $O(\lambda)$, we obtain

$$\begin{aligned} \mathbf{U}_p^{(1)} \cdot \int_{A_p} \boldsymbol{\tau} \cdot \mathbf{n}\, dA = -\int_{A_p} (\mathbf{T}^{(1)} \cdot \mathbf{v}) \cdot \mathbf{n}\, dA + \int_{V_f} \{\boldsymbol{\Sigma}(\hat{\mathbf{u}}_0 + \mathbf{V}) \\ - \boldsymbol{\Sigma}(\mathbf{V})\} : \nabla \mathbf{v}\, dV + \kappa \frac{\tilde{\lambda}}{\lambda} \int_{V_p} \tilde{\boldsymbol{\Sigma}}(\tilde{\mathbf{u}}_0) : \nabla \tilde{\mathbf{v}}\, dV. \end{aligned} \tag{27}$$

The integral $\int_{A_p} \boldsymbol{\tau} \cdot \mathbf{n}\, dA$ is just the hydrodynamic force on a spherical drop that is translating at zero Reynolds number through a quiescent

Newtonian fluid in the "lateral" (i.e. $\mathbf{e}_3$) direction

$$\int_{A_p} \boldsymbol{\tau}\cdot\mathbf{n}\,dA = \left[-\frac{2\pi(2+3\kappa)}{1+\kappa} + \text{wall corrections}\right]\mathbf{e}_3.$$

Thus the component of $\mathbf{U}_p^{(1)}$ in the lateral direction, $(U_p^{(1)})_3$, can be calculated directly from (27) using the solution of the disturbance and complementary Stokes-flow problems. The undisturbed stress at $O(\lambda)$, $\mathbf{T}^{(1)}$, is "known" insofar as Equation (27) is concerned.

The motion of a spherical drop in a Newtonian fluid with *weak inertia* can be obtained, at $O(\text{Re})$, using essentially the same procedure provided $\text{Re} \ll (a/d)^m$ as before so that the asymptotic expansion in Re is regular. In order to adapt (27) to this case, we need only note that 1. the undisturbed stress at $O(\text{Re})$, i.e. $\mathbf{T}^{(1)}$, is zero if the undisturbed velocity field is unidirectional, as we shall assume for simplicity; and 2. the *nonlinear* inertia terms must be "substituted" for the non-Newtonian stress terms. With these changes

$$-(U_p^{(1)})_3\left[\frac{2\pi(2+3\kappa)}{1+\kappa}\right] = \int_{V_f}\{\mathbf{f}(\hat{\mathbf{u}}_0+\mathbf{V})-f(\mathbf{V})\}\cdot\mathbf{v}\,dV + \int_{V_p}\mathbf{f}(\tilde{\mathbf{u}}_0)\cdot\tilde{\mathbf{v}}\,dV. \tag{28}$$

Thus, again, the lateral velocity of a spherical drop at $O(\text{Re})$ can be calculated directly using the solutions of the complementary and disturbance Stokes-flow problems. If the condition $\text{Re} \ll (a/d)^m$ is *not* satisfied, calculation of the migration velocity at $O(\text{Re})$ is most efficiently done via a full matched asymptotic solution for the velocity and pressure fields through $O(\text{Re})$.

Finally, we turn to the problem of a slightly deformed drop in a Newtonian fluid at zero Reynolds number. Even in this case, the basic approach is essentially unchanged. The boundary conditions of velocity and tangential stress continuity are applied at the surface of the *deformed* drop, with the shape determined from the normal stress condition

$$(\hat{\mathbf{S}}+\mathbf{T}):\mathbf{nn} = \kappa\tilde{\mathbf{S}}:\mathbf{nn} + \frac{1}{\delta}(\nabla\cdot\mathbf{n}), \tag{29}$$

where $\delta = a\mu G/\sigma$, G is a characteristic shear rate, and σ the interfacial tension. Let us consider the case $\kappa = O(1)$ and $\delta \ll 1$, where the deformation is small and quasi-steady. The shape of the drop is conveniently specified to $O(\delta)$ as

$$F = r - 1 - \delta f_1 = 0 \tag{30}$$

where f_1 is the "deformation" at $O(\delta)$. The unit normal $\mathbf{n}$ and the sum of the principle radii of curvature $\nabla \cdot \mathbf{n}$ are thus

$$\mathbf{n} = \mathbf{e}_r - \delta \nabla f_1 + O(\delta^2)$$

and

$$\nabla \cdot \mathbf{n} = 2 - \delta[2f_1 + \nabla^2 f_1] + O(\delta^2). \tag{31}$$

In order to simplify the application of the conditions of velocity and tangential stress continuity, it is advantageous to approximate all quantities that are to be evaluated at the surface of the *deformed* drop in terms of equivalent quantities at $r = 1$ using a Taylor series expansion about $r = 1$ for $\delta \ll 1$. Hence, in effect, we replace the original problem, which has boundary conditions at the *deformed* surface, with an equivalent problem, for $\delta \ll 1$, in which modified boundary conditions are applied at the surface of a *sphere.* Although the complementary problem must normally be solved for a particle (drop) of the same shape as the "real" one under consideration, the reduction of the full problem to the motion of a *spherical* drop with a series (in δ) of modified boundary conditions at the sphere surface means that we can also choose the drop to be spherical for the complementary problem, which is thus again (22)–(24). An expression for the lateral velocity $(\mathbf{U}_p^{(1)})_3$ at $O(\delta)$ can therefore be obtained by first combining (25) and (26) (with $\mathbf{\Sigma} \equiv 0$), and then applying the modified surface boundary conditions to evaluate the integrand of the combined integral over A_p that appears. The result in this case is somewhat more complicated than before,

$$\mathbf{U}_p^{(1)} \cdot \int_{A_p} \mathbf{t} \cdot \mathbf{n}\, dA = -\int_{A_p} \Bigg\{ \Bigg[-f_1 \frac{\partial}{\partial r} (\mathbf{S}_0 - \kappa \tilde{\mathbf{S}}_0) \cdot \mathbf{e}_r + (\mathbf{S}_0 - \kappa \tilde{\mathbf{S}}_0) \cdot \nabla f_1 + \nabla f_1 (2f_1 + \nabla^2 f_1) \Bigg] \cdot \mathbf{u} + (\boldsymbol{\tau} - \kappa \tilde{\boldsymbol{\tau}}) : \mathbf{e}_r \mathbf{e}_r \Bigg[f_1 \left(\frac{\partial}{\partial r} \mathbf{u}_0 \right) \cdot \mathbf{e}_r - \mathbf{u}_0 \cdot \nabla f_1 \Bigg] + \kappa \tilde{\boldsymbol{\tau}} \cdot \mathbf{e}_r \cdot \Bigg[f_1 \frac{\partial}{\partial r} (\mathbf{u}_0 - \tilde{\mathbf{u}}_0) \Bigg] \Bigg\} dA, \tag{32}$$

but is easily evaluated using the complementary and disturbance velocity field at $O(1)$, plus the shape function f_1 calculated from $\hat{\mathbf{u}}_0$, $\hat{p}_0$ and $\tilde{\mathbf{u}}_0$, $\tilde{p}_0$ using the normal stress condition (29).

D *Concluding Remarks*

We have seen, in this section, that an analysis of the effects *on particle motion* of weak inertia in a *bounded* domain, non-Newtonian rheology, or shape deformation of a drop can all be carried out in a common framework requiring only detailed solutions for creeping motion of the undeformed particle or drop in the same undisturbed flow, **V**, plus solution of a second "complementary" creeping-motion problem in the same domain but with the fluid stationary far from the particle. Although this latter problem may be formidable, it will never be more difficult than the disturbance-flow problem at $O(1)$ and will always be much simpler than the full second-order problem, i.e. $O(\lambda)$, $O(\text{Re})$, or $O(\delta)$, for the velocity and pressure fields in the fluid(s). Furthermore, the complementary problem depends only on the particle and wall geometry, plus boundary conditions at the particle surface, and may be solved once and for all independent of the type of nonlinear effect we wish to consider.

It is evident that any other type of weak nonlinear departure from the Stokes problem for a rigid particle in a Newtonian fluid could also be analyzed by the same techniques, provided that the corresponding asymptotic expansion is *regular*. Two problems come to mind immediately, which have received limited attention in the present context. One is the motion induced by weak deformation of an elastic particle, either from a sphere or some other simple axisymmetric form (Tam & Hyman 1973). The other is the "extra" motion induced in a particle that is translating or rotating in a quiescent fluid near a boundary, due to deformation of the boundary. One example of this type, presently being studied by Berdan (1980), is the lateral motion induced in a sphere that is translating under the action of an applied force, near a deforming fluid/fluid interface.

3 THEORETICAL RESULTS

Let us now turn to the theoretical results that have been obtained, using either the methods of the preceeding section, or full solutions of the asymptotic problems as is required when the expansion is singular. We focus our attention on the motion of a single particle, either in a quiescent fluid, or in a unidirectional shear flow. Relatively little has so far been done outside of this regime, though we should specifically mention the analyses of Vasseur & Cox (1977) for the effects of inertia on the relative motion of two rigid spheres that are sedimenting through a quiescent fluid, and of Brunn (1977b) for the effect of viscoelastic fluid properties in the same problem.

A *The Effects of Deformation on the Motion of a Liquid Drop*

We begin by considering the effects of shape deformation on the motion of a Newtonian liquid drop in a unidirectional shear flow. It is known (Bretherton 1962) that the lateral position of a *spherical* drop in a Newtonian fluid at zero Reynolds number is *fixed* for all time by its value at some initial instant. We have noted earlier, however, that the introduction of flow-induced asymmetry in the drop shape relative either to the axis of the undisturbed flow or, equivalently, to the walls in a bounded flow domain may lead to lateral "migration" of the drop in a direction normal to the walls (or to the undisturbed flow) even for Newtonian fluids at zero Reynolds number. The first quantitative experimental studies of this phenomenon were carried out by Goldsmith & Mason (1962) for neutrally buoyant drops of relatively low viscosity in Poiseuille flow through a circular tube, and by Karnis & Mason (1967) for a Couette flow. In both cases, a relatively rapid migration was found to occur toward the *centerline* of the undisturbed flow (even though *no* migration could be observed for rigid spherical particles under the same circumstances). It may be noted that the existence of transverse motion in shear flow appears to be characteristic of all deformable or flexible particles, including fiber "threads" (Goldsmith & Mason 1962, Cox 1970b) and red blood cells (Goldsmith 1968, 1971), and to be independent of whether the unidirectional flow is steady, oscillatory, or pulsatile (Goldsmith & Mason 1967). However, the only theory for deformable particles other than drops is by Tam & Hyman (1973), who considered the migration of a slightly deformed elastic sphere in a simple shear flow. No theoretical results are available for time-dependent, oscillatory, or pulsatile flows.

In the present review, we consider the lateral motion of a deformable drop only in a steady unidirectional undisturbed shear flow of a Newtonian fluid at zero Reynolds number. This problem is advantageous as a first example for application of the theory of the preceding section since no constitutive approximations (or other "ad hoc" assumptions) are required provided that we restrict our attention to "clean" systems where the interface can be adequately characterized in terms of a single constant value of interfacial tension. The presence of surface-active "dirt" would complicate the problem greatly. Not only would the interfacial tension be non-uniform and dependent on the local velocity field near the interface (cf Harper 1972), but there is also some evidence of a need to include dissipative processes at the interface when surfactant is present (i.e. interfacial viscosity or viscoelasticity) and this would require "ad hoc" constitutive assumptions. These complications have not been considered in any of the existing papers on "drop migration," but fortunately are apparently unimportant since there is good agreement between the existing theories (which neglect them) and experiments in real fluid systems (where surface-active materials are almost inevitably present).

From a qualitative point of view, it is known that migration is possible in a Stokes flow only if the geometry of the deformed drop is asymmetric relative to the undisturbed streamlines (or the walls) and this would seem to suggest that a drop must *always* move in the direction of the "center-line" where the deformed droplet will be symmetrically oriented relative to the walls. However, the mechanisms of migration are more subtle than suggested by this simple picture. In the case of a neutrally buoyant drop, there are two "competing" mechanisms for migration; one is a direct consequence of hydrodynamic interaction between the drop and the walls of the apparatus and the other is a result of interaction between drop deformation and gradients of shear rate in the undisturbed flow. If the drop is not neutrally buoyant (and the undisturbed flow is vertical) the details of both of these mechanisms change. Furthermore, there is almost certainly an additional lateral velocity component associated with *sedimentation* of the deformed drop in this case.

The most general theories of drop migration have considered either a steady two-dimensional or an axisymmetric unidirectional undisturbed flow with a quadratic dependence on lateral position. Initially, it is advantageous to focus on the simpler two-dimensional problem in which the flow is assumed to be bounded by two parallel infinite plane walls. In this case, the undisturbed flow relative to a point fixed at the centroid of the particle may be expressed in the form

$$\mathbf{V} = (\alpha + \beta x_3 + \gamma x_3^2)\,\mathbf{e}_1 - \mathbf{U}_p. \tag{33}$$

Examples of common flows described by this expression are linear shear flow where

$$\alpha = V_w s, \quad \beta = V_w \varepsilon, \quad \gamma = 0 \tag{34a}$$

and a plane Poiseuille flow where

$$\alpha = 4\,V_{max} s(1-s), \quad \beta = 4\,V_{max}(1-2s)\varepsilon, \quad \gamma = -4\,V_{max} s^2 \varepsilon^2. \tag{34b}$$

Here V_w is the linear velocity of one of the boundaries in shear flow (the second bounding wall is assumed to be *fixed*) and V_{max} is the centerline velocity of the Poiseuille flow, both V_w and V_{max} being measured relative to some fixed laboratory reference frame. The variable s is the distance from a wall scaled by the total distance between the walls, $0 \leqq s \leqq 1$, and ε is the ratio of drop radius to gap width, $\varepsilon \equiv a/d$, which is assumed to be small in all existing theories.

The migration of a neutrally buoyant drop in a linear shear flow, $\gamma \equiv 0$, was first analyzed by Chaffey et al (1965) who considered the case of a single plane wall. These authors used the method of reflections to approximate the hydrodynamic interactions between the drop and the wall, and a full solution of the first-order corrections to the velocity and pressure fields due to drop deformation, to calculate the lateral migration velocity. The restriction to small deformation, which is inherent in the analytical development of the previous section, was also adopted by Chaffey and co-workers (as well as in all subsequent migration theories) and was assumed to be a consequence of a large interfacial tension. In this case, the drop shape is quasi-steady in a steady unidirectional shear flow, changing only as the position of the drop changes. Chaffey et al (1965) found that a drop would migrate away from the wall, in apparent and qualitative agreement with experimental observations. In the other possible limit of very large internal viscosity (cf Taylor 1932, 1934 and Frankel & Acrivos 1970), the deformation may also be small, but the problem is inherently unsteady. What is more relevant, however, is that the migration velocity in the latter case could be at most second-order in the inverse viscosity ratio, $\kappa^{-1} \ll 1$, since the first-order approximation to the drop shape has the major axis of the deformed drop aligned with the axis of the undisturbed velocity vector.

The motion of a deformable drop in a unidirectional shear flow *with a shear gradient* (i.e. $\gamma \neq 0$ in the two-dimensional case cited above) was first considered by Haber & Hetsroni (1971) and Hetsroni et al (1970) for both neutrally buoyant and non-neutrally buoyant drops in a vertical tube. Later, the neutrally buoyant problem was repeated by Wohl (1976) and Wohl & Rubinow (1974) for plane and axisymmetric Poiseuille flow. All of these analyses of migration in flows with a shear gradient neglect hydrodynamic interactions between the drop and bounding walls. Further, all were done by calculating the full corrections to the velocity and pressure fields at first-order in the appropriate deformation parameter, instead of the simpler reciprocal theorem approach outlined in the previous section.

The theories mentioned above, with the exception of Hetsroni et al (1970) and Haber & Hetsroni (1971), predict migration of a neutrally buoyant drop in the direction of the centerline (i.e. away from the walls) in Poiseuille or Couette flow, and this appears qualitatively consistent with the experimental results cited earlier. Haber & Hetsroni predicted that a neutrally buoyant drop would move radially outward in Poiseuille flow, but their analysis appears to contain algebraic errors (cf Wohl & Rubinow 1974, Chan & Leal 1979). The theories of Hetsroni and co-workers also suggest that migration can occur in *either* direction when the drop is *not* neutrally buoyant depending on the particular conditions. However, this conclusion must be considered as tentative since it is not clear whether the errors evident in the neutrally buoyant calculations are also present in the non-neutrally buoyant case. With this exception, it appeared after Wohl (1976) that the lateral-migration problem for deformable drops was solved. Closer inspection, however, revealed a number of apparent problems with the previous theories, even for neutrally buoyant drops. First, none provided a good fit to any of the actual experimental trajectory data—in spite of statements to the contrary in the original papers. Second, the results of Wohl (1976) and Wohl & Rubinow (1974) showed a ten-fold difference in the predicted migration velocities for axisymmetric and planar Poiseuille flow, all other conditions being identical, and this caused us to doubt the accuracy of their analyses. Third, the linear shear-flow theory of Chaffey et al (1965) was only for a single plane boundary, whereas available experimental data pertain primarily to circumstances where there are either two walls or a circular tube as the bounding surface. For these reasons, the problem of migration of a neutrally buoyant drop was recently reinvestigated by Chan & Leal (1979).

The analysis of Chan & Leal (1979) considers two distinct cases. First is the lateral migration in an *unbounded* quadratic unidirectional shear flow of the type given by Equation (33) with $\gamma \neq 0$. Second is the migration in a linear shear flow with $\gamma = 0$ in a region bounded by two *plane* walls. In the first case the migration velocity was predicted to be

$$(U_p)_3 = -\frac{\beta\gamma}{(1+\kappa)^2(2+3\kappa)}\left\{\left[\frac{16+19\kappa}{42(2+3\kappa)(4+\kappa)}(13-36\kappa-73\kappa^2-24\kappa^3) + \frac{10+11\kappa}{105}(8-\kappa+3\kappa^2)\right]+O(\varepsilon)\right\}\delta. \tag{35}$$

It may be noted that migration in an unbounded flow is predicted to occur only if there is curvature in the undisturbed velocity profile. Indeed, it may be noted that contributions from the bounding walls appear for this case ($\gamma \neq 0$) as a correction to the expression (35) at higher order in ε. The *rate* of migration can be seen to depend on the *magnitude* of deformation, as indicated by the scaling factor δ, as well as the viscosity ratio κ and the length-scale ratio $\varepsilon = a/d$. If $\beta = O(\varepsilon)$ and $\gamma = O(\varepsilon^2)$, as in Equations (34), the predicted migration velocity is thus $O(\varepsilon^3)$. Surprisingly, the *direction* of migration depends on the viscosity ratio. For $\kappa < 0.5$ or $\kappa > 10$, $(U_p)_3$ is positive and migration is predicted to occur toward the "centerline" of the undisturbed flow. For intermediate values, $0.5 < \kappa < 10$, on the other hand, migration is predicted to occur in the direction *toward* the nearest wall. Migration toward the centerline, for small κ, is qualitatively consistent with existing observations. Migration toward the apparatus boundaries has *not* previously been reported; however, neither have any *experiments* yet been done for κ in the range $0.5 < \kappa < 10$. Equation (35) is different in detail from that of Wohl (1976) for the same problem. However, a pointwise numerical comparison for

$0.01 \leqq \kappa \leqq 100$ has shown that the quantitative behavior of the two expressions is very similar.

In the class of *unbounded* quadratic flows, Chan & Leal (1979) have also considered the migration of slightly deformed drops in an *axisymmetric* Poiseuille flow. The predicted migration velocity in this case is

$$(U_p)_3 = -\frac{\beta\gamma}{(1+\kappa)^2(2+3\kappa)}\left\{\left[\frac{3}{14}\cdot\frac{16+19\kappa}{2+3\kappa}(1-\kappa-2\kappa^2) + \frac{10+11\kappa}{140}(8-\kappa+3\kappa^2)\right]+O(\varepsilon)\right\}\delta. \tag{36}$$

This expression differs significantly from the result obtained by Wohl & Rubinow (1974), also for axisymmetric Poiseuille flow. In particular, apart from differences of detail, Wohl & Rubinow predict migration velocities that exceed (36) by a factor of approximately 10. Experimental evidence (which we shall discuss briefly below), as well as comparison with the two-dimensional Equation (35), suggests strongly that the more recent result of Chan & Leal (1979) is correct.

We have noted above that the effect of the walls is not of direct importance in the lateral migration of drops, provided the profile curvature γ is at least $O(\varepsilon^2)$, and the other conditions of the existing theories are satisfied, namely $\varepsilon \ll 1$ and $\delta \ll 1$. In a linear velocity field where $\gamma \equiv 0$, however, the expressions (35) and (36) vanish, thus indicating that migration must be weaker (if it is to occur at all) and due completely to hydrodynamic interaction between the deformed drop and the walls. Chan & Leal have shown that the migration velocity in this case takes the form

$$(U_p)_3 = \varepsilon^2\beta^2\left[\frac{3(16+19\kappa)(54+97\kappa+54\kappa^2)}{4480(1+\kappa)^3}\right]\left\{\frac{1}{s^2}-\frac{1}{(1-s)^2}+2(1-2s)\right\}. \tag{37}$$

Unlike the expression (35), the predicted *direction* of migration in a linear shear flow is toward the centerline for *all* values of the viscosity ratio κ. It may be noted, however, that the magnitude of this wall-induced migration is only $O(\varepsilon^4)$, which is asymptotically small relative to the contribution of profile curvature except when $\gamma < O(\varepsilon^2)$. The corresponding expression of Chaffey et al for *one* wall is similar to (37) in the limit $s \to 0$, except for an erroneous multiplicative factor of 11 in their work.

The only quantitative experiments have been carried out in a circular tube geometry, where the results can be compared directly with (36), and in a Couette flow device intended to approximate a simple linear shear flow. Interpretation of data in the latter case is complicated by the fact that the undisturbed velocity profile is neither exactly unidirectional nor linear due to the curvature of the boundaries. Since γ is generally small in this case, i.e. less than $O(\varepsilon^2)$, the contributions of the general types (35) and (37) to the migration velocity may be of comparable magnitude. Detailed analysis shows that the contribution due to profile curvature in the Couette device is

$$(U_{\mathrm{p}})_3 = \delta\left[-\frac{2(4+61\kappa+85\kappa^2+25\kappa^3)}{7(2+3\kappa)(1+\kappa)^2}\right]\frac{A_2^2 a^3}{R_0^5} \tag{38}$$

where

$$A_2 = \frac{R_1^2 R_2^2(\Omega_1-\Omega_2)}{(R_2^2-R_1^2)\,Ga}$$

and R_0 is the distance from the axis of the Couette device to the center of the sphere. This result predicts that the profile-curvature effect in a Couette device will always cause migration toward the *inner* cylinder. It may be noted that this rigorous result is *opposite* in direction to the prediction that would be obtained using (35) directly with a local two-dimensional approximation to the velocity field, cf Ho & Leal (1976). When the shear-gradient contribution, (38), is combined with the contribution from hydrodynamic interactions between the particle and wall, (37), a drop is predicted to migrate to an equilibrium position between the centerline and the inner wall of the Couette device. Under the conditions of *existing* Couette flow experiments, however, the wall-interaction contribution is *numerically* dominant and the *predicted* equilibrium position is quite near the centerline. This conclusion agrees very well with the experimental observations of Karnis & Mason (1967). Chan & Leal (1979) have in fact made a detailed comparison of the theoretically predicted trajectories, based on the results described above, and the observed trajectories both in Poiseuille flow (Goldsmith & Mason 1962), and in a Couette flow (Karnis & Mason 1967). In both cases, the theory and experiment show excellent agreement.

B *The Effects of Inertia on Particle Motions*

The second class of problems we shall discuss in some detail are those involving weak inertia effects on the motion of small particles. In many ways, this class of problems is more interesting than that described in the preceeding section. The range of phenomena is much wider, including lateral migration, orbital drift, and rotational effects in sedimentation. The subtlety of the necessary analysis is made much greater by the fact that the expansion procedure can be either regular or singular depending upon the positions of any boundaries. Furthermore, the number of possible mechanisms for each phenomenon is larger for rigid particles because they can be subjected both to external forces and couples. This wealth of possible variations in conditions, as well as the spectacular nature of some of the experimental observations, has resulted in a large number of fundamental studies, both experimental and theoretical. This is particularly true of the lateral-migration problem where the early observation by Segre & Silberberg (1962b) of an equilibrium position midway between the centerline and the walls for a sphere in Poiseuille flow sparked a flurry of activity. The problem of "orbital drift" for an axisymmetric particle in a linear shear flow has occupied a central position in low-Reynolds-number hydrodynamics since the original work of Jeffery (1922), who suggested that among all possible orbital motions, a particle subject to weak inertia (or other possible departures from Stokes-flow assumptions) would adopt that orbit corresponding to a *minimum* average rate of viscous dissipation.

It is somewhat surprising to discover that in spite of this high level of activity and visibility, there are still considerable gaps of information on both the experimental and theoretical sides. There has been no work, for example, on inertial effects in lateral migration for particles other than rigid spheres or for flows that are not steady. There has been little theory or experiment on sedimentation of nonspherical particles in the low-Reynolds-number range in spite of the well-known indeterminacy in orientation. Similarly, the theory of "orbit drift" due to inertia is still relatively limited although 57 years have elapsed since Jeffery's initial, thought-provoking proposal.

The goal of this section, following the precedent of the previous one, is to review the present state of knowledge of weak inertial effects on the motion of small particles, with particular emphasis on the theoretical side of the various problems. A detailed and still reasonably current review of the experimental side may be found in the article of Goldsmith & Mason (1967). Other more recent experimental works have generally been discussed in the references cited below.

1 SEDIMENTATION IN AN UNBOUNDED FLUID The simplest class of problems in which weak inertial effects are known to play a key role is the sedimentation of single particles through a quiescent Newtonian fluid. We confine our attention here to homogeneous particles of sufficiently simple shape that hydrodynamic coupling between translation and rotation is absent. Examples include an ellipsoid, a sphere, a body of revolution with fore-aft symmetry, and a transversely isotropic particle, such as the tri-dumbbell. Two types of indeterminacy occur for

such particles in sedimentation in the Stokes-flow regime: first, the lateral position of the particle relative to vertical bounding walls is determined for all time by its initial position (though not *fixed* at that position unless the particle is spherically isotropic or falling in a fixed orientation parallel to one of its symmetry axes); second, the orientation of a nonspherical rigid particle is either fixed at its initial value in an unbounded fluid, or is determined for all time by its initial orientation in the presence of bounding walls. The influence of weak inertial effects on the particle *position* will be discussed in the next subsection where we consider "lateral migration" phenomena induced by inertia. Here, we consider the influence of weak inertia on the particle orientation in an unbounded fluid.

It is well known that rigid particles settling in a quiescent Newtonian fluid at finite Reynolds number will eventually adopt an orientation that is independent of the initial orientation of the particle. In the case of axisymmetric rigid particles this equilibrium orientation has the axis of symmetry either horizontal (i.e. normal to the direction of motion) or vertical depending upon whether the particle is prolate or oblate in shape. Although there have been many experimental studies of particle settling, especially in the chemical engineering literature, that qualitatively corroborate this fact (cf Pettyjohn & Christiansen 1948, Becker 1959) there are none of which we are aware that are quantitative in the low-Reynolds-number regime.

The most important theoretical investigation of the influence of slight inertia on the orientation of sedimenting nonspherical particles is due to Cox (1965). Using the method of matched asymptotic expansions and a generalization of the reciprocal theorem approach of Section 2 (cf Brenner & Cox 1963), Cox was able to derive general formulae, in terms of "intrinsic" resistance coefficients, for the force and torque on a single particle of arbitrary shape moving with both translation and rotation in an infinite quiescent fluid, correct to terms of $O(\mathrm{Re}^2 \ln \mathrm{Re})$. In the case of a centrally symmetric body translating (*without* rotation) with velocity $\mathbf{V}$, Cox's results for the force and torque can be shown to reduce to the form

$$\mathbf{F} = \mathbf{A}\cdot\mathbf{V} + (\text{terms of order } \mathrm{Re}^2 \text{ and } \mathrm{Re} \ln \mathrm{Re})$$
$$\mathbf{G} = \frac{1}{6\pi}\,\mathrm{Re}\,\{\mathbf{K} : \mathbf{VV}\}. \tag{39}$$

Thus, the couple $\mathbf{G}$ on the particle is identically zero at $O(1)$ and the particle would translate with a fixed orientation in the absence of inertia, as noted above. However, there *is* a nonzero couple acting on the particle at $O(\text{Re})$, and this will induce rotation unless the particle is constrained by application of an "external" couple. The possible existence of one (or more) equilibrium orientations for an unconstrained particle is thus seen to depend on the existence of zeros in

$$\mathbf{K}:\mathbf{VV}$$

for some specific orientations. In order to determine whether any such orientations exist, it is necessary to consider a particle of specific shape for which **K** can be calculated. Cox (1965) examined the case of a spheroid of small eccentricity, i.e.

$$r_S = 1 + eS_2(\theta,\phi), \tag{40}$$

where $S_2(\theta,\phi)$ is a surface spherical harmonic of order 2, and $|e| \ll 1$. In terms of a coordinate system fixed with the 1 axis along the axis of symmetry and the 2 axis in the plane of particle motion, we can write

$$\mathbf{V} = (V \cos \alpha,\ V \sin \alpha,\ 0), \tag{41}$$

where α is the angle between the axis of symmetry and **V**. In this case, detailed evaluation of **K** yields

$$G_1 = G_2 = 0, \quad G_3 = -\tfrac{29}{40} e \,\text{Re}\, V^2 \sin \alpha \cos \alpha + O(\text{Re}^2).$$

Since any angular motion induced by G_3 is $O(\text{Re})$, it can be shown that a freely suspended particle would rotate with angular velocity

$$\Omega_3 = -\tfrac{87}{160} e \cdot \text{Re} \cdot V^2 \cdot \sin \alpha \cos \alpha. \tag{42}$$

Since $\Omega_3 = -d\alpha/dt$, it thus follows that

$$\alpha \to 0 \quad \text{as} \quad t \to \infty \quad \text{for} \quad e < 0$$

and

$$\alpha \to \pi/2 \quad \text{as} \quad t \to \infty \quad \text{for} \quad e > 0.$$

Two positions of stable equilibrium exist for a nearly spherical spheroid. A prolate spheroid ($e > 0$) is predicted to sediment at equilibrium with its axis of symmetry *normal* to **V**, whereas an oblate spheroid will rotate until the axis of symmetry is parallel to **V**. In both cases, the action of inertia is thus to yield the orientation corresponding to maximum dissipation for a given $|\mathbf{V}|$.

This theoretical result agrees with the available qualitative experimental observations. However, the restriction to a nearly spherical spheroid greatly restricts any possible practical utility of the theory. Furthermore, it is possible (though unlikely) that predictions for bodies of more general shape (though still centrally symmetric) might exhibit qualitatively different behavior. However, no additional work has been done on the problem so far as we are aware.

2 ORBIT DRIFT FOR TRANSVERSELY ISOTROPIC PARTICLES IN SIMPLE SHEAR FLOW A second problem in which inertia can exert a dominant role in the ultimate steady-state configuration is the rotational motion of rigid transversely isotropic particles in a simple shear flow. The motion of a general ellipsoidal particle at zero Reynolds number was first investigated 57 years ago in the now classical paper of Jeffery (1922). Jeffery's solution [and Bretherton's (1962) much later analysis for a more general class of particle shapes] shows that an isolated transversely isotropic particle in a Newtonian fluid at zero Reynolds number will rotate indefinitely on one of an infinite one-parameter family of periodic orbits in which the major axis of the particle traverses around the vorticity axis of the undisturbed flow. The orbit of rotation for a particular particle is denoted by the orbit constant C, $0 \leqq C \leqq \infty$, and is thus determined for all time (at zero Reynolds number) by the orientation of the particle at some initial instant. It is the lack of any intrinsically preferred orbit that constitutes the "indeterminacy" of the Stokes-flow solution in this case.

The first systematic experimental investigation of the motion of a rigid axisymmetric particle in shear flow at small but nonzero Reynolds numbers was carried out by Karnis, Goldsmith & Mason (1963, 1966), who examined the rotation of single rods and disks in both Poiseuille and Couette flows. Much earlier, Taylor (1923) had reported that prolate and oblate spheroids in Couette flow assumed the orbits of *minimum* energy dissipation, in apparent agreement with Jeffery's hypothesis. However, Mason and his co-workers showed conclusively that the effect of inertia was to cause rods and disks to gradually drift into the orbits of maximum dissipation, namely $C = \infty$ for rods in which the axis of the particle rotates completely in the plane of the flow, and $C = 0$ for disks in which the particle spins with the axis of symmetry collinear with the vorticity vector of the undisturbed flow. Later, Harper & Chang (1968) obtained the same result for rotations of dumbbell-shaped particles in a Couette flow.

The only satisfactory theory of inertia-induced orbit drift in shear flow was also reported by Harper & Chang (1968) for a neutrally buoyant dumbbell in a *linear* shear flow. Harper & Chang did *not* use the reciprocal theorem approach of Section 2, but instead calculated the motion of the dumbbell by generalizing the result obtained earlier by Saffman (1965) for the lift on a single sphere translating in an unbounded linear shear flow. When expressed in terms of the time rate of change of the polar angle θ_1 between the vorticity axis and the axis of the particle, and of the angle ϕ_1 between the plane defined by the vorticity and symmetry axes and the direction of the velocity gradient (cf Leal 1975), Harper & Chang's result was

$$\frac{d\phi_1}{dt} = \cos^2 \phi_1 + \tfrac{2}{3}e^2 + \sin \phi_1 \cos \phi_1 \sin^2 \theta_1 \{U \cos^2 \phi_1 - S \sin^2 \phi_1 + (A-D) \sin \phi_1 \cos \phi_1\} \text{ Re} \tag{43a}$$

and

$$\frac{d\theta_1}{dt} = \sin \theta_1 \cos \theta_1 \sin \phi_1 \cos \phi_1 \{1 + \sin^2 \theta_1 [(S+U) \sin \phi_1 \cos \phi_1 + A \sin^2 \phi_1 + D \cos^2 \phi_1 - B] \text{ Re}\}. \tag{43b}$$

Here e is the ratio of the bead (sphere) radius to the bead separation in the dumbbell, and A, B, D, S, U are components of the dimensionless lift tensor

$$\mathbf{L} = \begin{bmatrix} B & 0 & 0 \\ 0 & D & S \\ 0 & U & A \end{bmatrix}$$

defined by the expression for the force on one of the beads,

$$\mathbf{F} = 6\pi\, \mathbf{U} + \text{Re}(\mathbf{L} \cdot \mathbf{U}) \tag{44}$$

relative to axes in which the undisturbed shear flow is

$$\mathbf{V} = y\, \mathbf{k}.$$

The velocity of the bead *relative* to the local undisturbed velocity $\mathbf{V}$ is denoted as $\mathbf{U}$.

Transformation of (43a) and (43b) to the "natural" C, τ system of dependent variables, first introduced by Leal & Hinch (1971), allows a more straightforward examination of the effect of inertia on the orbital motion of the dumbbell. Thus we employ

$$\tau = \tan \phi_1, \quad C = \tan \theta_1 \cos \phi_1$$

and transform according to

$$\frac{dC}{dt} = \sec^2 \theta_1 \cos \phi_1 \dot{\theta}_1 - \tan \theta_1 \sin \phi_1 \dot{\phi}_1.$$

The result is

$$\frac{dC}{dt} = \text{Re}\, [\sin^2 \theta_1 \sin^2 \phi_1 \cdot S + (D-B) \sin^2 \theta_1 \sin \phi_1 \cos \phi_1]\, C. \tag{45}$$

It can be seen that the "stability" of the orbit is determined by the first term on the right. If $S = 0$, the orbit constant only oscillates periodically in magnitude but remains constant when averaged over ϕ_1 from 0 to 2π. When $S \neq 0$, on the other hand, the orbit tends asymptotically to either $C = 0$ or $C = \infty$ depending upon the sign of S. Harper & Chang have shown that $S > 0$ for the dumbbell, and in this case the orbit tends to $C = \infty$, which is the orbit of maximum dissipation. This result is in qualitative agreement with the experimental observations of Karnis, Goldsmith & Mason (1967) for rodlike particles in simple shear flow. However, it is clearly restricted insofar as particle geometry is concerned and it would be advantageous to examine other cases to verify the generality of the result. Two obvious choices for such an analysis would be the nearly spherical spheroid, examined by Cox (1965) for sedimentation orientation, and the slender rodlike particle for which "slender-body theory" could be applied to obtain the necessary creeping-motion solutions.

3 LATERAL MIGRATION PHENOMENA A third class of weak inertia effects on single particle dynamics, with potentially important consequences in the flow properties of suspensions, is the existence of lateral translational motions either *across* streamlines of an undisturbed flow, or normal to the direction of action of the body force in sedimentation through a quiescent fluid. Following the studies of Segre & Silberberg (1962a,b, 1963), who showed that a neutrally buoyant sphere would translate across streamlines of a Poiseuille flow to an equilibrium position 60% of the way from the centerline to the tube walls, many experimental investigations were carried out to study the effects of buoyancy (Repetti & Leonard 1964, 1966, Jeffery & Pearson 1965), restricted particle rotation (Oliver 1962), different types of undisturbed flows including Couette flow and plane Poiseuille flow (Goldsmith & Mason 1967), time-dependent flows (Takano & Mason 1966, Shizgal, Goldsmith & Mason 1967), migration in a quiescent fluid (Vasseur & Cox 1977), and particles of different shapes (Karnis, Goldsmith & Mason 1963, 1966). This work is effectively reviewed by a number of authors including Brenner (1966), Cox & Mason (1971), Goldsmith & Mason (1967), and Goldsmith & Skalak (1975) and need not be discussed here in detail. Qualitatively we may note

1. Non-neutrally buoyant spheres in a vertical Poiseuille flow (either axisymmetric or plane) will migrate either to the centerline or to the walls depending on whether the particle leads or lags the local undisturbed motion.

2. Neutrally buoyant spheres migrate to the centerline in Couette flow.
3. Sedimenting spheres migrate towards the centerline (or center plane) of the container.
4. Axisymmetric, but nonspherical, particles generally show the same behavior in all shear flows as a sphere provided that the data are averaged in time to remove any periodic lateral motion associated with the periodic rotation of the particle.
5. Particles migrate the same, in a time-averaged sense, in oscillatory or pulsatile flows as in a steady flow in the same device.
6. Particles inhibited in rotation by some external field coupling show similar qualitative behavior to a freely rotating particle, but with modified rates of migration and/or final equilibrium positions.

Thus, unlike some of the problems discussed earlier, there is a wealth of experimental information on inertia-induced particle migrations. This has, in turn, generated a large amount of theoretical work, though it is still limited to rigid spherical particles and steady undisturbed flows.

Two classes of theoretical analyses have, in fact, been reported: those for *bounded* and *unbounded* domains. The majority of the earliest work was concerned with the latter class of problems, but has been adequately reviewed by earlier authors (see Brenner 1966) and will only be briefly discussed here. The first analysis of this type was due to Rubinow & Keller (1961), who used the method of matched asymptotic expansions to calculate the lift force due to inertia on a rigid sphere that simultaneously translates and rotates (in orthogonal directions) in an unbounded stationary fluid. This analysis was subsequently generalized by Cox (1965) for translation in an arbitrary direction relative to the rotation axis. Later, Saffman (1965) considered the inertia-induced lift force on a sphere in a linear shear flow, when the sphere translates parallel to the undisturbed flow with relative velocity V and rotates with angular velocity $\mathbf{\Omega}_p$ parallel to the undisturbed vorticity. Saffman assumed

$$1 \gg \mathrm{Re}_G \geqq \mathrm{Re}_\Omega \gg \mathrm{Re}_v \tag{46}$$

where

$$\mathrm{Re}_G \equiv \frac{Ga^2}{\nu}, \quad \mathrm{Re}_\Omega \equiv \frac{|\Omega_p|\,a^2}{\nu}, \quad \mathrm{Re}_v = \frac{Va}{\nu}.$$

Here a is the sphere radius and G the mean velocity gradient. The migration velocity predicted by Saffman is larger by $O(\mathrm{Re}_G^{-\frac{1}{2}})$ than that predicted by Rubinow and Keller for $|\Omega_p|$ that is smaller than or comparable in magnitude to $G/2$. Thus, in the presence of a velocity gradient, under the condition (46), a lift term of the Rubinow-Keller type will appear, among others, at the next higher order in Saffman's theory. More recently, Harper & Chang (1968) generalized Saffman's result to bodies of arbitrary shape and also considered more general directions of translation relative to the undisturbed shear flow. Finally, Drew (1978) used the same matched asymptotic methods to consider the lift force on a sphere that translates and rotates relative to either a pure rotational plane linear flow or a pure straining plane linear flow (i.e. hyperbolic flow). The direction of translation is in the plane of the undisturbed flow, but is otherwise arbitrary, while the direction of rotation is normal to this plane. In the case of the pure rotational flow, Drew found that there is only a drag correction, but no lift. For the pure straining flow, on the other hand, there is both a correction to Stokes's drag and a lift force of $O(\mathrm{Re}^{\frac{1}{2}})$.

As interesting as the various results for an *unbounded* fluid may be, however, they take no account of the presence of boundaries or of the nonlinearities in the undisturbed velocity profiles. Since either (or both) of these effects may result in the appearance of contributions to the "lift" force on the particle that are larger than any of the contributions for an unbounded fluid, the majority of more recent studies have focused on predictions of migration velocities taking explicit account of the walls and profile curvature. It may be remarked that the ratio of the lateral force in an *unbounded* fluid to that calculated for any *bounded* shear flow with $\varepsilon \equiv a/d$ small (i.e. migration due to wall effect and profile curvature) is $O(\overline{\mathrm{Re}}^{\frac{1}{2}})$, where $\overline{\mathrm{Re}}$ is the Reynolds number of the mean flow. Thus, wall effects and profile curvature provide the dominant migration mechanisms when the Reynolds number of the *mean flow* is small. This condition has been satisfied in the majority of experimental, laboratory studies of lateral migration, but may not be true in many technological situations. In the latter case, the theoretical results for an *unbounded* fluid could be used as long as $\varepsilon \equiv a/d$ is small. Let us now consider the case of *bounded* flows for $\varepsilon \ll 1$ and $\overline{\mathrm{Re}} \ll 1$.

Lateral migration phenomena in bounded flows can be divided into two classes, one which occurs in an undisturbed shear flow and the other in sedimentation through a quiescent fluid. In addition, within each class there are two possibilities. First, the bounding walls may be sufficiently close to the particle that viscous forces dominate inertia everywhere in the flow domain. In this case, the reciprocal-theorem approach of Section 2 can be applied directly. The second subclass is then characterized by the fact that the walls are far enough away from the particle that inertia becomes significant in the region near the walls, and in this case the most straightforward basis for analysis is via matched asymptotic expansions. The general framework for the first of these two subclasses was originally outlined, for rigid spherical particles, by Cox & Brenner (1968). However, these authors only obtained general forms for the lateral velocity (or force) but did *not* evaluate the detailed functional form of the coefficients in these general expressions and could thus say nothing about the direction, rate, or spatial dependence of the migration velocity.

In the case of sedimenting spheres in a quiescent fluid, the detailed evaluation of Cox & Brenner's formulae was recently carried out by Cox & Hsu (1977) and Vasseur & Cox (1976). The corresponding Class-2 problem of migration in sedimentation when the inertial domain lies inside the walls was considered by Vasseur & Cox (1977). Cox & Hsu (1977) examined the inertia-induced lateral migration of a spherical particle sedimenting near a single vertical plane wall, under the conditions $\mathrm{Re}/\varepsilon \ll 1$, $\varepsilon \ll 1$, $\mathrm{Re} \ll 1$ and showed that the direction of migration is away from the wall. Here $\mathrm{Re} \equiv Va/\nu$, where $\mathbf{V}$ is the sedimentation velocity and $V = |\mathbf{V}|$. We note that the first of the conditions (46) is necessary in order to insure that the walls lie inside the viscous subdomain of the disturbance velocity field, as assumed by Cox & Brenner (1968). This analysis was generalized by Vasseur & Cox (1976), who studied the case of two parallel vertical plane walls. In both cases, the integral-transform technique, mentioned in Section 2, was used to solve the basic and complementary Stokes-flow problems. This method, which requires that the particle be replaced by a *point* force, is useful if the dominant contribution to the volume integrals in (32) comes from the region near the walls where $\tilde{r} = r/\varepsilon = O(1)$—and this is true for the inertia-induced lateral migration problem. The migration velocity obtained by Vasseur & Cox (1976) takes the form

$$U_{\mathrm{p}} = V \,\mathrm{Re}\, f(s),$$

where s is the dimensionless position of the sphere scaled with respect to the distance d between the walls. The function $f(s)$ is positive for $\frac{1}{2} < s \leqq 1$, negative for $0 \leqq s < \frac{1}{2}$, and exactly zero at $s = \frac{1}{2}$. Thus, the particle is predicted to migrate away from the walls to the midpoint between the two walls. This prediction seems to be in qualitative agreement with the observations of Karnis, Goldsmith & Mason (1966) for a sphere settling in a circular tube, though the theory has not in fact been generalized from the two- to three-dimensional geometry. Vasseur & Cox (1977) later considered inertial migration of a solid sphere in sedimentation when the condition $\mathrm{Re}/\varepsilon \ll 1$ is *not* satisfied so that the walls lie outside the viscous subdomain of the disturbance velocity field. Both one- and two-wall cases were examined. In the one-wall case, the lateral migration velocity was found to be *dependent* on both position and Re in the form

$$U_{\rm p} = \tfrac{3}{32} V \,{\rm Re} \left\{ 1 - \tfrac{11}{32} \left(\frac{lV}{\nu} \right)^2 + \ldots \right\} \tag{47}$$

for $lV/\nu \ll 1$, and

$$U_{\rm p} = \tfrac{3}{8} V \,{\rm Re} \left\{ \left(\frac{\nu}{lV} \right)^2 + 2.21901 \left(\frac{\nu}{lV} \right)^{\frac{5}{2}} + \ldots \right\} \tag{48}$$

for $lV/\nu \gg 1$. Intermediate values of lV/ν were evaluated numerically—note that the l is the dimensional separation distance between the particle and the wall. A curiosity of the latter result is that the migration velocity is independent of the sedimentation velocity V to first order in ν/V. The fact that $U_{\rm p} > 0$ for all l means that the sphere will migrate away from the wall at all positions. It may also be noted that the limit $(3/32)V$ Re for $lV/\nu \to 0$ "matches" the result of Cox & Hsu for a single plane wall in which the wall is close enough to the particle to be within the viscous subdomain of the disturbance flow while still keeping $\varepsilon \ll 1$ so that the point-force approximation can be used to determine the necessary Stokes solutions as described above. The migration velocity for a sphere between *two* plane walls, both in the *inertial* subdomain of the disturbance velocity field, takes a more complicated form

$$U_{\rm p} = V \,{\rm Re}\, g\left(\frac{lV}{\nu}, s \right) \tag{49}$$

where s is the distance from one wall scaled with d, i.e. $s = l/d$, as before, and this expression had to be evaluated numerically. It may be noted, however, that $g = 0$ for $s = \frac{1}{2}$ and is positive for $0 \leqq s < \frac{1}{2}$ and negative for $\frac{1}{2} < s \leqq 1$ as before. Furthermore g *increases* with decrease in s between 0 and $\frac{1}{2}$, for any fixed value of lV/ν, but *decreases* with increase in lV/ν for any fixed value of s. This theory of Vasseur & Cox (1977) reduces to their earlier theory in the limit $lV/\nu \to 0$ for all s when the particle is close enough to the wall to be inside the inner (viscous) region of expansion. Vasseur & Cox (1977) also report qualitative *experimental* results that verify the basic features of their theory.

The first detailed theoretical analysis of lateral migration in shear flows was reported by Ho & Leal (1974), who considered the motion of neutrally buoyant rigid spheres in a unidirectional quadratic undisturbed flow of the type (33), between two parallel plane boundaries. The analysis assumed

$${\rm Re} \ll \varepsilon^2$$

where, in this case, the appropriate Reynolds number is Ga^2/ν, so that the walls lie inside the inner (viscous) region of the small-Re expansion. It may be noted that the condition $\mathrm{Re} \ll \varepsilon^2$ is equivalent to requiring $\overline{\mathrm{Re}} \ll 1$, where $\overline{\mathrm{Re}}$ is the Reynolds number of the undisturbed velocity field and this condition, in turn, is the same as the condition that is required in order that the wall and profile curvature contributions to migration dominate the contributions that would exist in an unbounded fluid. Ho & Leal (1974) utilized the method of reflections to calculate the disturbance and complementary Stokes-flow solutions satisfying boundary conditions on the walls, thus introducing an additional requirement that the sphere not be too close to the wall. The same problem was later considered by Vasseur & Cox (1976) using the point-force integral-transform technique to represent the necessary Stokes-flow solutions. The two predicted results are similar in character, except near the walls where they differ, due possibly to the difference in assumptions inherent in the method of reflections and the integral-transform technique. Expressed in a general form, the predicted migration velocity in this case is

$$U_{\mathrm{p}} = \kappa^2 \,\mathrm{Re}\, V_{\mathrm{m}}[\beta^2 G_1(s) + \beta\gamma G_2(s)] \tag{50}$$

where V_{m} is the characteristic velocity of the undisturbed flow. The functions G_1 and G_2 depend only on the particle's position relative to the boundaries (*not* on the form of the undisturbed flow) and satisfy the symmetry conditions

$$G_1(s) = -G_1(1-s); \quad G_2(s) = G_2(1-s).$$

Further, $G_1(s)$ is positive for $0 < s < 0.5$, while $G_2(s)$ is *always* positive. The first term in (50), which represents the interaction of the disturbance stresslet and its wall correction with the undisturbed shear, thus produces an inward motion toward the centerline in all cases. In a *linear* shear flow, where $\gamma \equiv 0$, the particle is therefore predicted to migrate to the centerplane, which is its equilibrium position. The second term, on the other hand, represents the interaction between the disturbance flow (stresslet), without wall corrections, and the curvature of the undisturbed velocity profile. This term is *negative* for $s < \frac{1}{2}$ and *positive* for $s > \frac{1}{2}$, and thus produces migration *toward* the regions of highest shear rate—one or the other of the walls. In the case of a quadratic profile, like that in a plane Poiseuille flow, the two terms produce migration in opposite directions, and a more detailed analysis shows that the resulting equilibrium point is 60% of the way from the centerplane to the walls. This prediction is in qualitative agreement with the original observations of Segre & Silberberg (1962a,b) for tube flow. More importantly, however, Ho & Leal (1974) have shown that the predictions of equilibrium points and *trajectory* are in excellent quantitative agreement with the experimental observations of Halow & Wills (1970), which were carried out in a Couette flow device. Vasseur & Cox (1976) have gone on, under the same conditions on Re and ε, to consider non-neutrally buoyant particles, as well as neutrally buoyant ones that are constrained so that they cannot rotate. The detailed results, given by Vasseur & Cox (1976), appear to agree qualitatively with available experiments, the most important comparison being with the two-dimensional data of Repetti & Leonard (1964, 1966) for non-neutrally buoyant particles.

It should be noted that the problem of migration in a shear flow, including boundary effects and/or profile curvature, has *not* been done for the case in which the walls lie in the outer (inertial) domain of the perturbation solution. Neither has any analysis been reported for non-spherical particles, or any wall geometry other than infinite parallel plane boundaries. Finally, no theoretical work has yet been done on oscillatory or pulsatile flows. All of these problems deserve future attention.

C *Non-Newtonian Effects on Particle Motions*

Finally, we come to the third class of problems mentioned in the introductory section of this review, namely the effects of weak non-Newtonian fluid properties on the motion and ultimate configuration of particles and/or drops. This class of problems has been recently reviewed by Leal (1979) and we shall thus confine our discussion to a brief summary of previous studies, plus a slightly more detailed description of results for two "new" cases of particle motion that were not included in the earlier paper.

The focus of all theoretical work, to date, has been "weak, slow" flows of viscoelastic suspending fluids, where the non-Newtonian contributions to particle and fluid motions are restricted to small instantaneous departures from Newtonian behavior that can be described via the *n*th-order fluid constitutive approximation of Rivlin-Ericksen. The *n*th-order fluid "model" represents a *common* limiting form for almost all of the best known constitutive equations that are fully viscoelastic and non-linear, and this fact has generally been accepted as an indication that it will provide representative results for a wider range of materials than could *presently* be claimed for any of its more general parents. Indeed, existing theories of particle motions in the present class, which have generally been limited to only the *first* term of the *n*th-order-fluid approximation, i.e. to a "second-order" fluid, have nevertheless shown good qualitative, and in some cases quantitative, agreement with experimental results.

The existing theories, reviewed earlier by Leal (1979), where this is true include: 1. rotation of a slender axisymmetric particle in sedimentation to a vertical equilibrium position (Leal 1975), 2. orbital drift of both a slender axisymmetric particle (Leal 1975) and a rigid tri-dumbbell (Brunn 1977a) in simple shear flow, and finally 3. migration of neutrally buoyant rigid spheres and spherical drops in a quadratic unidirectional shear flow between parallel plane boundaries (Ho & Leal 1974 and Chan & Leal 1979), and in an unbounded but otherwise general quadratic undisturbed flow (Brunn 1976b and Chan & Leal 1977).

Two problems not discussed by Leal (1979), but recently solved for the second-order fluid, are the migration of a non-neutrally buoyant spherical particle in a vertically oriented unbounded quadratic shear flow, and the migration of a spherical particle that is sedimenting in a quiescent fluid between two parallel plane boundaries. The first of these problems was considered independently by Chan & Leal (1977), Brunn (1967a), and Caswell (1977) for the case when the force F on the particle is at least $O(1)$ compared with the small parameter λ that describes the ratio of non-Newtonian to Newtonian terms in the second-order fluid model. The parameter λ is the ratio of the intrinsic relaxation time scale for the fluid relative to the convective time scale of the fluid's motion. The method of analysis by Chan & Leal (1977) was precisely that described in Section 2, and the result obtained for the lateral migration velocity in a general unbounded quadratic undisturbed shear flow,

$$\mathbf{V} = \boldsymbol{\alpha}+\boldsymbol{\beta}\cdot\mathbf{x}+\gamma:\mathbf{xx},$$

is

$$\lambda\mathbf{U}_{\mathrm{p}}^{(1)} = \lambda\{(1+2\varepsilon_1)[\boldsymbol{\alpha}-\mathbf{U}_{\mathrm{p}}^{(0)}]\cdot\mathbf{e}+\varepsilon_1[\boldsymbol{\alpha}-\mathbf{U}_{\mathrm{p}}^{(0)}]\wedge\boldsymbol{\Omega}+O(\varepsilon^3)\} \tag{51}$$

where $\mathbf{U}_{\mathrm{p}}^{(1)}$ and $\mathbf{V}$ are both scaled with respect to V_{m}, as before. Here ε_1 is a dimensionless rheological parameter characteristic of the second-order fluid, which is generally assumed to have a value of approximately -0.6. It should also be noted that $\mathbf{e}$ is the rate-of-strain tensor of the undisturbed flow, while $\boldsymbol{\Omega}$ is the vorticity vector. The coefficient $\boldsymbol{\alpha}-\mathbf{U}_{\mathrm{p}}^{(0)}$ is the "slip" velocity for creeping motion of a particle in an unbounded Newtonian fluid

$$\boldsymbol{\alpha}-\mathbf{U}_{\mathrm{p}}^{(0)} = -\frac{\mathbf{F}_0}{6\pi}$$

and $\mathbf{F}_0$ is the $O(1)$ "buoyancy" force acting on the particle. As noted, the analysis leading to (51) is predicated on the assumption that $\lambda \ll 1$. The comparable results obtained independently by other authors for this same problem are slightly in error (Brunn omits the second term, whereas

Caswell has $1+\varepsilon_1$ instead of $1+2\varepsilon_1$ in the first term). Furthermore, it is noteworthy that the contribution to $\mathbf{U}_p^{(1)}$ for a pure straining flow ($\omega \equiv 0$) is predicted by (51) (and also by Brunn) to be proportional to $(1+2\varepsilon_1)$, which is itself proportional to the magnitude of the second normal stress difference in a simple shear flow. If we examine a plane ("unbounded") Poiseuille flow, the predicted direction of migration can be shown to depend upon the sign of the slip velocity. For $\alpha - U_p^{(0)} > 0$, the lateral component of $\mathbf{U}_p^{(1)}$ is negative for $0 \leqq s < \frac{1}{2}$ and positive for $\frac{1}{2} < s \leqq 1$. In this case, the particle is predicted to migrate toward the walls. For $\alpha - \mathbf{U}_p^{(0)} < 0$, on the other hand, the particle migrates toward the centerline. So far as we are aware, all of the available experiments of migration in a viscoelastic fluid were carried out for neutrally buoyant particles. Thus, we cannot presently determine whether these predictions are reasonable or not.

The second problem that has been solved recently (Chan 1979) is the migration of a Newtonian spherical drop in sedimentation through a second-order fluid between two parallel plane boundaries. The method of analyses was again taken directly from Section 2 of this paper. The result for the lateral velocity component, scaled with respect to the Stokes sedimentation velocity, is

$$U_p = \lambda \frac{3\varepsilon^2}{160(1+\kappa)^3} [(12+32\kappa+40\kappa^2+15\kappa^3) + \varepsilon_1(20+72\kappa+77\kappa^2+27\kappa^3)]\, F(s)$$

where

$$F(s) \sim \frac{1}{s^2} - \frac{1}{(1-s)^2} + 2(1-2s).$$

In this case, the *direction* of migration depends critically on the value of the rheological (second-order fluid) parameter ε_1, which is generally believed to be in the range

$$-0.5 > \varepsilon_1 > -0.6.$$

In the limit $\kappa \to 0$

$$U_p = \lambda \frac{3\varepsilon^2}{160} [12+20\varepsilon_1]\, F(s).$$

Thus, for $\varepsilon > -0.6$, an inviscid drop is predicted to migrate toward the centerplane for all values of $|\varepsilon_1 \lambda| \ll 1$. On the other hand, as $\kappa \to \infty$ we approach the result for a rigid sphere

$$U_p = \lambda \frac{3\varepsilon^2}{160} [15 + 27\varepsilon_1] F(s),$$

and in this case, the particle is predicted to migrate toward the centerplane if $\varepsilon_1 > -0.55$, but to migrate toward the nearest wall if $\varepsilon_1 < -0.55$. This suggests the interesting possibility that observation of the sedimentation of drops with various values of the viscosity ratio κ may provide the basis for a sensitive measurement of ε_1, which is remarkably elusive by other means. However, so far as we are aware, no experimental data are yet available for comparison with these predictions.

4 CONCLUSIONS

In this paper we have considered the effects of weak inertia, non-Newtonian rheology, or flow-induced shape deformation on the motions of small particles in a viscous fluid. In Section 2, a general approach, which has heretofore only appeared in part, was outlined for theoretical investigation of these effects and in Section 3 we briefly reviewed the work that has so far been published, albeit with a strong emphasis toward the existing theoretical analysis. Although much is known, even more remains to be done before all of the possible effects will be well understood. In the case of lateral migration of drops, experiments over a wider range of viscosity ratio and experiments with a viscoelastic drop would be particularly useful. Both experiment and theory are required for a wider range of particle shape in the case of inertia-induced orbit drift. The same is true with even more urgency in the case of sedimentation with inertia. Even in the heavily studied inertia-migration problem, no theory has yet been completed outside the case of steady flow between parallel plane boundaries. Apart from these and other omissions of detail, we may also add in every case the virtually unknown effects of finite particle size and of finite departures from the "standard conditions" of creeping-flow theory. Many problems, apart from the three discussed here, have hardly been studied at all, though they would clearly be accessible with the same theoretical tools. We have already mentioned, as one example, the interactions between two particles, or between a particle and an interface, the latter of which is currently under investigation in our group.

ACKNOWLEDGMENT

Preparation of this manuscript was supported, in part, by grants from the National Science Foundation and from the Office of Naval Research. The author wishes to thank Mr. Paul Chan, who freely offered his assistance in preparing the paper.

Literature Cited

Acrivos, A., Lo, T. 1978. Deformation and breakup of a single slender drop in an extensional flow. *J. Fluid Mech.* 86:641

Batchelor, G. K. 1970. Slender-body theory for particles of arbitrary cross-section in Stokes flow. *J. Fluid Mech.* 44:419

Batchelor, G. K. 1976. Developments in microhydrodynamics. *Proc. IUTAM Cong. (Delft).* Amsterdam: North-Holland

Becker, H. A. 1959. The effects of shape and Reynolds number on drag in the motion of a freely oriented body in an infinite fluid. *Can. J. Chem. Eng.* 37:85

Berdan, C. 1980. PhD thesis. Calif. Inst. Tech., Pasadena

Blake, J. R. 1971. A note on the image system for a Stokeslet in a no-slip boundary. *Proc. Cambridge Philos. Soc.* 70:303

Brenner, H. 1966. Hydrodynamic resistance of particles at small Reynolds numbers. *Adv. Chem. Eng.* 6:287

Brenner, H. 1970. Dynamics of neutrally buoyant particles in low Reynolds number flows. *Prog. Heat Mass Transfer* 6:509

Brenner, H., Cox, R. G. 1963. The resistance to a particle of arbitrary shape in translational motion at small Reynolds numbers. *J. Fluid Mech.* 17:561

Bretherton, F. P. 1962. The motion of rigid particles in a shear flow at low Reynolds number. *J. Fluid Mech.* 14:284

Brunn, P. 1976a. The slow motion of a sphere in a second-order fluid. *Rheol. Acta* 15:163

Brunn, P. 1976b. The behavior of a sphere in non-homogeneous flows of a viscoelastic fluid. *Rheol. Acta* 15:589

Brunn, P. 1977a. The slow motion of a rigid particle in a second-order fluid. *J. Fluid Mech.* 82:529

Brunn, P. 1977b. Interaction of spheres in a viscoelastic fluid. *Rheol. Acta* 16:461

Buckmaster, J. D. 1972. Pointed bubbles in slow viscous flow. *J. Fluid Mech.* 55:385

Buckmaster, J. D. 1973. The bursting of pointed drops in slow viscous flow. *J. Appl. Mech.* E40:18

Buckmaster, J. D., Flaherty, J. E. 1973. The bursting of two-dimensional drops in slow viscous flow. *J. Fluid Mech.* 60:625

Burgers, J. M. 1938. On the motion of small particles of elongated form suspended in a viscous liquid. *Second Report on Viscosity and Plasticity.* Amsterdam: North-Holland

Caswell, B. 1970. The stability of particle motion near a wall in Newtonian and non-Newtonian fluids. *Chem. Eng. Sci.* 27:373

Caswell, B. 1977. Sedimentation of particles in non-Newtonian fluids. *The Mechanics of Viscoelastic Fluids, ASME, AMD* 22:19

Chaffey, C., Brenner, H., Mason, S. G. 1965. Particle motions in sheared suspensions. Part 18. Wall migration (theoretical). *Rheol. Acta* 4:64

Chan, P. C.-H. 1979. PhD thesis. Calif. Inst. Tech., Pasadena

Chan, P. C.-H., Leal, L. G. 1977. A note on the motion of a spherical particle in a general quadratic flow of a second-order fluid. *J. Fluid Mech.* 82:549

Chan, P. C.-H., Leal, L. G. 1979. The motion of a deformable drop in a second-order fluid. *J. Fluid Mech.* 92:131

Chwang, A. 1975. Hydromechanics of low-Reynolds-number flow. Part 3. Motion of a spheroidal particle in quadratic flows. *J. Fluid Mech.* 72:17

Chwang, A., Wu, T. Y. T. 1974. Hydromechanics of low-Reynolds-number flow. Part 1. Rotation of axisymmetric prolate bodies. *J. Fluid Mech.* 63:607

Chwang, A., Wu, T. Y. T. 1975. Hydromechanics of low-Reynolds-number flow. Part 2. Singularity method for Stokes flow. *J. Fluid Mech.* 67:787

Chwang, A., Wu, T. Y. T. 1976. Hydromechanics of low-Reynolds-number flow. Part 4. Translation of spheroids. *J. Fluid Mech.* 75:677

Cox, R. G. 1965. The steady motion of a particle of arbitrary shape at small Reynolds numbers. *J. Fluid Mech.* 23:625

Cox, R. G. 1970a. The motion of long slender bodies in a viscous fluid. Part 1. General theory. *J. Fluid Mech.* 44:791

Cox, R. G. 1970b. *Rep. No. PGRL/25.* Pulp and Paper Res. Inst. Canada

Cox, R. G., Brenner, H. 1968. The lateral migration of solid particles in Poiseuille flow. Part 1. Theory. *Chem. Eng. Sci.* 23:147

Cox, R. G., Hsu, S. K. 1977. The lateral migration of solid particles in a laminar flow near a plane. *Int. J. Multiphase Flow* 3:201

Cox, R. G., Mason, S. G. 1971. Suspended particles in fluid flow through tubes. *Ann. Rev. Fluid Mech.* 3:291

Davis, A. M. J., O'Neill, M. E., Dorrepaal, J. M., Ranger, K. B. 1976. Separation from the surface of two equal spheres in Stokes flow. *J. Fluid Mech.* 77:625

de Mestre, N. J., Russell, W. B. 1975. Low-Reynolds-number translation of a slender cylinder near a plane wall. *J. Eng. Math.* 9:81

Drew, D. A. 1978. The force on a small sphere in slow viscous flow. *J. Fluid Mech.* 88:393

Frankel, W., Acrivos, A. 1970. The constitutive equation for a dilute emulsion. *J. Fluid Mech.* 44:65

Gierszewski, P. L., Chaffey, C. E. 1978. Rotation of an isolated triaxial ellipsoid suspended in slow viscous flow. *Can. J. Phys.* 55:6

Gluckman, M. J., Weinbaum, S., Pfeffer, R. 1972. Axisymmetric slow viscous flow past an arbitrary convex body of revolution. *J. Fluid Mech.* 55:677

Goldsmith, H. 1968. The microrheology of red blood cell suspensions. *J. Gen. Physiol.* 52:55

Goldsmith, H. 1971. Red cell motions and wall interactions in tube flow. *Federation Proc.* 30:1578

Goldsmith, H., Mason, S. G. 1962. The flow of suspensions through tubes. Part 1. Single spheres, rods and discs. *J. Colloid Sci.* 17:448

Goldsmith, H. L., Mason, S. G. 1967. The microrheology of dispersions. *Rheology* 4:85

Goldsmith, H. L., Skalak, R. 1975. Hemodynamics. *Ann. Rev. Fluid Mech.* 7:213

Haber, S., Hetsroni, G. 1971. The dynamics of a deformable drop suspended in an unbounded Stokes flow. *J. Fluid Mech.* 49:257

Halow, J. S., Wills, G. B. 1970. Radial migration of spherical particles in Couette systems. *AIChE J.* 16:281

Happel, J., Brenner, H. 1965. *Low Reynolds Number Hydrodynamics.* Englewood Cliffs, N.J.: Prentice-Hall

Harper, E. Y., Chang, I.-D. 1968. Maximum dissipation resulting from lift in a slow viscous shear flow. *J. Fluid Mech.* 33:209

Harper, J. F. 1972. The motion of bubbles and drops through liquids. *Adv. Appl. Mech.* 12:59

Hetsroni, G., Haber, S., Brenner, H., Greenstein, T. 1970. A second-order theory for a deformable drop suspended in a long conduit. *Prog. Heat Mass Transfer* 6:591

Hinch, E. J., Leal, L. G. 1979. Rotation of small non-axisymmetric particles in a simple shear flow. *J. Fluid Mech.* 92:591

Ho, B. P., Leal, L. G. 1974. Inertial migration of rigid spheres in two-dimensional unidirectional flows. *J. Fluid Mech.* 65:365

Ho, B. P., Leal, L. G. 1976. Migration of rigid spheres in a two-dimensional unidirectional shear flow of a second-order fluid. *J. Fluid Mech.* 76:783

Jeffery, G. B. 1922. The motion of ellipsoidal particles immersed in a viscous fluid. *Proc. R. Soc. London Ser. A* 102:161

Jeffery, R. C., Pearson, J. R. A. 1965. Particle motion in laminar vertical tube flow. *J. Fluid Mech.* 22:721

Karnis, A., Goldsmith, H., Mason, S. G. 1963. Axial migration of particles in Poiseuille flow. *Nature* 200:159

Karnis, A., Goldsmith, H., Mason, S. G. 1966. The flow of suspensions through tubes. Part 5. Inertial effects. *Can. J. Chem. Eng.* 44:181

Karnis, A., Mason, S. G. 1967. Particle motions in sheared suspension. Part 23. Wall migration of fluid drops. *J. Colloid Int. Sci.* 24:164

Leal, L. G. 1975. The slow motion of slender rod-like particles in a second-order fluid. *J. Fluid Mech.* 69:305

Leal, L. G. 1979. The motion of small particles in non-Newtonian fluids. *J. Non-Newtonian Fluid Mech.* 5:33

Leal, L. G., Hinch, E. J. 1971. The effect of weak Brownian rotations on particles in shear flow. *J. Fluid Mech.* 46:685

Leal, L. G., Hinch, E. J. 1972. Theoretical studies of a suspension of rigid particles affected by Brownian couples. *Rheol. Acta* 12:127

Lee, S. H., Chadwick, R. S., Leal, L. G. 1979. Motion of a sphere in the presence of a plane interface. Part 1. *J. Fluid Mech.* 93:705

Lee, S. H., Leal, L. G. 1979. Motion of a sphere in the presence of a plane interface. Part 2. *J. Fluid Mech.* In press

Lorentz, H. A. 1907. A general theory concerning the motion of a viscous fluid. *Abhandl. Theoret. Phys.* 1:23

Michael, D. H., O'Neill, M. E. 1977. The separation of Stokes flows. *J. Fluid Mech.* 80:785

Oliver, D. R. 1962. Influence of particle rotation on radial migration in the Poiseuille flow of suspensions. *Nature* 194:1269

Pettyjohn, E. S., Christiansen, E. B. 1948. Effect of particle shape on free-settling rates of isometric particles. *Chem. Eng. Prog.* 44:157

Rallison, J. R., Acrivos, A. 1978. A numerical study of the deformation and burst of a viscous drop in an extensional flow. *J. Fluid Mech.* 89:191

Repetti, R. V., Leonard, E. F. 1964. Segre-Silberberg annulus formation: a possible explanation. *Nature* 203:1346

Repetti, R. V., Leonard, E. F. 1966. Physical basis for the axial accumulation of red blood cells. *Chem. Eng. Prog. Symp. Ser.* 62:80

Rubinow, S. I., Keller, J. B. 1961. The transverse force on a spinning sphere moving in a viscous fluid. *J. Fluid Mech.* 11:447

Russell, W. B. 1976. Low-shear limit of the secondary electroviscous effect. *J. Colloid Interface Sci.* 55:590

Russell, W. B. 1977a. The rheology of suspensions of charged rigid spheres. *J. Fluid Mech.* 85:209

Russell, W. B. 1977b. Bulk stresses due to deformation of the electrical double layer around a charged sphere. *J. Fluid Mech.* 85:673

Russell, W. B., Hinch, E. J., Leal, L. G., Tiefenbruck, G. 1977. Rods falling near a vertical wall. *J. Fluid Mech.* 83:273

Saffman, P. 1965. The lift on a small sphere in a slow shear flow. *J. Fluid Mech.* 22:385

Saville, D. A. 1977. Electrokinetic effects with small particles. *Ann. Rev. Fluid Mech.* 9:321

Segre, G., Silberberg, A. 1962a. Behavior of macroscopic rigid spheres in Poiseuille flow. Part 1. *J. Fluid Mech.* 14:115

Segre, G., Silberberg, A. 1962b. Behavior of macroscopic rigid spheres in Poiseuille flow. Part 2. *J. Fluid Mech.* 14:136

Segre, G., Silberberg, A. 1963. Non-Newtonian behavior of dilute suspensions of macroscopic spheres in a capillary viscometer. *J. Colloid Sci.* 18:312

Shizgal, B., Goldsmith, H. L., Mason, S. G. 1965. The flow of suspensions through tubes. Part 4. Oscillatory flow of rigid spheres. *Can. J. Chem. Eng.* 43:97

Spielman, L. A. 1977. Particle capture from low-speed laminar flows. *Ann. Rev. Fluid Mech.* 9:297

Stokes, G. 1851. On the effect of the internal friction of fluids on the motion of pendulums. *Trans. Cambridge Philos. Soc.* 9:8

Takano, M., Mason, S. G. 1966. *Tech. Rep. 487*. Pulp and Paper Res. Inst., Canada

Tam, C. K. W., Hyman, W. A. 1973. Transverse motion of an elastic sphere in a shear field. *J. Fluid Mech.* 59:177

Taylor, G. I. 1923. The motion of ellipsoidal particles in a viscous fluid. *Proc. R. Soc. London Ser. A* 103:58

Taylor, G. I. 1932. The viscosity of a fluid containing small drops of another fluid. *Proc. R. Soc. London Ser. A* 138:41

Taylor, G. I. 1934. The formation of emulsions in definable fields of flow. *Proc. R. Soc. London Ser. A* 146:501

Tillett, J. P. K. 1970. Axial and transverse Stokes flow past slender axisymmetric bodies. *J. Fluid Mech.* 44:401

Vasseur, P., Cox, R. G. 1976. The lateral migration of a spherical particle in two-dimensional shear flows. *J. Fluid Mech.* 78:385

Vasseur, P., Cox, R. G. 1977. The lateral migration of spherical particles sedimenting in a stagnant bounded fluid. *J. Fluid Mech.* 80:561

Wohl, P. R. 1976. The lift force on a drop in unbounded plane Poiseuille flow. *Adv. Eng. Sci. NASA CP-2001* 4:1493

Wohl, P. R., Rubinow, S. I. 1974. The transverse force on a drop in an unbounded parabolic flow. *J. Fluid Mech.* 62:185

AUTHOR INDEX

CUMULATIVE INDEXES

CONTRIBUTING AUTHORS, VOLUMES 8–12

CHAPTER TITLES, VOLUMES 8–12